Grundlehren der mathematischen Wissenschaften 286

A Series of Comprehensive Studies in Mathematics

Anatolij N. Andrianov

Quadratic Forms and Hecke Operators

Springer-Verlag
Berlin Heidelberg New York
London Paris Tokyo

Professor Anatolij N. Andrianov
LOMI
Fontanka 27
Leningrad, 191011
USSR

Mathematics Subject Classification (1980): 10C15, 10D20, 10D24

ISBN-13:978-3-642-70343-0 e-ISBN-13:978-3-642-70341-6
DOI: 10.1007/978-3-642-70341-6

With 3 Figures

Library of Congress Cataloging-in-Publication Data
Andrianov, Anatolij N.
Quadratic forms and Hecke operators.
(Grundlehren der mathematischen Wissenschaften; 286)
Bibliography: p.
Includes indexes.
1. Forms, Modular. 2. Hecke operators. 3. Series, Theta. I. Title. II. Series.
QA243.A53 1987 512'.7 86-26300
ISBN 0-387-15294-6 (U.S.)

Softcover reprint of the hardcover 1st edition 1987

Printing and bookbinding: K. Triltsch, Graphischer Betrieb, Würzburg
2141/3140-543210

To my wife Olga

Preface

The numerous explicit formulae of the classical theory of quadratic forms revealed remarkable multiplicative properties of the numbers of integral representations of integers by positive definite integral quadratic forms. These properties were explained by the original theory of Hecke operators. As regards the integral representations of quadratic forms in more than one variable by quadratic forms, no multiplicative properties were known at that time, and so there was nothing to explain. However, the idea of Hecke operators was so natural and attractive that soon attempts were made to cultivate it in the neighbouring field of modular forms of several variables. The approach has proved to be fruitful; in particular, a number of multiplicative properties of integral representations of quadratic forms by quadratic forms were eventually discovered. By now the theory has reached a certain maturity, and the time has come to give an up-to-date report in a concise form, in order to provide a solid ground for further development.

The purpose of this book is to present in the form of a self-contained text-book the contemporary state of the theory of Hecke operators on the spaces of holomorphic modular forms of integral weight (the Siegel modular forms) for congruence subgroups of integral symplectic groups. The book can also be used for an initial study of modular forms of one or several variables and theta-series of positive definite integral quadratic forms. In this book Hecke operators are mainly considered as a tool for the investigation of multiplicative properties of Fourier coefficients of modular forms, in accordance with the initial approach of Hecke. I do not discuss other directions which consider the relations of Hecke operators to algebraic geometry, Galois theory or representation theory, since in the case of several variables they are still in embryo.

The book is designed for those who wish to work in the arithmetical theory of automorphic forms, for those who are working in the field, and for those who merely want to look into it. No special knowledge is assumed beyond the standard university courses in algebra (general and linear) and analysis (real and complex). The classical case of one variable is included.

I believe that new concepts are best assimilated in a concrete context. However sometimes one must admit the greater usefulness and simplicity of a general treatment. So I have often made the first choice, but sometimes the second.

For an overview of the contents the reader may consult the Table of Contents and the remarks at the beginning of each chapter. Here I shall make only a few comments. The first three chapters are independent of each other, except for some general lemmas. Chapter 4 is based on Chapters 2 and 3, Chapter 5 makes use of all previous chapters. The exercises usually indicate some interesting topics which could not be included in the main text, partly for reasons of space and partly because of their incompleteness. The Notes at the end of the book contain essential references. The main text contains no references, but this does not mean that all of the results are due to the author.

There are three kinds of indexing throughout the book: first, the indexing of chapters, sections, and subsections which is clear from the Table of Contents; second, the indexing of theorems, propositions, lemmas, and exercises which goes through each of the sections with the first two indices indicating the number of the chapter and that of the section respectively; and, finally, the similar indexing of formulae.

I hope that this book will help to attract attention of young mathematicians to this beautiful and promising area. I also hope that some of my own enjoyment of the subject will be communicated to the reader.

I should like to express my gratitude to Professor H. Maaß and Professor S. Raghavan for their detailed comments and criticism of the manuscript; it is hardly necessary to mention that all of the imperfections that remain are the sole responsibility of the author. I am grateful to Springer-Verlag in Heidelberg for their kind invitation to write this book and their patience.

Leningrad, January 1987 A. Andrianov

Table of Contents

Chapter 1. Theta-Series

In this chapter we introduce theta-series of positive definite integral quadratic forms and study their automorphic properties.

§ 1.1. Definition of Theta-Series

1.1.1. Representations of Quadratic Forms by Quadratic Forms

Let

$$\mathsf{q}(x_1,\ldots,x_m) = \sum_{1\le\alpha,\beta\le m} q_{\alpha\beta}x_\alpha x_\beta, \qquad \mathsf{a}(y_1,\ldots,y_n) = \sum_{1\le\gamma,\delta\le n} a_{\gamma\delta}y_\gamma y_\delta$$

be two quadratic forms with coefficients in a commutative ring K_0. By *a representation of* a *by* q *over a subring* K *of* K_0 we mean a matrix $C = (c_{\alpha\gamma}) \in \mathsf{M}_{m,n}(\mathsf{K})$ such that the change of variables

$$x_\alpha = \sum_{\gamma=1}^{n} c_{\alpha\gamma}y_\gamma \qquad (\alpha = 1,\ldots,m)$$

transforms q into a; the number of representations will be denoted by $r_{\mathsf{K}}(\mathsf{q},\mathsf{a})$. If $\mathsf{a}(y_1) = ay_1^2$, then a matrix C with the transpose ${}^tC = (c_1,\ldots,c_m)$ is a representation of a by q if and only if $(c_1,\ldots,c_m)$ is a solution of the equation

$$\sum_{1\le\alpha,\beta\le m} q_{\alpha\beta}x_\alpha x_\beta = a. \qquad (1.1.1)$$

Thus, the number $r_{\mathsf{K}}(\mathsf{q}, ay_1^2)$ is equal to the number of solutions of (1.1.1) over K.

If 2 is not a zero divisor in K_0, it is convenient to use matrix notation. For a quadratic form

$$\mathsf{q}(x_1,\ldots,x_m) = \sum_{1\le\alpha,\beta\le m} q_{\alpha\beta}x_\alpha x_\beta \qquad (1.1.2)$$

the 'even' symmetric matrix

$$Q = Q(\mathsf{q}) = (q_{\alpha\beta}) + {}^t(q_{\alpha\beta}) \qquad (1.1.3)$$

will be called *the matrix of* q (it would be more traditional but less convenient to call it the matrix of $2\mathsf{q}$). In matrix notation one can write

$$\mathsf{q}(x_1,\ldots,x_m) = \tfrac{1}{2}{}^t\mathbf{x}Q\mathbf{x}, \tag{1.1.4}$$

where $\mathbf{x}$ is the column with entries $x_1, \ldots, x_m$; a matrix $C \in \mathsf{M}_{m,n}(\mathsf{K})$ is a representation of a by q if and only if C is a solution of the matrix equation

$${}^tXQX = A, \tag{1.1.5}$$

where Q and A are the matrices of q and a respectively, and X is the variable $(m \times n)$-matrix; in particular, we have

$$r_{\mathsf{K}}(\mathsf{q}, \mathsf{a}) = r_{\mathsf{K}}(Q, A),$$

where $r_{\mathsf{K}}(Q, A)$ is the number of solutions of (1.1.5) over K.

We say that two forms q, q', in the same number m of variables are *equivalent over* K (or K-*equivalent*) if there exists a representation V of one of them by the other with V belonging to the group $\mathsf{GL}_m(\mathsf{K})$ of invertible matrices of order m over K; we write $\mathsf{q} \sim_{\mathsf{K}} \mathsf{q}'$. Clearly K-equivalence is an equivalence relation in the technical sense (reflexive, symmetric and transitive). We may therefore speak of *the* K-*equivalence class of* q; we write $\{\mathsf{q}\}_{\mathsf{K}}$. If Q and Q' are the matrices of q and q' respectively, the equivalence $\mathsf{q} \sim_{\mathsf{K}} \mathsf{q}'$ means that the matrices satisfy

$${}^tVQV = Q' \qquad \text{with } V \in \mathsf{GL}_m(\mathsf{K}). \tag{1.1.6}$$

In this case we say that Q and Q' *are equivalent over* K (or K-*equivalent*); we write $Q \sim_{\mathsf{K}} Q'$ and denote by $\{Q\}_{\mathsf{K}}$ *the* K-*equivalence class of* Q.

If $V \in \mathsf{GL}_m(\mathsf{K})$ and $V' \in \mathsf{GL}_n(\mathsf{K})$, then the mapping $C \to VCV'$ of $\mathsf{M}_{m,n}(\mathsf{K})$ is clearly one-to-one. So we get the following

Proposition 1.1.1. *The function* $r_{\mathsf{K}}(Q, A)$ *depends only on* K-*classes of* Q *and* A.

The theory of quadratic forms over rings is almost as old and varied as mathematic itself. Its problems and approaches are different for different rings and forms. In this book the basic ring will be the ring $\mathbb{Z}$ of rational integers; the quadratic forms under consideration will be mainly positive definite or semi-definite. Other rings and forms play an auxiliary part. The numbers of representations over $\mathbb{Z}$ by a positive definite form are finite. In this case the theory studies various properties of the numbers, including 'multiplicative' properties, explicit and asymptotic formulae. We shall see that a natural setting for the study is given by the theory of modular forms. A link between quadratic forms and modular forms is provided by theta-series of quadratic forms.

Exercise 1.1.2. *Let* Q, A *be two symmetric matrices of order* m *and* n *respectively over the field* $\mathbb{R}$ *of real numbers. Suppose that* $Q > 0$. *Show that the equation* ${}^tXQX = A$ *has a solution* $C = (c_{\alpha\beta})$ *over* $\mathbb{R}$ *if and only if* $A \geq 0$ *and* $\operatorname{rank} A \leq m$, *and the entries of the solution satisfy*

$$\max_{\alpha,\beta} |c_{\alpha\beta}| \leq \lambda^{-1/2} \mu^{1/2},$$

where λ is the least characteristic value of Q, and μ is the largest characteristic value of A.

[Hint: Prove first that $Q \geq \lambda E_m$ and $A \leq \mu E_n$.]

1.1.2. Definition of Theta-Series

Let

$$\mathbb{A}_n = \{A = (A_{\alpha\beta}) \in \mathsf{M}_n; {}^tA = A \geq 0, A_{\alpha\alpha} \in 2\mathbb{Z}\} \tag{1.1.7}$$

be the set of symmetric integral *even* (i.e. with even diagonal entries) and semi-definite matrices of order n, and let

$$\mathbb{A}_n^+ = \{A \in \mathbb{A}_n; A > 0\}. \tag{1.1.8}$$

It is clear that $r(Q, A) = r_{\mathbb{Z}}(Q, A)$ is finite if $Q \in \mathbb{A}_m^+$ and $A \in \mathbb{A}_n$. Thus, for a fixed $Q \in \mathbb{A}_m^+$, we get a function

$$r(Q, \quad): \mathbb{A}_n \to \mathbb{Z}.$$

An analytical approach to the function is based on the consideration of some generating functions. There are different ways to construct a generating function; for example, it can be defined by the power series

$$\sum_{A=((1+e_{\alpha\beta})a_{\alpha\beta}) \in \mathbb{A}_n} r(Q, A) \prod_{1 \leq \alpha \leq \beta \leq n} t_{\alpha\beta}^{a_{\alpha\beta}}$$

in $\langle n \rangle = n(n+1)/2$ variables $t_{\alpha\beta}$, where $e_{\alpha\beta}$ are the entries of the unit matrix $E_n = (e_{\alpha\beta})$. Setting $t_{\alpha\beta} = \exp(2\pi i z_{\alpha\beta})$, we get the Fourier series

$$\begin{aligned} \theta^n(Z, Q) &= \sum_{A=((1+e_{\alpha\beta})a_{\alpha\beta}) \in \mathbb{A}_n} r(Q, A) \exp\Big(\pi i \cdot 2 \sum_{1 \leq \alpha \leq \beta \leq n} a_{\alpha\beta} z_{\alpha\beta}\Big) \\ &= \sum_{A \in \mathbb{A}_n} r(Q, A) \exp(\pi i \sigma(AZ)), \end{aligned} \tag{1.1.9}$$

where Z is the symmetric matrix of order n with the entries $z_{\alpha\beta}$ on and above the principal diagonal, and where σ denotes the trace. The last form is more convenient in many respects; in particular, it will help us to find the domain of convergence of the series. Let us write the matrix Z in the form $Z = X + iY$ with real matrices X and Y. If Y does not satisfy the condition $Y \geq 0$, then there is an integral row $\mathbf{c} = (c_1, \ldots, c_n)$ such that $\mathbf{c} Y {}^t\mathbf{c} < 0$ (a real solution of the inequlaity exists by the definition; a rational solution can be found by approximation; an integral solution can be obtained from the rational one by homogeneity). Let C be the $m \times h$-matrix whose first row is $\mathbf{c}$ and all other rows of which are zero-rows. Then the matrices $A_d = d^2\, {}^tCQC = {}^t(dC)Q(dC)$ with $d \in \mathbb{Z}$ belong to $\mathbb{A}_n$, satisfy $r(Q, A_d) \geq 1$ and

$$|r(Q, A_d) \exp(\pi i \sigma(A_d Z))| = r(Q, A_d) \exp(-\pi d^2 \sigma(QCY\,{}^tC)) \to \infty \quad \text{as } d \to \infty.$$

Thus, the series (1.1.9) is divergent. Hence if the series (1.1.9) is convergent on an open subset of the $\langle n \rangle$-dimensional complex space, the subset must be contained in the domain

$$\mathbb{H}_n = \{Z = X + iY \in \mathsf{M}_n(\mathbb{C}); {}^tZ = Z, Y > 0\} \tag{1.1.10}$$

which is called *the Siegel upper half-plane of genus n*. Obviously, $\mathbb{H}_n$ is an open submanifold of the $\langle n \rangle$-dimensional complex Euclidean space.

Proposition 1.1.3. *The series $\theta^n(Z, Q)$ with $Q \in \mathbb{A}_m^+$ and $n \in \mathbb{N}$ converges absolutely and uniformly on each subset of $\mathbb{H}_n$ of the form*

$$\mathbb{H}_n(\varepsilon) = \{Z = X + iY \in \mathbb{H}_n; Y \geq \varepsilon E_n\} \tag{1.1.11}$$

with $\varepsilon > 0$.

Proof. If ε_1 is the least characteristic value of Q, then every matrix N of $\mathsf{M}_{m,n}(\mathbb{R})$ satisfies ${}^tNQN \geq \varepsilon_1^t NN$. Thus, for a $Z \in \mathbb{H}_n(\varepsilon)$, the series

$$\sum_{N \in \mathsf{M}_{m,n}} \exp(\pi i \sigma({}^tNQNZ)) \tag{1.1.12}$$

is dominated term by term by the convergent series

$$\sum_{N \in \mathsf{M}_{m,n}} \exp(-\pi\varepsilon\varepsilon_1 \sigma({}^tNN)) = \left(\sum_{t \in \mathbb{Z}} \exp(-\pi\varepsilon\varepsilon_1 t^2)\right)^{mn}.$$

Therefore the series (1.1.12) converges absolutely and uniformly on $\mathbb{H}_n(\varepsilon)$. Putting together all the terms of (1.1.12) with ${}^tNQN = A$ for $A \in \mathbb{A}_n$, we see that the number of the terms is $r(Q, A)$. So the series (1.1.12) is equal to $\theta^n(Z, Q)$. □

The series

$$\begin{aligned}\theta^n(Z, Q) &= \sum_{A \in \mathbb{A}_n} r(Q, A) \exp(\pi i \sigma(AZ)) \\ &= \sum_{N \in \mathsf{M}_{m,n}} \exp(\pi i \sigma({}^tNQNZ))\end{aligned} \tag{1.1.13}$$

is called *the theta-series of genus n of the matrix Q* (or the corresponding *quadratic form* (1.1.4)). From Proposition 1.1.3 we obtain

Theorem 1.1.4. *The theta-series $\theta^n(Z, Q)$ defines a holomorphic function on $\mathbb{H}_n$. The function is bounded on each subset of $\mathbb{H}_n$ of the form $\mathbb{H}_n(\varepsilon)$ with $\varepsilon > 0$.*

§ 1.2. Symplectic Transformations

1.2.1. The Symplectic Group

Here we shall consider certain analytic transformations on $\mathbb{H}_n$. First, we note that the mappings

$$U(V)\colon Z \to {}^tV^{-1}ZV^{-1} \qquad \text{with } V \in \mathsf{GL}_n(\mathbb{R})$$

and

$$T(S)\colon Z \to Z + S \qquad \text{with } S \in \mathsf{S}_n(\mathbb{R}),$$

where $\mathsf{S}_n(\mathbb{R})$ is the set of symmetric matrices of order n over $\mathbb{R}$, are clearly analytic automorphisms of $\mathbb{H}_n$. Now we show that the mapping $J\colon Z \to -Z^{-1}$ is an analytic automorphism of $\mathbb{H}_n$ as well. It suffices to check that $-Z^{-1}$ exists and belongs to $\mathbb{H}_n$ when $Z = X + iY \in \mathbb{H}_n$. Since Y is positive definite, there is a matrix $V \in \mathsf{GL}_n(\mathbb{R})$ such that $VY\,{}^tV = E_n$. Hence, $VZ\,{}^tV = T + iE_n$ with $T = VX\,{}^tV$. Since the matrix $T^2 + E_n = T\,{}^tT + E_n$ is positive definite, it is invertible. Thus, the matrix $T + iE$ is invertible too, and

$$(T + iE)^{-1} = (T - iE)(T^2 + E)^{-1}.$$

It follows that Z is invertible and

$$-Z^{-1} = -{}^tV(T + iE)^{-1}V = {}^tV(-T + iE)(T^2 + E)^{-1}V \in \mathbb{H}_n.$$

We shall say that a one-to-one mapping $f\colon \mathbb{H}_n \to \mathbb{H}_n$ is *represented by a matrix* $M = \begin{pmatrix} A & B \\ C & D \end{pmatrix}$, where A, B, C and D are blocks of dimension $n \times n$, if the matrix $CZ + D$ is invertible for all $Z \in \mathbb{H}_n$, and

$$f(Z) = M\langle Z\rangle = (AZ + B)(CZ + D)^{-1} \quad (Z \in \mathbb{H}_n). \tag{1.2.1}$$

If f is represented by M and f_1 is represented by M_1, it follows that the mapping $f_1 f\colon Z \to f_1(f(Z))$ is represented by $M_1 M$. Since the mappings $U(V)$, $T(S)$ and J are represented by the matrices

$$U(V) = \begin{pmatrix} V^* & 0 \\ 0 & V \end{pmatrix} \qquad \text{with } V \in \mathsf{GL}_n(\mathbb{R}), \tag{1.2.2}$$

where, for an invertible matrix V, we set $V^* = {}^tV^{-1}$,

$$T(S) = \begin{pmatrix} E_n & S \\ 0 & E_n \end{pmatrix} \qquad \text{with } S \in \mathsf{S}_n(\mathbb{R}), \tag{1.2.3}$$

and

$$J = J_n = \begin{pmatrix} 0 & E_n \\ -E_n & 0 \end{pmatrix}, \tag{1.2.4}$$

respectively, we obtain the following

Proposition 1.2.1. *Let* S *be the subgroup of* $\mathsf{GL}_{2n}(\mathbb{R})$ *generated by the matrices of the form* (1.2.2)–(1.2.4). *Then, for every* $M = \begin{pmatrix} A & B \\ C & D \end{pmatrix} \in \mathsf{S}$ *and* $Z \in \mathbb{H}_n$, *the matrix* $CZ + D$ *is invertible; the mapping*

$$f = f(M)\colon Z \to M\langle Z\rangle = (AZ + B)(CZ + D)^{-1} \tag{1.2.5}$$

is an analytic automorphism of $\mathbb{H}_n$; *and the mapping* $M \to f(M)$ *is a homomorphism of* S *into the group of analytic automorphisms of* $\mathbb{H}_n$.

Now we describe the group S as an algebraic group. First we note that each of the generators M of S satisfies ${}^tMJ_nM = J_n$. Thus, S is a subgroup of *the (real) symplectic group of genus n*:

$$\mathsf{Sp}_n(\mathbb{R}) = \{M \in \mathsf{M}_{2n}(\mathbb{R}); {}^tMJ_nM = J_n\}. \tag{1.2.6}$$

It follows from the definition that a matrix $M = \begin{pmatrix} A & B \\ C & D \end{pmatrix} \in \mathsf{M}_{2n}(\mathbb{R})$ is *symplectic* (i.e. $M \in \mathsf{Sp}_n(\mathbb{R})$) if and only if its $n \times n$ blocks A, B, C and D satisfy the relations

$${}^tAC = {}^tCA, \qquad {}^tBD = {}^tDB, \qquad {}^tAD - {}^tCB = E_n. \tag{1.2.7}$$

It is easy to see that M is symplectic if and only if the transpose ${}^tM = JM^{-1}J^{-1}$ is. It follows that the conditions (1.2.7) are equivalent with the conditions

$$A\,{}^tB = B\,{}^tA, \qquad C\,{}^tD = D\,{}^tC, \qquad A\,{}^tD - B\,{}^tC = E_n. \tag{1.2.8}$$

If $M = \begin{pmatrix} A & B \\ C & D \end{pmatrix} \in \mathsf{Sp}_n(\mathbb{R})$, then

$$M^{-1} = J^{-1}\,{}^tMJ = \begin{pmatrix} {}^tD & -{}^tB \\ -{}^tC & {}^tA \end{pmatrix}. \tag{1.2.9}$$

Theorem 1.2.2. *The symplectic group of genus n is generated by the matrices* (1.2.2)–(1.2.4), *i.e.* $\mathsf{S} = \mathsf{Sp}_n(\mathbb{R})$.

Proof. Let $M = \begin{pmatrix} A & B \\ C & D \end{pmatrix}$ be a symplectic matrix. The A — block of $U(V^*)MU(V_1^*)$ is equal to VAV_1. We can choose $V, V_1 \in \mathsf{GL}_n(\mathbb{R})$ such that $VAV_1 = \begin{pmatrix} E_r & 0 \\ 0 & 0 \end{pmatrix}$, where r is the rank of A. So we can assume that $A = \begin{pmatrix} E_r & 0 \\ 0 & 0 \end{pmatrix}$. If we write C in the form $\begin{pmatrix} C_1 & C_2 \\ C_3 & C_4 \end{pmatrix}$ with the block C_1 of dimension $r \times r$ and so on, then it follows from ${}^tAC = {}^tCA$ that $C_2 = 0$ and $C_1 = {}^tC_1$. Besides $\det C_4 \neq 0$, since the columns of M are linearly independent. Let us consider the matrix $T(\lambda E_n)M$ with a $\lambda \in \mathbb{R}$. The A-block of the matrix is equal to

$$A' = A + \lambda C = \begin{pmatrix} E_r + \lambda C_1 & 0 \\ \lambda C_3 & \lambda C_4 \end{pmatrix}.$$

Thus, rank $A' = n$, if λ is sufficiently small. So we can assume that $A = E_n$. Then C is symmetric and we have

$$J^{-1}T(C)JM = \begin{pmatrix} E & 0 \\ -C & E \end{pmatrix}\begin{pmatrix} E & B \\ C & D \end{pmatrix} = \begin{pmatrix} E & B_1 \\ 0 & D_1 \end{pmatrix}.$$

The last matrix is symplectic. Then, by (1.2.7), it follows that $D_1 = E$ and ${}^tB_1 = B_1$. So the matrix is equal to $T(B_1)$. □

Exercise 1.2.3. *Let* $M = \begin{pmatrix} A & B \\ C & D \end{pmatrix}$ *be a real matrix of order 2n, which satisfies*

$$ {}^tMJ_nM = \mu J_n \quad \text{with } \mu \neq 0. $$

Show that the matrix $CZ + D$ *is invertible for* $Z \in \mathbb{H}_n$*; the mapping*

$$ Z \to (AZ + B)(CZ + D)^{-1} $$

is an analytic automorphism of $\mathbb{H}_n$ *if* $\mu > 0$*, and is an analytic isomorphism of* $\mathbb{H}_n$ *onto*

$$ \{Z = X + iY \in \mathsf{M}_n(\mathbb{C}); {}^tZ = Z, Y < 0\}, $$

if $\mu < 0$.

1.2.2. The Siegel Upper Half-Plane

We have seen that $\mathsf{Sp}_n(\mathbb{R})$ acts on $\mathbb{H}_n$ as a group of analytic automorphism, the action being defined by

$$ M = \begin{pmatrix} A & B \\ C & D \end{pmatrix} : Z \to M\langle Z \rangle = (AZ + B)(CZ + D)^{-1}. \tag{1.2.10} $$

Proposition 1.2.4. *The action of* $\mathsf{Sp}_n(\mathbb{R})$ *on* $\mathbb{H}_n$ *is transitive.*

Proof. We only have to show that each point $Z = X + iY \in \mathbb{H}_n$ is of the form $M\langle iE_n \rangle$ with $M \in \mathsf{Sp}_n(\mathbb{R})$. Since $Y > 0$, we can write $Y = A\,{}^tA$ with $A \in \mathsf{GL}_n(\mathbb{R})$. It follows from (1.2.7) that the matrix

$$ M = \begin{pmatrix} A & B \\ 0 & A^* \end{pmatrix} \quad \text{with } B = XA^* $$

is symplectic. Also $M\langle iE \rangle = iA\,{}^tA + B\,{}^tA = X + iY$. □

The proposition enables us to identify $\mathbb{H}_n$ with a homogeneous space of $\mathsf{Sp}_n(\mathbb{R})$. For example, if

$$ \mathsf{U} = \{M \in \mathsf{Sp}_n(\mathbb{R}); M\langle iE_n \rangle = iE_n\} $$

is the stabilizer of iE_n in $\mathsf{Sp}_n(\mathbb{R})$, then the mapping $M \to M\langle iE_n \rangle$ defines a one-to-one mapping $\mathsf{Sp}_n(\mathbb{R})/\mathsf{U} \to \mathbb{H}_n$, which is compatible with the action of $\mathsf{Sp}_n(\mathbb{R})$, where on the left side the group acts by multiplication from the left.

Lemma 1.2.5. *The group* U *is given by*

$$\mathsf{U} = \left\{ \begin{pmatrix} A & B \\ -B & A \end{pmatrix} \in \mathsf{Sp}_n(\mathbb{R}) \right\};$$

the mapping $\begin{pmatrix} A & B \\ -B & A \end{pmatrix} \to A + iB$ *is an isomorphism of* U *onto the unitary group of order n; in particular,* U *is compact.*

Proof. The relation $(Ai + B)(Ci + D)^{-1} = iE$ is equivalent with $Ai + B = i(Ci + D) = Di - C$, which means that $D = A$ and $C = -B$. The relations (1.2.7) for a matrix of the form $\begin{pmatrix} A & B \\ -B & A \end{pmatrix}$ turn into ${}^tAB = {}^tBA$, ${}^tAA + {}^tBB = E$, which is equivalent with ${}^t(A - iB)(A + iB) = E$. □

Let $Z = X + iY$ and $Z_1 = X_1 + iY_1$ be two matrices in $\mathbb{H}_n$, and let $t \in \mathbb{R}$ with $0 \leq t \leq 1$. Then the matrix $tY + (1 - t)Y_1$ is clearly positive definite. Thus, $tZ + (1 - t)Z_1 \in \mathbb{H}_n$. It follows:

Proposition 1.2.6. *The domain* $\mathbb{H}_n$ *is simply connected and convex.*

The upper half-plane $\mathbb{H}_n$ is clearly an open subset of $\mathbb{R}^{n(n+1)}$. We shall show that there exists an element of volume on $\mathbb{H}_n$, which is invariant under the group $\mathsf{Sp}_n(\mathbb{R})$, and compute the element.

Lemma 1.2.7. *Let*

$$\mathrm{d}Z = \prod_{1 \leq \alpha \leq \beta \leq n} \mathrm{d}x_{\alpha\beta}\,\mathrm{d}y_{\alpha\beta} \qquad (Z = (x_{\alpha\beta} + iy_{\alpha\beta}) \in \mathbb{H}_n)$$

be the Euclidean element of volume on $\mathbb{H}_n$*. Then, for each matrix* $M = \begin{pmatrix} A & B \\ C & D \end{pmatrix} \in \mathsf{Sp}_n(\mathbb{R})$*, the following relation holds:*

$$\mathrm{d}M\langle Z\rangle = |\det(CZ + D)|^{-2n-2}\,\mathrm{d}Z.$$

Proof. For $Z = (z_{\alpha\beta}) = (x_{\alpha\beta} + iy_{\alpha\beta}) \in \mathbb{H}_n$, we set $Z' = (z'_{\gamma\delta}) = (x'_{\gamma\delta} + iy'_{\gamma\delta}) = M\langle Z\rangle$. We have to compute the Jacobian of the variables $x'_{\gamma\delta}$, $y'_{\gamma\delta}$ with respect to the variables $x_{\alpha\beta}$, $y_{\alpha\beta}$. First we shall consider the transformation of the differentials of the complex variables $z_{\alpha\beta}$. For $Z_1, Z_2 \in \mathbb{H}_n$, since Z'_2 is symmetric, we get

$$\begin{aligned} Z'_2 - Z'_1 &= (Z_2\,{}^tC + {}^tD)^{-1}(Z_2\,{}^tA + {}^tB) - (AZ_1 + B)(CZ_1 + D)^{-1} \\ &= (Z_2\,{}^tC + {}^tD)^{-1}\{(Z_2\,{}^tA + {}^tB)(CZ_1 + D) - (Z_2\,{}^tC + {}^tD)(AZ_1 + B)\} \\ &\quad \times (CZ_1 + D)^{-1} \qquad (1.2.11) \\ &= (Z_2\,{}^tC + {}^tD)^{-1}(Z_2 - Z_1)(CZ_1 + D)^{-1}, \end{aligned}$$

where we have also used the relations (1.2.7). It follows that

$$\mathbf{D}Z' = (Z\,{}^tC + {}^tD)^{-1}\mathbf{D}Z(CZ + D)^{-1} \tag{1.2.12}$$

with $\mathbf{D}Z = (\mathrm{d}z_{\alpha\beta})$ and $\mathbf{D}Z' = (\mathrm{d}z'_{\gamma\delta})$, $\mathrm{d}z_{\alpha\beta}$, $\mathrm{d}z'_{\gamma\delta}$ being the differentials of the variables $z_{\alpha\beta}$, $z'_{\gamma\delta}$ respectively. Now, if $\rho(U)$ with $U \in \mathrm{GL}_n(\mathbb{C})$ is the transformation $(v_{\alpha\beta}) \to U(v_{\alpha\beta})\,{}^tU$ of variables $v_{\alpha\beta} = v_{\beta\alpha}$, $1 \le \alpha, \beta \le n$, then

$$\det \rho(U) = (\det U)^{n+1}. \tag{1.2.13}$$

To prove this, we can replace U by a matrix of the form $W^{-1}UW$. Thus we can assume that U is an upper triangular matrix. If the variables $v_{\alpha\beta}$ $(1 \le \alpha \le \beta \le n)$ are ordered lexicographically, then the matrix $\rho(U)$ again is upper triangular with the diagonal entries $u_i u_j$ $(1 \le i \le j \le n)$, where $u_1, \ldots, u_n$ are the diagonal entries of U. The relation (1.2.13) follows. Let $\mathbf{d}Z(\mathbf{d}Z')$ be the column with the entries $\mathrm{d}z_{\alpha\beta}(1 \le \alpha \le \beta \le n)(\mathrm{d}z'_{\gamma\delta}(1 \le \gamma \le \delta \le n))$ arranged in a fixed order. Then the above considerations imply the relation

$$\mathbf{d}Z' = \rho((CZ + D)^*)\,\mathbf{d}Z.$$

Taking $\mathbf{d}Z = \mathbf{d}X + i\,\mathbf{d}Y$, $\mathbf{d}Z' = \mathbf{d}X' + i\,\mathbf{d}Y'$ and $\rho((CZ + D)^*) = R + iS$, we conclude that

$$\mathbf{d}X' = R\,\mathbf{d}X - S\,\mathbf{d}Y, \qquad \mathbf{d}Y' = S\,\mathbf{d}X + R\,\mathbf{d}Y.$$

Thus, the Jacobian is equal to

$$\begin{aligned}\det\begin{pmatrix} R & -S \\ S & R\end{pmatrix} &= \det\begin{pmatrix} E & iE \\ 0 & E\end{pmatrix}\begin{pmatrix} R & -S \\ S & R\end{pmatrix}\begin{pmatrix} E & -iE \\ 0 & E\end{pmatrix}\\ &= \det\begin{pmatrix} R + iS & 0 \\ 0 & R - iS\end{pmatrix} = |\det(R + iS)|^2 \\ &= |\det(CZ + D)|^{-2n-2}. \end{aligned}$$

□

Lemma 1.2.8. *Let* $Z' = X' + iY' = M\langle Z\rangle$ *with* $Z = X + iY \in \mathbb{H}_n$ *and* $M = \begin{pmatrix} A & B \\ C & D\end{pmatrix} \in \mathrm{Sp}_n(\mathbb{R})$. *Then*

$$Y' = {}^t(C\bar{Z} + D)^{-1}Y(CZ + D)^{-1};$$

in particular,

$$\det Y' = |\det(CZ + D)|^{-2}\det Y. \tag{1.2.14}$$

Proof. The lemma follows from (1.2.11), since $Y' = -1/2i(\bar{Z}' - Z')$. □

By comparing the last two lemmas, we get

Proposition 1.2.9. *Let* d^*Z *be the element of volume on* $\mathbb{H}_n$ *given by*

$$\mathrm{d}^*Z = (\det Y)^{-n-1}\,\mathrm{d}Z = \det(y_{\alpha\beta})^{-n-1} \prod_{1 \le \alpha \le \beta \le n} \mathrm{d}x_{\alpha\beta}\,\mathrm{d}y_{\alpha\beta}$$

with $Z = X + iY = (x_{\alpha\beta}) + i(y_{\alpha\beta}) \in \mathbb{H}_n$. *Then* d^*Z *is invariant under all symplectic transformation on* $\mathbb{H}_n$, *i.e.*

$$\mathrm{d}^*M\langle Z\rangle = \mathrm{d}^*Z \qquad (M \in \mathrm{Sp}_n(\mathbb{R})).$$

Exercise 1.2.10. *Show that the Cayley mapping*

$$Z \to W = (Z - iE_n)(Z + iE_n)^{-1} \qquad (Z \in \mathbb{H}_n)$$

is an analytical isomorphism of $\mathbb{H}_n$ *onto the bounded domain*

$$\{W \in \mathrm{M}_n(\mathbb{C}); {}^tW = W, \overline{W}W < E_n\},$$

where the inequality is understood in the sense of Hermitian matrices. Show that the inverse mapping is given by

$$W \to Z = i(E_n + W)(E_n - W)^{-1}.$$

Exercise 1.2.11. *Show that the element of volume*

$$\mathrm{d}^*Y = (\det Y)^{-(n+1)/2} \prod_{1 \le \alpha \le \beta \le n} \mathrm{d}y_{\alpha\beta}$$

on the space

$$\mathrm{P}_n = \{Y \in \mathrm{M}_n(\mathbb{R}); {}^tY = Y, Y > 0\}$$

is invariant under all transformation of the form

$$Y \to {}^tgYg \qquad \text{with } g \in \mathrm{GL}_n(\mathbb{R}).$$

§ 1.3. Symplectic Transformations of Theta-Series

1.3.1. Transformations of Theta-Functions

A theta-series is "almost" invariant under a big group of symplectic transformations of the variables. This remarkable property is a corner-stone of applications of theta-series. The group appears usually as a subgroup of the symplectic group, which is generated by certain generators, whose action on the theta-series can be determined by a direct computation. However, generators of a subgroup of the symplectic group are generally unknown. So the direct computation of the transformation group often proves to be complicated. Instead we first determine the transformation group of the simplest theta-series, the so-called theta-functions. Then we express all theta-series in terms of the theta-functions and obtain the transformation group of a theta-series as a subgroup of the transformation group of the corresponding theta-function.

If Q is a symmetric matrix of order k and N is a $k \times l$-matrix, we shall write

$$Q[N] = {}^tNQN. \tag{1.3.1}$$

It is easy to see that the series

$$\theta^k(Z; W, W') = \sum_{N \in \mathsf{M}_{k,1}} \exp(\pi i(Z[N - W'] + 2\,{}^tNW - {}^tW'W)) \tag{1.3.2}$$

converges absolutely, when $Z \in \mathbb{H}_k$, W, $W' \in \mathsf{M}_{k,1}(\mathbb{C})$. It converges uniformly, when $Z \in \mathbb{H}_n(\varepsilon)$ with $\varepsilon > 0$, and W, W' are in a fixed compact subset of $\mathsf{M}_{k,1}(\mathbb{C})$. The same is also true for the series which can be obtained from (1.3.2) by means of any partial differentiation. The holomorphic function on $\mathbb{H}_k \times \mathsf{M}_{k,1}(\mathbb{C}) \times \mathsf{M}_{k,1}(\mathbb{C})$ defined by (1.3.2) is called *the theta-function of genus k*.

Let $V \in \Lambda^k = \mathsf{GL}_k(\mathbb{Z})$. Since (1.3.2) converges absolutely, it follows by a direct computation that

$$\theta^k(Z[{}^tV]; W, W') = \theta^k(Z; V^{-1}W, {}^tVW').$$

Replacing here W by VW, and W' by V^*W', we get the identity

$$\theta^k(Z[{}^tV]; VW, V^*W) = \theta^k(Z; W, W') \qquad (V \in \Lambda^k). \tag{1.3.3}$$

Now let $S = (s_{\alpha\beta})$ be an integral symmetric matrix of order k. Replacing Z by $Z + S$ in (1.3.2), we get

$$\begin{aligned}\theta^k(Z + S; W, W') = \sum_{N \in \mathsf{M}_{k,1}} &\exp(\pi i(Z[N - W'] + S[N - W'] \\ &+ 2\,{}^tNW - {}^tW'W)).\end{aligned}$$

Since

$$S[N - W'] = S[N] - 2{}^tNSW' + S[W']$$

and

$$\begin{aligned}S[N] &= \sum_{\alpha,\beta=1}^{k} s_{\alpha\beta}n_\alpha n_\beta = \sum_{\alpha=1}^{k} s_{\alpha\alpha}n_\alpha^2 + 2\sum_{1 \le \alpha < \beta \le k} s_{\alpha\beta}n_\alpha n_\beta \\ &\equiv \sum_{\alpha=1}^{k} s_{\alpha\alpha}n_\alpha = {}^tN \cdot \mathrm{dc}(S) \pmod 2,\end{aligned}$$

where $\mathrm{dc}(S)$ is the column with entries $s_{\alpha\alpha}$ $(1 \le \alpha \le k)$, it follows that

$$\begin{aligned}&\theta^k(Z + S; W, W') \\ &\quad = \exp\left(\frac{\pi i}{2}{}^tW'\,\mathrm{dc}(S)\right)\theta^k\left(Z; W - SW' + \frac{1}{2}\mathrm{dc}(S), W'\right).\end{aligned}$$

Replacing W by $W + SW' - \frac{1}{2}\mathrm{dc}(S)$ and dividing both sides by $\exp((\pi i/2) \times {}^tW'\,\mathrm{dc}(S))$, we obtain the identity

$$\begin{aligned}\exp\left(-\frac{\pi i}{2}{}^tW' \cdot \mathrm{dc}(S)\right)\theta^k\left(Z + S; W + SW' - \frac{1}{2}\mathrm{dc}(S), W'\right) \\ = \theta^k(Z; W, W').\end{aligned} \tag{1.3.4}$$

The final one and the only non-trivial transformation formula of the theta-function, the so-called *inversion formula*, goes back to Jacobi.

Lemma 1.3.1. (*The inversion formula for the theta-function*). *The following identity holds*

$$\theta^k(-Z^{-1}; W', -W) = (\det(-iZ))^{1/2}\theta^k(Z; W, W'). \tag{1.3.5}$$

The square root in (1.3.5) *is to be taken positive when* $Z = iY$ *i.e. is purely imaginary; in general, the sign is found by analytic continuation in the simply connected variety* $\mathbb{H}_n$.

Proof. The function

$$\exp(-\pi i\,{}^tW'W)\theta^k(Z; W, W')$$
$$= \sum_{N \in \mathsf{M}_{k,1}} \exp(\pi i Z[N - W'] + 2\pi i\,{}^t(N - W')W)$$

is holomorphic and periodic of period 1 in all entries w_t' of W'. Let us introduce the variables u_t by

$$u_t = \exp(2\pi i w_t'), \qquad w_t = (1/2\pi i)\log u_t.$$

Then the function is an analytic function in $u_1, \ldots, u_k$ in

$$0 \le |u_1|, \ldots, |u_k| < \infty.$$

Thus, it can be expanded in the Laurent series:

$$\exp(-\pi i\,{}^tW'W)\theta^k(Z; W, W')$$
$$= \sum_{l_1,\ldots,l_k=-\infty}^{+\infty} c(L)u_1^{l_1}\cdots u_k^{l_k} = \sum_{L \in \mathsf{M}_{k,1}} c(L)\exp(2\pi i\,{}^tW'L), \tag{1.3.6}$$

where the coefficients $c(L)$ depend only on Z and W. The series converges uniformly in W' on any set $\mathsf{M}_{k,1}(\mathbb{R}) + W_0$ with a fixed W_0. Thus, it can be integrated term by term over its subsets. Let us multiply both sides of (1.3.6) by $\exp(-2\pi i\,{}^tW'L)$, set $W' = H + W_0$, where W_0 will be choosen later, and integrate the series term by term over the cube

$$\mathsf{C} = \{H = (h_t) \in \mathsf{M}_{k,1}(\mathbb{R}); 0 \le h_t \le 1\}$$

with respect to the Euclidean measure $\mathrm{d}H = \mathrm{d}h_1 \ldots \mathrm{d}h_k$. We get

$$c(L) = \int_{\mathsf{C}} \theta^k(Z; W, H + W_0)\exp(-\pi i\,{}^t(H + W_0)W - 2\pi i\,{}^t(H + W_0)L)\,\mathrm{d}H$$
$$= \int_{\mathsf{C}} \sum_{N \in \mathsf{M}_{k,1}} \exp(\pi i(Z[N - H - W_0] + 2\,{}^t(N - H - W_0)W$$
$$+ 2\,{}^t(N - H - W_0)L))\,\mathrm{d}H$$

(note ${}^tNL$ is an integer). Since the integrand converges uniformly, we can integrate it term by term. So we get

$$\begin{aligned} c(L) &= \sum_{N \in \mathsf{M}_{k,1}} \int_{-N+C} \exp(\pi i(Z[H + W_0] - 2{}^t(H + W_0)(W + L)))\, \mathrm{d}H \\ &= \int_{\mathsf{M}_{k,1}(\mathbb{R})} \exp(\pi i(Z[H + W_0] - 2\,{}^t(H + W_0)(W + L)))\, \mathrm{d}H. \end{aligned}$$

Applying the identity

$$\begin{aligned} & Z[H + W_0] - 2\,{}^t(H + W_0)(W + L) \\ & \qquad = Z[H + W_0 - Z^{-1}(W + L)] - Z^{-1}[W + L] \end{aligned}$$

and setting $W_0 = Z^{-1}(W + L)$, we have

$$c(L) = \exp(-\pi i Z^{-1}[L + W]) \cdot \int_{\mathsf{M}_{k,1}(\mathbb{R})} \exp(\pi i Z[H])\, \mathrm{d}H.$$

Substituting the expression in (1.3.6), we obtain

$$\begin{aligned} & \theta^k(Z; W, W') \\ & \quad = I(Z) \sum_{L \in \mathsf{M}_{k,1}} \exp(\pi i(-Z^{-1}[L + W] + 2\,{}^tLW' + {}^tWW')) \\ & \quad = I(Z)\theta^k(-Z^{-1}; W', -W), \end{aligned}$$

where

$$I(Z) = \int_{\mathsf{M}_{k,1}(\mathbb{R})} \exp(\pi i Z[H])\, \mathrm{d}H.$$

If $Z = iY$, then, substituting $H = VH'$ with $V \in GL_k(\mathbb{R})$ such that $Y[V] = E_k$, we get

$$\begin{aligned} I(Z) &= \int_{\mathsf{M}_{k,1}(\mathbb{R})} \exp(-\pi\, {}^tH'H') |\det V|\, \mathrm{d}H' \\ &= (\det Y)^{-1/2} \left(\int_{\mathbb{R}} \exp(-\pi h^2)\, dh \right)^k = \det(-iZ)^{-1/2}. \end{aligned}$$

The two sides are analytic functions on $\mathbb{H}_k$. So the relation holds on the whole of $\mathbb{H}_k$, by the analytic continuation principle with respect to each of the complex variables $z_{\alpha\beta}$ with $Z = (z_{\alpha\beta}) \in \mathbb{H}_k$. To go further, we need some notation. First we set

$$\Omega = \begin{pmatrix} W \\ W' \end{pmatrix} \in \mathsf{M}_{2k,1}(\mathbb{C}), \qquad \text{if } W, W' \in \mathsf{M}_{k,1}(\mathbb{C}),$$

and define

$$\theta(Z, \Omega) = \theta^k(Z; W, W'). \tag{1.3.7}$$

Then we denote by Γ^k the set of all integral symplectic matrices of order k, i.e.

$$\Gamma^k = \mathrm{Sp}_k(\mathbb{Z}) = \mathrm{Sp}_k(\mathbb{R}) \cap \mathsf{M}_{2k}. \tag{1.3.8}$$

It follows, by (1.2.9), that Γ^k is a group. The group Γ^k is called *the Siegel modular group of genus k*. For a matrix $M = \begin{pmatrix} A & B \\ C & D \end{pmatrix} \in \Gamma^k$, we shall denote by $\xi(M) = \mathrm{dc}(B\,{}^tA)$ (resp. $\eta(M) = \mathrm{dc}(C\,{}^tD)$, the k – column whose entries are the diagonal entries of $B\,{}^tA$ (resp. $C\,{}^tD$) and set

$$\zeta(M) = \begin{pmatrix} \xi(M) \\ \eta(M) \end{pmatrix} = \begin{pmatrix} \mathrm{dc}(B\,{}^tA) \\ \mathrm{dc}(C\,{}^tD) \end{pmatrix} \in \mathsf{M}_{2k,1}. \tag{1.3.9}$$

Finally, for $M = \begin{pmatrix} A & B \\ C & D \end{pmatrix} \in \Gamma^k$ and a complex-valued function F on $\mathbb{H}_k \times \mathsf{M}_{2k,1}(\mathbb{C})$, we define

$$(F|M)(Z,\Omega) \tag{1.3.10}$$
$$= \det(CZ+D)^{-1/2} \exp\left(-\frac{\pi i}{2} {}^t\zeta(M) J_k M\Omega \right) F\left(M\langle Z\rangle, M\Omega - \frac{1}{2}\zeta(M) \right).$$

The function $F|M$ is defined up to a sign, which depends on the sign of the first factor.

Remark. If φ is a holomorphic function on $\mathbb{H}_k$ which does not vanish, we denote by $\varphi^{1/2}$ one of two single-valued holomorphic functions which can be obtained by the analytic continuation of a local element $\pm\varphi^{1/2}$ over the simply connected space $\mathbb{H}_k$.

In the notation, the formulae (1.3.3)–(1.3.5) imply the relations

$$(\theta|M)(Z,\Omega) = \chi(M)\theta(Z,\Omega), \tag{1.3.11}$$

where M is one of the matrices

$$U(V) \text{ with } V \in \Lambda^k, \qquad T(S) \text{ with } S \in \mathsf{S}_k(\mathbb{Z}), \text{ or } J_k, \tag{1.3.12}$$

and where $\chi(M)$ is a complex number such that the product $\chi(M)\det(CZ+D)^{1/2}$ equals 1 in the first two cases and equals $\det(-iZ)^{1/2}$ in the third.

Proposition 1.3.2. *Let Γ'_k be the subgroup of Γ^k generated by all matrices* (1.3.12). *Then* (1.3.11) *holds for any $M \in \Gamma'_k$ with a multiplier $\chi(M)$. $\chi(M)$ satisfies $\chi(M)^8 = 1$ and depends on the choice of the sign on the left.*

Proof. Since the relation (1.3.11) holds for the generators of Γ', it suffices to show that

$$(\theta|M|M_1)(Z,\Omega) = \varepsilon\cdot(\theta|MM_1)(Z,\Omega) \qquad (M, M_1 \in \Gamma'), \tag{1.3.13}$$

where ε satisfies $\varepsilon^8 = 1$.

It is easy to see that the replacement of $\zeta(M)$ by $\zeta(M) + 2L$ with $L = \begin{pmatrix} L_1 \\ L_2 \end{pmatrix} \in$ $\mathsf{M}_{2k,1}$ multiplies $\theta|M$ by

$$\exp\left(-\frac{\pi i}{2}{}^tLJ_k\zeta(M) - \pi i\,{}^tL_1L_2\right)$$

which is a fourth root of unity.

Further, if $M = \begin{pmatrix} A & B \\ C & D \end{pmatrix}$ and $M_1 = \begin{pmatrix} A_1 & B_1 \\ C_1 & D_1 \end{pmatrix}$, it follows from the definition that

$$\begin{aligned}
&(\theta|M|M_1)(Z,\Omega)\\
&\quad= \det(C_1Z + D_1)^{-1/2}\exp\left(-\frac{\pi i}{2}{}^t\zeta(M_1)JM_1\Omega\right)\\
&\qquad\times(\theta|M)\left(M_1\langle Z\rangle, M_1\Omega - \frac{1}{2}\zeta(M_1)\right)\\
&\quad= \det(C_1Z + D_1)^{-1/2}\exp\left(-\frac{\pi i}{2}{}^t\zeta(M_1)JM_1\Omega\right)\det(CM_1\langle Z\rangle + D)^{-1/2}\\
&\qquad\times\exp\left(-\frac{\pi i}{2}{}^t\zeta(M)JM\left(M_1\Omega - \frac{1}{2}\zeta(M_1)\right)\right)\\
&\qquad\times\theta\left(M\langle M_1\langle Z\rangle\rangle, M\left(M_1\Omega - \frac{1}{2}\zeta(M_1)\right) - \frac{1}{2}\zeta(M)\right).
\end{aligned}$$

Let $MM_1 = \begin{pmatrix} A_2 & B_2 \\ C_2 & D_2 \end{pmatrix}$. Since

$$\begin{aligned}
CM_1\langle Z\rangle + D &= (C(A_1Z + B_1) + D(C_1Z + D_1))(C_1Z + D_1)^{-1}\\
&= (C_2Z + D_2)(C_1Z + D_1)^{-1},
\end{aligned} \tag{1.3.14}$$

and $\exp((\pi i/4)\,{}^t\zeta(M)JM\zeta(M_1))$ is an eighth root of unity, it follows that the last expression is equal to

$$\det(C_2Z + D_2)^{-1/2}\exp\left(-\frac{\pi i}{2}{}^t(M\zeta(M_1) + \zeta(M))JMM_1\Omega\right)$$

$$\cdot\,\theta\left(MM_1\langle Z\rangle, MM_1\Omega - \frac{1}{2}(M\zeta(M_1) + \zeta(M)\right)$$

up to a factor, an eighth root of unity (note that $J[M] = J$). If we prove that $M\zeta(M_1) + \zeta(M) \equiv \zeta(MM_1) \pmod 2$, it will follow that the last expression is $(\theta|MM_1)(Z,\Omega)$ up to a fourth root of unity. It will prove (1.3.13) and the proposition.

Lemma 1.3.3. *Let* $M, M_1 \in \Gamma^k$. *Then*

$$\zeta(MM_1) \equiv M\zeta(M_1) + \zeta(M) \pmod 2.$$

Proof. Writing $M = \begin{pmatrix} A & B \\ C & D \end{pmatrix}$ and $M_1 = \begin{pmatrix} A_1 & B_1 \\ C_1 & D_1 \end{pmatrix}$, we conclude that $MM_1 = \begin{pmatrix} AA_1 + BC_1 & AB_1 + BD_1 \\ CA_1 + DC_1 & CB_1 + DD_1 \end{pmatrix}$. Then, by the definition, we have

$$\zeta(MM_1) = \begin{pmatrix} \mathrm{dc}(AB_1 + BD_1)^t(AA_1 + BC_1) \\ \mathrm{dc}(CA_1 + DC_1)^t(CB_1 + DD_1) \end{pmatrix}.$$

It is easy to see that

$$\mathrm{dc}(MS\,{}^tM) \equiv M \cdot \mathrm{dc}\, S \pmod 2,$$

if M and S are integral matrices of the same order and S is symmetric. It follows that

$$\zeta(MM_1) \equiv \begin{pmatrix} A\,\mathrm{dc}(B_1\,{}^tA_1) + B\cdot\mathrm{dc}(C_1\,{}^tD_1) + \mathrm{dc}(A(B_1\,{}^tC_1 + A_1\,{}^tD_1)\,{}^tB) \\ C\,\mathrm{dc}(B_1\,{}^tA_1) + D\cdot\mathrm{dc}(C_1\,{}^tD_1) + \mathrm{dc}(C(A_1\,{}^tD_1 + B_1\,{}^tC_1)\,{}^tD) \end{pmatrix}$$
$$\equiv M\zeta(M_1) + \zeta(M) \pmod 2,$$

since, by (1.2.8), $A_1\,{}^tD_1 + B_1\,{}^tC_1 = E + 2B_1\,{}^tC_1 \equiv E \pmod 2$. Lemma 1.3.3 and Proposition 1.3.2 are proved. □

Exercise 1.3.4. *Show that the theta-function of genus k satisfies*

$$\theta^k(Z; W + L, W') = \exp(-\pi i\,{}^tLW')\theta^k(Z; W, W'),$$
$$\theta^k(Z; W + ZL, W') = \exp(\pi i(-Z[L] - 2\,{}^tLW + {}^tLZW'))\theta^k(Z; W, W')$$

for any $L \in \mathsf{M}_{k,1}$. *Show that each function* $F(Z; W, W')$, *which satisfies the relations and is holomorphic in* W, *is of the form*

$$F(Z; W, W') = F_0(Z, W')\theta^k(Z; W, W')$$

with a function F_0 *independent of* W.

Exercise 1.3.5. *Show that the theta-function of genus k satisfies*

$$\theta^k(Z; W, W' + L) = \exp(\pi i\,{}^tLW)\theta^k(Z; W, W'),$$
$$\theta^k(Z; W, W' - Z^{-1}L) = \exp(\pi i(Z^{-1}[L] - 2\,{}^tLW' + {}^tLZ^{-1}W))\theta^k(Z; W, W')$$

for any $L \in \mathsf{M}_{k,1}$. *Show that each function* $F(Z, W, W')$, *which satisfies the relations and is holomorphic in* W', *is of the form*

$$F(Z; W, W') = F_1(Z, W)\theta^k(-Z^{-1}; W', -W)$$

with a function F_1 *independent of* W'.

1.3.2. The Siegel Modular Group and the Theta-Group

Here we shall prove that the Siegel group Γ^k is generated by the matrices (1.3.12), i.e. $\Gamma'_k = \Gamma^k$. Hence, the functional equation (1.3.11) holds for any $M \in \Gamma^k$. The equation turns out to be especially simple, when M is in a subgroup of Γ^k, the so-called theta group.

Theorem 1.3.6. *The Siegel modular group of genus k,*

$$\Gamma^k = \mathrm{Sp}_k(\mathbb{Z}),$$

is generated by the matrices (1.3.12).

We shall prove a more general statement.

Proposition 1.3.7. *Let M be a matrix of order $2k$ with entries in $\mathbb{Q}$. If M satisfies*

$$^tMJ_kM = \mu J_k \tag{1.3.15}$$

with a non-zero scalar μ, then there exists a matrix $g \in \Gamma'_k$ such that

$$gM = \begin{pmatrix} A_1 & B_1 \\ 0 & D_1 \end{pmatrix},$$

where 0 is the zero matrix of order k.

Proof of Theorem 1.3.6 modulo Proposition 1.3.7. By the proposition, if $M \in \Gamma^k$, then there exists a matrix $g \in \Gamma'_k$ such that

$$M_1 = gM = \begin{pmatrix} A_1 & B_1 \\ 0 & D_1 \end{pmatrix}.$$

Since M_1 is symplectic, it follows that ${}^tA_1 D_1 = E_k$ and ${}^tB_1 D_1 = {}^tD_1 B_1$. Since M_1 is integral, it follows that $A_1, D_1 \in \Lambda^k$ and $S = B_1 D_1^{-1} = {}^t(B_1 D_1^{-1}) \in \mathrm{S}_k(\mathbb{Z})$. Thus, $M_1 = T(S)U(D_1)$, and $M = g^{-1}M_1 \in \Gamma'_k$. □

In order to prove the proposition, we first prove two lemmas.

Lemma 1.3.8. *Let $\mathbf{t} = {}^t(t_1, \ldots, t_k)$ be a non-zero integral column with $k \geq 2$. Then there exists a matrix $V \in \mathrm{SL}_k(\mathbb{Z})$ such that*

$$V\mathbf{t} = {}^t(d, 0, \ldots, 0), \tag{1.3.16}$$

where $d = g.c.d.(t_1, \ldots, t_k)$ (g.c.d. = the greatest common divisor).

Proof. Let $k = 2$. Since the numbers $\gamma = -t_2/d$, $\delta = t_1/d$ are coprime, it follows that there exist integers α, β such that $\alpha\delta - \beta\gamma = 1$. Thus,

$$\begin{pmatrix} \alpha & \beta \\ \gamma & \delta \end{pmatrix}\begin{pmatrix} t_1 \\ t_2 \end{pmatrix} = \begin{pmatrix} d \\ 0 \end{pmatrix} \quad \text{and} \quad \begin{pmatrix} \alpha & \beta \\ \gamma & \delta \end{pmatrix} \in \mathrm{SL}_2(\mathbb{Z}).$$

An obvious induction on k completes the proof. □

Lemma 1.3.9. *Let* $\mathbf{t}$ *be a non-zero integral* $2k$ – *column with* $k \geq 1$. *Then there exists a matrix* $g \in \Gamma'_k$ *such that*

$$g\mathbf{t} = {}^t(d, 0, \ldots, 0), \tag{1.3.17}$$

where d *is the greatest common divisor of the entries of* $\mathbf{t}$.

Proof. We shall write $\mathbf{t} = \begin{pmatrix} \mathbf{a} \\ \mathbf{c} \end{pmatrix}$, where $\mathbf{a}$ and $\mathbf{c}$ are k – columns. First we prove that the set $\{g\mathbf{t}; g \in \Gamma'\}$ contains a column $\mathbf{t}' = \begin{pmatrix} \mathbf{a}' \\ \mathbf{c}' \end{pmatrix}$ with $\mathbf{a}' = \mathbf{0}$ or $\mathbf{c}' = \mathbf{0}$. If the statement is false, we have $\mathbf{a}' \neq \mathbf{0}$ and $\mathbf{c}' \neq \mathbf{0}$ for every $\mathbf{t}' = g\mathbf{t}$. Let a' (resp. c') be the greatest common divisor of the entries of $\mathbf{a}'$ (resp. $\mathbf{c}'$). Let us choose $\mathbf{t}'$ so that the product $a'c'$ is minimal and $a' \geq c'$ (if $c' > a'$, we replace $\mathbf{t}'$ by $J_k\mathbf{t}'$). By Lemma 1.3.8, on replacing of $\mathbf{t}'$ by $U(V)\mathbf{t}'$ with a $V \in \mathrm{GL}_k(\mathbb{Z})$, we may assume that $\mathbf{c}' = {}^t(c', 0, \ldots, 0)$. The replacement of $\mathbf{t}'$ by $T(S)\mathbf{t}'$ carries $\mathbf{a}'$ to $\mathbf{a}' + S\mathbf{c}'$, and does not change $\mathbf{c}'$. It is clear that an integral symmetric matrix S of order k can be chosen in such a way that all entries of $\mathbf{a}' + S\mathbf{c}'$ are among the numbers $0, 1, \ldots, c' - 1$. Then the greatest common divisor δ of the entries satisfies $\delta < c' \leq a'$ and $\delta c' < a'c'$, contrary to the choice of a', c'. Thus, the set contains a column $\mathbf{t}'$ with $\mathbf{a}' = \mathbf{0}$ or $\mathbf{c}' = \mathbf{0}$. If $\mathbf{a}' = \mathbf{0}$, we replace $\mathbf{t}'$ by $J_k\mathbf{t}'$. So we can assume that $\mathbf{c}' = \mathbf{0}$. Then, by Lemma 1.3.8, the column $U(V)\mathbf{t}'$ with a matrix $V \in \mathrm{GL}_k(\mathbb{Z})$ is of the form (1.3.17). □

Proof of Proposition 1.3.7. We can assume that M is an integral matrix. We can also assume that the first column of M is of the form (1.3.17), by Lemma 1.3.9. If $k = 1$, this proves the proposition. Suppose that the proposition is proved for the matrices of orders $2k' < 2k$. The relation (1.3.15) for $M = \begin{pmatrix} A & B \\ C & D \end{pmatrix}$ is equivalent to the conditions

$${}^tAC = {}^tCA, \quad {}^tBD = {}^tDB, \quad {}^tAD - {}^tCB = \mu E_k. \tag{1.3.18}$$

By our assumption, we can write

$$A = \begin{pmatrix} a_{11} & a_{12} \cdots a_{1k} \\ 0 & \\ \vdots & A_0 \\ 0 & \end{pmatrix} \quad \text{with } a_{11} \neq 0, \qquad C = \begin{pmatrix} 0 & c_{12} \cdots c_{1k} \\ 0 & \\ \vdots & C_0 \\ 0 & \end{pmatrix}. \quad \square \tag{1.3.19}$$

Since ${}^tAC = {}^tCA$, it follows that

$$c_{12} = \cdots = c_{1k} = 0, \quad {}^tA_0C_0 = {}^tC_0A_0.$$

Then, by ${}^tAD = {}^tCB + \mu E_k$, we conclude that

$$D = \begin{pmatrix} d_{11} & 0 \cdots 0 \\ d_{21} & \\ \vdots & D_0 \\ d_{k1} & \end{pmatrix} \quad \text{and} \quad {}^t\!A_0 D_0 - {}^t C_0 B_0 = \mu E_{k-1},$$

where B_0 is the corresponding block of B. Finally, by ${}^tBD = {}^tDB$, we get ${}^tB_0 D_0 = {}^tD_0 B_0$. It follows that the matrix $M_0 = \begin{pmatrix} A_0 & B_0 \\ C_0 & D_0 \end{pmatrix}$ satisfies ${}^tM_0 J_k M_0 = \mu J_{k-1}$. By the inductive hypothesis, there exists $g_0 \in \Gamma'_{k-1}$ such that $g_0 M_0 = \begin{pmatrix} * & * \\ 0 & * \end{pmatrix}$. For a matrix $M' = \begin{pmatrix} A' & B' \\ C' & D' \end{pmatrix}$ of order $2k - 2 \geq 2$, we let $\hat{M}' = \begin{pmatrix} A'' & B'' \\ C'' & D'' \end{pmatrix}$ denote the matrix of order $2k$ with blocks

$$A'' = \begin{pmatrix} 1 & 0 \\ 0 & A' \end{pmatrix}, \quad B'' = \begin{pmatrix} 0 & 0 \\ 0 & B' \end{pmatrix}, \quad C'' = \begin{pmatrix} 0 & 0 \\ 0 & C' \end{pmatrix}, \quad \text{and} \quad D'' = \begin{pmatrix} 1 & 0 \\ 0 & D' \end{pmatrix}.$$

Then the C-block of $\hat{g}_0 M$ is the zero-block. Hence, to prove the proposition it suffices to show that the mapping $g_0 \to \hat{g}_0$ maps Γ'_{k-1} into Γ'_k. It is clear that the mapping defines a homomorphic imbedding of Γ^{k-1} into Γ^k. So it suffices to show that the mapping carries the generators of Γ'_{k-1} into Γ'_k. This is clear for all generators except J_{k-1}. For J_{k-1}, we have

$$\hat{J}_{k-1} = \begin{pmatrix} E^1 & E_k - E^1 \\ E^1 - E_k & E^1 \end{pmatrix} = -(J_k T(E^1))^3 \in \Gamma'_k \tag{1.3.20}$$

where E^1 is the matrix of order k of the form $E^1 = \operatorname{diag}(1, 0, \ldots, 0)$. □

It follows that the functional equation (1.3.11) holds for each $M \in \Gamma^k$. If $\zeta(M) \equiv 0 \pmod 2$, we can assume that $\zeta(M) = 0$, by a remark in the beginning of the proof of Proposition 1.3.2. Then the functional equation (1.3.11) turns into

$$\det(CZ + D)^{-1/2} \theta(M\langle Z\rangle, M\Omega) = \chi(M)\theta(Z, \Omega), \tag{1.3.21}$$

where $\chi(M)$ is an eighth root of unity. It follows from Lemma 1.3.3 that the set

$$\theta^k = \{M \in \Gamma^k; \zeta(M) \equiv 0 \pmod 2\}$$

is a subgroup of Γ^k. Returning to the original notation, we obtain

Theorem 1.3.10. *The set*

$$\Theta^k = \left\{ M = \begin{pmatrix} A & B \\ C & D \end{pmatrix} \in \Gamma^k; \operatorname{dc}(B\,{}^t\!A) \equiv \operatorname{dc}(C\,{}^tD) \equiv 0 \pmod 2 \right\}$$

is a subgroup of Γ^k. *For any* $M = \begin{pmatrix} A & B \\ C & D \end{pmatrix} \in \Theta^k$, *the theta-function* $\theta^k(Z; W, W')$ *satisfies*

$$\det(CZ + D)^{-1/2}\theta^k((AZ + B)(CZ + D)^{-1}; AW + BW', CW + DW')$$
$$= \chi(M)\theta^k(Z; W, W'),$$

where $\chi(M)$ is an eighth root of unity.

The group Θ^k is called *the theta-group of genus k.*

Exercise 1.3.11 (Witt). *Show that the group Θ^k is generated by the matrices $U(V)$ with $V \in \Lambda^k$, $T(S)$ with $S \in \mathbb{E}_k$ (integral symmetric matrices of order k with even diagonal entries), J_k, and matrices of the form* $\begin{pmatrix} E^i & E_k - E^i \\ E^i - E_k & E^i \end{pmatrix}$ *with* $E^i = \operatorname{diag}(\underbrace{1,\dots,1}_{i},\underbrace{0,\dots,0}_{k-i})$ *for $i = 1, 2, \dots, k-1$.*

1.3.3. Symplectic Transformations of Theta-Series

Here we shall consider the action of symplectic transformations on the theta-series. First it is convenient to generalize the definition of theta-series.

Let $Q \in \mathsf{S}_m(\mathbb{R})$ with $Q > 0$, $Z \in \mathbb{H}_n$, and $W, W' \in \mathsf{M}_{m,n}(\mathbb{C})$, where $m, n \in \mathbb{N}$. Then the series

$$\theta^n(Z, Q; (W, W')) = \theta^n(Z, Q, \Omega)$$
$$= \sum_{N \in \mathsf{M}_{m,n}} e\{Q[N - W']Z + 2\,{}^tNQW - {}^tWQW'\} \qquad (1.3.22)$$

with $\Omega = (W, W') \in \mathsf{M}_{m,2n}(\mathbb{C})$, where for a square matrix T we set

$$e\{T\} = \exp(\pi i \sigma(T)), \qquad (1.3.23)$$

and $\sigma(T)$ is the trace of T, converges absolutely and uniformly for $Z \in \mathbb{H}_n(\varepsilon)$ with $\varepsilon > 0$ and Ω in a fixed compact subset of $\mathsf{M}_{m,2n}(\mathbb{C})$. This can be proved in the same way as Proposition 1.1.3. The series (1.3.22) will be called *the theta-function of genus n of* $\mathbb{Q}$.

Since clearly $\theta^n(Z, (1), (W, W')) = \theta^n(Z; {}^tW, {}^tW')$, the theta-function (1.3.2) is a special case of (1.3.22). Conversely, it turns out that every theta-function (1.3.22) is a restriction of a certain theta-function (1.3.2). This will allow us to reduce the study of symplectic transformations of general theta-functions to the special case considered above.

First we recall the definition and the simplest properties of the tensor product of two matrices. Let A and $B = (b_{\alpha\beta})$ be two square matrices of order m and n respectively with entries in a field F. The matrix

$$A \otimes B = \begin{pmatrix} Ab_{11} \cdots Ab_{1n} \\ \cdots\cdots\cdots \\ \cdots\cdots\cdots \\ Ab_{n1} \cdots Ab_{nn} \end{pmatrix} \in \mathsf{M}_{mn}(\mathsf{F})$$

is called *the tensor product of A and B.* The tensor product is clearly linear in each argument and satisfies

$$(A \otimes B) \cdot (A_1 \otimes B_1) = AA_1 \otimes BB_1.$$

It follows that the matrix $A \otimes B$ is invertible if and only if the matrices A and B are, and

$$(A \otimes B)^{-1} = A^{-1} \otimes B^{-1}.$$

Also

$$\det(A \otimes B) = \det(A \otimes E_n) \det(E_m \otimes B) = (\det A)^n (\det B)^m,$$

and

$${}^t(A \otimes B) = {}^tA \otimes {}^tB.$$

Finally, if $\mathsf{F} = \mathbb{R}$ and the matrices A, B are symmetric and positive definite, then $A \otimes B$ is also positive definite.

Lemma 1.3.12. *With the above notation and assumptions, the following identity holds:*

$$\theta^n(Z, Q, (W, W')) = \theta^{mn}(Q \otimes Z; \mathbf{c}(QW), \mathbf{c}(W')) \tag{1.3.24}$$

where the function on the right is the theta-function (1.3.2) *and for a matrix* $T = (\mathbf{t}_1, \ldots, \mathbf{t}_n) \in M_{m,n}(\mathbb{C})$ *with the columns* $\mathbf{t}_\alpha$, $\mathbf{c}(T)$ *denotes the column*

$$\mathbf{c}(T) = \begin{pmatrix} \mathbf{t}_1 \\ \vdots \\ \mathbf{t}_n \end{pmatrix} \in \mathsf{M}_{mn,1}(\mathbb{C}).$$

Besides, if $M = \begin{pmatrix} A & B \\ C & D \end{pmatrix} \in \mathsf{Sp}_n(\mathbb{R})$, *then the matrix*

$$M_Q = \begin{pmatrix} A_Q & B_Q \\ C_Q & D_Q \end{pmatrix} = \begin{pmatrix} E_m \otimes A & Q \otimes B \\ Q^{-1} \otimes C & E_m \otimes D \end{pmatrix}$$

belongs to $\mathsf{Sp}_{mn}(\mathbb{R})$ *and the following identities hold:*

$$\begin{aligned} &\theta^n(M\langle Z \rangle; Q, (W, W')\,{}^tM) &(1.3.25)\\ &\quad = \theta^{mn}(M_Q\langle Q \otimes Z \rangle; A_Q\mathbf{c}(QW) + B_Q\mathbf{c}(W'), C_Q\mathbf{c}(QW) + D_Q\mathbf{c}(W')), \end{aligned}$$

$$\det(CZ + D)^m = \det(C_Q \cdot (Q \otimes Z) + D_Q). \tag{1.3.26}$$

Proof. First, in the notation, we have

$$(Q \otimes Z)[\mathbf{c}(T)] = ({}^t\mathbf{t}_1, \ldots, {}^t\mathbf{t}_n) \begin{pmatrix} Qz_{11} \cdots Qz_{1n} \\ \cdots\cdots\cdots \\ \cdots\cdots\cdots \\ Qz_{n1} \cdots Qz_{nn} \end{pmatrix} \begin{pmatrix} \mathbf{t}_1 \\ \vdots \\ \mathbf{t}_n \end{pmatrix}$$

$$= \sum_{\alpha, \beta = 1}^{n} z_{\alpha\beta}\, {}^t\mathbf{t}_\alpha Q \mathbf{t}_\beta = \sigma(Q[T]Z).$$

Also, for $T, V \in \mathsf{M}_{m,n}(\mathbb{C})$, we have

$$ {}^t\mathbf{c}(T)\mathbf{c}(V) = \sum_{\alpha=1}^{n} {}^t\mathbf{t}_\alpha \mathbf{v}_\alpha = \sigma({}^tTV). $$

Hence we get

$$ \begin{aligned} &\theta^n(Z, Q, (W, W')) \\ &\quad = \sum_{\mathbf{c}(N) \in \mathsf{M}_{mn,1}} \exp(\pi i((Q \otimes Z)[\mathbf{c}(N) - \mathbf{c}(W')] + 2\,{}^t\mathbf{c}(N) \\ &\qquad \cdot \mathbf{c}(QW) - {}^t\mathbf{c}(W')\mathbf{c}(QW))) \\ &\quad = \theta^{mn}(Q \otimes Z; \mathbf{c}(QW), \mathbf{c}(W')), \end{aligned} $$

which proves the first assertion of the lemma.

To prove the inclusion $M_Q \in \mathsf{Sp}_{mn}(\mathbb{R})$, it suffices to show that the blocks of M_Q satisfy (1.2.7). By the proverties of the tensor product, we have

$$ \begin{aligned} {}^tC_Q A_Q &= (Q^{-1} \otimes {}^tC)(E_m \otimes A) = Q^{-1} \otimes {}^tCA \\ &= Q^{-1} \otimes {}^tAC = {}^t(E_m \otimes A)(Q^{-1} \otimes C) = {}^tA_Q \cdot C_Q. \end{aligned} $$

In the same way, we have

$$ {}^tB_Q D_Q = Q \otimes {}^tBD = Q \otimes {}^tDB = {}^tD_Q B_Q. $$

Finally,

$$ {}^tA_Q D_Q - {}^tC_Q B_Q = (E_m \otimes A\,{}^tD) - (E_m \otimes {}^tCD) = E_m \otimes E_n = E_{mn}. $$

In order to prove (1.3.25), we only have to show that

$$ \begin{gathered} Q \otimes M\langle Z\rangle = M_Q\langle Q \otimes Z\rangle, \\ \mathbf{c}(Q(W\,{}^tA + W'\,{}^tB)) = A_Q\mathbf{c}(QW) + B_Q\mathbf{c}(W'), \end{gathered} \tag{1.3.27} $$

and

$$ \mathbf{c}(W\,{}^tC + W'\,{}^tD) = C_Q\mathbf{c}(QW) + D_Q\mathbf{c}(W'). $$

We have

$$ \begin{aligned} A_Q(Q \otimes Z) + B_Q &= (E_m \otimes A)(Q \otimes Z) + (Q \otimes B) = Q \otimes (AZ + B), \\ C_Q(Q \otimes Z) + D_Q &= (Q^{-1} \otimes C)(Q \otimes Z) + (E_m \otimes D) = E_m \otimes (CZ + D), \end{aligned} $$

whence we get (1.3.27) and the relation

$$ \det(C_Q(Q \otimes Z) + D_Q) = \det(CZ + D)^m. $$

Finally, if $A = (a_{\alpha\beta})$, $B = (b_{\alpha\beta})$, $W = (\mathbf{w}_1, \ldots, \mathbf{w}_n)$, and $W' = (\mathbf{w}'_1, \ldots, \mathbf{w}'_n)$, then we have

$$A_Q\mathbf{c}(QW) + B_Q\mathbf{c}(W')$$
$$= \begin{pmatrix} E_m a_{11} \cdots E_m a_{1n} \\ \cdots\cdots\cdots \\ \cdots\cdots\cdots \\ E_m a_{n1} \cdots E_m a_{nn} \end{pmatrix} \begin{pmatrix} Q\mathbf{w}_1 \\ \vdots \\ Q\mathbf{w}_n \end{pmatrix} + \begin{pmatrix} Qb_{11} \cdots Qb_{1n} \\ \cdots\cdots\cdots \\ \cdots\cdots\cdots \\ Qb_{n1} \cdots Qb_{nn} \end{pmatrix} \begin{pmatrix} \mathbf{w}'_1 \\ \vdots \\ \mathbf{w}'_n \end{pmatrix}$$
$$= \begin{pmatrix} Q(\mathbf{w}_1 a_{11} + \cdots + \mathbf{w}_n a_{1n}) \\ \vdots \\ Q(\mathbf{w}_1 a_{n1} + \cdots + \mathbf{w}_n a_{nn}) \end{pmatrix} + \begin{pmatrix} Q(\mathbf{w}'_1 b_{11} + \cdots + \mathbf{w}'_n b_{1n}) \\ \vdots \\ Q(\mathbf{w}'_1 b_{n1} + \cdots + \mathbf{w}'_n b_{nn}) \end{pmatrix}$$
$$= \mathbf{c}(Q(W\,{}^t\!A + W'\,{}^t\!B)).$$

The last relation is proved in the same way. □

Let Q be the matrix of a non-degenerate integral quadratic form q in m variables, i.e. $Q \in \mathbb{E}_m$, $\det Q \neq 0$. The least positive integer $q = q(Q) = q(\mathsf{q})$ such that $qQ^{-1} \in \mathbb{E}_m$ is called *the level of Q* (or of q)

Theorem 1.3.13. *Let q be the level of a matrix $Q \in \mathbb{A}_m^+$, and let $M = \begin{pmatrix} A & B \\ C & D \end{pmatrix}$ be a matrix of the group*

$$\Gamma_0^n(q) = \left\{\begin{pmatrix} A & B \\ C & D \end{pmatrix} \in \Gamma^n; C \equiv 0 \pmod q\right\} \tag{1.3.28}$$

with $m, n \in \mathbb{N}$. Then the theta-function (1.3.22) *of genus n of Q satisfies the functional equation*

$$\det(CZ + D)^{-m/2}\theta^n(M\langle Z\rangle, Q, \Omega\,{}^tM) = \chi_Q^n(M)\theta^n(Z, Q, \Omega), \tag{1.3.29}$$

where $\chi_Q^n(M)$ is an eighth root of unity depending on the choice of the square root on the left if m is odd; in particular, the theta-series (1.1.13) *of genus n of Q satisfies*

$$\det(CZ + D)^{-m/2}\theta^n(M\langle Z\rangle, Q) = \chi_Q^n(M)\theta^n(Z, Q) \tag{1.3.30}$$

for any $M = \begin{pmatrix} A & B \\ C & D \end{pmatrix} \in \Gamma_0^n(q)$.

Proof. Let M_Q be as in Lemma 1.3.12. Thus, by the lemma, $M_Q \in \mathrm{Sp}_{mn}(\mathbb{R})$. It follows from the definitions that M_Q is an integral matrix. Hence, $M_Q \in \Gamma^{mn}$. Finally, all diagonal entries of $B_Q\,{}^t\!A_Q = Q \otimes B\,{}^t\!A$ and $C_Q\,{}^tD_Q = Q^{-1} \otimes C\,{}^tD = qQ^{-1} \otimes (q^{-1}C\,{}^tD)$ are even, for the first factors of the tensor products have even diagonal entries, and the second factors are integral matrices. It follows that $M_Q \in \Theta^{mn}$. Thus, by Lemma 1.3.12 and Theorem 1.3.10, we get

$$\det(CZ + D)^{-m/2}\theta^n(M\langle Z\rangle, Q, \Omega\,{}^tM)$$
$$= \det(C_Q(Q \otimes Z) + D_Q)^{-1/2}\theta^{mn}(M_Q\langle Q \otimes Z\rangle; A_Q\mathbf{c}(QW)$$

$$+ B_Q\mathbf{c}(W'), C_Q\mathbf{c}(QW) + D_Q\mathbf{c}(W'))$$
$$= \chi(M_Q)\theta^{mn}(Q \otimes Z; \mathbf{c}(QW), \mathbf{c}(W'))$$
$$= \chi_Q^n(M)\theta^n(Z, Q, (W, W')),$$

where $\chi_Q^n(M) = \chi(M_Q)$, which proves (1.3.29). The relation (1.3.30) follows from (1.3.29) by setting $\Omega = 0_{m,n}$. □

When we transform a theta-series by an integral symplectic matrix which does not belong to the corresponding group $\Gamma_0^n(q)$, in general, the image is not proportional to the original. However, it stays in a finite-dimensional space, the space of generalized theta-series. Let $Q \in \mathbb{A}_m^+$, and let q be the level of Q. For a matrix T in the set

$$\mathsf{T}^n(Q) = \{T \in \mathsf{M}_{m,n}; QT \equiv 0 \ (\mathrm{mod}\ q)\} \tag{1.3.31}$$

we define *the generalized theta-series of genus n of Q by*

$$\begin{aligned}\theta^n(Z, Q \mid T) &= \theta^n(Z, Q, (0, -q^{-1}T)) \\ &= \sum_{N \in \mathsf{M}_{m,n}} e\{Q[N + q^{-1}T]Z\}.\end{aligned} \tag{1.3.32}$$

It is clear that the generalized theta-series depends only on T modulo q, and

$$\theta^n(Z, Q \mid T) = \theta^n(Z, Q) \quad \text{if } T \equiv 0 \ (\mathrm{mod}\ q)$$

It follows that the set of all linear combinations of the generalized theta-series of genus n of Q is a finite-dimensional space containing the theta-series $\theta^n(Z, Q)$.

Proposition 1.3.14. *The generalized theta-series of genus n of $Q \in \mathbb{A}_m^+$ satisfy*

$$\theta^n(U(V^*)\langle Z\rangle, Q \mid T) = \theta^n(Z, Q \mid TV), \qquad \textit{for } V \in \Lambda^n,$$
$$\theta^n(T(S)\langle Z\rangle, Q \mid T) = e\{q^{-2}Q[T]S\}\theta^n(Z, Q \mid T), \qquad \textit{for } S \in \mathsf{S}_n(\mathbb{Z})$$

and

$$\theta^n(J_n\langle Z\rangle, Q \mid T)$$
$$= (\det Q)^{-n/2}(\det(-iZ))^{m/2} \sum_{T' \in \mathsf{T}^n(Q)/\mathrm{mod}\ q} e\{2q^{-2}\,{}^tTQT'\}\theta^n(Z, Q \mid T'),$$

where $U(V)$, $T(S)$ and J_n are the generators (1.3.12) *of Γ^n. The square root $(\det(-iZ))^{1/2}$ is to be taken positive when $Z = iY$; in general, the sign is found by analytic continuation in the simply connected space $\mathbb{H}_n$.*

First we prove a generalization of Lemma 1.3.1.

Lemma 1.3.15 (*The inversion formula for the theta-function of genus n of Q*). *The theta-function* (1.3.22) *satisfies*

$$\begin{aligned}\theta^n(-Z^{-1}, Q^{-1}, (QW', -QW)) \\ = (\det Q)^{n/2}(\det(-iZ))^{m/2}\theta^n(Z, Q, (W, W')),\end{aligned} \tag{1.3.33}$$

where $(\det(-iZ))^{1/2}$ *is defined as above.*

Proof. We shall deduce (1.3.33) from (1.3.5) by the use of Lemma 1.3.12. By (1.3.24) and the properties of the tensor product, we have

$$\begin{aligned}\theta^n(-Z^{-1}, Q^{-1}, (QW', -QW)) &= \theta^{mn}(Q^{-1} \otimes (-Z^{-1}); \mathbf{c}(W'), -\mathbf{c}(QW)) \\ &= \theta^{mn}(-(Q \otimes Z)^{-1}; \mathbf{c}(W'), -\mathbf{c}(QW)).\end{aligned}$$

By Lemma 1.3.1, the last expression equals

$$\det(-i(Q \otimes Z))^{1/2}\theta^{mn}(Q \otimes Z; \mathbf{c}(QW), \mathbf{c}(W')),$$

which can be written as

$$(\det Q)^{n/2}(\det(-iZ))^{m/2}\theta^n(Z, Q, (W, W')),$$

by (1.3.24) and the properties of the tensor product. □

Proof of Proposition 1.3.14. By the definition, we have

$$\begin{aligned}\theta^n(Z[{}^tV]; Q|T) &= \sum_{N \in \mathsf{M}_{m,n}} e\{{}^tVQ[N + q^{-1}T]VZ\} \\ &= \sum_{N \in \mathsf{M}_{m,n}} e\{Q[NV + q^{-1}TV]Z\} \\ &= \sum_{N' \in \mathsf{M}_{m,n}} e\{Q[N' + q^{-1}TV]Z\},\end{aligned}$$

since $\mathsf{M}_{m,n}V = \mathsf{M}_{m,n}$ for $V \in \Lambda^n$, which proves the first formula. To prove the second, we note that

$$\begin{aligned}\sigma(Q[N + q^{-1}T]S) &= \sigma(Q[N]S) + 2\sigma({}^tNq^{-1}QTS) + \sigma(Q[q^{-1}T]S) \\ &\equiv \sigma(Q[q^{-1}T]S) \pmod 2\end{aligned}$$

with $N \in \mathsf{M}_{m,n}$ and $S \in \mathsf{S}_n(\mathbb{Z})$, since all diagonal entries of $Q[N]$ together with those of Q are even, and $q^{-1}QT$ is integral. It follows that

$$\begin{aligned}\theta^n(Z + S, Q|T) &= \sum_{N \in \mathsf{M}_{m,n}} e\{Q[N + q^{-1}T]Z\}e\{Q[N + q^{-1}T]S\} \\ &= \sum_{N \in \mathsf{M}_{m,n}} e\{Q[N + q^{-1}T]Z\} \cdot e\{Q[q^{-1}T]S\}.\end{aligned}$$

To prove the third formula, we apply (1.3.33) with Q^{-1} instead of Q, $W' = 0$, and $W = q^{-1}QT$ to the theta-series $\theta^n(Z, Q|T) = \theta^n(Z, Q, (0, -q^{-1}T))$. We have

$$\begin{aligned}\theta^n(-Z^{-1}, Q|T) &= (\det Q)^{-n/2}(\det(-iZ))^{m/2}\theta^n(Z, Q^{-1}, (q^{-1}QT, 0)) \\ &= (\det Q)^{-n/2}(\det(-iZ))^{m/2} \sum_{N \in \mathsf{M}_{m,n}} e\{Q^{-1}[N]Z + 2q^{-1}\,{}^tNT\}.\end{aligned}$$

For every $L \in \mathsf{M}_{m,n}$, the matrix $qQ^{-1}L$ is clearly in $\mathsf{T}^n(Q)$. Conversely, any $T' \in \mathsf{T}^n(Q)$ is of the form $qQ^{-1}L$ with $L \in \mathsf{M}_{m,n}$. Thus, the mapping $L \to qQ^{-1}L$ is an isomorphism of $\mathsf{M}_{m,n}$ onto $\mathsf{T}^n(Q)$. It maps the subgroup $Q\,\mathsf{M}_{m,n}$ onto $q\,\mathsf{M}_{m,n}$. Hence, the mapping induces an isomorphism of the factor groups:

$$\mathsf{M}_{m,n}/Q\,\mathsf{M}_{m,n} \to \mathsf{T}^n(Q)/\mathrm{mod}\, q.$$

Extending the above equality, we get

$$= (\det Q)^{-n/2}(\det(-iZ))^{m/2} \sum_{\substack{N \in \mathsf{M}_{m,n} \\ L \in \mathsf{M}_{m,n}/Q\mathsf{M}_{m,n}}} e\{Q^{-1}[QN + L]Z + 2q^{-1}\,{}^t(QN + L)T\}$$

$$= (\det Q)^{-n/2}(\det(-iZ))^{m/2} \sum_{N,L} e\{Q[N + q^{-1}(qQ^{-1}L)]Z$$

$$+ 2q^{-1}\,{}^t(N + q^{-1}(qQ^{-1}L))QT\}$$

$$= (\det Q)^{-n/2}(\det(-iZ))^{m/2} \sum_{\substack{N \in \mathsf{M}_{m,n} \\ T' \in \mathsf{T}^n(Q)/\mathrm{mod}\, q}} e\{Q[N + q^{-1}T']Z + 2q^{-2}\,{}^tT'QT\},$$

for $q^{-1}\,{}^tNQT = {}^tN(q^{-1}QT)$ is an integral matrix. □

Exercise 1.3.16. *Show that the set* $\mathsf{T}^n(Q)/\mathrm{mod}\, q$ *consists of* $(\det Q)^n$ *elements.*

Exercise 1.3.17. *Let* $Q \in \mathbb{A}_m^+$, *and* $T \in \mathsf{T}^n(Q)$. *Suppose that* $M = \begin{pmatrix} A & B \\ C & D \end{pmatrix} \in \Gamma_0^n(q)$, *where* q *is the level of* Q. *Show that* $\theta^n(Z, Q \mid T)$ *satisfies*

$$\det(CZ + D)^{-m/2}\theta^n(M\langle Z\rangle, Q \mid T) = \chi_Q^n(M)e\{q^{-2}A\,{}^tBQ[T]\}\theta^n(Z, Q \mid TA)$$

with the same scalar $\chi_Q^n(M)$ *as in Theorem* 1.3.13; *in particular, if* $M \equiv E_{2n} \pmod q$, *then*

$$\det(CZ + D)^{-m/2}\theta^n(M\langle Z\rangle, Q \mid T) = \chi_Q^n(M)\theta^n(Z, Q \mid T).$$

Exercise 1.3.18. *In the above notation, let* $T_1, \ldots, T_d$ *be a set of representatives of* $\mathsf{T}^n(Q)$ *modulo* q, *and let* $\Theta^n(Z, Q)$ *be the column with the entries* $\theta^n(Z, Q \mid T_1), \ldots, \theta^n(Z, Q \mid T_d)$. *Show that* $\Theta^n(Z, Q)$ *satisfies*

$$\det(CZ + D)^{-m/2}\Theta^n(M\langle Z\rangle, Q) = X_Q^n(M)\Theta^n(Z, Q)$$

for each $M \in \Gamma^n$, *where* $X_Q^n(M)$ *is an unitary matrix.*

§ 1.4. Computation of the Multiplier

The scalar factor $\chi_Q^n(M)$, which has appeared in the functional equations for the theta-function and the theta-series of genus n of Q, will be called *the multiplier*. Here we shall compute the multiplier.

1.4.1. Factors of Automorphy

We assume that $Q \in \mathbb{A}_m^+$, q is the level of Q, and $n \in \mathbb{N}$. For $M \in \Gamma_0^n(q)$ and $Z \in \mathbb{H}_n$, we set

$$j_Q^n(M, Z) = \theta^n(M\langle Z\rangle, Q)\theta^n(Z, Q)^{-1}. \tag{1.4.1}$$

By the definition, the function $j_Q^n\colon \Gamma_0^n(q) \times \mathbb{H}_n \to \mathbb{C}$ is meromorphic in the second argument. By Theorem 1.3.13, the function can be written in the form

$$j_Q^n(M, Z) = \det(CZ + D)^{m/2}\chi_Q^n(M) \qquad \left(M = \begin{pmatrix} A & B \\ C & D \end{pmatrix}\right).$$

It follows that $j_Q^n(M, Z)$ does not vanish on $\mathbb{H}_n$ and is holomorphic. Finally, it satisfies

$$\begin{aligned} j_Q^n(MM_1, Z) &= \theta^n(MM_1\langle Z\rangle, Q)\theta^n(Z, Q)^{-1} \\ &= \theta^n(M\langle M_1\langle Z\rangle\rangle, Q)\theta^n(M_1\langle Z\rangle, Q)^{-1}\theta^n(M_1\langle Z\rangle, Q)\theta^n(M, Z)^{-1} \\ &= j_Q^n(M, M_1\langle Z\rangle)j_Q^n(M_1, Z) \qquad \text{for } M,\ M_1 \in \Gamma_0^n(q) \text{ and } Z \in \mathbb{H}_n. \end{aligned}$$

It is now convenient to consider a general situation. Let S be a semigroup which operates on a set H, say $g\colon h \to g(h)$ for $g \in \mathsf{S}$ and $h \in \mathsf{H}$, as a subsemigroup of the group of all one-to-one mappings of H onto itself. A mapping $\varphi\colon \mathsf{S} \times \mathsf{H} \to \mathsf{T}$, where T is a group, is called a T-*valued factor of automorphy of* S *on* H if it satisfies

$$\varphi(gg_1, h) = \varphi(g, g_1(h))\varphi(g_1, h)$$

for all $g, g_1 \in \mathsf{S}$, and $h \in \mathsf{H}$.

Lemma 1.4.1. *Let* $\varphi\colon \mathsf{S} \times \mathsf{H} \to \mathsf{T}$ *be a* T-*valued factor of automorphy of* S *on* H. *Then the following assertions hold*:

(1) *If* $f\colon \mathsf{T} \to \mathsf{T}_1$ *is a homomorphism of* T *into a group* T_1, *then the mapping*

$$\mathsf{S} \times \mathsf{H} \ni (g, h) \to f(\varphi(g, h)) \in \mathsf{T}_1$$

is a T_1-*valued factor of automorphy of* S *on* H;

(2) *Let* $\chi\colon \mathsf{S} \to \mathsf{T}$ *be a mapping, and let the image of* S *be contained in the centre of* T. *Then the mapping* $(g, h) \to \chi(g)\varphi(g, h)$ *is a factor of automorphy of* S *on* H *if and only if* χ *is a homomorphism.*

(3) *For a mapping* $F\colon \mathsf{H} \to \mathsf{V}$, *where* V *is a left* T-*module, and* $g \in \mathsf{S}$, *let us define the mapping* $F|g\colon \mathsf{H} \to \mathsf{V}$ *by*

$$(F|g)(h) = (F|_\varphi g)(h) = \varphi(g, h)^{-1}F(g(h)).$$

Then it satisfies

$$F|g|g_1 = F|gg_1 \qquad (g, g_1 \in \mathsf{S}).$$

Proof. The assertions directly follow from the definitions. For example, to prove (3), we have

$$\begin{aligned}(F|g|g_1)(h) &= \varphi(g_1,h)^{-1}(F|g)(g_1(h))\\ &= \varphi(g_1,h)^{-1}\varphi(g,g_1(h))^{-1}F(g(g_1h))\\ &= \varphi(gg_1,h)^{-1}F(gg_1(h))\\ &= (F|gg_1)(h).\end{aligned}$$

□

The considerations in the beginning of the section show that j_Q^n is a $\mathbb{C}^*$-valued factor of automorphy of the group $\Gamma_0^n(q)$ on $\mathbb{H}_n$, where q is the level of Q. Other examples are given by the following lemma.

Lemma 1.4.2. (1) *Let* $M = \begin{pmatrix} A & B \\ C & D \end{pmatrix}$ *be a matrix of the group*

$$\mathsf{G} = \mathsf{GSp}_n^+(\mathbb{R}) = \{M \in \mathsf{M}_{2n}(\mathbb{R}); {}^tMJ_nM = \mu(M)J_n, \mu(M) > 0\}. \tag{1.4.2}$$

Then the matrix $CZ + D$ is invertible for all $Z \in \mathbb{H}_n$, and the mapping

$$f_M\colon Z \to M\langle Z\rangle = (AZ + B)(CZ + D)^{-1}$$

is an analytic automorphism of $\mathbb{H}_n$. The mapping $M \to f_M$ is a homomorphism of G into the group of analytic automorphisms of $\mathbb{H}_n$.

(2) *Let $\rho\colon \mathsf{GL}_n(\mathbb{C}) \to \mathsf{GL}_m(\mathbb{C})$ be a homomorphism. Then the mapping*

$$(M, Z) \to \rho(CZ + D) \in \mathsf{GL}_m(\mathbb{C})$$

is a $\mathsf{GL}_m(\mathbb{C})$-valued factor of automorphy of G on $\mathbb{H}_n$; in particular

$$j(M, Z)^k = \det(CZ + D)^k \quad \text{with } k \in \mathbb{Z} \tag{1.4.3}$$

is a $\mathbb{C}^*$-valued factor of automorphy of G on $\mathbb{H}_n$.

Proof. It is obvious that $\mu(\lambda M) = \lambda^2\mu(M)$, for $\lambda \in \mathbb{R}$ and $M \in \mathsf{G}$. Hence, $\lambda M \in \mathsf{Sp}_n(\mathbb{R})$, if $\lambda = \mu(M)^{-1/2}$. Then the first statement follows from Proposition 1.2.1 and Theorem 1.2.2.

Now, if $M = \begin{pmatrix} A & B \\ C & D \end{pmatrix}$, $M_1 = \begin{pmatrix} A_1 & B_1 \\ C_1 & D_1 \end{pmatrix}$, and $MM_1 = \begin{pmatrix} A_2 & B_2 \\ C_2 & D_2 \end{pmatrix}$, we have

$$\begin{aligned}&\rho(C(A_1Z + B_1)(C_1Z + D_1)^{-1} + D)\rho(C_1Z + D_1)\\ &\quad = \rho((CA_1 + DC_1)Z + (CB_1 + DD_1)) = \rho(C_2Z + D_2).\end{aligned}$$

□

1.4.2. Quadratic Forms of Level 1

Proposition 1.4.3. *Let $Q \in \mathbb{A}_m^+$, where $m \in \mathbb{N}$. Suppose that the level of Q is 1. Then $m \equiv 0 \pmod 8$, and $\chi_Q^n(M) = 1$ for every $n \in \mathbb{N}$ and $M \in \Gamma^n$.*

Proof. Since the level of Q is 1, it follows that the matrix Q^{-1} is integral. Hence, $\det Q = \pm 1$. Since $Q > 0$, it follows that $\det Q = 1$.

By the inversion formula (1.3.33) with $W = W' = 0$, we have

$$\theta^n(-Z^{-1}, Q^{-1}) = \det(-iZ)^{m/2}\theta^n(Z, Q).$$

On the other hand, since Q^{-1} is unimodular, it follows that

$$\begin{aligned}\theta^n(Z', Q^{-1}) &= \sum_{N \in \mathbb{M}_{m,n}} e\{Q^{-1}[N]Z'\} \\ &= \sum_{N \in \mathbb{M}_{m,n}} e\{Q[Q^{-1}N]Z'\} = \theta^n(Z'; Q) \qquad (Z' \in \mathbb{H}_n).\end{aligned}$$

Thus,

$$\theta^n(-Z^{-1}, Q) = \det(-iZ)^{m/2}\theta^n(Z, Q). \tag{1.4.4}$$

In particular, if $n = 1$, we have

$$\theta^1(-z^{-1}, Q) = (-iz)^{m/2}\theta^1(z, Q).$$

On the other hand, it is obvious that

$$\theta^1(z + 1, Q) = \theta^1(z, Q).$$

It follows from the relations that

$$\theta^1(-(z + 1)^{-1}, Q) = (-i(z + 1))^{m/2}\theta^1(z, Q).$$

Let us set $A = \begin{pmatrix} 0 & 1 \\ -1 & -1 \end{pmatrix}$. Then, for a $z \in \mathbb{H}_1$ we have

$$A\langle z\rangle = -(z + 1)^{-1}, \quad A^2\langle z\rangle = -(z + 1)z^{-1}, \quad \text{and} \quad A^3\langle z\rangle = z.$$

It follows that

$$\begin{aligned}\theta^1(z, Q)^\nu &= \theta^1(A\langle A^2\langle z\rangle\rangle, Q)^\nu = (-i(-(z + 1)z^{-1} + 1))^{\nu m/2}\theta^1(A^2\langle z\rangle, Q)^\nu \\ &= (iz^{-1})^{\nu m/2}(-i(-(z + 1)^{-1} + 1))^{\nu m/2}\theta^1(A\langle z\rangle, Q)^\nu \\ &= (iz^{-1})^{\nu m/2}(-iz(z + 1)^{-1})^{\nu m/2}(-i(z + 1))^{\nu m/2}\theta^1(z, Q),\end{aligned}$$

where we take $\nu = 1$, if m is even, and $\nu = 2$, if m is odd. Let $z_0 \in \mathbb{H}_1$ be such that $\theta^1(z_0, Q) \neq 0$. Then we obtain

$$(iz_0^{-1})^{\nu m/2}(-iz_0(z_0 + 1)^{-1})^{\nu m/2}(-i(z_0 + 1))^{\nu m/2} = 1.$$

Since νm is even, it follows that

$$i^{\nu m/2}(-i)^{\nu m/2}(-i)^{\nu m/2} = (-i)^{\nu m/2} = 1$$

which implies the congruence

$$\nu m/2 \equiv 0 \pmod 4. \tag{1.4.5}$$

Hence, $m \equiv 0 \pmod 4$. Then $\nu = 1$, and (1.4.5) shows that $m \equiv \pmod 8$.

Since $m \equiv 0 \pmod 8$, we can rewrite (1.4.4) in the form

$$\theta^n(J_n\langle Z\rangle, Q) = \det(-Z)^{m/2}\theta^n(Z, Q).$$

For other generators (1.3.12) of Γ^n, we directly obtain

$$\theta^n(U(V)\langle Z\rangle, Q) = \theta^n(Z, Q), \qquad \theta^n(T(S)\langle Z\rangle, Q) = \theta^n(Z, Q).$$

It follows that the factor of automorphy $j_Q^n(M, Z)$ equals $\det(-Z)^{m/2}$, 1, and 1, when $M = J_n$, $U(V)$, and $T(S)$ respectively. Clearly the factor of automorphy $j(M, Z)^{m/2}$ takes the same values at the generators. On the other hand, by Theorem 1.3.13, we have

$$j_Q^n(M, Z) = \chi_Q^n(M) j(M, Z)^{m/2},$$

where $\chi_Q^n\colon \Gamma^n \to \mathbb{C}^*$ is a homomorphism, according to Lemmas 1.4.2 and 1.4.1 (2). Comparing the relations, we conclude that the homomorphism χ_Q^n is trivial on the generators of Γ^n and so is trivial on the whole of Γ^n. □

Exercise 1.4.4. *Show that the matrix Q_8 of the quadratic form*

$$x_1^2 + x_2^2 + 2x_3^2 + 2x_4^2 + 20x_5^2 + 12x_6^2 + 4x_7^2 + 2x_8^2$$
$$+ x_1x_2 + x_2x_3 + 3x_3x_4 + 5x_4x_5 + 3x_5x_6 + x_6x_7 + x_7x_8$$

is contained in $\mathbb{A}_8^+$ and satisfies $\det Q_8 = 1$. *Show that there exist matrices $Q_m \in \mathbb{A}_m^+$ with* $\det Q_m = 1$ *for any $m \equiv 0 \pmod 8$.*

1.4.3. The Multiplier as a Gaussian Sum

First we fix some *square roots.* If α is a real number, we can write $\alpha = r\exp(i\varphi)$ with $r \geq 0$ and $\varphi = 0$ or $\varphi = \pi$. Then, for $k \in \mathbb{Z}$, we set

$$\alpha^{k/2} = (\alpha^{1/2})^k = r^{k/2}\exp\left(\frac{i\varphi k}{2}\right), \quad \text{where} \quad r^{k/2} \geq 0. \tag{1.4.6}$$

Now let $\begin{pmatrix} A & B \\ C & D \end{pmatrix} \in \mathrm{GSp}_n^+(\mathbb{R})$. The function $f(Z) = \det(CZ + D)$ does not vanish on $\mathbb{H}_n$, by Lemma 1.4.2. If $\det D \neq 0$, then we will denote by $f(Z)^{1/2}$ the branch of the square root which is holomorphic on $\mathbb{H}_n$, and satisfies the condition

$$\lim_{Z = i\lambda E,\, \lambda \to +0.} \det(CZ + D)^{1/2} = (\det D)^{1/2}, \tag{1.4.7}$$

the right side being defined by (1.4.6).

Finally, for $k \in \mathbb{Z}$, we set

$$\det(CZ + D)^{k/2} = (\det(CZ + D)^{1/2})^k. \tag{1.4.8}$$

Proposition 1.4.5. *In the notation of Theorem* 1.3.13, *let us assume that* $q \neq 1$ *and* $\det(CZ+D)^{-m/2}$ *is defined by* (1.4.7)–(1.4.8). *Then, for any* $M = \begin{pmatrix} A & B \\ C & D \end{pmatrix} \in \Gamma_0^n(q)$ *the multiplier* $\chi_Q^n(M)$ *is given by*

$$\chi_Q^n(M) = (\det D)^{-m/2} |\det D|^m G(-D^{-1}C, Q) \tag{1.4.9}$$

where, for $S \in \mathsf{S}_n(\mathbb{Q})$, $G(S,Q)$ *is the Gaussian sum given by*

$$G(S,Q) = d^{-mn} \sum_{L \in \mathsf{M}_{m,n}(\mathbb{Z}/d\mathbb{Z})} e\{Q[L]S\} \tag{1.4.10}$$

with a positive integer d such that dS is integral.

Remark. Note that, for $M = \begin{pmatrix} A & B \\ C & D \end{pmatrix} \in \Gamma_0^n(q)$, $\det D$ and q are coprime. Hence, $\det D \neq 0$, provided that $q > 1$. Note also that the Gaussian sum (1.4.10) does not depend on the choice of d. □

Proof. We shall compute the limit

$$\lim_{\lambda \to +0} \lambda^{mn/2} \det(i\lambda C + D)^{-m/2} \theta^n(M\langle i\lambda E\rangle, Q), \tag{1.4.11}$$

where $E = E_n$, in two ways. On one hand, by Theorem 1.3.13, it equals

$$\lim_{\lambda \to +0} \lambda^{mn/2} \chi_Q^n(M) \theta^n(i\lambda E, Q)$$

which, by the inversion formula (1.3.33), can be written in the form

$$\begin{aligned} &\lim_{\lambda \to +0} \lambda^{mn/2} \chi_Q^n(M) \det(\lambda E)^{-m/2} (\det Q)^{-n/2} \theta(i\lambda^{-1}E, Q^{-1}) \\ &= \chi_Q^n(M)(\det Q)^{-n/2} \lim_{\lambda \to +0} \theta(i\lambda^{-1}E, Q^{-1}) \\ &= \chi_Q^n(M)(\det Q)^{-n/2}. \end{aligned}$$

On the other hand, if we set

$$M\langle i\lambda E\rangle = BD^{-1} + Z_0 \tag{1.4.12}$$

then, by (1.2.8), we get

$$\begin{aligned} Z_0(i\lambda C + D) &= (i\lambda A + B) - BD^{-1}(i\lambda C + D) \\ &= i\lambda(A - BD^{-1}C) = i\lambda({A}^{t}D - BD^{-1}C^{t}D)D^* \\ &= i\lambda(A^{t}D - B^{t}C)D^* = i\lambda D^*, \end{aligned}$$

hence,

$$Z_0 = i\lambda D^*(i\lambda C + D)^{-1}. \tag{1.4.13}$$

By the definition, we can write

$$\theta^n(M\langle i\lambda E\rangle, Q) = \sum_{N \in \mathsf{M}_{m,n}} e\{Q[N](BD^{-1} + Z_0)\}.$$

Let d be a positive integer such that the matrix dBD^{-1} is integral. Every matrix $N \in \mathsf{M}_{m,n}$ can be written in the form

$$N = L + dN_1 \qquad \text{with } L \in \mathsf{M}_{m,n}/\text{mod}\, d \text{ and } N_1 \in \mathsf{M}_{m,n}.$$

Since

$$Q[N] = Q[L] + d^2Q[N_1] + d\,{}^tLQN_1 + d\,{}^tN_1QL,$$

and the matrix

$$(d^2Q[N_1] + d\,{}^tLQN_1 + d\,{}^tN_1QL)BD^{-1}$$

is integral with even trace, it follows that

$$\begin{aligned}\theta(M\langle i\lambda E\rangle, Q) &= \sum_{L \in \mathsf{M}_{m,n}/\text{mod}\, d} e\{Q[L](BD^{-1} + Z_0)\} \sum_{N_1 \in \mathsf{M}_{m,n}} e\{d^2Q[N_1]Z_0 + 2d\,{}^tN_1QLZ_0\} \\ &= \sum_{L \in \mathsf{M}_{m,n}/\text{mod}\, d} e\{Q[L](BD^{-1} + Z_0)\}\theta(d^2Z_0, Q, (dLZ_0, 0)).\end{aligned}$$

By the inversion formula (1.3.33), the last expression is equal to

$$(\det Q)^{-n/2}(\det(-id^2Z_0))^{-m/2} \tag{1.4.14}$$
$$\times \sum_{L \in \mathsf{M}_{m,n}/\text{mod}\, d} e\{Q[L](BD^{-1} + Z_0)\}\theta(-(d^2Z_0)^{-1}, Q^{-1}, (0, -dQLZ_0)).$$

It follows from (1.4.13) that $Z_0 \to 0$ as $\lambda \to 0$, and $-Z_0^{-1} = -C\,{}^tD + i\lambda^{-1}D\,{}^tD$. Since the series $\theta(Z, Q, (W, W'))$ converges uniformly, when W, W' are in a fixed compact set and $Z \in \mathbb{H}_n(\varepsilon)$ with $\varepsilon > 0$, we can interchange the limit and the summation in

$$\lim_{\lambda \to 0} \theta(-(d^2Z_0)^{-1}, Q^{-1}, (0, -dQLZ_0)).$$

It follows that the limit exists and equals 1. Thus, by (1.4.14), we see that the limit (1.4.11) is equal to

$$(\det Q)^{-n/2} \sum_{L \in \mathsf{M}_{m,n}/\text{mod}\, d} e\{Q[L] \cdot BD^{-1}\}$$
$$\lim_{\lambda \to +0} \lambda^{mn/2}(\det(-i\lambda C + D))^{-m/2} \cdot (\det(-id^2Z_0))^{-m/2}$$
$$= (\det Q)^{-n/2}G(BD^{-1}, Q) \cdot (\det D)^{-m/2} \lim_{\lambda \to +0} (\det(-i(iD^*(i\lambda C + D)^{-1})))^{-m/2}.$$

By Lemma 1.3.15, the function under the last limit is continuous in $Z = iD^*(i\lambda C + D)^{-1} \in \mathbb{H}_n$. Hence,

$$\lim_{\lambda \to +0} (\det(-i(iD^*(i\lambda C + D)^{-1})))^{1/2} = (\det(-i(iD^*D^{-1})))^{1/2} = |\det D|^{-1}.$$

Thus, the limit (1.4.11) is equal to

$$(\det Q)^{-n/2}(\det D)^{-m/2}\,|\det D|^m G(BD^{-1}, Q).$$

Comparing it with (1.4.12), we obtain

$$\chi_Q^n(M) = (\det D)^{-m/2} |\det D|^m G(BD^{-1}, Q). \tag{1.4.15}$$

Hence, to prove the proposition, it suffices to check that

$$G(BD^{-1}, Q) = G(-D^{-1}C, Q) \quad \text{if} \begin{pmatrix} A & B \\ C & D \end{pmatrix} \in \Gamma^n \text{ and } \det D \neq 0. \tag{1.4.16}$$

To prove the relation, we shall modify the definition of Gaussian sums. Let S and Q be as before, and let D satisfies

$$D \in \mathsf{M}_n, \quad \det D \neq 0, \text{ and } DS \in \mathsf{M}_n. \tag{1.4.17}$$

Then we set

$$G_D(S, Q) = |\det D|^{-m} \sum_{L \in \mathsf{M}_{m,n}/\mathsf{M}_{m,n}D} e\{Q[L]S\}. \tag{1.4.18}$$

It is easy to see that

$$G_{MD}(S, Q) = G_D(S, Q)$$

for any integral non-singular matrix M of order n. Suppose that D and D_1 both satisfy (1.4.17), and let $D' = \det D \det D_1 \cdot E_n$. Since the matrices $D'D^{-1}$ and $D'D_1^{-1}$ are integral and non-singular, it follows that

$$G_D(S, Q) = G_{D'}(S, Q) = G_{D_1}(S, Q).$$

Thus, $G_D(S, Q)$ does not depend on the choice of D, and we can take $D = dE_n$, where $d \in \mathbb{N}$, $dS \in \mathsf{M}_n$. It gives us the relation

$$G_D(S, Q) = G(S, Q). \tag{1.4.19}$$

Hence, we obtain the relations

$$G(-D^{-1}C, Q) = G_D(-D^{-1}C, Q) \quad \text{and} \quad G(BD^{-1}, Q) = G_{{}^tD}(BD^{-1}, Q)$$

(note ${}^tDBD^{-1} = {}^tDD^* \, {}^tB = {}^tB$ is integral). Since ${}^tDB = {}^tBD$, the mapping $L \to LB$ defines a homomorphism

$$B\colon \mathsf{M}_{m,n}/\mathsf{M}_{m,n}{}^tD \to \mathsf{M}_{m,n}/\mathsf{M}_{m,n}D.$$

Similarly, the mapping $L \to L\,{}^tC$ defines a homomorphism

$${}^tC\colon \mathsf{M}_{m,n}/\mathsf{M}_{m,n}D \to \mathsf{M}_{m,n}/\mathsf{M}_{m,n}{}^tD.$$

Since $B\,{}^tC = A\,{}^tD - E_n$, the product $B\,{}^tC$ of the homomorphisms is an automorphism. Since ${}^tCB = {}^tAD - E_n$, the product ${}^tCB$ is an automorphism too. Thus, the mappings ${}^tC$ and B are isomorphisms, and we can write

$$G_D(-D^{-1}C, Q) = \sum_{L \in \mathsf{M}_{m,n}/\mathsf{M}_{m,n}{}^tD} e\{Q[LB](-D^{-1}C)\}$$

$$= \sum_{L \in \mathsf{M}_{m,n}/\mathsf{M}_{m,n}{}^tD} e\{Q[L](-BD^{-1}C\,{}^tB)\} = G_{{}^tD}(BD^{-1}, Q)$$

since $C\,{}^tB = {}^t(A\,{}^tD - E) = D\,{}^tA - E$. □

Exercise 1.4.6. *Show that*

$$G(S,Q) = d^{-mn} \sum_{L \in \mathsf{M}_{mn,1}(\mathbb{Z}/d\mathbb{Z})} \exp((\pi i/d)(Q \otimes S')[L])$$

if $d \in \mathbb{N}$ *and* $S' = dS$ *is integral.*

Exercise 1.4.7. (*Gaussian sum as an automorphic form*). *For* $n \in \mathbb{N}$ *and* $q > 1$, *let*

$$\mathsf{S} = \left\{ S = BD^{-1}; \begin{pmatrix} A & B \\ C & D \end{pmatrix} \in \Gamma_0^n(q) \right\}.$$

Show that

(1) *If* $M = \begin{pmatrix} A & B \\ C & D \end{pmatrix} \in \Gamma_0^n(q)$ *and* $S \in \mathsf{S}$, *then* $\det(CS + D) \neq 0$, *and the transformation rule*

$$M: S \to (AS + B)(CS + D)^{-1} = M\langle S \rangle$$

gives a transitive action of $\Gamma_0^n(q)$ *on* S;

(2) *If m is even,* $Q \in \mathbb{A}_m^+$, *and q is the level of Q, then the Gaussian sum* (1.4.10) *satisfies*

$$\det(CS + D)^{m/2} G(M\langle S \rangle, Q) = \chi_Q^n(M) G(S, Q),$$

for any $M = \begin{pmatrix} A & B \\ C & D \end{pmatrix} \in \Gamma_0^n(q)$ *and* $S \in \mathsf{S}$, *where* $\chi_Q^n(M)$ *the same multiplier as in Theorem* 1.3.13.

[Hint: Use the fact that χ_Q^n is a character of $\Gamma_0^n(q)$.]

1.4.4. Quadratic Forms in an Even Number of Variables

Let $Q \in \mathbb{A}_m^+$. Assume that m is even, say $m = 2k$. By Theorem 1.3.13, the factor of automorphy (1.4.1) can be written in the form

$$j_Q^n(M,Z) = \chi_Q^n(M) \det(CZ + D)^k = \chi_Q^n(M) j(M,Z)^k,$$

where $M = \begin{pmatrix} A & B \\ C & D \end{pmatrix} \in \Gamma_0^n(q)$ and q is the level of Q. By Lemmas 1.4.2 (2) and 1.4.1 (2), the mapping $\chi_Q = \chi_Q^n: \Gamma_0^n(q) \to \mathbb{C}^*$ satisfies

$$\chi_Q(MM_1) = \chi_Q(M)\chi_Q(M_1) \qquad \text{for } M, M_1 \in \Gamma_0^n(q). \tag{1.4.20}$$

If $q = 1$, then χ_Q is trivial, by Proposition 1.4.3. So we can assume that $q > 1$. Then Proposition 1.4.5 shows that

$$\chi_Q(M) = (\det D)^k G(-D^{-1}C, Q). \tag{1.4.21}$$

Let us denote by $K = K^n$ the subgroup of $\Gamma_0^n(q)$ generated by the matrices $U(V)$ with $V \in \mathsf{SL}_n(\mathbb{Z})$, $T(S)$ with $S \in \mathsf{S}_n(\mathbb{Z})$, and ${}^tT(S)$ with $S \in q\mathsf{S}_n(\mathbb{Z})$ (see (1.3.12)). It follows from (1.4.21) that χ_Q is trivial on the generators of K. Hence, it is trivial on the whole of K, and so the function χ_Q is constant on any double coset KMK with $M \in \Gamma_0^n(q)$.

Lemma 1.4.8. *Any double coset KMK with $M = \begin{pmatrix} A & B \\ C & D \end{pmatrix} \in \Gamma_0^n(q)$ contains a matrix of the form $M_0 = \begin{pmatrix} A_0 & B_0 \\ C_0 & D_0 \end{pmatrix}$ with $A_0 = \begin{pmatrix} E_{n-1} & 0 \\ 0 & \alpha \end{pmatrix}$, $B_0 = \begin{pmatrix} 0 & 0 \\ 0 & \beta \end{pmatrix}$, $C_0 = \begin{pmatrix} 0 & 0 \\ 0 & \gamma \end{pmatrix}$, and $D_0 = \begin{pmatrix} E_{n-1} & 0 \\ 0 & \delta \end{pmatrix}$, where $\begin{pmatrix} \alpha & \beta \\ \gamma & \delta \end{pmatrix} \in \Gamma_0^1(q)$, and*

$$\delta \equiv \det D \pmod q. \tag{1.4.22}$$

Proof. If $M' = \begin{pmatrix} A' & B' \\ C' & D' \end{pmatrix} = MU(V)$, then $D' = DV$. By Lemma 1.3.8, we can choose a $V \in \mathrm{SL}_n(\mathbb{Z})$ such that the first row of D' is $(d, 0, \dots, 0)$, where d is the greatest common divisor of the entries of the first row of D. Let $(c'_{11}, \dots, c'_{1n})$ be the first row of C' and $c = \text{g.c.d}(c'_{11}, \dots, c'_{1n})$. Then there are integers $s_{21}, \dots, s_{2n}$ such that $c'_{11}s_{21} + \cdots + c'_{1n}s_{2n} = c$. Let us denote by $S = (s_{\alpha\beta})$ an integral symmetric matrix with the second row $(s_{21}, \dots, s_{2n})$, and let $M'' = \begin{pmatrix} A'' & B'' \\ C'' & D'' \end{pmatrix} = M'T(S)$. The first entry of the first row of $D'' = C'S + D'$ is $c'_{11}s_{11} + \cdots + c'_{1n}s_{1n} + d$, and the second entry is c. Since c divides $c'_{11}, \dots, c'_{1n}$ and the numbers c, d are coprime, it follows that the entries are coprime. Hence, replacing M'' by $M''U(V')$ with a $V' \in \mathrm{SL}_n(\mathbb{Z})$, we can assume that the first row of the D-block of M'' is $(1, 0, \dots, 0)$. Since

$$\begin{pmatrix} 1 & 0\cdots 0 \\ -d_{21} & 1\cdots 0 \\ \cdots & \cdots \\ -d_{n1} & 0\cdots 1 \end{pmatrix} \begin{pmatrix} 1 & 0,\dots,0 \\ d_{21} & \\ \vdots & * \\ d_{n1} & \end{pmatrix} = \begin{pmatrix} 1 & 0\cdots 0 \\ 0 & \\ \vdots & * \\ 0 & \end{pmatrix},$$

we can replace M'' by $U(V'')M''$ with a $V'' \in \mathrm{SL}_n(\mathbb{Z})$ to obtain a matrix with D-block of the form

$$\begin{pmatrix} 1 & 0\cdots 0 \\ 0 & \\ \vdots & * \\ 0 & \end{pmatrix}. \tag{1.4.23}$$

So we can assume that the D-block of M is of the form (1.4.23). If now

$$M' = \begin{pmatrix} A' & B' \\ C' & D' \end{pmatrix} = T(S)M\,{}^tT(S_1) = \begin{pmatrix} A' & B + SD \\ C + DS_1 & D \end{pmatrix},$$

then clearly we can choose matrices $S \in \mathrm{S}_n(\mathbb{Z})$ and $S_1 \in q\mathrm{S}_n(\mathbb{Z})$ such that the first column of $B + SD$ is the zero-column and the first row of $C + DS_1$ is the zero-row. By $C'\,{}^tD' = D'\,{}^tC'$ and ${}^tB'D' = {}^tD'B'$, we then conclude that the same is true for the first column of C' and the first row of B'. Now, from ${}^tA'D' - {}^tC'B' = E$ and $A'\,{}^tD' - B'\,{}^tC' = E$, it follows that A' and D' are of the form (1.4.23). So we have found a representative of the form

$$\begin{pmatrix} 1 & 0 & 0 & 0 \\ 0 & A_1 & 0 & B_1 \\ 0 & 0 & 1 & 0 \\ 0 & C_1 & 0 & D_1 \end{pmatrix} = \begin{pmatrix} A_1 & B_1 \\ C_1 & D_1 \end{pmatrix}^{\wedge}$$

in KMK. It is clear that $\begin{pmatrix} A_1 & B_1 \\ C_1 & D_1 \end{pmatrix} \in \Gamma_0^{n-1}(q)$. Since the mapping $M_1 \to \hat{M}_1$ is clearly a homomorphic imbedding of $\Gamma_0^{n-1}(q)$ into $\Gamma_0^n(q)$, and $\hat{K}^{n-1} \subset K^n$, we can proceed by an induction on n to get a representative of the form M_0. If $M' = \begin{pmatrix} A' & B' \\ C' & D' \end{pmatrix} \in KMK$, it is clear that $\det D' \equiv \det D \pmod q$. The congruence (1.4.22) follows. □

We continue the computation of χ_Q. Since χ_Q is constant on KMK, it follows, by Lemma 1.4.8 and (1.4.21), that

$$\begin{aligned} \chi_Q(M) &= \chi_Q(M_0) = \delta^k G(-D_0^{-1}C_0, Q) \\ &= \delta^k d^{-mn} \sum_{L=(\mathbf{l}_1,\ldots,\mathbf{l}_n)\in \mathrm{M}_{m,n}(\mathbb{Z}/d\mathbb{Z})} e\left\{\begin{pmatrix} 0 & 0 \\ 0 & -\delta^{-1}\gamma \end{pmatrix}\begin{pmatrix} * & * \\ * & Q[\mathbf{l}_n] \end{pmatrix}\right\}, \end{aligned}$$

where $d \in |\delta|\,\mathbb{N}$, and $\mathbf{l}_1, \ldots, \mathbf{l}_n$ are the columns of L. It follows that

$$\begin{aligned} \chi_Q(M) &= \delta^k d^{-mn} \sum_{L=(\mathbf{l}_1,\ldots,\mathbf{l}_n)\in \mathrm{M}_{m,n}(\mathbb{Z}/d\mathbb{Z})} e\{-\delta^{-1}\gamma Q[\mathbf{l}_n]\} \\ &= \delta^k d^{-m} \sum_{\mathbf{l}\in \mathrm{M}_{m,1}(\mathbb{Z}/d\mathbb{Z})} e\{-\delta^{-1}\gamma Q[\mathbf{l}]\}. \end{aligned}$$

The formula (1.4.21) shows that the last expression is $\chi_Q^1\left(\begin{pmatrix} \alpha & \beta \\ \gamma & \delta \end{pmatrix}\right)$. So, in the notation of Lemma 1.4.8, we get the relation

$$\chi_Q^n(M) = \chi_Q^1\left(\begin{pmatrix} \alpha & \beta \\ \gamma & \delta \end{pmatrix}\right). \tag{1.4.24}$$

Proposition 1.4.9. *Let $Q \in \mathbb{A}_m^+$. Suppose that m is even, $m = 2k$, and the level q of Q is greater than 1. Then, for any $\begin{pmatrix} \alpha & \beta \\ \gamma & \delta \end{pmatrix} \in \Gamma_0^1(q)$, the following formula holds*

$$\chi_Q^1\left(\begin{pmatrix} \alpha & \beta \\ \gamma & \delta \end{pmatrix}\right) = \chi_Q(\delta), \tag{1.4.25}$$

where χ_Q is the character of the quadratic form with matrix Q, that is a real Dirichlet character modulo q given on integers δ prime to q by

$$\chi_Q(\delta) = \delta^{-k} \sum_{\mathrm{l} \in \mathrm{M}_{m,1}(\mathbb{Z}/\delta\mathbb{Z})} \exp((\pi i/\delta) Q[\mathrm{l}]); \tag{1.4.26}$$

in particular,

$$\chi_Q(-1) = (-1)^k.$$

If p is an odd prime number, then

$$\chi_Q(p) = \left(\frac{(-1)^k \det Q}{p}\right)$$

(the Legendre symbol; see A.2.3).

Proof. Formula (1.4.21) shows that the number $\xi = \chi_Q^1\left(\begin{pmatrix} \alpha & \beta \\ \gamma & \delta \end{pmatrix}\right)$ is contained in the cyclotomic field $\mathbb{Q}_{|\delta|} = \mathbb{Q}(\exp 2\pi i/\delta)$. Since χ_Q^1 is a character of $\Gamma_0^1(q)$ and $\chi_Q^1\left(\begin{pmatrix} 1 & b \\ 0 & 1 \end{pmatrix}\right) = 1$ for any $b \in \mathbb{Z}$, it follows that

$$\chi_Q^1\left(\begin{pmatrix} \alpha & \beta \\ \gamma & \delta \end{pmatrix}\right) = \chi_Q^1\left(\begin{pmatrix} \alpha & \beta \\ \gamma & \delta \end{pmatrix}\begin{pmatrix} 1 & b \\ 0 & 1 \end{pmatrix}\right) = \chi_Q^1\left(\begin{pmatrix} \alpha & \beta + b\alpha \\ \gamma & \delta + b\gamma \end{pmatrix}\right).$$

Thus, the ξ is also contained in any field $\mathbb{Q}_{|\delta + b\gamma|}$ with $b \in \mathbb{Z}$. It is easy to see that the arithmetic progression $\delta + b\gamma$ contains a pair of coprime numbers, say a, c. Then $\mathbb{Q}_{|a|} \cap \mathbb{Q}_{|c|} = \mathbb{Q}$, for the field generated by $\mathbb{Q}_{|a|}$ and $\mathbb{Q}_{|c|}$ is $\mathbb{Q}_{|ac|}$, and $[\mathbb{Q}_{|ac|} : \mathbb{Q}] = [\mathbb{Q}_{|a|} : \mathbb{Q}][\mathbb{Q}_{|c|} : \mathbb{Q}]$. It follows that ξ is a rational number. So the number is invariant under any automorphism of $\mathbb{Q}_{|\delta|}$ of the form

$$\exp(2\pi i/\delta) \to \exp(2\pi i t/\delta), \qquad \text{where} \qquad \text{g.c.d.}(t, \delta) = 1.$$

Setting $t = -\beta$, we have

$$\chi_Q^1\left(\begin{pmatrix} \alpha & \beta \\ \gamma & \delta \end{pmatrix}\right) = \delta^k d^{-m} \sum_{\mathrm{l} \in \mathrm{M}_{m,1}(\mathbb{Z}/d\mathbb{Z})} e\{\delta^{-1} Q[\mathrm{l}]\},$$

for $-\gamma\beta \equiv -1 \pmod{\delta}$. The last expression, say $\chi_Q(\delta)$, depends only on δ and Q. Setting here $d = \delta$, we obtain (1.4.26).

If δ and q are coprime, there exist integers a, b such that $a\delta - qb = 1$. Then $\begin{pmatrix} a & b \\ q & \delta \end{pmatrix} \in \Gamma_0^1(q)$, and

$$\begin{aligned} \chi_Q(\delta) &= \chi_Q^1\left(\begin{pmatrix} a & b \\ q & \delta \end{pmatrix}\right) = \chi_Q\left(\begin{pmatrix} a & b \\ q & \delta \end{pmatrix}\begin{pmatrix} 1 & t \\ 0 & 1 \end{pmatrix}\right) \\ &= \chi_Q^1\left(\begin{pmatrix} a & b + tq \\ q & \delta + tq \end{pmatrix}\right) = \chi_Q(\delta + tq), \end{aligned}$$

for any $t \in \mathbb{Z}$. Thus, $\chi_Q(\delta)$ is defined for all integers δ with g.c.d.$(\delta, q) = 1$, and depends only on δ modulo q. If δ_1 is also prime to q and $\begin{pmatrix} a_1 & b_1 \\ q & \delta_1 \end{pmatrix} \in \Gamma_0^n(q)$, it follows that

$$\chi_Q(\delta)\chi_Q(\delta_1) = \chi_Q^1\left(\begin{pmatrix} a & b \\ q & \delta \end{pmatrix}\right)\chi_Q^1\left(\begin{pmatrix} a_1 & b_1 \\ q & \delta_1 \end{pmatrix}\right)$$

$$= \chi_Q^1\left(\begin{pmatrix} * & * \\ * & qb_1 + \delta\delta_1 \end{pmatrix}\right) = \chi_Q(qb_1 + \delta\delta_1) = \chi_Q(\delta\delta_1).$$

Thus, χ_Q is a real Dirichlet character modulo q. Now let p be an odd prime number prime to q. Setting $\delta = p$ in (1.4.26), we can write

$$\chi_Q(p) = p^{-k} \sum_{\mathfrak{l} \in \mathsf{M}_{m,1}(\mathbb{Z}/p\mathbb{Z})} \exp((2\pi i/p)(1/2)Q[\mathfrak{l}]).$$

For $M \in \mathsf{M}_m$ with $p \nmid \det M$, the mapping $\mathfrak{l} \to M\mathfrak{l}$ induces a one-to-one mapping of $\mathsf{M}_{m,1}/p\mathsf{M}_{m,1}$ onto itself. Thus, for any such M, we have

$$\chi_Q(p) = p^{-k} \sum_{\mathfrak{l} \in \mathsf{M}_{m,1}(\mathbb{Z}/p\mathbb{Z})} \exp((2\pi i/p)(1/2)Q[M\mathfrak{l}]).$$

The matrix M can be choosen in such a way that the quadratic form $(1/2)Q[MX]$ is a diagonal form modulo p, i.e.

$$(1/2)Q[MX] \equiv a_1x_1^2 + \cdots + a_mx_m^2 \pmod p$$

(see A.1.1). Then it is clear that

$$a_1 \cdots a_m \equiv \det(\tfrac{1}{2}Q[M]) = 2^{-m}(\det M)^2 \det Q \pmod p,$$

and we can write

$$\chi_Q(p) = p^{-k} \sum_{l_1,\dots,l_m \in \mathbb{Z}/p\mathbb{Z}} \exp((2\pi i/p)(a_1l_1^2 + \cdots + a_ml_m^2))$$

$$= p^{-k}G_p(a_1)\cdots G_p(a_m),$$

where

$$G_p(a) = \sum_{l \in \mathbb{Z}/p\mathbb{Z}} \exp(2\pi i a l^2/p) \tag{1.4.27}$$

is the ordinary Gaussian sum. By the multiplicativity of the Legendre symbol modulo p (see (A.2.16)), we get

$$G_p(a) = \sum_{v \in \mathbb{Z}/p\mathbb{Z}} \left(1 + \left(\frac{v}{p}\right)\right)\exp(2\pi i a v/p)$$

$$= \sum_{v \in \mathbb{Z}/p\mathbb{Z}} \left(\frac{v}{p}\right)\exp(2\pi i a v/p) \tag{1.4.28}$$

$$= \left(\frac{a}{p}\right) \sum_{v \in \mathbb{Z}/p\mathbb{Z}} \left(\frac{av}{p}\right)\exp(2\pi i a v/p) = \left(\frac{a}{p}\right) G_p(1),$$

where a is prime to p. On the other hand, since p is an odd prime, we have

$$\begin{aligned} G_p(1)^2 &= \left(\frac{-1}{p}\right) G_p(1) G_p(-1) \\ &= \left(\frac{-1}{p}\right) \sum_{l_1, l_2 \in \mathbb{Z}/p\mathbb{Z}} \exp(2\pi i (l_1 - l_2)(l_1 + l_2)/p) \qquad (1.4.29) \\ &= \left(\frac{-1}{p}\right) \sum_{v_1 \in \mathbb{Z}/p\mathbb{Z}} \sum_{v_2 \in \mathbb{Z}/p\mathbb{Z}} \exp(2\pi i v_1 v_2/p) = \left(\frac{-1}{p}\right) p. \end{aligned}$$

Coming back to $\chi_Q(p)$, we obtain

$$\begin{aligned} \chi_Q(p) &= p^{-k} \left(\frac{a_1 \cdots a_m}{p}\right) G_p(1)^m \\ &= \left(\frac{(-1)^k a_1 \cdots a_m}{p}\right) = \left(\frac{(-1)^k 2^{-m} (\det M)^2 \det Q}{p}\right) = \left(\frac{(-1)^k \det Q}{p}\right). \quad \square \end{aligned}$$

The number $(-1)^k \det Q$ is called *the discriminant of the quadratic form with matrix Q.*

Summing up, we obtain

Theorem 1.4.10. *Let $Q \in \mathbb{A}_m^+$ be the matrix of a positive definite integral quadratic form of level q in an even number of variables, say $m = 2k$. Then, for each matrix $M = \begin{pmatrix} A & B \\ C & D \end{pmatrix} \in \Gamma_0^n(q)$, the multiplier $\chi_Q^n(M)$ in the functional equations* (1.3.29) *and* (1.3.30) *of Theorem* 1.3.13 *is given by*

$$\chi_Q^n(M) = 1, \qquad (1.4.30)$$

if $q = 1$, and

$$\chi_Q^n(M) = \chi_Q(\det D), \qquad (1.4.31)$$

if $q > 1$, where χ_Q is the character of the quadratic form with matrix Q, i.e. a real Dirichlet character modulo q satisfying

$$\chi_Q(-1) = (-1)^k, \qquad (1.4.32)$$

$$\chi_Q(p) = \left(\frac{(-1)^k \det Q}{p}\right) \quad \textit{(the Legendre symbol)}, \qquad (1.4.33)$$

if p is an odd prime with $p \nmid q$, and

$$\chi_Q(2) = 2^{-k} \sum_{\mathrm{l} \in \mathrm{M}_{m,1}(\mathbb{Z}/2\mathbb{Z})} \exp(\pi i Q[\mathrm{l}]/2), \qquad (1.4.34)$$

if q is odd.

Proof. The formula (1.4.30) was proved in Proposition 1.4.3; (1.4.31) follows from formulae (1.4.24), (1.4.25), and (1.4.22). The formulae (1.4.32)–(1.4.34) were proved in Proposition 1.4.9. □

Exercise 1.4.11. *Show that the discriminant d of an integral quadratic form in an even number of variables satisfies* $d \equiv 0$ *or* $d \equiv 1 \pmod 4$.

1.4.5. Quadratic Forms in an Odd Number of Variables

We first prove

Proposition 1.4.12. *The determinant* $\det Q$ *and the level* $q(Q)$ *of a non-singular even matrix* Q *of odd order satisfy*

$$\det Q \equiv 0 \pmod 2, \qquad q(Q) \equiv 0 \pmod 4. \tag{1.4.35}$$

Proof. Let $a_{\alpha\beta}$ with $\alpha, \beta = 1, \dots, m$ be entries of Q. If $Q^{-1} = {}^tQ^{-1} = (\det Q)^{-1}(A_{\alpha\beta})$, then, for $\alpha = 1, \dots, m$, we can write

$$\det Q = \sum_{\beta=1}^{m} a_{\alpha\beta} A_{\alpha\beta},$$

whence

$$m \det Q = \sum_{\alpha,\beta=1}^{m} a_{\alpha\beta} A_{\alpha\beta}$$

$$= \sum_{\alpha=1}^{m} a_{\alpha\alpha} A_{\alpha\alpha} + 2 \sum_{1 \le \alpha < \beta \le m} a_{\alpha\beta} A_{\alpha\beta}.$$

Since $a_{\alpha\alpha}$ are even, it follows that $m \det Q$ is even, whence $\det Q$ is even too.

Since the matrix $q(Q)Q^{-1}$ is integral, it follows that $\det Q$ divides $q(Q)^m$. Thus, $q = q(Q)$ is even. The congruence $q \equiv 0 \pmod 4$ will be proved by induction on odd m. For $m = 1$, it is clear. Assuming that the congruence is proved for matrices of odd orders less than an odd $m > 1$, we consider a non-degenerate even Q of order m. First assume that $Q = 2Q_1$ with an integral matrix Q_1. Since the matrix $Q_2 = qQ^{-1} = (q/2)Q_1^{-1}$ is a non-degenerate even matrix of odd order, it follows that $\det Q_2$ is even. On the other hand, $\det Q_2$ divides $(q/2)^m$. Thus, q is divisible by 4. If the matrix $Q = (a_{\alpha\beta})$ has an odd entry, we can assume without loss of generality that $a_{12} = a_{21}$ is odd, for the replacement of Q by $Q[V]$ with a permutation matrix $V \in \mathsf{GL}_m(\mathbb{Z})$ does not change the level and other properties of Q. Let us write $Q = \begin{pmatrix} Q_{11} & Q_{12} \\ Q_{21} & Q_{22} \end{pmatrix}$, where Q_{11} is the block of dimension 2×2. Since

$$\det Q_{11} = a_{11}a_{22} - a_{12}^2 \equiv -a_{12}^2 \equiv 1 \pmod 2,$$

it follows that Q_{11} is non-singular, and Q_{11}^{-1} is 2-integral with "even" diagonal entries. From the identity

$$Q[U] = \begin{pmatrix} Q_{11} & 0 \\ 0 & Q_{22} - Q_{11}^{-1}[Q_{12}] \end{pmatrix}, \quad \text{where} \quad U = \begin{pmatrix} E_2 & -Q_{11}^{-1}Q_{12} \\ 0 & E_{m-2} \end{pmatrix},$$

we get

$$Q^{-1}[U^*] = \begin{pmatrix} Q_{11}^{-1} & 0 \\ 0 & (Q_{22} - Q_{11}^{-1}[Q_{12}])^{-1} \end{pmatrix}.$$

The matrix $Q_{22} - Q_{11}^{-1}[Q_{12}]$ is 2-integral and has even diagonal entries. Thus, it can be written in the form $d^{-1}Q'$ with an odd positive integer d and even matrix Q'. By the last identity, the matrix $q \cdot d \cdot (Q')^{-1}$ is 2-integral and has even diagonal entries. By the inductive hypothesis, it follows that qd is divisible by 4, and hence so is q. □

Theorem 1.4.13. *Let $Q \in \mathbb{A}_m^+$ be the matrix of a positive definite integral quadratic form of level q in an odd number of variables, say $m = 2k + 1$. Then, for each matrix $M = \begin{pmatrix} A & B \\ C & D \end{pmatrix}$ of $\Gamma_0^n(q)$, the multiplier $\chi_Q^n(M)$ in the functional equations* (1.3.29) *and* (1.3.30) *of Theorem* 1.3.13 *with* $\det(CZ + D)^{-m/2}$ *being defined by* (1.4.7)–(1.4.8) *can be written in the form*

$$\chi_Q^n(M) = \chi_Q(\det D)\chi_{(2)}^n(M), \tag{1.4.36}$$

where χ_Q is the real Dirichlet character modulo q satisfying

$$\chi_Q(-1) = (-1)^k \tag{1.4.37}$$

and

$$\chi_Q(p) = \left(\frac{(-1)^k 2 \det Q}{p}\right) \quad \textit{(the Legendre symbol)},$$

if p is a prime, $p \nmid q$, and where $\chi_{(2)}^n$ is the multiplier on $\Gamma_0^n(4)$ associated to the matrix $(2) \in \mathbb{A}_1^+$.

Remark. By Proposition 1.4.5, we have

$$\chi_{(2)}^n(M) = (\det D)^{-1/2} |\det D|\, G(-D^{-1}C, (2))$$

with

$$G(-D^{-1}C, (2)) = d^{-n} \sum_{\mathbf{l} \in \mathsf{M}_{m,1}(\mathbb{Z}/d\mathbb{Z})} \exp(-2\pi i (D^{-1}C)[\mathbf{l}]),$$

where d is a positive integer such that $dD^{-1}C$ is integral. Since q is even, we can assume that d is odd. Then the quadratic form $(dD^{-1}C)[\mathbf{X}]$ can be diagonalized modulo d. Then the Gaussian sum turns into a product of one-dimensional Gaussian sums, whose values are well known.

Proof of the theorem. For $Q_1 \in \mathbb{A}^+_{m_1}$ and $Q_2 \in \mathbb{A}^+_{m_2}$ we set

$$Q_1 \oplus Q_2 = \begin{pmatrix} Q_1 & 0 \\ 0 & Q_2 \end{pmatrix} \in \mathbb{A}^+_{m_1+m_2}$$

(*the direct sum of* Q_1 *and* Q_2). Since clearly

$$\theta^n(Z, Q_1 \oplus Q_2) = \theta^n(Z, Q_1) \cdot \theta^n(Z, Q_2),$$

and

$$G(S, Q_1 \oplus Q_2) = G(S, Q_1) \cdot G(S, Q_2),$$

it follows, by the definition of multipliers and (1.4.8), or by (1.4.9), that

$$\chi^n_{Q_1 \oplus Q_2}(M) = \chi^n_{Q_1}(M)\chi^n_{Q_2}(M), \quad \text{for} \quad M \in \Gamma^n_0(q_1) \cap \Gamma^n_0(q_2)$$

where q_1 and q_2 are the levels of Q_1 and Q_2 respectively. Thus, by Theorem 1.4.10, we can write

$$\chi^n_Q(M)^2 = \chi^n_{Q \oplus Q}(M) = \chi_{Q \oplus Q}(\det D),$$

if $M = \begin{pmatrix} A & B \\ C & D \end{pmatrix} \in \Gamma^n_0(q)$, and

$$\chi^n_Q(M)\chi^n_{(2)}(M) = \chi^n_{Q \oplus (2)}(M) = \chi_{Q \oplus (2)}(\det D),$$

if $M = \begin{pmatrix} A & B \\ C & D \end{pmatrix} \in \Gamma^n_0(q) \cap \Gamma^n_0(4) = \Gamma^n_0(q)$ (see Proposition 1.4.12). Dividing the first relation by the second, we get

$$\chi^n_Q(M)\chi^n_{(2)}(M)^{-1} = \chi_{Q \oplus Q}(\det D)\chi_{Q \oplus (2)}(\det D)^{-1}.$$

Since the level of $Q \oplus Q$ as well as that of $Q + (2)$ is clearly q, it follows from Theorem 1.4.10 that $\chi_Q = \chi_{Q \oplus Q}\chi^{-1}_{Q + (2)}$ is a real Dirichlet character modulo q satisfying (1.4.37). □

Exercise 1.4.14. *Show that*

$$\chi^n_{(2)}(M) = \pm\left(\operatorname{sign}(\det D) \cdot \left(\frac{-4}{|\det D|}\right)\right)^{1/2},$$

for any $n \in \mathbb{N}$ *and* $M = \begin{pmatrix} A & B \\ C & D \end{pmatrix} \in \Gamma^n_0(4)$, *where* $\left(\frac{-4}{d}\right)$ *is the Jacobi symbol. Use this to prove that the multiplier* χ^n_Q *satisfies*

$$\chi^n_Q(M)^4 = 1,$$

for any $m, n \in \mathbb{N}$, $Q \in \mathbb{A}^+_m$, *and* $M \in \Gamma^n_0(q)$, q *being the level of* Q.

Chapter 2. Modular Forms

The properties of the theta-series of positive definite integral quadratic forms presented in the first chapter make it reasonable to look at the sets of all functions with similar properties. This way we come to the spaces of modular forms. Although, generally speaking, the spaces are not spanned by the theta-series, they have an advantage of being defined in an invariant way. A number of known properties of theta-series rest solely on the fact that the theta-series are modular forms.

§ 2.1. Fundamental Domains for Subgroups of the Modular Group

The functional equations obtained in the first chapter show that all values of a theta-series at points of *the orbit*

$$K\langle Z\rangle = \{M\langle Z\rangle; M \in K\} \tag{2.1.1}$$

of a point $Z \in \mathbb{H}_n$ *under the corresponding subgroup* K *of* Γ^n are determined by one of the values. Thus, the theta-series is uniquely determined by its restriction to any subset of $\mathbb{H}_n$ which meets each K-orbit on $\mathbb{H}_n$. In this section we shall construct *a fundamental domain* on $\mathbb{H}_n$ for each subgroup of finite index in Γ^n, that is a closed subset of $\mathbb{H}_n$ which meets each K-orbit and has no distinct inner points belonging to the same orbit.

2.1.1. The Modular Triangle

For short, the imaginary part of a point $z = x + iy \in \mathbb{H}_1$ will be called *the height of* z and denoted by $h(z)$. By Lemma 1.2.8, we get

$$h(M\langle z\rangle) = |cz + d|^{-2} h(z), \qquad \text{for } z \in \mathbb{H}_1 \text{ and } M = \begin{pmatrix} a & b \\ c & d \end{pmatrix} \in \Gamma^1.$$

For a fixed $z = x + iy \in \mathbb{H}_1$, the inequality

$$|cz + d|^2 = (cx + d)^2 + (cy)^2 < 1$$

has obviously only a finite number of solutions in integers c, d. It follows that, on the orbit $\Gamma^1\langle z\rangle$, the height takes only finitely many values greater than a fixed one. Thus, each Γ^1-orbit on $\mathbb{H}_1$ contains points of maximal height, and the points can be characterized by the inequalities $|cz + d| \geq 1$ where (c, d) runs over the set of second rows of all matrices in Γ^1, i.e. the set of all pairs of coprime integers. A transformation of the form $z = x + iy \to \begin{pmatrix} 1 & b \\ 0 & 1 \end{pmatrix}\langle z\rangle = z + b = (x + b) + iy$ with $b \in \mathbb{Z}$ does not change the height and can be chosen so that $|x + b| \leq \frac{1}{2}$. We conclude that each Γ^1-orbit on $\mathbb{H}_1$ meets the set

$$\mathsf{D}_1' = \{z = x + iy \in \mathbb{H}_1;\ |x| \leq \tfrac{1}{2},\ |cz + d| \geq 1 \quad \text{if } c, d \in \mathbb{Z} \text{ and g.c.d.}(c, d) = 1\}. \tag{2.1.2}$$

In fact, the set D_1', can be defined by a finite number of these inequalities: Let us consider the set

$$\mathsf{D}_1 = \{z = x + iy \in \mathbb{H}_1;\ |x| \leq \tfrac{1}{2},\ |z| = x^2 + y^2 \geq 1\}. \tag{2.1.3}$$

Since $1, 0 \in \mathbb{Z}$ and g.c.d.$(1, 0) = 1$, it follows that $\mathsf{D}_1' \subset \mathsf{D}_1$. On the other hand, if $z = x + iy \in \mathsf{D}_1$, $c, d \in \mathbb{Z}$, and g.c.d.$(c, d) = 1$, then

$$|cz + d|^2 = c^2(x^2 + y^2) + 2cdx + d^2 \geq c^2 - |cd| + d^2 \geq 1;$$

hence $z \in \mathsf{D}_1'$. Thus, $\mathsf{D}_1' = \mathsf{D}_1$. So each Γ^1-orbit on $\mathbb{H}_1$ meets D_1. D_1 can be considered as a triangle with the vertices at $\rho = \exp(\pi i/3)$, ρ^2 and $i\infty$, the so-called modular triangle. We shall show that D_1 is a fundamental domain for Γ^1 on $\mathbb{H}_1$.

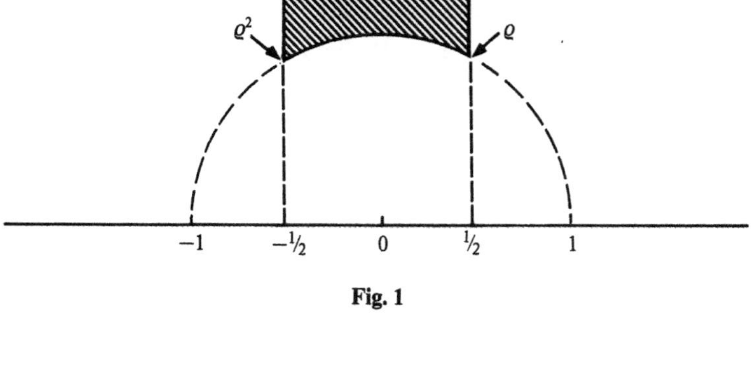

Fig. 1

Theorem 2.1.1.

(1) *Given $z \in \mathbb{H}_1$ there exists a matrix $M \in \Gamma^1$ such that $M\langle z\rangle \in \mathsf{D}_1$.*

(2) *If $z = x + iy$ and z' are two different points in D_1 belonging to the same orbit, then either $x = \pm\frac{1}{2}$ and $z' = z \mp 1$, or $|z| = 1$ and $z' = -1/z$; in particular, no diffirent inner points of D_1 belong to the same orbit.*

Proof. The first part has been proved. For the second part, let $z = x + iy \in D_1$, $M = \begin{pmatrix} a & b \\ c & d \end{pmatrix} \in \Gamma^1$, and $z' = x' + iy' = M\langle z\rangle \in D_1$, where $z' \neq z$. Since $\mathsf{D}_1 = \mathsf{D}_1'$, it follows that $h(z) = h(M\langle z\rangle)$, whence $|cz + d| = 1$. If $c = 0$, then $d = \pm 1$. So $M = \begin{pmatrix} \pm 1 & b \\ 0 & \pm 1 \end{pmatrix}$ and $z' = z \pm b$. Since $-\frac{1}{2} \leq x,\ x' \leq \frac{1}{2}$ and $x' = x + b'$, we have $x = \pm\frac{1}{2}$, $b' = \mp 1$, and $x' = \mp\frac{1}{2}$. If $d = 0$, then $c = \pm 1$ and $|z| = 1$. So $M = \begin{pmatrix} a & \mp 1 \\ \pm 1 & 0 \end{pmatrix}$ and $z' = \pm a - 1/z$. Note that in this case we have $-1/z \in \mathsf{D}_1$ and $|-1/z| = 1$. It follows that $a = 0$, unless $z = \rho = \exp(\pi i/3)$ and $a = -1$, or $z = \rho^2$ and $a = 1$. But in the last cases we have $z' = z$, contrary to the assumption. Finally, if $c \neq 0$ and $d \neq 0$, then it follows from the inequality

$$1 = |cz + d|^2 = c^2(x^2 + y^2) + 2cdx + d^2 \geq c^2 - |cd| + d^2 \geq 1$$

that

$$x^2 + y^2 = 1, \qquad x = \pm\tfrac{1}{2}, c = \pm 1, d = \pm 1, \qquad \text{and } 2cdx = -1.$$

Thus, $(c, d) = \pm(1, 1)$ and $z = \rho^2$, or $(c, d) = \pm(1, -1)$ and $z = \rho$. In the first case M is $\begin{pmatrix} a & a-1 \\ 1 & 1 \end{pmatrix}$ up to a sign, and

$$z' = (a\rho^2 + (a - 1)/\rho^2 + 1 = a - 1/\rho^2 + 1 = a + \rho^2.$$

Since $z' \neq z$, it follows that $a = 1$ and $z' = 1 + \rho^2 = \rho = -1/\rho^2$. The second case is considered in the same way. □

Exercise 2.1.2. *Show that the stabilizer*

$$\mathsf{S}(z) = \{M \in \Gamma^1; M\langle z\rangle = z\}$$

of a point $z \in \mathbb{H}_1$ in Γ^1 is $\{\pm E_2\}$, unless z belongs to $\Gamma^1\langle i\rangle$ or $\Gamma^1 < \rho>$. Show that

$$\mathsf{S}(i) = \left\{\pm E_2, \pm\begin{pmatrix} 0 & 1 \\ -1 & 0 \end{pmatrix}\right\}, \mathsf{S}(\rho) = \left\{\pm E_2, \pm\begin{pmatrix} 0 & -1 \\ 1 & -1 \end{pmatrix}, \pm\begin{pmatrix} -1 & 1 \\ -1 & 0 \end{pmatrix}\right\}.$$

Exercise 2.1.3. *Two binary quadratic forms $\mathsf{q}(x_1, x_2) = q_{11}x_1^2 + q_{12}x_1x_2 + q_{22}x_2^2$, $\mathsf{q}'(y_1, y_2) = q_{11}'y_1^2 + q_{12}'y_1y_2 + q_{22}'y_2^2$ are said to be properly equivalent* (over $\mathbb{Z}$) *if*

$$\mathsf{q}'(y_1, y_2) = \mathsf{q}(\alpha y_1 + \beta y_2, \gamma y_1 + \delta y_2) \quad \text{with} \begin{pmatrix} \alpha & \beta \\ \gamma & \delta \end{pmatrix} \in \Gamma^1. \tag{2.1.4}$$

Show that any real positive definite form q *is properly equivalent to a form* q′, *whose coefficients satisfy* $|q'_{12}| \le q'_{11} \le q'_{22}$. *Show that the domain given by the inequalities in the space of the coefficients of the forms contains no distinct inner points corresponding to properly equivalent forms.*

[Hint: Let ω and ω', be the roots of $\mathrm{q}(x, 1) = 0$ and $\mathrm{q}'(x, 1) = 0$, respectively, belonging to $\mathbb{H}_1$. Show that (2.1.4) is equivalent with the conditions $(q'_{12})^2 - 4q'_{11}q'_{22} = q_{12}^2 - 4q_{11}q_{22}$ and $\omega' = M^{-1}\langle\omega\rangle$, where $M = \begin{pmatrix} \alpha & \beta \\ \gamma & \delta \end{pmatrix}$; then use Theorem 2.1.1].

Exercise 2.1.4. *Show that the number of classes of properly equivalent integral positive definite quadratic forms* $q_{11}x_1^2 + q_{12}x_1x_2 + q_{22}x_2^2$ *of fixed discriminant* $q_{12}^2 - 4q_{11}q_{22} = \Delta < 0$ *is finite.*

2.1.2. The Minkowski Domain

The construction of a fundamental domain for Γ^n for arbitrary n is similar to that of D_1: again each orbit of Γ^n on $\mathbb{H}_n$ contains points $Z = X + iY$ of maximal *height* $h(Z) = \det Y$, and further reduction is carried out by the transformation of Γ^n which do not change the height. However, if $n = 1$, then the only such transformations are the translations $z \to z + b$ with $b \in \mathbb{Z}$; but if $n > 1$, then besides the translations $Z \to Z + B$ with $B \in \mathrm{S}_n(\mathbb{Z})$ there are also the mappings $Z \to Z[V]$ with $V \in \mathrm{GL}_n(\mathbb{Z})$. Here we shall determine a fundamental domain for the group

$$\Lambda = \Lambda^n = \mathrm{GL}_n(\mathbb{Z}) \tag{2.1.5}$$

acting on the space

$$\mathrm{P} = \mathrm{P}_n = \{Y \in \mathrm{S}_n(\mathbb{R}); Y > 0\} \tag{2.1.6}$$

by

$$\Lambda \ni V: Y \to Y[V] = {}^tVYV,$$

and consider some properties of the domain.

The columns of a matrix $U \in \Lambda$ we shall denote by $\mathbf{u}_1, \ldots, \mathbf{u}_n$, so that $U = (\mathbf{u}_1, \ldots, \mathbf{u}_n)$. To choose the reduced representative $Y[U]$ of *the orbit*

$$\Lambda(Y) = \{Y[V]; V \in \Lambda\} \tag{2.1.7}$$

of a point $Y \in \mathrm{P}$, we determine the matrix U column by column, using some minimal conditions. Let

$$\Lambda_k^n = \{(\mathbf{u}_1, \ldots, \mathbf{u}_k); (\mathbf{u}_1, \ldots, \mathbf{u}_k, *, \ldots, *) \in \Lambda^n\} \tag{2.1.8}$$

be the set of integral $n \times k$-metrices, composed of the first columns of all matrices in Λ^n. For a given $Y \in \mathrm{P}$, we choose $\mathbf{u}_1 \in \Lambda_1^n$ so that $Y[\mathbf{u}_1]$ is minimal; this can be done, since Y is positive definite. Next we determine $\mathbf{u}_2$ so that $(\mathbf{u}_1, \mathbf{u}_2) \in \Lambda_2^n$ and $Y[\mathbf{u}_2]$ is minimal. Replacing $\mathbf{u}_2$ by $-\mathbf{u}_2$, if necessary, we may assume that ${}^t\mathbf{u}_1 Y \mathbf{u}_2 \geq 0$. Proceeding in this way, at the k-th step we choose $\mathbf{u}_k$ so that $(\mathbf{u}_1, \ldots, \mathbf{u}_k) \in \Lambda_k^n$, $Y[\mathbf{u}_k]$ is minimal and ${}^t\mathbf{u}_{k-1} Y \mathbf{u}_k \geq 0$. Finally, when $k = n$, we obtain a certain unimodular matrix $U = (\mathbf{u}_1, \ldots, \mathbf{u}_n) \in \Lambda_n^n = \Lambda$ and a matrix $T = (t_{\alpha\beta}) = Y[U] \in \Lambda(Y)$. The matrix T will be called *reduced.*

Let us determine explicitly the condition for a matrix to be reduced in the terms of its entries.

Lemma 2.1.5. *Let* $\mathbf{v} \in \mathsf{M}_{r,1}$, *where* $r \geq 1$. *Then* $\mathbf{v} \in \Lambda_1^r$ *if and only if the entries of* $\mathbf{v}$ *are coprime.*

Proof. The "only if" part is clear. Conversely, if the entries of $\mathbf{v}$ are coprime, then, by Lemma 1.3.8, there exists a matrix $V \in \Lambda^r$ such that

$$V\mathbf{v} = {}^t(1, 0, \ldots, 0), \qquad \text{hence} \qquad \mathbf{v} = V^{-1}\, {}^t(1, 0, \ldots, 0).$$

It follows that the first column of V^{-1} is $\mathbf{v}$; so $\mathbf{v} \in \Lambda_1^r$. □

Lemma 2.1.6. *Let* $U, U' \in \Lambda^n$. *Then the first* r *columns of* U' *are the same as those of* U *if and only if*

$$U' = U\begin{pmatrix} E_r & B \\ 0 & D \end{pmatrix} \qquad \text{with } D \in \Lambda^{n-r} \text{ and } B \in \mathsf{M}_{r,n-r}.$$

Proof. The "if" part is clear. Now let $\mathbf{u}_1, \ldots, \mathbf{u}_n$ be the columns of U. Assume that the first r columns of U' are $\mathbf{u}_1, \ldots, \mathbf{u}_r$. Set

$$U^{-1}U' = V = \begin{pmatrix} A & B \\ C & D \end{pmatrix} \in \Lambda^n,$$

where $A = (a_{\alpha\beta})$, B, $C = (C_{\alpha\beta})$, and D are blocks of dimension $r \times r$, $r \times (n-r)$, $(n-r) \times r$ and $(n-r) \times (n-r)$, respectively. Looking at a β-th column with $1 \leq \beta \leq r$ in

$$U' = U\begin{pmatrix} A & B \\ C & D \end{pmatrix},$$

we get

$$\mathbf{u}_\beta = \sum_{\alpha=1}^{r} a_{\alpha\beta}\mathbf{u}_d + \sum_{\alpha=1}^{n-r} c_{\alpha\beta}\mathbf{u}_{r+\alpha}.$$

Since $\mathbf{u}_1, \ldots, \mathbf{u}_n$ are linearly independent, it follows that $a_{\alpha\beta} = 1$ for $\alpha = \beta$, $a_{\alpha\beta} = 0$ for $\alpha \neq \beta$ and $c_{\alpha\beta} = 0$. Thus, $A = E_r$, $C = 0$, and hence $D \in \Lambda^{n-r}$. □

Let $U = (\mathbf{u}_1, \ldots, \mathbf{u}_n) \in \Lambda^n$ and $1 \leq k \leq n$. By Lemma 2.1.6, the set of k-th columns of all matrices $U' \in \Lambda^n$ with the first columns $\mathbf{u}_1, \ldots, \mathbf{u}_{k-1}$ coincides with the set of the columns of the form

$$U\mathbf{v} \quad \text{where} \quad \mathbf{v} = {}^t(v_1, \ldots, v_n) \in \mathsf{M}_{n,1}$$

and $v_k, \ldots, v_n$ are the entries of the first column of a matrix $D \in \Lambda^{n-k+1}$. By Lemma 2.1.5, the last condition means that $v_k, \ldots, v_n$ are coprime. Thus, if $U = (\mathbf{u}_1, \ldots, \mathbf{u}_n) \in \Lambda^n$ and $1 \le k \le n$, then

$$\{u \in \mathsf{M}_{n,1}; (\mathbf{u}_1, \ldots, \mathbf{u}_{k-1}, \mathbf{u}) \in \Lambda_k^n\} = U\mathsf{V}_{k,n}, \tag{2.1.9}$$

where

$$\mathsf{V}_{k,n} = \{\mathbf{v} = {}^t(v_1, \ldots, v_n) \in \mathsf{M}_{n,1}; \text{g.c.d.}(v_k, \ldots, v_n) = 1\}.$$

By the definition and relations (2.1.9), it follows that a matrix $T = (t_{\alpha\beta}) = Y[U]$ is reduced if and only if it satisfies the conditions

$$Y[U\mathbf{v}] \ge Y[\mathbf{u}_k] \quad \text{for all } \mathbf{v} \in \mathsf{V}_{k,n} \text{ and } 1 \le k \le n,$$

and

$${}^t\mathbf{u}_{k-1} Y \mathbf{u}_k \ge 0 \quad \text{for } 1 < k \le n,$$

where $U = (\mathbf{u}_1, \ldots, \mathbf{u}_n)$. Since $Y[U] = T$, $Y[\mathbf{u}_k] = t_{kk}$ and ${}^t\mathbf{u}_{k-1} Y \mathbf{u}_k = t_{k-1,k}$, the conditions mean precisely that $T \in \mathsf{F}_n$, where

$$\begin{aligned}\mathsf{F}_n = \{T = (t_{\alpha\beta}) \in \mathsf{P}_n;\ & t_{kk} \le T[V] \text{ for } \mathbf{v} \in \mathsf{V}_{k,n} \text{ and} \\ & 1 \le k \le n;\ t_{k-1,k} \ge 0 \text{ for } 1 < k \le n\}\end{aligned} \tag{2.1.10}$$

is *the Minkowski reduction domain.*

Theorem 2.1.7. *Every orbit $\Lambda(Y)$ of Λ^n on P_n contains at least one and at most finitely many matrices of F_n. If T, T' are two inner points of F_n, and $T' = T[U]$ with $U \in \Lambda$, then $U = \pm E_n$; in particular, no different inner points of F_n belong to the same orbit.*

Proof. The above considerations show that, for given $Y \in \mathsf{P}_n$, there exists a $U \in \Lambda^n$ such that $Y[U] \in \mathsf{F}_n$, and every column of such U can be chosen in finitely many ways.

Let us denote by $\mathbf{e}_1, \ldots, \mathbf{e}_n$ the columns of the unit matrix E_n and set

$$\begin{aligned}\mathsf{F}_n^0 = \{T = (t_{\alpha\beta}) \in \mathsf{P}_n;\ & t_{k,k} < T[\mathbf{v}] \text{ for } \mathbf{v} \in \mathsf{V}_{k,n}, \\ & \mathbf{v} \ne \pm \mathbf{e}_k (1 \le k \le n);\ t_{k-1,k} > 0 \text{ for } 1 < k \le n\}.\end{aligned}$$

It is clear that $\mathsf{F}_n^0 \subset \mathsf{F}_n$ and each inner point of F_n is in F_n^0. If $T = (t_{\alpha\beta})$ and $T' = (t'_{\alpha\beta})$ are in F_n^0 and $T' = T[U]$ with $U = (\mathbf{u}_1, \ldots, \mathbf{u}_n) \in \Lambda$, then $t_{kk} = t'_{kk} = T[\mathbf{u}_k]$ for $k = 1, \ldots, n$. Since $\mathbf{u}_1 \in \mathsf{V}_{1,n}$, it follows that $\mathbf{u}_1 = \pm \mathbf{e}_1$. Then $\mathbf{u}_2 \in \mathsf{V}_{2,n}$, and we obtain that $\mathbf{u}_2 = \pm \mathbf{e}_2$. Proceeding in this way, we see that $\mathbf{u}_k = \pm \mathbf{e}_k$ for all $k = 1, \ldots, n$. Now, the conditions

$$t_{k-1,k} > 0, \quad t'_{k-1,k} = {}^t\mathbf{u}_{k-1} T \mathbf{u}_k > 0 \quad (1 < k \le n)$$

imply that $\mathbf{u}_1 = \mathbf{e}_1, \ldots, \mathbf{u}_n = \mathbf{e}_n$ or $\mathbf{u}_1 = -\mathbf{e}_1, \ldots, \mathbf{u}_n = -\mathbf{e}_n$. So $U = \pm E_n$ and $T = T'$. □

The entries of a reduced matrix $T = (t_{\alpha\beta})$ satisfy some useful inequalities. First of all, since

$$t_{kk} \leq T[\mathbf{e}_{k+1}] = t_{k+1,k+1} (1 \leq k < n),$$

it follows that

$$t_{11} \leq t_{22} \leq \cdots \leq t_{nn}. \tag{2.1.11}$$

Then, by

$$t_{ll} \leq T[\mathbf{e}_k \pm \mathbf{e}_l] = t_{kk} \pm 2t_{kl} + t_{ll} \quad \text{with } 1 \leq k < l \leq n,$$

we have

$$|2t_{kl}| \leq t_{kk} \quad \text{for } k \neq l. \tag{2.1.12}$$

Finally, we shall prove

Theorem 2.1.8. *If* $T = (t_{\alpha\beta}) \in \mathsf{F}_n$, *then*

$$t_{11}t_{22} \cdots t_{nn} \leq c_n \det T, \tag{2.1.13}$$

where c_n *depends only on* n.

Proof. For $\alpha = 1, 2, \ldots, n$, we define *the* α*-th minimum* $\mu_\alpha = \mu_\alpha(T)$ *of* T to be the positive number such that

(i) the set of integral columns $\mathbf{m}$ with $T[\mathbf{m}] \leq \mu_\alpha$ spans a space of dimension $\geq \alpha$ over $\mathbb{Q}$;

(ii) the set of integral columns $\mathbf{m}$ with $T[\mathbf{m}] < \mu_\alpha$ spans a space of dimension $< \alpha$ over $\mathbb{Q}$.

It is clear that $\mu_1, \ldots, \mu_n$ are well defined and satisfy $\mu_1 \leq \mu_2 \leq \cdots \mu_n$, and that there exists a set of linearly independent columns $\mathbf{m}_1, \ldots, \mathbf{m}_n$ with $T[\mathbf{m}_\alpha] = \mu_\alpha$.

Now we shall prove three lemmas

Lemma 2.1.9. *Let* $T = (t_{\alpha\beta}) \in \mathsf{F}_n$, *and let* $\mu_1, \ldots, \mu_n$ *be the minima of* T. *Then*

$$t_{\alpha\alpha} \leq c(\alpha)\mu_\alpha \quad \text{for } 1 \leq \alpha \leq n, \tag{2.1.14}$$

where $c(\alpha)$ *depends only on* α.

Proof of the lemma. Let $\mathbf{m}_1, \ldots, \mathbf{m}_n$ be linearly independent columns with $T[\mathbf{m}_\alpha] = \mu_\alpha$, and let $\mathbf{e}_1, \ldots, \mathbf{e}_n$ be the column of E_n. For a fixed α at least one of $\mathbf{m}_1, \ldots, \mathbf{m}_\alpha$ must be linearly independent of $\mathbf{e}_1, \ldots, \mathbf{e}_{\alpha-1}$. Suppose that it is $\mathbf{m}_k$. Then there is a column $\mathbf{e}'_\alpha$ such that $(\mathbf{e}_1, \ldots, \mathbf{e}_{\alpha-1}, \mathbf{e}'_\alpha) \in \Lambda^n_\alpha$; and such that

$$\mathbf{m}_k = h_1\mathbf{e}_1 + \cdots + h_{\alpha-1}\mathbf{e}_{\alpha-1} + s\mathbf{e}'_\alpha,$$

where $s \geq 1$ and $h_1, \ldots, h_{\alpha-1}$ are integers. On replacing of $\mathbf{e}'_\alpha$ by $\mathbf{e}'_\alpha + b_1\mathbf{e}_1 + \cdots + b_{\alpha-1}\mathbf{e}_{\alpha-1}$ with suitable integral $b_1, \ldots, b_{\alpha-1}$, we may suppose without loss of generality that $|h_\beta| \leq s/2$ for $1 \leq \beta < \alpha$. Now

$$\mathbf{e}'_\alpha = (1/s)\mathbf{m}_k - (h_1/s)\mathbf{e}_1 - \cdots - (h_{\alpha-1}/s)\mathbf{e}_{\alpha-1},$$

and so, by the triangle inequality for the norm $\|x\| = (T[\mathbf{x}])^{1/2}$ on $\mathsf{M}_{n,1}(\mathbb{R})$, we have

$$\begin{aligned} T[\mathbf{e}'_\alpha]^{1/2} &\leq \frac{1}{s} T[\mathbf{m}_k]^{1/2} + \sum_{\beta=1}^{\alpha-1} |h_\beta/s|\, T[\mathbf{e}_\beta]^{1/2} \\ &\leq T[\mathbf{m}_k]^{1/2} + \frac{1}{2}\sum_{\beta=1}^{\alpha-1} T[\mathbf{e}_\beta]^{1/2}. \end{aligned} \tag{2.1.15}$$

(2.1.14) is clearly true for $\alpha = 1$ with $c(1) = 1$. By induction we may suppose that (2.1.14) holds for $\beta < \alpha$, and so

$$T[\mathbf{e}_\beta] = t_{\beta\beta} \leq c(\beta)\mu_\beta \leq c(\beta)\mu_\alpha.$$

Since $k \leq \alpha$, we get $T[\mathbf{m}_k] = \mu_k \leq \mu_\alpha$. Hence, (2.1.15) implies that $T[\mathbf{e}'_\alpha] \leq c(\alpha)\mu_\alpha$, where

$$c(\alpha) = \left(1 + \frac{1}{2}\sum_{\beta=1}^{\alpha-1} c(\beta)^{1/2}\right)^2.$$

Since $(\mathbf{e}_1, \ldots, \mathbf{e}_{\alpha-1}, \mathbf{e}'_\alpha) \in \Lambda^n_\alpha$ and T is reduced, it follows that $t_{\alpha\alpha} \leq T[\mathbf{e}'_\alpha]$. □

Lemma 2.1.10. *The first minimum $\mu_1 = \mu_1(T)$ of a matrix $T \in \mathsf{P}_n$ satisfies*

$$\mu_1 \leq \gamma_n (\det T)^{1/n},$$

where γ_n depends only on n.

Proof of the lemma. We consider $\mathsf{M}_{n,1}(\mathbb{R})$ as n-dimensional real space. For $\mu > 0$ the set

$$\{X \in \mathsf{M}_{n,1}(\mathbb{R});\, T[X] \leq \mu\}$$

is clearly convex and symmetric, and has volume

$$v = s_n \mu^{n/2} (\det T)^{-1/2},$$

where s_n is the volume of the unit n-dimensional sphere. By Minkowski's convex body theorem this sphere contains an integral point other than the origin provided that $v \geq 2^n$, i.e. if

$$\mu > 4 s_n^{-2/n} (\det T)^{1/n}.$$

This shows that

$$\mu_1 = \inf_{\mathbf{m} \in \mathsf{M}_{n,1}(\mathbb{Z}),\, \mathbf{m} \neq 0} T[\mathbf{m}] \leq 4 s_n^{-2/n} (\det T)^{1/n}.$$ □

Lemma 2.1.11. *Let $T \in \mathsf{P}_n$ and $\mu_1, \ldots, \mu_n$ be the minima of T. Then*

$$\mu_1 \cdots \mu_n \leq (\gamma_n)^n \det T,$$

where γ_n is the same constant as in Lemma 2.1.10.

Proof of the lemma. Let $\mathbf{m}_1, \ldots, \mathbf{m}_n$ be the linearly independent columns with $T[\mathbf{m}_\alpha] = \mu_\alpha$. Then the matrix $M = (\mathbf{m}_1, \ldots, \mathbf{m}_n)$ is non-singular. By Theorem A.1.5 the matrix $T[M]$ can be written in the form

$$T[M] = {}^tLL \qquad \text{with } L = \begin{pmatrix} l_{11} & l_{12} & \cdots & l_{1n} \\ 0 & l_{22} & \cdots & l_{2n} \\ \cdots & \cdots & \cdots & \cdots \\ 0 & 0 & \cdots & l_{nn} \end{pmatrix}.$$

Set

$$Q = D[LM^{-1}] \qquad \text{with} \qquad D = \operatorname{diag}(\mu_1^{-1}, \ldots, \mu_n^{-1}),$$

we shall show that $Q[\mathbf{m}] \geq 1$ for all $\mathbf{m} \in \mathsf{M}_{n,1}$ other than $\mathbf{m} = \mathbf{0}$. For let $\mathbf{m} = M\mathbf{h}$ with ${}^t\mathbf{h} = (h_1, \ldots, h_n)$, and let α be the largest index such that $h_\alpha \neq 0$. Then $\mathbf{m}$ is of the form $\mathbf{m} = h_1\mathbf{m}_1 + \cdots + h_\alpha\mathbf{m}_\alpha$. Since $h_\alpha \neq 0$, it follows that $\mathbf{m}$ is linearly independent of $\mathbf{m}_1, \ldots, \mathbf{m}_{\alpha-1}$. Then, by the definition of μ_α, we have $T[\mathbf{m}] \geq \mu_\alpha$. Since the entries $(L\mathbf{h})_\beta$ of $L\mathbf{h}$ with $\beta > \alpha$ are equal to zero, we have

$$\begin{aligned} Q[\mathbf{m}] &= D[L\mathbf{h}] = \sum_{\beta=1}^{\alpha} \mu_\beta^{-1}(L\mathbf{h})_\beta^2 \\ &\geq \mu_\alpha^{-1} \sum_{\beta=1}^{\alpha} (L\mathbf{h})_\beta^2 = \mu_\alpha^{-1} E_n[L\mathbf{h}] = \mu_\alpha^{-1} T[\mathbf{m}] \geq 1. \end{aligned}$$

This inequality and Lemma 2.1.10 imply that

$$\gamma_n(\det Q)^{1/n} = \gamma_n((\mu_1 \cdots \mu_n)^{-1} \det T)^{1/n} \geq 1. \qquad \square$$

Returning to the proof of Theorem 2.1.8, we see that (2.1.13) follows from Lemmas 2.1.9 and 2.1.11. □

Corollary. *Let $T = (t_{\alpha\beta}) \in \mathsf{F}_n$. Then*

$$n^{1-n}c_n^{-1}T_0 \leq T \leq nT_0, \qquad \textit{where} \qquad T_0 = \operatorname{diag}(t_{11}, t_{22}, \ldots, t_{nn}), \tag{2.1.16}$$

and c_n is the same constant as in Theorem 2.1.8.

Proof. Let $\rho_1, \ldots, \rho_n$ be the characteristic values of $T[T_0^{-1/2}]$, where $T_0^{1/2} = \operatorname{diag}(t_{11}^{1/2}, \ldots, t_{nn}^{1/2})$. Then

$$\rho_1 + \cdots + \rho_n = \sigma(T[T_0^{-1/2}]) = n,$$

and, by Theorem 2.1.8, we have

$$\rho_1 \cdots \rho_n = \det T\,(t_{11} \cdots t_{nn})^{-1} \geq 1/c_n.$$

It follows that $\rho_\alpha \leq n$ and $\rho_\alpha \geq n^{1-n}c_n^{-1}$ for $\alpha = 1, \ldots, n$. If V is an orthogonal matrix satisfying

$$T[T_0^{-1/2}][V] = \operatorname{diag}(\rho_1, \ldots, \rho_n),$$

then

$$n^{1-n}c_n^{-1}E_n \leq T[T_0^{-1/2}][V] \leq nE_n,$$

which implies (2.1.16). □

The inequalities, we have proved, imply

Theorem 2.1.12. *The number of classes*

$$\{A\}_{\mathbb{Z}} = \{A[U]; U \in \Lambda^n\}$$

of matrices $A \in \mathbb{A}_n^+$ *of a fixed determinant* $\det A = d$ *is finite.*

Proof. By Theorem 2.1.7, each such class contains a reduced representative $A' = (a_{\alpha\beta}) \in \mathsf{F}_n$. Then the inequalities (2.1.11)–(2.1.13) imply that all $|a_{\alpha\beta}|$ are bounded above by a constant depending on n and d. But all $a_{\alpha\beta}$ are integral numbers. □

Corollary. *The number of classes* $\{A\}_{\mathbb{Z}}$ *of matrices* $A \in \mathbb{A}_n^+$ *of a fixed level* q *is finite.*

Proof. Let q be the level of $A \in \mathbb{A}_n^+$. Since $qA^{-1} \in \mathbb{A}_n$, it follows that q^n is divisible by $\det A$. Thus, $\det A$ can take only finitely many values, if q is fixed. □

Proposition 2.1.13. *Let*

$$c_n = \sup_{T=(t_{\alpha\beta}) \in \mathsf{F}_n} t_{11} \cdot t_{22} \cdots t_{nn} \cdot (\det T)^{-1}$$

Then, for $n > 1$, $c_n \geq c_{n-1}$.

Proof. If $T' = (t_{\alpha\beta}) \in \mathsf{F}_{n-1}$, and $t = t_{nn}$ is a real number with $t \geq \max(t_{11}, t_{22}, \ldots, t_{nn})$, then the matrix $T = \begin{pmatrix} T' & 0 \\ 0 & t \end{pmatrix}$ is contained in F_n. For if $\mathbf{v} = \begin{pmatrix} \mathbf{v}' \\ v_n \end{pmatrix} \in \mathsf{V}_{k,n}$ and $v_n \neq 0$, then

$$T[\mathbf{v}] = T'[\mathbf{v}'] + tv_n^2 \geq t \geq t_{kk};$$

but if $v_n = 0$, then $k < n$, $\mathbf{v}' \in \mathsf{V}_{k,n-1}$ and so $T[\mathbf{v}] = T'[\mathbf{v}'] \geq t_{kk}$. This implies the inequality

$$c_n \geq \sup_{T = \begin{pmatrix} T' & 0 \\ 0 & t \end{pmatrix} \in \mathsf{F}_n} t_{11} \cdots t_{n-1,n-1} \cdot t(t \det T')^{-1} = c_{n-1}. \qquad \square$$

Exercise 2.1.14. *Show that*

$$\mathsf{F}_2 = \left\{\begin{pmatrix} t_{11} & t_{12} \\ t_{12} & t_{22} \end{pmatrix} \in \mathsf{P}_2; 0 \le 2t_{12} \le t_{11} \le t_{22}\right\}.$$

[Hint: see Exercise 2.1.3.]

Exercise 2.1.15. *Show that in Lemma* 2.1.10 *for* $n = 2$ *one can take* $\gamma_2 = 2/3^{1/2}$, *and the value in minimal.*

$$\left[\text{Hint: For the minimality, consider } T = \begin{pmatrix} 1 & \frac{1}{2} \\ \frac{1}{2} & 1 \end{pmatrix}\right].$$

Exercise 2.1.16. *Show that in Theorem* 2.1.8 *for* $n = 2$ *one can take* $c_2 = \frac{4}{3}$, *and the value is minimal.*

2.1.3. The Fundamental Domain for the Siegel Modular Group

Quite similarly to the case $n = 1$, we first choose representatives Z of Γ^n-orbits satisfying the inequalities $|\det(CZ + D)| \ge 1$, where (C, D) runs over the set of the "second rows" of all matrices $\begin{pmatrix} A & B \\ C & D \end{pmatrix}$ in Γ^n. To prove that the choice is possible, we start from the set.

A pair (C, D) of square matrices of order n will be called *symmetric* if it satisfies $C\,{}^tD = D\,{}^tC$; it will be called *coprime* if the matrices C and D are integral, and the matrices GC and GD with a matrix G of order n are both integral only if G is integral.

Lemma 2.1.17. *Let* (C, D) *be a pair of integral matrices of order n. Then the following conditions are equivalent*:

(i) *There are matrices A and B such that* $M = \begin{pmatrix} A & B \\ C & D \end{pmatrix} \in \Gamma^n$,

(ii) *The pair* (C, D) *is symmetric and coprime.*

Proof. If (C, D) satisfies (i), then, by the second relation of (1.2.8), the pair is symmetric; it is also coprime, for if GC and GD are integral, then the matrix

$$G = G\,{}^t(-B\,{}^tC + A\,{}^tD) = -(GC)\,{}^tB + (GD)\,{}^tA$$

is integral as well (see (1.2.8)). Now suppose that (C, D) satisfies (ii). We note first that every pair of the form

$$(C', D') = (C, D)M'' = (C, D)\begin{pmatrix} A'' & B'' \\ C'' & D'' \end{pmatrix} = (CA'' + DC'', CB'' + DD'')$$

with $M'' = \begin{pmatrix} A'' & B'' \\ C'' & D'' \end{pmatrix} \in \Gamma^n$ is symmetric and coprime: By (1.2.8), the matrix

$$\begin{aligned} C'\,{}^tD' &= CA''\,{}^tB''\,{}^tC + DC''\,{}^tB''\,{}^tC + CA''\,{}^tD''\,{}^tD + DC''\,{}^tD''\,{}^tD \\ &= CA''\,{}^tB''\,{}^tC + DC''\,{}^tD''\,{}^tD + DC''\,{}^tB''\,{}^tC + CB''\,{}^tC''\,{}^tD + C\,{}^tD \end{aligned}$$

is symmetric. Thus, the pair (C', D') is symmetric; it is obviously coprime. We now show that the matrix $M'' \in \Gamma^n$ can be choosen in such a way that $(C', D') = (E, 0)$. Let $\mathbf{t}$ be the first row of (C, D). Since (C, D) is coprime, it follows that $\mathbf{t} \neq \mathbf{0}$. By Lemma 1.3.9, there is a $M_0 \in \Gamma^n$ such that $M_0\,{}^t\mathbf{t} = (t, 0, \ldots, 0)$ with $t \in \mathbb{N}$. Then

$$(C, D)\,{}^tM_0 = (C', D'), \quad \text{where} \quad C' = \begin{pmatrix} t & 0 \cdots 0 \\ c'_{21} & \\ \vdots & C_1 \\ c'_{n1} & \end{pmatrix}, \quad D' = \begin{pmatrix} 0 & \cdots & 0 \\ d'_{21} & & \\ \vdots & & D_1 \\ d'_{n1} & & \end{pmatrix}$$

Since the matrix $C'\,{}^tD'$ is symmetric, we conclude that $c'_{21} = \cdots = c'_{n1} = d'_{21} = \cdots = d'_{n1} = 0$ and $C_1\,{}^tD_1$ is symmetric. Since (C', D') is coprime, it follows that $t = 1$ and the pair (C_1, D_1) is coprime. This proves the assertion for $n = 1$. If $n > 1$, by induction we may suppose that the assertion holds for the pair (C_1, D_1), i.e. there is a $M_2 = \begin{pmatrix} A_2 & B_2 \\ C_2 & D_2 \end{pmatrix} \in \Gamma^{n-1}$ such that $(C_1, D_1) \cdot M_2 = (E_{n-1}, 0)$. Then the matrix $\hat{M}_2$ with the blocks $\begin{pmatrix} 1 & 0 \\ 0 & A_2 \end{pmatrix}, \begin{pmatrix} 0 & 0 \\ 0 & B_2 \end{pmatrix}, \begin{pmatrix} 0 & 0 \\ 0 & C_2 \end{pmatrix}$, and $\begin{pmatrix} 1 & 0 \\ 0 & D_2 \end{pmatrix}$ is clearly in Γ^n and we have $(C, D)\,{}^tM_0\hat{M}_2 = (E, 0)$. This shows that (C, D) can be presented in the form

$$(C, D) = (E, 0)(M'')^{-1} = (0, E)J_n^{-1}(M'')^{-1} \qquad \text{with } M'' \in \Gamma^n$$

Thus, (C, D) satisfies (i) with $M = J_n^{-1}(M'')^{-1} \in \Gamma^n$. □

We shall say that two coprime symmetric pairs (C, D) and (C', D') are *equivalent* if

$$(C', D') = U(C, D) = (UC, UD) \qquad \text{with a } U \in \Lambda^n. \tag{2.1.17}$$

Then it follows that

$$C\,{}^tD' = D\,{}^tC'. \tag{2.1.18}$$

Conversely, let (C, D) and (C', D') satisfy (2.1.18) and let M (resp M') be a matrix of Γ^n with the "second row" (C, D) (resp. (C', D')). Then

$$M' = (M'M^{-1})M \quad \text{and} \quad M'M^{-1} = \begin{pmatrix} U_1 & V \\ 0 & U \end{pmatrix} \in \Gamma^n.$$

It follows that $U \in \Lambda^n$ and the pairs satisfy (2.1.17). Thus, the conditions (2.1.17) and (2.1.18) are equivalent.

Lemma 2.1.18. *Each coprime symmetric pair* (C, D) *with* $\operatorname{rank} C = r$, *where* $0 \leq r \leq n$, *is equivalent to a pair of the form*

$$\left(\begin{pmatrix} C_1 & 0 \\ 0 & 0 \end{pmatrix} {}^tU_1, \begin{pmatrix} D_1 & 0 \\ 0 & E_{n-r} \end{pmatrix} U_1^{-1}\right) \tag{2.1.19}$$

($(0, E_n)$ if $r = 0$), *where* (C_1, D_1) *is a coprime symmetric pair of matrices of order* r *with* $\operatorname{rank} C_1 = r$ *and* $U_1 \in \Lambda^n$.

Two coprime symmetric pairs of the form (2.1.19) *associated to* C_1, D_1, U_1 *and* C_2, D_2, U_2 *respectively are equivalent if and only if*

$$U_2 = U_1 \begin{pmatrix} V & B \\ 0 & V' \end{pmatrix} \quad \textit{with } V \in \Lambda^r \tag{2.1.20}$$

and the pairs (C_1, D_1) *and* $(C_2\, {}^tV, D_2 V^{-1})$ *are equivalent.*

Remark. If $U_1 = (Q_1, Q_1')$ and $U_2 = (Q_2, Q_2')$ are two matrices in Λ^n, where Q_1 and Q_2 are blocks of dimension $n \times r$, then condition (2.1.20) is equivalent to the condition

$$Q_2 = Q_1 V \quad \text{with } V \in \Lambda^r. \tag{2.1.21}$$

For (2.1.21) clearly implies (2.1.20). Conversely, it follows from (2.1.21) that the first r columns of the matrix

$$U_2' = U_2 \begin{pmatrix} V & 0 \\ 0 & E_{n-r} \end{pmatrix}^{-1} \in \Lambda^n$$

are the same as the corresponding columns of U_1. Then, by Lemma 2.1.6, we have

$$U_1' = \begin{pmatrix} E_r & B' \\ 0 & V' \end{pmatrix} U_1,$$

which implies (2.1.20).

Proof of the lemma. If $r = 0$ or $r = n$, the lemma is trivial. Let $0 < r < n$. Then the homogeneous system of linear equations

$$C\mathbf{x} = 0, \quad \text{where} \quad {}^t\mathbf{x} = (x_1, \ldots, x_n)$$

has a non-zero integral solution $\mathbf{v}$. Clearly we can suppose that the entries of $\mathbf{v}$ are coprime. By Lemma 2.1.5, there is a matrix $V \in \Lambda^n$ with the last column $\mathbf{v}$. Then the last column of CV is the zero-column. The same consideration with respect to the last row of CV shows that there is $V_1 \in \Lambda^n$ such that

$$V_1 C V = \begin{pmatrix} C' & 0 \\ 0 & 0 \end{pmatrix},$$

where C' is of order $n - 1$. If $r = \operatorname{rank} C = \operatorname{rank} C' < n - 1$ then in the same way we can find $V_1', V' \in \Lambda^{n-1}$ such that the last column (resp. row) of $V_1' C' V'$ is the zero-column (resp. zero-row). Then we have

$$\begin{pmatrix} V_1' & 0 \\ 0 & 1 \end{pmatrix} V_1 C V \begin{pmatrix} V' & 0 \\ 0 & 1 \end{pmatrix} = \begin{pmatrix} C'' & 0 \\ 0 & 0 \end{pmatrix},$$

where C'' is of order $n-2$. Continuing in the same way, we finally obtain two unimodular matrices, call them U' and U_1^*, such that

$$U'CU_1^* = \begin{pmatrix} C_1 & 0 \\ 0 & 0 \end{pmatrix}, \quad \text{i.e.} \quad U'C = \begin{pmatrix} C_1 & 0 \\ 0 & 0 \end{pmatrix} {}^tU_1,$$

where C_1 is of order r and rank $C_1 = r$. Set

$$U'D = \begin{pmatrix} D_1 & D_2 \\ D_3 & D_4 \end{pmatrix} U_1^{-1},$$

where D_1 is the square block of order r. It is clear that the pair

$$\left(\begin{pmatrix} C_1 & 0 \\ 0 & 0 \end{pmatrix}, \begin{pmatrix} D_1 & D_2 \\ D_3 & D_4 \end{pmatrix} \right)$$

together with $(U'C, U'D)$, is symmetric and coprime. It follows that (C_1, D_1) is symmetric, $D_3 = 0$ and $D_4 \in \Lambda^{n-r}$. Now, for

$$U = \begin{pmatrix} E_r & -D_2 D_4^{-1} \\ 0 & D_4^{-1} \end{pmatrix} U' \in \Lambda^n$$

we have

$$UC = \begin{pmatrix} C_1 & 0 \\ 0 & 0 \end{pmatrix} {}^tU_1, \quad UD = \begin{pmatrix} D_1 & 0 \\ 0 & E_{n-r} \end{pmatrix} U_1^{-1}.$$

It follows that the pair (C_1, D_1) is coprime. The first part of the lemma is proved.

Consider now two coprime symmetric pairs of the form (2.1.19) associated to C_1, D_1, U_1 and C_2, D_2, U_2 respectively. If they are equivalent, then, by (2.1.18), we have

$$\begin{pmatrix} C_1 & 0 \\ 0 & 0 \end{pmatrix} {}^tU_1 {}^tU_2^{-1} \begin{pmatrix} {}^tD_2 & 0 \\ 0 & E \end{pmatrix} = \begin{pmatrix} D_1 & 0 \\ 0 & E \end{pmatrix} U_1^{-1} U_2 \begin{pmatrix} {}^tC_2 & 0 \\ 0 & 0 \end{pmatrix}.$$

Let us set

$${}^tU_1 {}^tU_2^{-1} = \begin{pmatrix} W & H' \\ H & W' \end{pmatrix}, \quad U_1^{-1} U_2 = \begin{pmatrix} V & B \\ B' & V' \end{pmatrix},$$

where W and V are blocks of dimension $r \times r$, then the last relation becomes

$$\begin{pmatrix} C_1 W {}^tD_2 & C_1 H' \\ 0 & 0 \end{pmatrix} = \begin{pmatrix} D_1 V {}^tC_2 & 0 \\ B' {}^tC_2 & 0 \end{pmatrix},$$

which implies that

$$C_1 W {}^tD_2 = D_1 V {}^tC_2 \quad \text{and} \quad B' = 0,$$

since C_2 is non-singular. In particular, we have

$$U_2 = U_1 \begin{pmatrix} V & B \\ 0 & V' \end{pmatrix}.$$

Also, $W = V^*$, for ${}^tU_1 U_2^* = (U_1^{-1} U_2)^*$. Thus,

$$C_1\,{}^t(D_2 V^{-1}) = D_1\,{}^t(C_2\,{}^tV),$$

and so $(C_2\,{}^tV, D_2 V^{-1})$ is equivalent to (C_1, D_1). The converse is clear. □

We now consider orbits $\Gamma^n\langle Z\rangle\,(Z \in \mathbb{H}_n)$ of the Siegel modular group on $\mathbb{H}_n$. For $Z = X + iY \in \mathbb{H}_n$, the number $h(Z) = \det Y$ will be called *the height of* Z. By Lemma 1.2.8, the height satisfies

$$h(M\langle Z\rangle) = |\det(CZ + D)|^{-2} h(Z) \quad \text{if } M = \begin{pmatrix} A & B \\ C & D \end{pmatrix} \in \Gamma^n. \tag{2.1.22}$$

Lemma 2.1.19. *Each orbit of* Γ^n *on* $\mathbb{H}^n$ *contains points* Z *of maximal height. The points are characterized as those satisfying*

$$|\det(CZ + D)| \geq 1 \tag{2.1.23}$$

for every coprime symmetric pair (C, D).

Proof. If $h(Z) \geq h(M\langle Z\rangle)$ for every $M \in \Gamma^n$, then, by (2.1.22), the inequality (2.1.23) is true for any (C, D) which is the second row of a matrix of Γ^n, i.e. for any coprime symmetric pair, by Lemma 2.1.17. Conversely, the inequalities (2.1.23) imply that $h(M\langle Z\rangle) \leq h(Z)$ for any $M \in \Gamma^n$. So we only have to show that each orbit contains a point of maximal height. To prove this, it suffices to show that, for a fixed $Z \in \mathbb{H}_n$, the inequality

$$|\det(CZ + D)| < 1 \tag{2.1.24}$$

has only finitely many solutions in non-equivalent coprime symmetric pairs (C, D), for then the function $h(M\langle Z\rangle)$ on Γ^n takes only finitely many different values greater than $h(Z)$.

Let $\operatorname{rank} C = r$. Then, by Lemma 2.1.18, we can take the pair (C, D) in the form (2.1.19). Writing $U_1 = (Q, Q')$, where Q is the $n \times r$-block, we have

$$CZ + D = \left(\begin{pmatrix} C_1 & 0 \\ 0 & 0 \end{pmatrix} \begin{pmatrix} {}^tQ \\ {}^tQ' \end{pmatrix} Z(Q, Q') + \begin{pmatrix} D_1 & 0 \\ 0 & E \end{pmatrix} \right) U_1^{-1}$$

$$= \begin{pmatrix} C_1 Z[Q] + D_1 & C_1\,{}^tQZQ' \\ 0 & E \end{pmatrix}.$$

It follows that

$$|\det(CZ + D)| = |\det C_1|\,|\det(Z[Q] + P)|,$$

where $P = C_1^{-1} D_1$ is a rational symmetric matrix of order r. By Theorem 2.1.7, on replacing of Q by QV with a $V \in \Lambda^n$ (see Lemma 2.1.18 and Remark), we may assume that $Y[Q] \in \mathsf{F}_r$. Note also that the class of (C_1, D_1) under (2.1.17) or (2.1.18) is uniquely determined by the symmetric matrix $P = C_1^{-1}D_1$, for if (C_2, D_2) is another coprime symmetric pair of matrices of order r with $\det C_2 \neq 0$ and $C_1^{-1}D_1 = C_2^{-1}D_2$, then $C_1^{-1}D_1 = {}^tD_2 C_2^*$, and hence $C_1\,{}^tD_2 = D_1\,{}^tC_2$. We write $T = Y[Q]$ and $S = X[Q] + P$. Since $T > 0$ and S is symmetric, it follows that there is a real matrix F of order r with $T[F] = E_r$ and $S[F] = H = \mathrm{diag}(h_1, \ldots, h_r)$. Since $(\det F)^{-2} = \det T$, we have

$$|\det(Z[Q] + P)| = |\det(S + iT)| = |\det(H + iE)[F^{-1}]| = |\det T \cdot \prod_{\alpha=1}^{r} (h_\alpha + i)|.$$

Thus, (2.1.24) is equivalent to

$$|\det C_1| \det T \prod_{\alpha=1}^{r} (1 + h_\alpha^2)^{1/2} < 1. \tag{2.1.25}$$

This inequality implies that $\det T < 1$. Let $\mathbf{q}_1, \ldots, \mathbf{q}_r$ be the columns of Q. Since the matrix $T = ({}^t\mathbf{q}_\alpha Y \mathbf{q}_\beta)$ is reduced, it follows that

$$\prod_{\alpha=1}^{r} Y[\mathbf{q}_\alpha] \leq c_r \det T < c_r,$$

by Theorem 2.1.8. On the other hand, we have

$$Y[\mathbf{q}_\alpha] \geq \lambda\,{}^t\mathbf{q}_\alpha \mathbf{q}_\alpha \geq \lambda$$

for $1 \leq \alpha \leq r$, where λ is the least characteristic value of Y. It follows that

$$Y[\mathbf{q}_\alpha] < \lambda^{1-r} c_r \qquad (1 \leq \alpha \leq r).$$

Hence, all $\mathbf{q}_\alpha$ are contained in a finite set of integral columns. In particular, there are only finitely many Q which are not equivalent under (2.1.21) and such that pair (2.1.19) satisfies (2.1.24). Now, $\det T = \det Y[Q]$ takes only finitely many values. So the inequality (2.1.25) implies that all of the values $|\det C_1|, h_1, \ldots, h_r$ are bounded. Also, since $T = F^* F^{-1}$, it follows that all entries of F^{-1} are bounded. Then the same is true for $S = H[F^{-1}]$, and therefore for $P = S - X[Q]$. Note that all denominators of entries of P divide $\det C_1$ and therefore are bounded. Now P is a bounded rational matrix with bounded denominators of all entries. Hence, P belongs to a finite set and so the same is true for the classes of pairs (C_1, D_1), by the correspondence established above. The second part of Lemma 2.1.18 completes the proof. □

Theorem 2.1.20. *Let D_n be the subset of $\mathbb{H}_n$ consisting of those $Z = X + iY \in \mathbb{H}_n$ which satisfy the following conditions:*

(i) $|\det(CZ + D)| \geq 1$ *for all coprime symmetric pairs (C, D) of matrices of order n;*

(ii) $Y \in \mathsf{F}_n$, *where F_n is the Minkowski domain;*

(iii) $X \in \mathsf{X}_n = \{X = (x_{\alpha\beta}) \in \mathsf{S}_n(\mathbb{R});\ |x_{\alpha\beta}| \leq \frac{1}{2} (1 \leq \alpha, \beta \leq n)\}$.

Then D_n *meets each* Γ^n*-orbit on* $\mathbb{H}_n$. *If* Z *and* Z' *are two inner points of* D_n *and* $Z' = M\langle Z\rangle$ *with* $M \in \Gamma^n$, *then* $M = \pm E_{2n}$; *in particular, no different inner points of* D_n *belong to the same orbit.*

Proof. Let us consider the orbit $\Gamma^n\langle Z''\rangle$ of a point $Z'' \in \mathbb{H}_n$. By Lemma 2.1.19, the orbit contains a point $Z' = X' + iY'$ of maximal height and the point satisfies the condition (i). Any transformation of the form

$$Z' \to \begin{pmatrix} {}^tV & SV^{-1} \\ 0 & V^{-1} \end{pmatrix}\langle Z'\rangle = Z'[V] + S = X'[V] + S + iY'[V]$$

with $V \in \Lambda^n$ and $S \in \mathsf{S}_n(\mathbb{Z})$ is represented by a matrix of Γ^n and does not change the height of Z'. By Theorem 2.1.7, there exists a $V \in \Lambda^n$ such that $Y = Y'[V]$ is in F_n. Clearly, there exists $S \in \mathsf{S}_n(\mathbb{Z})$ such that $X = X'[V] + S \in \mathsf{X}_n$. Then

$$Z = X + iY \in \Gamma^n\langle Z''\rangle \cap \mathsf{D}_n.$$

Suppose, now, that Z and Z' are two points of D_n with $Z' = M\langle Z\rangle$, where $M = \begin{pmatrix} A & B \\ C & D \end{pmatrix} \in \Gamma^n$. Then $h(Z) = h(Z')$, by Lemma 2.1.19. It follows from (2.1.22) that $|\det(CZ + D)| = 1$. On the other hand, since $Z = M^{-1}\langle Z\rangle$, it follows that $|\det(-{}^tCZ' + {}^tA)| = 1$ (see (1.2.9)). If $C \neq 0$, then the equations are non-trivial, and therefore the points Z, Z' belong to the boundary of D_n. If $C = 0$, then M is of the form

$$M = \begin{pmatrix} {}^tV & SV^{-1} \\ 0 & V^{-1} \end{pmatrix} \quad \text{with } V \in \Lambda^n \text{ and } S \in \mathsf{S}_n(\mathbb{Z}).$$

So we have

$$Z' = X' + iY' = X[V] + S + iY[V], \quad \text{where} \quad X + iY = Z;$$

in particular, $Y' = Y[V]$. Since Y and Y' are both in F_n. Theorem 2.1.7 implies that Y and Y' are boundary points of F_n or $V = \pm E_n$. In the last case we have $X' = X + S$, whence $S = 0$, unless X and X' belong to the boundary of X_n. Finally, we conclude that $M = \pm E_{2n}$, unless Z and Z' are boundary points of D_n. □

Theorem 2.1.21. *If* $Z = X + iY$ *belongs to* D_n, *then*

$$Y \geq c_n' E_n \quad \textit{with } c_n' = (2c_n)^{-1} n^{1-n} 3^{1/2}, \tag{2.1.26}$$

where c_n *is the same constant as in Theorem 2.1.8; in particular,*

$$\sigma(Y^{-1}) \leq \sigma_n \quad \textit{with } \sigma_n = 3^{-1/2} \cdot 2 \cdot n^n c_n. \tag{2.1.27}$$

Proof. Write $Z = (z_{\alpha\beta})$ with $z_{\alpha\beta} = x_{\alpha\beta} + iy_{\alpha\beta}$. The inequality $|\det(CZ + D)| \geq 1$ for the coprime symmetric pair

$$(C, D) = \left(\begin{pmatrix} 1 & 0 \\ 0 & 0 \end{pmatrix}, \begin{pmatrix} 0 & 0 \\ 0 & E_{n-1} \end{pmatrix} \right)$$

implies the inequality $|z_{11}| = (x_{11}^2 + y_{11}^2)^{1/2} \geq 1$. Since $|x_{11}| \leq \frac{1}{2}$, it follows that $y_{11}^2 \geq 1 - \frac{1}{4}$, i.e. $y_{11} \geq 3^{1/2}/2$. The last inequality and the inequalities (2.1.11) imply that $y_{\alpha\alpha} \geq 3^{1/2}/2$ for $\alpha = 1, \ldots, n$. These inequalities together with the inequalities (2.1.16) prove (2.1.26), for $Y \in \mathsf{F}_n$. It follows from (2.1.26) that each characteristic value of Y^{-1} is not greather than $1/c_n'$. Hence, $\sigma(Y^{-1}) \leq n/c_n'$. □

It follows from Theorem 2.1.21 that the fundamental domain is closed in the space of all complex symmetric matrices. Siegel has proved that D_n is connected and the boundary of D_n consists of a finite number of algebraic hypersurfaces.

2.1.4. Subgroups of Finite Index

Let K be a subgroup of finite index in the modular group Γ^n. Let us set

$$K' = (-E_{2n})K \cup K. \tag{2.1.28}$$

It is clear, that K' is again a subgroup of Γ^n. Let $\{M_1, \ldots, M_\nu\}$ be a set of representatives for the left cosets of Γ^n modulo K', so that

$$\Gamma^n = \bigcup_{1 \leq \alpha \leq \nu} K' M_\alpha, \quad \text{and} \quad K' M_\alpha \cap K' M_\beta = \varnothing \quad \text{for } \alpha \neq \beta \tag{2.1.29}$$

we define

$$\mathsf{D}(K) = \bigcup_{\alpha=1}^{\nu} M_\alpha \langle \mathsf{D}_n \rangle, \tag{2.1.30}$$

where D_n is the fundamental domain for Γ^n. Then we have the following theorem.

Theorem 2.1.22. *The set $\mathsf{D}(K)$ meets each orbit of K on $\mathbb{H}_n$. If Z and Z' are two inner points of $\mathsf{D}(K)$ and $Z' = g\langle Z \rangle$ with $g \in K$, then $g = \pm E_{2n}$; in particular, no different inner points of $\mathsf{D}(K)$ belong to the same K-orbit.*

Proof. Since $K\langle Z \rangle = K'\langle Z \rangle$ for each $Z \in \mathbb{H}_n$, and $\mathsf{D}(K) = \mathsf{D}(K')$, we may assume that $K = K'$. By Theorem 2.1.20, given $Z \in \mathbb{H}_n$, there exists a matrix $M \in \Gamma^n$ such that $M\langle Z \rangle \in \mathsf{D}_n$. Let KM_α be the coset in (2.1.29) containing M^{-1}. Then $M^{-1} = g^{-1}M_\alpha$ with $g \in K$, and we have $Z \in M^{-1}\langle \mathsf{D}_n \rangle = g^{-1}M_\alpha\langle \mathsf{D}_n \rangle$, hence $g\langle Z \rangle \in M_\alpha\langle \mathsf{D}_n \rangle \subset \mathsf{D}(K)$. If Z and $Z' = g\langle Z \rangle$ are inner points of $\mathsf{D}(K)$, we can replace Z by a close point which is an inner point of a subset $M_\alpha\langle \mathsf{D}_n \rangle$ of $\mathsf{D}(K)$. Then $Z = M_\alpha\langle Z_1 \rangle$, where Z_1 is an inner point of D_n. In the same way, we may assume that $g\langle Z \rangle = M_\beta\langle Z_2 \rangle$ with an inner point Z_2 of D_n. Since

$$M_\beta^{-1} g M_\alpha \langle Z_1 \rangle = M_\beta^{-1} g \langle Z \rangle = Z_2,$$

it follows from Theorem 2.1.20 that $M_\beta^{-1} g M_\alpha = \pm E_{2n}$, which clearly implies that $\alpha = \beta$ and $g = \pm E_{2n}$. □

It follows that

$$\mathbb{H}_n = \bigcup_{M \in K'/\{\pm E_{2n}\}} M\langle \mathsf{D}(K)\rangle, \tag{2.1.31}$$

and the sets $M\langle \mathsf{D}(K)\rangle$ have no inner points in common.

Proposition 2.1.23. (1). *Let $K \subset \Gamma^n$ be a subgroup of finite index,* and *let $D(K)$ be a fundamental domain for K on $\mathbb{H}_n$. Then the volume*

$$v(K) = v(\mathsf{D}(K)) = \int_{\mathsf{D}(K)} \mathrm{d}^* Z,$$

where $\mathrm{d}^* Z$ *is the invariant element of volume defined in Proposition* 1.2.9, *does not depend on the choice of fundamental domain and is finite.*

(2) *Let $K_1 \subset K$ be two subgroup of finite index of Γ^n. Then*

$$v(K_1) = [K' : K_1']v(K),$$

the groups K', K_1' being defined by (2.1.28).

Proof. It follows from Proposition 1.2.9 that $v(\mathsf{D}(K))$ is independent of the choice of $\mathsf{D}(K)$. Taking $\mathsf{D}(K)$ in the form (2.1.30), we get

$$\begin{aligned} v(D(K)) &= \sum_{\alpha=1}^{\nu} \int_{M_\alpha\langle \mathsf{D}_n\rangle} \mathrm{d}^* Z = \sum_{\alpha=1}^{\nu} \int_{\mathsf{D}_n} \mathrm{d}^* M_\alpha\langle Z\rangle \\ &= \sum_{\alpha=1}^{\nu} \int_{\mathsf{D}_n} \mathrm{d}^* Z = [\Gamma^n : K']v(\mathsf{D}_n). \end{aligned} \tag{2.1.32}$$

Hence, in order to prove that $v(\mathsf{D}(K))$ is finite, we only have to prove that $v(\mathsf{D}_n)$ is. It follows from the definition of D_n, Theorems 2.1.21 and 2.1.8 and the inequalities (2.1.11), (2.1.12) that

$$\begin{aligned} v(\mathsf{D}_n) &\le \int_{\{Y \in \mathsf{F}_n;\, Y \ge c_n' E\}} (\det Y)^{-(n+1)} \mathrm{d}Y \\ &\le c \int_{\substack{c_n' \le y_{11} \le \cdots \le y_{nn};\\ |y_{\alpha\beta}| \le y_{\alpha\alpha}/2\ (\alpha \ne \beta)}} (y_{11} \cdot y_{22} \cdots y_{nn})^{-(n+1)} \prod_{1 \le \alpha \le \beta \le n} \mathrm{d}y_{\alpha\beta} \\ &\le c \int_{y_{11}, y_{22}, \ldots, y_{nn} \ge c_n'} (y_{11} \cdot y_{22} \cdots y_{nn})^{-(n+1)} \prod_{\alpha=1}^{n} y_{\alpha\alpha}^{n-\alpha} \prod_{\alpha=1}^{n} \mathrm{d}y_{\alpha\alpha} \\ &= c \prod_{\alpha=1}^{n} \int_{c_n'}^{\infty} y_{\alpha\alpha}^{-(\alpha+1)} \mathrm{d}y_{\alpha\alpha} < \infty, \end{aligned}$$

where $c = c_n^{-(n+1)}$ and c_n is the constant of Theorem 2.1.8. The first part is proved.

The second part follows from (2.1.32), for

$$\begin{aligned} v(K_1) &= v(\mathsf{D}(K_1)) = [\Gamma^n : K_1']v(\mathsf{D}_n) \\ &= [\Gamma^n : K_1'][\Gamma^n : K']^{-1}[\Gamma^n : K']v(\mathsf{D}_n) = [K' : K_1']v(\mathsf{D}(K)). \qquad \square \end{aligned}$$

Exercise 2.1.24. *Draw a connected fundamental domains for $\Gamma_0^1(2)$.*

Exercise 2.1.25. *Compute $v(\Gamma^1)$.*

§ 2.2. Definition of Modular Forms

2.2.1. Congruence Subgroups of the Modular Group

For $n, q \in \mathbb{N}$, we shall denote by $\Gamma^n(q)$ *the principal congruence subgroup of the modular group Γ^n of level q*, i.e. the group

$$\Gamma^n(q) = \{M \in \Gamma^n; M \equiv E_{2n} \pmod q\}. \tag{2.2.1}$$

Considering matrices of Γ^n modulo q, we obtain a homomorphism of Γ^n into the finite group $\mathsf{GL}_{2n}(\mathbb{Z}/q\mathbb{Z})$, whose kernel is $\Gamma^n(q)$. It follows that $\Gamma^n(q)$ is a normal subgroup of Γ^n of finite index.

A subgroup K of Γ^n is called *a congruence subgroup*, if it contains a principal congruence subgroup $\Gamma^n(q)$:

$$\Gamma^n(q) \subset K \subset \Gamma^n. \tag{2.2.2}$$

Any such K satisfies

$$v(K) = [\Gamma^n : K] < \infty \tag{2.2.3}$$

Exercise 2.2.1. *Show that*

$$v(\Gamma^1(q)) = q^3 \prod_{p|q} (1 - p^{-2}),$$

where p runs over the prime divisors of q.

2.2.2. Modular Forms of Integral Weight

Let K be a congruence subgroup of Γ^n with $n \in \mathbb{N}$, and let χ be a *congruence character of K*, i.e. a homomorphism of K into the group of roots of unity, which is trivial on a principal congruence subgroup of Γ^n. A function $F: \mathbb{H}_n \to \mathbb{C}$ is called *a modular form (of genus n) of weight k*, where k is an integer, *and character*

χ *for* K if the following conditions hold:

(i) F is a regular function in the $\langle n \rangle$ complex variables on $\mathbb{H}_n$;

(ii) For every matrix $M = \begin{pmatrix} A & B \\ C & D \end{pmatrix}$ of K, F satisfies the functional equation

$$\det(CZ + D)^{-k} F(M\langle Z \rangle) = \chi(M) F(Z); \tag{2.2.4}$$

(iii) If $n = 1$, then every function

$$(cz + d)^{-k} F((az + b)(cz + d)^{-1}) \quad \text{with} \begin{pmatrix} a & b \\ c & d \end{pmatrix} \in \Gamma^1$$

is bounded on each subset $\mathbb{H}_1(\varepsilon) \subset \mathbb{H}_1$ with $\varepsilon > 0$ (see (1.1.11)).

The set $\mathfrak{M}_k(K, \chi)$ of all modular forms of weight k and character χ for K is clearly a linear space over $\mathbb{C}$.

2.2.3. Modular Forms of Half-Integral Weight

Let K be a congruence subgroup of Γ^n, χ a congruence character of K and k an integer. Suppose that

$$K \subset \Gamma_0^n(4). \tag{2.2.5}$$

A function $F : \mathbb{H}_n \to \mathbb{C}$ is called *a modular form* (*of genus n*) *of weight* $k + \frac{1}{2}$ *and character* χ *for* K if the following conditions hold:

(i) F is a regular function in the $\langle n \rangle$ complex variables on $\mathbb{H}_n$;

(ii) For every matrix $M = \begin{pmatrix} A & B \\ C & D \end{pmatrix}$ of K, F satisfies the functional equation

$$\det(CZ + D)^{-k} j_{(2)}^n(M, Z)^{-1} F(M\langle Z \rangle) = \chi(M) F(Z), \tag{2.2.6}$$

where $j_{(2)}^n(M, Z)$ is given by (1.4.1) for $Q = (2) \in \mathbb{A}_1^+$, $M \in \Gamma_0^n(4)$ and $Z \in \mathbb{H}_n$;

(iii) If $n = 1$, then the function

$$(cz + d)^{-(k+1/2)} F((az + b)(cz + d)^{-1}) \quad \text{for every} \begin{pmatrix} a & b \\ c & d \end{pmatrix} \in \Gamma^1$$

is bounded on each subset $\mathbb{H}_1(\varepsilon) \subset \mathbb{H}_1$ with $\varepsilon > 0$.

Remark. By Theorem 1.3.13, we can write

$$j_{(2)}^n(M, Z) = \chi_{(2)}^n(M) \det(CZ + D)^{1/2}, \tag{2.2.7}$$

where $\chi_{(2)}^n(M)$ is an eighth root of unity given by (1.4.9), in view of Proposition 1.4.5.

The set $\mathfrak{M}_{k+1/2}(K, \chi)$ of all modular forms of weight $k + \frac{1}{2}$ and character χ for K is clearly a linear space over $\mathbb{C}$.

2.2.4. Theta-Series as Modular Forms

Let $Q \in \mathbb{A}_m^+$ be the matrix of an integral positive definite quadratic form in m variables, q the level of Q and $n \in \mathbb{N}$. Then it follows from Theorem 1.1.4 and 1.3.13, Theorem 1.4.10 if m is even, and Theorem 1.4.13 together with Proposition 1.4.12 if m is odd that the theta-series $\theta^n(Z, Q)$ of genus n of Q satisfies the conditions (i), (ii) in the definition of modular forms of weight $m/2$ and the character χ_Q^n for the group $\Gamma_0^n(q)$. Furthermore, by Proposition 1.3.14 and Theorem 1.3.6, every function of the form

$$\det(CZ + D)^{-m/2}\theta^n(M\langle Z\rangle, Q) \quad \text{with } M = \begin{pmatrix} A & B \\ C & D \end{pmatrix} \in \Gamma^n \tag{2.2.8}$$

is a linear combination with constant coefficients of the generalized theta-series $\theta^n(Z, Q|T)$. Since the generalized theta-series converge absolutely and uniformly on each set $\mathbb{H}_n(\varepsilon)$ with $\varepsilon > 0$, it follows that every function (2.2.8) is bounded on each of the sets. We have proved

Theorem 2.2.2. *The theta-series $\theta^n(Z, Q)$ of genus n of a matrix $Q \in \mathbb{A}_m^+$, where n, $m \in \mathbb{N}$, is a modular form of weight $m/2$ and character χ_Q^n for the group $\Gamma_0^n(q)$:*

$$\theta^n(Z, Q) \in \mathfrak{M}_{m/2}(\Gamma_0^n(q), \chi_Q^n), \tag{2.2.9}$$

where q is the level of Q and χ_Q^n is the character of Q defined in Theorem 1.4.10 *if m is even, and in Theorem* 1.4.13 *if m is odd.*

To some extent, the theorem reduces the study of theta-series to that of modular forms. The last have an advantage of invariant definition.

Exercise 2.2.3. *Let n, $m \in \mathbb{N}$, $Q \in \mathbb{A}_m^+$ and $T \in \mathrm{T}^n(Q)$. Show that the generalized theta-series $\theta^n(Z, Q|T)$ is a modular form of weight $m/2$ and the unit character for $\Gamma^n(q)$, where q is the level of Q.*

[Hint: see Exercise 1.3.17].

§ 2.3. Fourier Expansions

2.3.1. Modular Forms for Triangular Subgroups

It turns out that some important properties of modular forms depend only on their regularity and the functional equations for a subgroup of infinite index in the modular group.

The subgroup

$$\Gamma_0^n = \left\{\begin{pmatrix} A & B \\ O & D \end{pmatrix} \in \Gamma^n\right\} \tag{2.3.1}$$

will be called *the triangular subgroup of* Γ^n. Let $T = T^n$ be a subgroup of Γ_0^n. A function $F : \mathbb{H}_n \to \mathbb{C}$ will be called *a modular form for* T if the following conditions hold:

(i) is a regular function in the $\langle n \rangle$ complex variables on $\mathbb{H}_n$;

(ii) For every matrix $M = \begin{pmatrix} A & B \\ O & D \end{pmatrix} \in T$, F satisfies the functional equation

$$F(M\langle Z\rangle) = F(AZD^{-1} + BD^{-1}) = F(Z); \tag{2.3.2}$$

(iii) If $n = 1$, F is bounded on each subset $\mathbb{H}_1(\varepsilon) \subset \mathbb{H}_1$ with $\varepsilon > 0$.

The space of all modular forms for T will be denoted by $\mathfrak{M}(T)$. If K is a congruence subgroup of Γ^n and χ is a character of K trivial on the principal congruence subgroup $\Gamma^n(q) \subset K$, then it follows from the definitions that

$$\mathfrak{M}_k(K, \chi), \quad \mathfrak{M}_{k+1/2}(K, \chi) \subset \mathfrak{M}(T_q^n), \tag{2.3.3}$$

where $k \in \mathbb{Z}$ and

$$T_q^n = \left\{ \begin{pmatrix} A & B \\ O & D \end{pmatrix} \in \Gamma^n(q); \det D = 1 \right\}. \tag{2.3.4}$$

2.3.2. Koecher's Effect

Here we shall prove the following

Theorem 2.3.1. *Let $T \subset \Gamma_0^n$ be a subgroup of finite index. Then each modular form $F \subset \mathfrak{M}(T)$ has an expansion of the form*

$$F(Z) = \sum_{A \in \mathbb{A}_n} f(A) e\{q^{-1} AZ\}, \tag{2.3.5}$$

where $\mathbb{A}_n$ is the set (1.1.7), *$e\{\ \}$ is defined by* (1.3.23) *and where $q = q(T)$ is the least $q \in \mathbb{N}$ such that*

$$T(qS) \in T \qquad \textit{for all } S \in \mathrm{S}_n(\mathbb{Z}). \tag{2.3.6}$$

The series (2.3.5) *converges absolutely on $\mathbb{H}_n$ and uniformly on any subset of $\mathbb{H}_n$ of the form $\mathbb{H}_n(\varepsilon)$ with $\varepsilon > 0$; in particular, $F(Z)$ is bounded on each of the subset. For each matrix*

$$U({}^tV) \in T, \tag{2.3.7}$$

the coefficients $f(A)$ in (2.3.5) *satisfy*

$$f(A[V]) = f(A) \qquad (A \in \mathbb{A}_n). \tag{2.3.8}$$

The expansion (2.3.5) is called *the Fourier expansion of* F and the numbers $f(A)(A \in \mathbb{A}_n)$ are *the Fourier coefficients of* F.

Proof. The functional equation (2.3.2) for the matrices of the form (2.3.6) turns into

$$F(Z + qS) = F(Z) \quad (Z = (z_{\alpha\beta}) \in \mathbb{H}_n, S \in \mathsf{S}_n(\mathbb{Z})).$$

This means that F is periodic of period q in all of the $z_{\alpha\beta} = z_{\beta\alpha}$. Since F is also regular, it can be expanded in the Fourier series

$$F(Z) = \sum_{a_{\alpha\beta} \in \mathbb{Z}(1 \le \alpha \le \beta \le n)} f'(\ldots, a_{\alpha\beta}, \ldots; Y)\exp\left(\frac{2\pi i}{q} \sum_{1 \le \alpha \le \beta \le n} a_{\alpha\beta} x_{\alpha\beta}\right),$$

where $Z = X + iY$ and $X = (x_{\alpha\beta})$, which can be differentiated with respect to all variables. Since

$$2 \sum_{1 \le \alpha \le \beta \le n} a_{\alpha\beta} x_{\alpha\beta} = \sigma(AX) \quad \text{with } A = \begin{pmatrix} 2a_{11} & a_{12} & \cdots & a_{1n} \\ a_{12} & 2a_{22} & \cdots & a_{2n} \\ \cdots & \cdots & \cdots & \cdots \\ a_{1n} & a_{2n} & \cdots & 2a_{nn} \end{pmatrix},$$

the expansion can be written in the form

$$F(Z) = \sum_A f(A, Y) e\{q^{-1} AZ\}$$

with $f(A, Y) = f'(\ldots, a_{\alpha\beta}, \ldots; Y)e\{-iq^{-1}AY\}$, where A runs over the set $\mathbb{E}_n$ of all matrices of $\mathsf{S}_n(\mathbb{Z})$ with even diagonal entries. Since $F(Z)$ is holomorphic in each $z_{\alpha\beta}$, it satisfies the Riemann equations with respect to $z_{\alpha\beta}$, which can be written in the form

$$\frac{\partial F}{\partial \bar{z}_{\alpha\beta}} = 0 \quad \text{with } \frac{\partial}{\partial \bar{z}} = \frac{1}{2}\left(\frac{\partial}{\partial x} + i\frac{\partial}{\partial y}\right).$$

Since the last expansion can be differentiated term by term, it follows that

$$\frac{\partial f(A, Y)}{\partial \bar{z}_{\alpha\beta}} = \frac{i}{2} \frac{\partial f(A, Y)}{\partial y_{\alpha\beta}} = 0, \quad \text{for } 1 \le \alpha \le \beta \le n.$$

Hence the coefficients $f(A, Y)$ are independent of Y, and so we get an expansion of the form

$$F(Z) = \sum_{A \in \mathbb{E}_n} f(A) e\{q^{-1} AZ\} \tag{2.3.9}$$

with constant coefficients $f(A)$.

The last expansion can be considered as a Laurent expansion of the regular function F in the variables $t_{\alpha\beta} = \exp(2\pi i z_{\alpha\beta}/q)$ $(1 \le \alpha \le \beta \le n)$. Thus, series (2.3.9) converges absolutely in $\mathbb{H}_n$.

The functional equation (2.3.2) for a matrix of the form (2.3.7) gives the relation

$$\sum_A f(A) e\{q^{-1} AV^{-1} ZV^*\} = \sum_A f(A) e\{q^{-1} AZ\}.$$

On replacing of A by $A[V]$ on the left, we obtain

$$\sum_A f(A[V])e\{q^{-1}AZ\} = \sum_A f(A)e\{q^{-1}AZ\}.$$

Comparing the coefficients, we get (2.3.8).

To complete the proof, we only have to show that $f(A) = 0$, unless $A \in \mathbb{A}_n$, and that the series converges uniformly on $\mathbb{H}_n(\varepsilon)$ with $\varepsilon > 0$.

Let us consider first the case $n = 1$. In this case the expansion (2.3.9) is

$$F(z) = \sum_{a \in \mathbb{Z}} f(2a)\exp(2\pi iaz/q) = \sum_{a \in \mathbb{Z}} f(2a)t^a \quad \left(t = \exp\left(\frac{2\pi iz}{q}\right)\right)$$

which can be considered as a Laurent series of a regular function in t for

$$|t| = \exp(-2\pi y/q) < 1 \qquad (z = x + iy \in \mathbb{H}_1),$$

exept, possibly, at $t = 0$. The condition (iii) of the definition of modular forms for T^1 shows that F is bounded in

$$|t| = \exp(-2\pi y/a) < \exp(-2\pi\varepsilon/q) \qquad \text{with } \varepsilon > 0.$$

Thus, F is regular at $t = 0$. Hence, $f(2a) = 0$ if $a < 0$. It follows that the series converges uniformly on any $\mathbb{H}_1(\varepsilon)$ with $\varepsilon > 0$.

Now let $n \geq 2$. Since the series (2.3.9) converges absolutely, and the coefficients $f(A)$ satisfy (2.3.8), we can write

$$F(Z) = \sum_{\{A\}_T} f(A)\varepsilon'(Z, \{A\}_T),$$

where the sum is taken over a set of representatives for the classes

$$\{A\}_T = \{A[V]; U({}^tV) \in T\} \tag{2.3.10}$$

of $A \in \mathbb{E}_n$, $\varepsilon'(Z, \{A\}_T)$ being defined by

$$\varepsilon'(Z, \{A\}_T) = \sum_{A' \in \{A\}_T} e\{q^{-1}A'Z\}. \tag{2.3.11}$$

If $f(A) \neq 0$, then the series $f(A)\varepsilon'(Z, \{A\}_T)$ converges for every $Z \in \mathbb{H}_n$, for it is a partial sum of an absolutely convergent series; in particular, the series

$$\varepsilon'(iE, \{A\}_T) = \sum_{A' \in \{A\}_T} \exp(-\pi\sigma(A')/q)$$

in convergent. Since the trace $\sigma(A')$ of any $A' \in \{A\}_T$ is an integer, it follows that the class $\{A\}_T$ cannot contain more than a finite number of A' with $\sigma(A') < 0$. On the other hand, we shall show that the function $\sigma(A')$ takes infinitely many negative values on the class $\{A\}_T$ of a $A \in \mathsf{S}_n(\mathbb{Z})$ with $n \geq 2$, unless $A \geq 0$. If A is not semi-definite, then there is an integral column $\mathbf{h}$ with ${}^t\mathbf{h} = (h_1, \ldots, h_n)$ such that $A[\mathbf{h}] < 0$. For $s \in \mathbb{Z}$ we set

$$V_s = E_n + sH \qquad \text{with } H = (t_1\mathbf{h}, t_2\mathbf{h}, \ldots, t_n\mathbf{h}) \in \mathsf{M}_n,$$

where $t_1, \ldots, t_n$ are some integers. Since the rank of sH equals 1 or 0, it follows that

$$\det V_s = 1 + \sigma(sH) = 1 + s(t_1 h_1 + \cdots + t_n h_n).$$

Since $n \geq 2$, the integers $t_1, \ldots, t_n$ can be choosen in such a way that

$$t_1 h_1 + \cdots + t_n h_n = 0, \quad \text{and} \quad t_1^2 + \cdots + t_n^2 > 0.$$

Then we have

$$V_s \in \mathrm{SL}_n(\mathbb{Z}), \quad \text{and} \quad H^2 = (h_\alpha t_\beta(t_1 h_1 + \cdots + t_n h_n)) = 0;$$

in particular, $V_s = V_1^s$. Since the index of T in Γ_0^n is finite, it follows that $U({}^tV_a) = U({}^tV_1)^a$ is contained in T for some $a \in \mathbb{N}$. Then for every $b \in \mathbb{Z}$ we have $U({}^tV_{ab}) = U({}^tV_a)^b \in T$, and so, by the definition, $A_b = A[V_{ab}]$ is in $\{A\}_T$. One computes that

$$\begin{aligned} \sigma(A_b) &= \sigma(A) + 2ab\sigma(AH) + a^2 b^2 \sigma(A[H]) \\ &= \sigma(A) + 2ab\sigma(AH) + a^2 b^2 A[\mathbf{h}](t_1^2 + \cdots + t_n^2). \end{aligned}$$

The last expression is a polynomial in b of degree 2 with the leading coefficient negative. Hence, the polynomial takes infinitely many negative values on $\mathbb{Z}$.

The above consideration shows that a coefficient $f(A)$ in (2.3.9) is equal to zero, unless A is in $\mathbb{A}_n$, which proves (2.3.5).

Finally, if $Z = X + iY \in \mathbb{H}_n(\varepsilon)$ with $\varepsilon > 0$, then, by (A.1.6), we have

$$\sigma(YA) \geq \varepsilon\sigma(A), \tag{2.3.12}$$

for every $A \in \mathbb{A}_n$. Since the series (2.3.5) converges absolutely at $Z = Z_0 = \dfrac{i\varepsilon}{2} E_n$, it follows that

$$|f(A)e\{q^{-1} Z_0 A\}| \leq c,$$

and hence

$$|f(A)| \leq c|e\{-q^{-1} Z_0 A\}| = c\exp(\pi\varepsilon\sigma(A)/2q),$$

for all A of $\mathbb{A}_n$, where c depends only on ε. The inequalities imply that

$$\sum_{A \in \mathbb{A}_n} |f(A)e\{q^{-1}AZ\}| \leq c \sum_{A \in \mathbb{A}_n} \exp(-\pi\varepsilon\sigma(A)/2q).$$

Now if $A = (A_{\alpha\beta}) \in \mathbb{A}_n$ and $\sigma(A) \leq N$, it follows from (A.1.5) that each of the entries $A_{\alpha\beta}$ satisfies $|A_{\alpha\beta}| \leq N$. Thus, the number of different matrices A of $\mathbb{A}_n$ with $\sigma(A) \leq N$ does not exceed the number

$$\left(\frac{N}{2} + 1\right)^n (2N + 1)^{n(n-1)/2} \tag{2.3.13}$$

(note that $A_{\alpha\alpha}$ are non-negative and even), and so we have

$$\sum_{A \in \mathbb{A}_n} |f(A)e\{q^{-1}AZ\}| \leq c \sum_{N=0}^{\infty} \left(\frac{N}{2} + 1\right)^n (2N + 1)^{n(n-1)/2} \exp(-\pi\varepsilon N/2q).$$

The last series converges, and therefore the series (2.3.5) converges uniformly on $\mathbb{H}_n(\varepsilon)$. □

Exercise 2.3.2. *Let n, $q \in \mathbb{N}$. Show that every series of the form* (2.3.5) *with the properties listed in Theorem* 2.3.1 *defines a modular form for $T = T_q^n$.*

Exercise 2.3.3. *Let $T = T_q^n$. Show that each series $\varepsilon'(Z, \{A\}_T)$ with $A \in \mathbb{A}_n$ is a modular form for T.*

2.3.3. Fourier Expansions of Modular Forms

By (2.3.3) and Theorem 2.3.1, each modular form for a congruence subgroup of Γ^n has a Fourier expansion of the form (2.3.5). However, in this case one can say more than that.

In the theory and applications of modular forms of weight k one often has to consider the transformations of the form

$$F \to F|_k M = \det(CZ + D)^{-k} F(M\langle Z\rangle) \qquad \text{with } M = \begin{pmatrix} A & B \\ C & D \end{pmatrix} \in \mathsf{G} = \mathsf{GSp}_n^+(\mathbb{R}). \tag{2.3.14}$$

By Lemma 1.4.2, $F|_k M$ is a regular function on $\mathbb{H}_n$ if F is. Further properties of the transformations depends on k. If k is an integer, then it follows from Lemma 1.4.2 that $\det(CZ + D)^{-k}$ is an factor of automorphy of G on $\mathbb{H}_n$, and so, by Lemma 1.4.1 (3) we have the relations

$$F|_k M|_k M_1 = F|_k MM_1, \qquad \text{for every } M, M_1 \in \mathsf{G}. \tag{2.3.15}$$

But if k is half-integral, we have only the relations

$$F|_k M|_k M_1 = \pm F|_k MM_1, \qquad \text{for every } M, M_1 \in \mathsf{G},$$

where the sign depends on the choice of the signs in $F|_k M|_k M_1$ and $F|_k MM_1$. Also, for an integral weight k, the condition (ii) of the definition of modular forms can be written in the form

$$F|_k M = \chi(M)F \qquad \text{if } F \in \mathfrak{M}_k(K, \chi) \text{ and } M \in K, \tag{2.3.16}$$

where $\chi(M)$ depends only on M. For a half-integral k, from (2.2.6) and Theorem 1.3.13 one can only conclude that

$$F|_k M = \eta F \qquad \text{if } F \in \mathfrak{M}_k(K, \chi) \text{ and } M \in K, \tag{2.3.17}$$

where η depends not only on M but also on the choice of the sign in $F|_k M$. Technical as they are, these differences make a joint treatment of the two cases rather awkward. Therefore, from now on, we consider only the case of integral k, formulating the similar results for half-integral k as exercises.

Theorem 2.3.4. *Let $F \in \mathfrak{M}_k(K, \chi)$ be a modular form of integral weight k and character χ for a congruence subgroup K of Γ^n. Assume that χ is trivial on the principal congruence subgroup $\Gamma^n(q) \subset K$. Then, for every $M \in \Gamma^n$, the following expansion holds*

$$F|_k M = \sum_{A \in \mathbb{A}_n} f_M(A) e\{q^{-1} A Z\}. \tag{2.3.18}$$

Each series (2.3.18) *converges absolutely on $\mathbb{H}_n$ and uniformly on any subset of $\mathbb{H}_n$ of the form $\mathbb{H}_n(\varepsilon)$ with $\varepsilon > 0$; in particular, each function $F|_k M$ with $M \in \Gamma^n$ is bounded on each of the subsets.*

The coefficients $f_M(A)$ of (2.3.18) *satisfy*

$$f_M(A[V]) = f_M(A), \qquad \text{for } A \in \mathbb{A}_n \text{ and } V \in \mathsf{SL}_n(\mathbb{Z}) \text{ with } V \equiv E_n \ (\operatorname{mod} q). \tag{2.3.19}$$

If $k \geq 0$, then

$$|f_M(A)| \leq c (\det A)^k, \qquad \text{for } A \in \mathbb{A}_n^+, \tag{2.3.20}$$

where c depends only on F.

Remark. We shall see later that $\mathfrak{M}_k(K, \chi) = \{0\}$, if $k < 0$. So the condition $k \geq 0$ in (2.3.20) is not a restriction.

Proof. Since clearly $F \in \mathfrak{M}_k(\Gamma^n(q), 1)$, we may replace $\mathfrak{M}_k(K, \chi)$ by $\mathfrak{M} = \mathfrak{M}_k(\Gamma^n(q), 1)$.

We have already noted that the functions $F|_k M$ are regular together with F. Also, if $M_0 \in \Gamma^n(q)$, then $M M_0 M^{-1} \in \Gamma^n(q)$, and so

$$F|_k M|_k M_0 = F|_k M M_0 M^{-1}|_k M = F|_k M,$$

by (2.3.15). Finally, any function $F|_k M|_k M'$ with $M' \in \Gamma^n$ is again of the form $F|_k M_1$, with $M_1 \in \Gamma^n$. It follows that $F|_k M \in \mathfrak{M}$ if $F \in \mathfrak{M}$ and $M \in \Gamma^n$. Thus, by (2.3.3), we get

$$F|_k M \in \mathfrak{M}(T_q^n) \qquad \text{if } F \in \mathfrak{M} \text{ and } M \in \Gamma^n.$$

Then the theorem follows from Theorem 2.3.1, except the estimate (2.3.20).

In order to prove (2.3.20), we consider the function

$$G = G_F = \sum_{M_i \in \Gamma^n(q) \backslash \Gamma^n} |F|_k M_i|. \tag{2.3.21}$$

It follows from (2.3.15) and (2.3.16) that G is independent of the choice of representatives in the classes $\Gamma^n(q) \backslash \Gamma^n$. Also, if $M \in \Gamma^n$, then we have

$$|G|_k M| = \sum_{M_i} |F|_k M_i|_k M| = \sum_{M_i} |F|_k M_i M| = G,$$

for $M_i M$ runs over a system of representatives for $\Gamma^n(q) \backslash \Gamma^n$, when M_i does. Since the functions $F|_k M_i$ are bounded on the sets $\mathbb{H}_n(\varepsilon)$ with $\varepsilon > 0$, it follows from Theorem 2.1.21 that they are bounded on the fundamental domain D_n for Γ^n. Hence, G is bounded on D_n.

Lemma 2.3.5. *Let G be a real non-negative function on $\mathbb{H}_n$ which satisfies*

$$|G|_k M| = G \qquad \text{for any } M \in \Gamma^n \tag{2.3.22}$$

and is bounded on D_n. Then it satisfies

$$G(X + iA^{-1}) \leq \gamma(\det A)^k$$

uniformly in $X \in \mathsf{S}_n(\mathbb{R})$ and $A \in \mathbb{A}_n^+$, where γ depends only on G.

Proof of the lemma. Let $Z = X + iY \in \mathbb{H}_n$. Since D_n is a fundamental domain for Γ^n on $\mathbb{H}_n$ there exists a matrix $M = \begin{pmatrix} A & B \\ C & D \end{pmatrix} \in \Gamma^n$ Such that $Z_0 = M\langle Z \rangle$ is in D_n. If $\det C \neq 0$, then $|\det C| \geq 1$, and we have

$$\begin{aligned} G(Z) &= |\det(CZ + D)|^{-k} G(M\langle Z \rangle) \\ &\leq |\det C|^{-k} |\det(X + C^{-1}D + iY)|^{-k} \gamma \leq \gamma(\det Y)^{-k} \end{aligned} \tag{2.3.23}$$

for G satisfies (2.3.22) and is bounded on D_n (see also (A.1.10)). If $\det C = 0$, we replace Z by a matrix

$$Z_1 = X_1 + iY_1 = M_1\langle Z \rangle \qquad \text{with } M_1 = \begin{pmatrix} 0 & E \\ -E & S \end{pmatrix} \text{ and } S \in \mathsf{S}_n(\mathbb{Z}).$$

Then we have

$$Z_0 = M\langle Z \rangle = MM_1^{-1}\langle Z_1 \rangle$$

and

$$MM_1^{-1} = \begin{pmatrix} A & B \\ C & D \end{pmatrix} \begin{pmatrix} S & -E \\ E & 0 \end{pmatrix} = \begin{pmatrix} AS + B & -A \\ CS + D & -D \end{pmatrix}.$$

Let us show that there exists a $S \in \mathsf{S}_n(\mathbb{Z})$ such that

$$\det(CS + D) \neq 0. \tag{2.3.24}$$

Denote by r the rank of C. It follows from Lemmas 2.1.17, 2.1.18 that the matrices C, D can be written in the form

$$C = U\begin{pmatrix} C_1 & 0 \\ 0 & 0 \end{pmatrix} {}^tU_1, \qquad D = U\begin{pmatrix} D_1 & 0 \\ 0 & E_{n-r} \end{pmatrix} U_1^{-1},$$

where $U, U_1 \in \Lambda^n$ and (C_1, D_1) is a coprime symmetric pair of matrices of order r with $\det C_1 \neq 0$. Then, clearly, there is an integer t such that $\det(tC_1 + D_1) \neq 0$, and the matrix

$$S = U_1^* \begin{pmatrix} tE_r & 0 \\ 0 & E_{n-r} \end{pmatrix} U_1^{-1}$$

satisfies (2.3.24). For any such S, by (2.3.23) and Lemma 1.2.8, we get

$$G(Z_1) \le \gamma(\det Y_1)^{-k} = \gamma(\det Y)^{-k}|\det(-Z+S)|^{2k},$$

hence

$$\begin{aligned} G(Z) &= |\det(-Z+S)|^{-k} G(M_1\langle Z\rangle) \\ &\le \gamma(\det Y)^{-k}|\det(-Z+S)|^{k}. \end{aligned} \tag{2.3.25}$$

The expression $\det(CS+D)$ is a polynomial in the entries $s_{\alpha\beta}$ of S, of degree at most 2 in each variable $s_{\alpha\beta}$. Since the polynomial does not vanish identically, it follows by induction that $\det(CS+D) \neq 0$ for a $S = (s_{\alpha\beta}) \in \mathsf{S}_n(\mathbb{Z})$ with

$$-2 < s_{\alpha\beta} - x_{\alpha\beta} < 2 \qquad (\alpha, \beta = 1, \ldots, n),$$

where $x_{\alpha\beta}$ are the entries of the real part $X = (x_{\alpha\beta})$ of Z. Then the expression $|\det(-Z+S)|^2 = |\det(S - X + iY)|^2$ is a polynomial in the entries $y_{\alpha\beta}$ of Y of degree $2n$ with bounded coefficients. By (A.1.5), we have the inequalities $|y_{\alpha\beta}| \le \sigma(Y)$ for $\alpha, \beta = 1, \ldots, n$, since $Y > 0$. Hence

$$|\det(-Z+S)| \le \delta(1+\sigma(Y))^n,$$

where δ depends only on n. Applying this to (2.3.25), we get

$$G(Z) \le \gamma\delta(1+\sigma(Y))^{nk}(\det Y)^{-k}.$$

Now let $Y = A^{-1}$ with $A \in \mathbb{A}_n^+$. We can write A in the form $A = A_0[U^{-1}]$, where A_0 belongs to the Minkowski reduction domain and $U \in \Lambda^n$. Then, by (2.3.22) with $M = U(V^*) \in \Gamma^n$ and the last inequality, we get

$$\begin{aligned} G(X + iA^{-1}) &= G(U(U^{-1}XU^* + iA_0^{-1})\,{}^tU) \\ &= G(U^{-1}XU^* + iA_0^{-1}) \le \gamma\delta(1+\sigma(A_0^{-1}))^{nk}(\det A_0)^k. \end{aligned}$$

Let $B_{\alpha\alpha}$ be the matrix obtained of A_0 by crossing out its α-th row and α-th column, and let a_α be the α-th diagonal entry of A_0. Since $B_{\alpha\alpha} > 0$, we can apply the inequality (A.1.8). This together with Theorem 2.1.8 for shows that

$$\begin{aligned} \sigma(A_0^{-1}) &= \sum_{\alpha=1}^{n} \det B_{\alpha\alpha}/\det A_0 \le \sum_{\alpha=1}^{n} a_\alpha^{-1} a_1 \cdots a_n / c_n^{-1} a_1 \cdots a_n \\ &= c_n \sum_{\alpha=1}^{n} a_\alpha^{-1} \le c_n n. \end{aligned}$$

Finally, we have

$$G(X + iA^{-1}) \le \gamma'(\det A_0)^k = \gamma'(\det A)^k$$

with $\gamma' = \gamma\delta(1 + nc_n)^{nk}$. □

Returning to the proof of (2.3.20), we can apply the lemma to the function (2.3.21), which gives the inequality

$$|(F|_k M)(X + iA^{-1})| \le G_F(X + iA^{-1}) \le \gamma(\det A)^k,$$

for $F|_k M$ is one of $F|_k M_i$ in (2.3.21). Hence we have

$$|f_M(A)| = |q^{-\langle n\rangle} \int_{0\le\cdots,x_{\alpha\beta},\cdots\le q} (F|_k M)((x_{\alpha\beta}) + iA^{-1})$$
$$\times e\{-q^{-1}A\cdot((x_{\alpha\beta}) + iA^{-1})\} \prod_{1\le\alpha\le\beta\le n} \mathrm{d}x_{\alpha\beta}$$
$$\le q^{-\langle n\rangle}\gamma\cdot(\det A)^k\cdot\exp(q^{-1}\pi n)q^{\langle n\rangle} = c(\det A)^k. \qquad \square$$

Let us return to the operators (2.3.14). In the beginning of the proof of Theorem 2.3.4 we saw that each function $F|_k M$ with $M \in \Gamma^n$ is contained in $\mathbb{M}_k(\Gamma^n(q), 1)$, if the F is. The general situation is more complicated; however, if M is rational (or rational up to a scalar factor), the operator $|_k M$ still maps modular forms of weight k to modular forms of the same weight but for another group:

Proposition 2.3.6. *Let K be a congruence subgroup of Γ^n, and let χ be a congruence character of K. Then, for any matrix $M \in \mathsf{GSp}_n^+(\mathbb{R}) \cap \mathsf{M}_{2n}(\mathbb{Q}) = \mathsf{GSp}_n^+(\mathbb{Q})$, the group*

$$K_M = M^{-1}KM \cap \Gamma^n \tag{2.3.26}$$

is a congruence subgroup of Γ^n, the character

$$\chi_M : W \to \chi_M(W) = \chi(MWM^{-1})\ (W \in K_M) \tag{2.3.27}$$

is a congruence character of K_M, and

$$F|_k M \in \mathfrak{M}_k(K_M, \chi_M)$$

if $F \in \mathfrak{M}_k(K, \chi)$.

Proof. Let $q \in \mathbb{N}$ be such that $\Gamma^n(q) \subset K$ and χ is trivial on $\Gamma^n(q)$. We write $M = tM_0$ with an integral matrix M_0 and set $q_0 = \mu(M_0)$. Then $\Gamma^n(qq_0) \subset K_M$ and χ_M is trivial on $\Gamma^n(qq_0)$: if $W \in \Gamma^n(qq_0)$, then $MWM^{-1} \in \Gamma^n(q) \subset K$, for the matrix

$$MWM^{-1} = M(E + (W - E))M^{-1} = E + qM((qq_0)^{-1}(W - E))q_0M^{-1}$$

is an integral symplectic matrix congruent to E modulo q, and $\chi_M(W) = \chi(MWM^{-1}) = 1$.

Now let $F \in \mathfrak{M}_k(K, \chi)$. It follows from Lemma 1.4.2 that the function $G = F|_k M$ is regular on $\mathbb{H}_n$. If $W \in K_M$, then $W = M^{-1}W_1M$ with $W_1 \in K$, and so, by (2.3.15) and (2.3.16), we have

$$G|_k W = F|_k M|_k M^{-1}W_1M = F|_k W_1|_k M$$
$$= \chi(W_1)F|_k M = \chi_M(W)G.$$

Thus, G satisfies all of the functional equations which define the space $\mathfrak{M}_k(K_M, \chi_M)$. Finally, if $n = 1$, each function $F|_k W'$ with $W' \in \Gamma^1$ is bounded on any subset $\mathbb{H}_1(\varepsilon)$ of $\mathbb{H}_1$ with $\varepsilon > 0$, by the definition of modular forms. Let $W \in \Gamma^1$, then we have

$$G|_k W = F|_k M|_k W = F|_k MW.$$

By Proposition 1.3.7, we can write $MW = W'M'$, where $W' \in \Gamma^1$ and $M' = \begin{pmatrix} a & b \\ 0 & d \end{pmatrix}$ with $ad = \mu(M) > 0$, hence

$$G|_k W = F|_k(W'M') = d^{-k}(F|_k W')(az/d + b/d).$$

Since the function $F|_k W'$ is bounded on $\mathbb{H}_1(a\varepsilon/d)$, it follows that $G|_k W$ is bounded on $\mathbb{H}_1(\varepsilon)$. We have proved that $G \in \mathfrak{M}_k(K_M, \chi_M)$. □

Exercise 2.3.7. *Let k be a half-integral number, and let $F \in \mathfrak{M}_k(K, \chi)$, where K is a congruence subgroup of Γ^n, χ is a character of K trivial on a principal congruence subgroup $\Gamma^n(q) \subset K$, and $n \in \mathbb{N}$. Show that Theorem 2.3.4 holds for each function $F|_k M$ with $M \in \Gamma^n$, provided that q in (2.3.18) is replaced by $4q$, and the relations (2.3.19) are replaced by*

$$f_M(A[V]) = \tau(V) f_M(A) \quad \text{for } A \in \mathbb{A}_n \text{ and } V \in \mathrm{SL}_n(\mathbb{Z}),\ V \equiv E_n \pmod q,$$

where τ is a character with $\tau^4 = 1$.

[Hint: Show first that, for a fixed $M \in \Gamma^n$ and every $M' \in T_q^n$,

$$(F|_k M)(M'\langle Z\rangle) = \tau(M') F|_k M,$$

where τ is a character of T_q^n with $\tau^4 = 1$; then apply Theorem 2.3.1 to the kernel of τ].

Exercise 2.3.8. *Let $F \in \mathfrak{M}_{k+1/2}(K, \chi)$ be a modular form of half-integral weight for a congruence subgroup $K \subset \Gamma_0^n(4)$. Set, for $M = \begin{pmatrix} A & B \\ C & D \end{pmatrix} \in \Gamma_0^n(4)$,*

$$F|_j M = (\det(CZ + D))^{-k} j_{(2)}^n(M, Z)^{-1} F(M\langle Z\rangle).$$

Show that

$$F|_j M|_j M_1 = F|_j MM_1 \quad \text{for every } M,\ M_1 \in \Gamma_0^n(4),$$

and

$$F|_j M = \chi(M) F \quad \text{for every } M \in K.$$

In the notation of Proposition 2.3.6, show that

$$F|_j M \in \mathfrak{M}_{k+1/2}(K_M, \chi_M) \quad \text{if } M \in \Gamma_0^n(4).$$

2.3.4. The Siegel Operator

Here we shall look at relations between the modular forms on $\mathbb{H}_n$ and those on $\mathbb{H}_{n-1}$. The relations are based on the properties of Fourier expansions of modular forms described in Theorems 2.3.1, 2.3.4. We let $\mathfrak{F}\mathfrak{S}_q^n$ denote the set of all Fourier series of the form (2.3.5), which converge absolutely on $\mathbb{H}_n$ and uniformly on any subset of $\mathbb{H}_n$ of the form $\mathbb{H}_n(\varepsilon)$ with $\varepsilon > 0$. Let $F \in \mathfrak{F}\mathfrak{S}_q^n$. If $Z \in \mathbb{H}_{n-1}(\varepsilon)$ with $\varepsilon > 0$, and $\lambda \geq \varepsilon$, then clearly

$$Z_\lambda = \begin{pmatrix} Z & 0 \\ 0 & i\lambda \end{pmatrix} \in \mathbb{H}_n(\varepsilon).$$

Since the series for F converges uniformly, we have

$$\lim_{\lambda \to +\infty} F(Z_\lambda) = \sum_{A \in \mathbb{A}_n} f(A) \lim_{\lambda \to +\infty} \{q^{-1} A Z_\lambda\}.$$

If $A = \begin{pmatrix} A' & * \\ * & A_{nn} \end{pmatrix}$, then $\sigma(AZ_\lambda) = \sigma(A'Z) + i\lambda A_{nn}$, hence

$$\begin{aligned} \lim_{\lambda \to +\infty} e\{q^{-1} A Z_\lambda\} &= \lim_{\lambda \to +\infty} \exp(-q^{-1} \pi A_{nn} \lambda) e\{q^{-1} A' Z\} \\ &= \begin{cases} e\{q^{-1} A' Z\}, & \text{if } A_{nn} = 0 \\ 0, & \text{if } A_{nn} > 0. \end{cases} \end{aligned} \tag{2.3.28}$$

Since $A \geq 0$, the equality $A_{nn} = 0$ implies that $A_{1n} = A_{n1} = \cdots = A_{n-1,n} = A_{n,n-1} = 0$, i.e. $A = \begin{pmatrix} A' & 0 \\ 0 & 0 \end{pmatrix}$. Thus, for $Z \in \mathbb{H}_{n-1}$, we have

$$(F|\Phi)(Z) = \lim_{\lambda \to +\infty} F(Z_\lambda) = \sum_{A' \in \mathbb{A}_{n-1}} f\left(\begin{pmatrix} A' & 0 \\ 0 & 0 \end{pmatrix}\right) e\{q^{-1} A' Z\}. \tag{2.3.29}$$

The last series is a partial series for the expansion (2.3.5) of F, and so it converges absolutely on $\mathbb{H}_{n-1}$ and uniformly on $\mathbb{H}_{n-1}(\varepsilon)$ with $\varepsilon > 0$. Thus, we have $F|\Phi \in \mathfrak{F}\mathfrak{S}_q^{n-1}$. If $n = 1$, we set

$$F|\Phi = \lim_{\lambda \to +\infty} F(i\lambda). \tag{2.3.30}$$

As above, the limit exists and is equal to the constant term of the Fourier expansion of F. Taking

$$\mathfrak{F}\mathfrak{S}_q^0 = \mathbb{C}, \tag{2.3.31}$$

we obtain a linear operator

$$\Phi : \mathfrak{F}\mathfrak{S}_q^n \to \mathfrak{F}\mathfrak{S}_q^{n-1},$$

for all $n, q \in \mathbb{N}$, which is called *the Siegel operator*.

Let us consider the action of the Siegel operator on modular forms for congruence subgroups of modular group. For $n > 1$ and a matrix $M' = \begin{pmatrix} A' & B' \\ C' & D' \end{pmatrix}$ with $(n-1) \times (n-1)$-blocks A', B', C', and D' we set

$$\check{M}' = \begin{pmatrix} A & B \\ C & D \end{pmatrix}, \text{where } A = \begin{pmatrix} A' & 0 \\ 0 & 1 \end{pmatrix}, B = \begin{pmatrix} B' & 0 \\ 0 & 0 \end{pmatrix}, C = \begin{pmatrix} C' & 0 \\ 0 & 0 \end{pmatrix}, D' = \begin{pmatrix} D' & 0 \\ 0 & 1 \end{pmatrix}.$$

Let

$$\Gamma^{n,n-1} = \{M \in \Gamma^n; M = \check{M}'\}. \tag{2.3.32}$$

Then $\Gamma^{n,n-1}$ is clearly a subgroup of Γ^n, and the mapping i defined by

$$i(\check{M}') = M' \tag{2.3.33}$$

is an isomorphism of $\Gamma^{n,n-1}$ onto Γ^{n-1}. For a subgroup $K = K^n \subset \Gamma^n$, we set

$$K^{n,n-1} = K \cap \Gamma^{n,n-1} \quad \text{and} \quad K^{(n-1)} = \imath(K^{n,n-1}) \subset \Gamma^{n-1}. \tag{2.3.34}$$

Let K be a congruence group of Γ^n, and let χ be a character of K which is trivial on $\Gamma^n(q) \subset K$. We have $K^{n,n-1} \supset \Gamma^{n,n-1} \cap \Gamma^n(q) = \Gamma^{n,n-1}(q)$, hence

$$\Gamma^{n-1}(q) = \imath(\Gamma^{n,n-1}(q)) \subset K^{(n-1)}.$$

Thus, $K^{(n-1)}$ is a congruence subgroup of Γ^{n-1}. Then the character $\chi^{(n-1)}$ of $K^{(n-1)}$ given by

$$\chi^{(n-1)}(M') = \chi(\check{M}') \qquad (M' \in K^{(n-1)}) \tag{2.3.35}$$

is trivial on $\Gamma^{n-1}(q)$. Now let $F \in \mathfrak{M}_k(K, \chi)$ be a modular form of integral weight k and character χ for K. If $Z \in \mathbb{H}_{n-1}$ and $M' = \begin{pmatrix} A' & B' \\ C' & D' \end{pmatrix} \in \Gamma^{n-1}$, then, for $\lambda > 0$, we have

$$Z_\lambda = \begin{pmatrix} Z & 0 \\ 0 & i\lambda \end{pmatrix} \in \mathbb{H}_n; \qquad M = \begin{pmatrix} A & B \\ C & D \end{pmatrix} = \check{M}' \in \Gamma^{n,n-1},$$

$$M\langle Z_\lambda \rangle = \begin{pmatrix} A'Z + B' & 0 \\ 0 & i\lambda \end{pmatrix} \begin{pmatrix} C'Z + D' & 0 \\ 0 & 1 \end{pmatrix}^{-1} = M'\langle Z \rangle_\lambda,$$

and

$$\det(CZ_\lambda + D) = \det \begin{pmatrix} C'Z + D' & 0 \\ 0 & 1 \end{pmatrix} = \det(C'Z + D').$$

It follows that

$$\begin{aligned} (F|\Phi)|_k M' &= \lim_{\lambda \to +\infty} \det(C'Z + D')^{-k} F(M'\langle Z \rangle_\lambda) \\ &= \lim_{\lambda \to +\infty} \det(CZ_\lambda + D)^{-k} F(M\langle Z_\lambda \rangle) = (F|_k M)|\Phi. \end{aligned} \tag{2.3.36}$$

In particular, if $M' \in K^{(n-1)}$, we have

$$(F|\Phi)|_k M' = \chi(M) F|\Phi = \chi^{(n-1)}(M') F|\Phi.$$

Thus, $F|\Phi$ satisfies the functional equations for functions of $\mathfrak{M}_k(K^{(n-1)}, \chi^{(n-1)})$. Further, it follows from (2.3.36) that any function $(F|\Phi)|_k M'$ with $M' \in \Gamma^{n-1}$ is bounded on each $\mathbb{H}_{n-1}(\varepsilon)$ with $\varepsilon > 0$, since the function $F|_k \check{M}'$ is bounded on $\mathbb{H}_n(\varepsilon)$. Finally, since $F|\Phi \in \mathfrak{F}\mathfrak{S}_q^{n-1}$, it follows that $F|\Phi$ is regular on $\mathbb{H}_{n-1}$. Thus, we have proved the following

Proposition 2.3.9. *Let K be a congruence subgroup of Γ^n with $n > 1$, and let χ be a congruence character of K. Then the Siegel operator Φ maps the space $\mathfrak{M}_k(K, \chi)$,* *where* $k \in \mathbb{Z}$, *into* $\mathfrak{M}_k(K^{(n-1)}, \chi^{(n-1)})$:

$$\Phi : \mathfrak{M}_k(K, \chi) \to \mathfrak{M}_k(K^{(n-1)}, \chi^{(n-1)}). \tag{2.3.37}$$

If we set

$$\mathfrak{M}_k(K^{(0)}, \chi^{(0)}) = \mathfrak{M}_k(K^{(1)}, \chi^{(1)})|\Phi \subset \mathbb{C}, \tag{2.3.38}$$

then Proposition 2.3.9 holds also for $n = 1$.

The action of the Siegel operator on the theta-series is especially simple:

Proposition 2.3.10. *Let* $Q \in \mathbb{A}_m^+$ *and* $T \in T^n(Q)$, *where* $m, n \in \mathbb{N}$. *Then the image of the generalized theta-series* $\theta^n(Z, Q|T)$ *under the Siegel operator* Φ *is equal to*

$$\begin{cases} \theta^{n-1}(Z', Q|T'), & \text{if } \mathbf{t} \equiv 0 \pmod q \\ 0, & \textit{otherwise}, \end{cases} \tag{2.3.39}$$

where q *is the level of* Q *and* $\mathbf{t}$ *is the last column of* $T = (T', \mathbf{t})$. (*We set* $\theta^0 = 1$). *In particular,* Φ *maps the theta-series* $\theta^n(Z, Q)$ *on* $\theta^{n-1}(Z', Q)$.

Proof. Since the series for $\theta^n(Z, Q|T)$ converges uniformly on $\mathbb{H}_n(\varepsilon)$ with $\varepsilon > 0$, we have

$$\lim_{\lambda \to \infty} \theta^n\left(\begin{pmatrix} Z' & 0 \\ 0 & i\lambda \end{pmatrix}, Q|T\right) = \sum_{N \in \mathbb{M}_{m,n}} \lim_{\lambda \to \infty} e\{Q[N + q^{-1}T]Z'_\lambda\}.$$

If $N = (N', \mathbf{n})$, where $\mathbf{n}$ is the last column of N, then the entry in the lower right corner of $Q[N + q^{-1}T]$ is clearly $Q[\mathbf{n} + q^{-1}\mathbf{t}]$. Hence, by (2.3.28), since Q is positive, we have

$$\lim_{\lambda \to \infty} e\{Q[N + q^{-1}T]Z'_\lambda\} = \begin{cases} e\{Q[N' + q^{-1}T']Z'\}, & \text{if } \mathbf{n} + q^{-1}\mathbf{t} = \mathbf{0} \\ 0, & \text{otherwise.} \end{cases}$$

Exercise 2.3.11. *Let* $F \in \mathfrak{M}_k(K, \chi)$ *be a modular form of integral or half-integral weight* k *for a congruence subgroup* K *of* Γ^n. *The constant term* $f_M(0)$ *of the Fourier expansion* (2.3.18) *of* $F|_k M$ *with* $M \in \Gamma^n$ *will be called a boundary value of* F. *Show that*

(1) *The boundary value* $f_M(0)$ *with* $M \in \Gamma^n$ *is given by*

$$f_M(0) = \lim_{\lambda \to +\infty} (F|_k M)(i\lambda E_n) = (F|_k M)|\underbrace{\Phi|\cdots|\Phi}_{n \text{ times}};$$

(2) *If* χ *is trivial on* $\Gamma^n(q) \subset K$ *and* k *is integral, then* $f_M(0)$ *for* $M = \begin{pmatrix} A & B \\ C & D \end{pmatrix} \in \Gamma^n$ *depends only on* (A, C) *modulo* q;

(3) *If* $F = \theta^n(Z, Q|T)$ *with* $Q \in \mathbb{A}_m^+$, $T \in T^n(Q)$, *and* $n \in \mathbb{N}$ *is a generalized theta-series, then, for each* $M = \begin{pmatrix} A & B \\ C & D \end{pmatrix} \in \Gamma^n$ *with* $\det C \neq 0$, *the boundary value* $f_M(0)$

of F is given by

$$f_M(0) = (\det Q)^{-n/2}(\det(iC))^{-m/2}|\det C|^m d^{-mn} \sum_{N \in \mathsf{M}_{m,n}(\mathbb{Z}/d\mathbb{Z})} e\{Q[N + q^{-1}T]AC^{-1}\}$$

where $d \in \mathbb{N}$ *is such that* dC^{-1} *is integral,* $(\det iC)^{-m/2}$ *being defined by*

$$(\det iC)^{-m/2} = \lim_{\lambda \to +\infty} (\det(iC + \lambda^{-1}D))^{-m/2},$$

and the last root has to be taken with the same sign as in $F|_{m/2}M$.

[Hint for (3): Write

$$M\langle i\lambda E_n \rangle = AC^{-1} + Z_0 \quad \text{with} \quad Z_0 = -C^*(i\lambda C + D)^{-1},$$

and then compute $\lim_{\lambda \to \infty} (F|_{m/2}M)(i\lambda E_n)$ in the same way as in the proof of Proposition 1.4.5].

2.3.5. Cusp Forms

Let us call a form $F \in \mathfrak{M}_k(K, \chi)$ *a cusp form* (*of weight k and character* χ *for K*) if it satisfies the condition

$$F|_k M | \Phi = 0 \tag{2.3.40}$$

for every $M \in \Gamma^n$. The definition means that $F(Z)$ tends to zero as Z approaches any "rational boundary point" of $\mathbb{H}_n$ in a certain way. In particular, this allows us to strengthen the estimates of Fourier coefficients of such F.

Theorem 2.3.12. *Let* $F \in \mathfrak{M}_k(K, \chi)$ *be a cusp form of integral weight k and character* χ *for a congruence subgroup K of* Γ^n *with* $n \in \mathbb{N}$. *Suppose that* χ *is trivial on* $\Gamma^n(q) \subset K$. *Then the following statements hold:*

(1) *For any* $M \in \Gamma^n$, *the Fourier expansion* (2.3.18) *of the function* $F|_k M \in \mathfrak{M}_k(K_M, \chi_M)$ *is of the form*

$$F|_k M = \sum_{A \in \mathbb{A}_n^+} f_M(A) e\{q^{-1}AZ\}, \tag{2.3.41}$$

i.e. $f_M(A) = 0$, *unless* $A > 0$;

(2) *If* $k \geq 0$, *then each function* $F|_k M$ *with* $M \in \Gamma^n$ *satisfies*

$$|(F|_k M)(X + iY)| \leq \delta_F'(\det Y)^{-k/2} \quad \text{for } X + iY \in \mathbb{H}_n, \tag{2.3.42}$$

and its Fourier coefficients satisfy

$$|f_M(A)| \leq \delta_F(\det A)^{k/2} \quad \text{for } A \in \mathbb{A}_n^+ \tag{2.3.43}$$

with δ_F', δ_F *depending only on F*;

(3) *For every* $M \in \mathsf{GSp}_n^+(\mathbb{Q})$, *the function* $F|_k M \in \mathfrak{M}_k(K_M, \chi_M)$ *is a cusp form.*

We first prove a lemma.

Lemma 2.3.13. *Let $A \in \mathbb{A}_m$ with $m > 1$. Assume that* $\det A = 0$ *Then there exists a $V \in \mathsf{SL}_m(\mathbb{Z})$ such that*

$$A[V] = \begin{pmatrix} A' & 0 \\ 0 & 0 \end{pmatrix} \quad \text{with } A' \in \mathbb{A}_{m-1}. \tag{2.3.44}$$

Proof of the lemma. Since A integral and $\det A = 0$, there exists a non-zero integral column $\mathbf{v}$ with $A\mathbf{v} = 0$. We may assume that the entries of $\mathbf{v}$ are coprime. Then, by Lemma 2.1.5, we can find a matrix $V \in \mathsf{SL}_m(\mathbb{Z})$ with the last column $\mathbf{v}$. The matrix V satisfies (2.3.44). □

Proof of the theorem. In order to prove the first statement, we only have to show that a coefficient $f_M(A)$ of the expansion (2.3.18) of a cusp form equals zero if $\det A = 0$. If $n = 1$, this immediately follows from (2.3.40), for $(F|_k M)|\Phi = f_M(0)$, in the case. Now we assume that $n > 1$. Let $V \in \mathsf{SL}_m(\mathbb{Z})$ be a matrix satisfying (2.3.44) for a given $A = A_0$ with $\det A_0 = 0$. Then $M_0 = U(V^*) \in \Gamma^n$, and we have

$$\begin{aligned} F|_k M|_k M_0 &= \sum_{A \in \mathbb{A}_n} f_M(A) e\{q^{-1} A V Z\,{}^t V\} \\ &= \sum_{A \in \mathbb{A}_n} f_M(A[V^{-1}]) e\{q^{-1} A Z\}. \end{aligned}$$

On the other hand, the Fourier coefficients of $F|_k M|_k M_0 = F|_k M M_0$ are $f_{MM_0}(A)$. By comparing the coefficients, we get $f_M(A[V^{-1}] = f_{MM_0}(A)$, for $A \in \mathbb{A}_n$. Taking here $A = A_0[V]$, we see that

$$f_M(A_0) = f_{MM_0}(A_0[V]) = f_{MM_0}\left(\begin{pmatrix} A' & 0 \\ 0 & 0 \end{pmatrix}\right) = 0,$$

since $f_{MM_0}\left(\begin{pmatrix} A' & 0 \\ 0 & 0 \end{pmatrix}\right)$ is a Fourier coefficient of the function $(F|_k M M_0)|\Phi = 0$. The first statement is proved. To prove the estimates, we first prove the following lemma.

Lemma 2.3.14. *Let*

$$\psi(Y) = \sum_{A \in \mathbb{A}_n^+} \psi(A) \exp(-\xi \sigma(AY)),$$

where $Y \in \mathsf{P}_n$ and $\xi > 0$, be a series with non-negative coefficients $\psi(A)$, which converges for all $Y \in \mathsf{P}_n$. Then, for every matrix Y of the Minkowski reduction domain F_n satisfying $Y \geq cE_n$ with $c > 0$, the following estimates hold

$$\begin{aligned} \psi(Y) &\leq \delta_1 \exp(-\delta_2 \sigma(Y)), \\ \psi(Y) &\leq \delta_1 \exp(-\delta_2 n (\det Y)^{1/n}), \end{aligned}$$

where $\delta_1 > 0$ depends only on ψ and c, and $\delta_2 > 0$ depends only on n and ξ.

Proof of the lemma. Since $Y \in \mathsf{F}_n$, it follows from (2.1.16) and (A.1.6) that

$$\begin{aligned}\sigma(AY) &\geq b_n \sigma(A \operatorname{diag}(y_{11}, y_{22}, \ldots, y_{nn})) \\ &= b_n \sum_{\alpha=1}^{n} 2a_{\alpha\alpha} y_{\alpha\alpha} \geq 2b_n \sigma(Y)\end{aligned}$$

for $A \in \mathbb{A}_n^+$, where $b_n = n^{1-n} c_n^{-1}$ depends only on n, and $2a_{\alpha\alpha} \geq 2$, are the diagonal entries of A. On the other hand, since $Y \geq cE_n$, we have $\sigma(AY) \geq c\sigma(A)$ for $A \geq 0$. Combining the inequalities, we get

$$\sigma(AY) \geq b_n \sigma(Y) + \tfrac{1}{2} c\sigma(A),$$

hence

$$\begin{aligned}\psi(Y) &\leq \exp(-\xi b_n \sigma(Y)) \sum_{A \in \mathbb{A}_n^+} \psi(A) \exp(-\xi c\sigma(A)/2) \\ &= \exp(-\xi b_n \sigma(Y)) \psi(\tfrac{1}{2} c E_n),\end{aligned}$$

which proves the first estimate. The second estimate follows from the first one, since, by (A.1.8), we have

$$\sigma(Y) = y_{11} + \cdots + y_{nn} \geq n(y_{11} \cdots y_{nn})^{1/n} \geq n(\det Y)^{1/n} \quad \text{for} \quad Y \in \mathsf{P}_n. \quad \square$$

Returning to the proof of Theorem 2.3.12, we consider the function

$$G = G_F = \sum_{M \in K \backslash \Gamma^n} |F|_k M|. \tag{2.3.45}$$

It follows from (2.3.15) and (2.3.16) that G does not depend on the choice of representatives in the classes $K \backslash \Gamma^n$. Hence G satisfies

$$\begin{aligned}|\det(CZ + D)|^{-k} G(M'\langle Z \rangle) &= \sum_M |F|_k M|_k M'| \\ &= \sum_M |F|_k MM'| = G(Z),\end{aligned} \tag{2.3.46}$$

for every $M' = \begin{pmatrix} A & B \\ C & D \end{pmatrix} \in \Gamma^n$. Hence, by Lemma 1.2.8, we see that the function

$$G'(Z) = G'(X + iY) = (\det Y)^{k/2} G(Z)$$

is invariant under Γ^n:

$$G'(M'\langle Z \rangle) = G'(Z) (Z \in \mathbb{H}_n, M' \in \Gamma^n).$$

Thus, the values of G' on $\mathbb{H}_n$ are the same as those on D_n, the fundamental domain for Γ^n on $\mathbb{H}_n$. If $Z = X + iY \in \mathsf{D}_n$, then, by (2.1.26) and the definition of D_n, we can apply Lemma 2.3.14 to each of the series

$$\sum_{A \in \mathbb{A}_n^+} |f_M(A)| \exp(-\pi\sigma(AY)/q) \geq |F|_k M|,$$

which gives the estimate

$$G'(X+iY) \leq \delta_3(\det Y)^{k/2}\exp(-\delta_2\sigma(Y)),$$

where $\delta_2, \delta_3 > 0$ depend only on F. Applying (A.1.8), we get

$$G'(X+iY) \leq \delta_3 \prod_{\alpha=1}^{n} y_{\alpha\alpha}^{k/2}\exp(-\delta_2 y_{\alpha\alpha}),$$

where $y_{\alpha\alpha}$ are the diagonal entries of Y (note that now $k \geq 0$). Since the last expression is clearly bounded on P_n, it follows that G' is bounded on D_n. Thus, G' is bounded on $\mathbb{H}_n$:

$$G'(Z) \leq \delta'_F \ (Z \in \mathbb{H}_n).$$

Since each of the functions $|F|_k M|$ with $M \in \Gamma^n$ is a summand in (2.3.45), we obtain

$$\begin{aligned}|(F|_k M)(X+iY)| &\leq G_F(X+iY) = (\det Y)^{-k/2} G'(X+iY)\\ &\leq \delta'_F(\det Y)^{-k/2},\end{aligned}$$

which is the estimate (2.3.42). Now we have

$$\begin{aligned}|f_M(A)| &= \Big|q^{-\langle n\rangle} \int_{0\leq\cdots,x_{\alpha\beta},\cdots\leq q} (F|_k M)((x_{\alpha\beta}) + iA^{-1})\\ &\quad \times e\{-q^{-1}A((x_{\alpha\beta}) + iA^{-1})\} \prod_{1\leq\alpha\leq\beta\leq n} dx_{\alpha\beta}\Big|\\ &\leq q^{-\langle n\rangle}\delta'_F(\det A)^{k/2}\exp(q^{-1}\pi n)q^{\langle n\rangle} = \delta_F(\det A)^{k/2},\end{aligned}$$

which proves (2.3.43).

Finally, by Proposition 2.3.6, in order to prove the last statement, we only have to show that

$$((F|_k M)|_k M')|\Phi = 0$$

if $M \in \mathsf{GSp}_n^+(\mathbb{Q})$ and $M' \in \Gamma^n$. By (2.3.15), we have $F|_k M|_k M' = F|_k MM'$. By Proposition 1.3.7, we can write the matrix MM' in the form

$$MM' = M_1\begin{pmatrix} A_1 & B_1 \\ 0 & D_1 \end{pmatrix} \quad \text{with} \quad M_1 \in \Gamma^n.$$

Since F is a cusp form, we have the expansion

$$F|_k M_1 = \sum_{A \in \mathbb{A}_n^+} f_1(A) e\{q^{-1}AZ\},$$

by the first part of the theorem; hence

$$\begin{aligned}F|_k MM' &= F|_k M_1\begin{pmatrix} A_1 & B_1 \\ 0 & D_1 \end{pmatrix} = F|_k M_1|_k\begin{pmatrix} A_1 & B_1 \\ 0 & D_1 \end{pmatrix}\\ &= (\det D_1)^{-k}\sum_{A\in\mathbb{A}_n^+} f_1(A)e\{q^{-1}AB_1D_1^{-1}\}e\{q^{-1}D_1^{-1}AA_1Z\}.\end{aligned}$$

Note that each matrix $q^{-1}D_1^{-1}AA_1$ is positive definite, since A is. Then it follows from the definition of the Siegel operator and (2.3.28) that $F|_k MM'|\Phi = 0$. □

We shall denote by $\mathfrak{N}_k(K,\chi)$ the space of all cusp forms of weight k and character χ for K.

Exercise 2.3.15. *A modular form of half-integral weight k is called a cusp form, if it meets* (2.3.40), *the operators* $|_k M$ *being defined by* (2.3.14). *Let* $\mathfrak{N}_k(K,\chi)$ *be the subspace of all cusp forms of* $\mathfrak{M}_k(K,\chi)$. *Show that the expansion* (2.3.41) *with* $4q$ *instead of* q, *the estimate* (2.3.42) *and the estimate* (2.3.43) *hold for each form of* $\mathfrak{N}_k(K,\chi)$.

[Hint: See Exercise 2.3.7].

2.3.6. Singular Forms

We have seen that Fourier coefficient $f(A)$ of a cusp form satisfy $f(A) = 0$ if $\det A = 0$. Modular forms with the opposite property, i.e. the forms whose Fourier coefficients $f(A)$ satisfy $f(A) = 0$ if $\det A \neq 0$, are also of considerable interest, especially in the theory of Hecke operators.

A modular form $F \in \mathfrak{M}_k(K,\chi)$ with the Fourier expansion

$$F(Z) = \sum_{A \in \mathbb{A}_n} f(A)e\{q^{-1}AZ\}$$

is called *singular* if

$$f(A) = 0 \qquad \text{of all } A \in \mathbb{A}_n^+. \tag{2.3.47}$$

Examples of singular forms are given by certain theta-series: Let $Q \in \mathbb{A}_{2k}^+$ and $n > 2k$. Then the theta-series

$$\theta^n(Z,Q) = \sum_{A \in \mathbb{A}_n} r(Q,A)e\{AZ\}, \tag{2.3.48}$$

where $r(Q,A)$ is the number of solutions of the equation $Q[X] = A$ in integral $(2k \times n)$-matrices X, is a singular form, since the rank of any A of the form $Q[X]$ is not greater than $2k = \operatorname{rank} Q$, and so $r(Q,A) = 0$ if $\operatorname{rank} A = n$. In some cases it happens that every singular form is a linear combination of the theta-series (2.3.48).

Let χ be a Dirichlet character modulo q. Then the mapping

$$\Gamma_0^n(q) \ni \begin{pmatrix} A & B \\ C & D \end{pmatrix} \to [\chi]\left(\begin{pmatrix} A & B \\ C & D \end{pmatrix}\right) = \chi(\det D) \tag{2.3.49}$$

is clearly a congruence character of $\Gamma_0^n(q)$. A character of $\Gamma_0^n(q)$ of the form $[\chi]$ will be called *one-dimensional*. For such a character $[\chi]$, a modular form $F \in \mathfrak{M}_k(\Gamma_0^n(q), [\chi]) = \mathfrak{M}_k^n(q,\chi)$ has the Fourier expansion of the form

$$F(Z) = \sum_{A \in \mathbb{A}_n} f(A)e\{AZ\}, \tag{2.3.50}$$

where the Fourier coefficients $f(A)$ satisfy

$$f(A[V]) = (\det V)^k \chi(\det V) f(A) \qquad \text{with } A \in \mathbb{A}_n,\ V \in \Lambda^n \tag{2.3.51}$$

((2.3.50) follows from (2.3.18) with $M = E_{2n}$, since F is invariant under $Z \to Z + S$ for $S \in \mathsf{S}_n(\mathbb{Z})$; (2.3.51) follows from

$$(F|_k U({}^tV))(Z) = (\det V)^{-k} F(V^{-1} Z V^*) = \chi(\det V) F(Z)$$

by the uniqueness of the Fourier expansion).

Theorem 2.3.16. *Let $\mathfrak{M} = \mathfrak{M}_k^n(q, \chi)$ be a space of modular forms of integral weight k and one-dimensional character $[\chi]$ for the group $\Gamma_0^n(q)$, where $n, q \in \mathbb{N}$. Assume that $\mathfrak{M}$ contains a non-constant singular form F. Then the following statements hold:*

(1) $n > 2k > 0$;

(2) *χ is a real Dirichlet character modulo q satisfying $\chi(-1) = (-1)^k$;*

(3) *F is a finite linear combination of the theta-series $\theta^n(Z, Q)$ with*

$$Q \in \mathbb{A}_{2k}^+,\ q(Q) | q, \quad \textit{and} \quad \chi_Q(a) = \chi(a), \quad \textit{if } \text{g.c.d.}(a, q) = 1, \tag{2.3.52}$$

where $q(Q)$ is the level of Q and χ_Q is the character of Q (see Theorem 1.4.10).

Corollary. *If $F \in \mathfrak{M}_k^n(q, \chi)$ is a singular form, then each form $F|_k M$ with $M \in \Gamma^n$ is also singular.*

Proof of the theorem. Let $f(A)$ be a non-zero coefficient of the Fourier expansion (2.3.50) with maximal $r = \operatorname{rank} A$. It follows from Lemma 2.3.13 that there is a $V \in \mathsf{SL}_n(\mathbb{Z})$ satisfying

$$A[V] = \begin{pmatrix} Q & 0 \\ 0 & 0 \end{pmatrix} \quad \text{with } Q \in \mathbb{A}_r^+.$$

Let us choose A and V so that $\det Q$ be minimal. Then we show that

$$r = 2k \quad \text{and} \quad Q \text{ satisfies (2.3.52).} \tag{2.3.53}$$

For this we consider the restriction of F on the subset of $\mathbb{H}_n$ consisting of the matrices

$$Z = \begin{pmatrix} Z_1 & 0 \\ 0 & Z_2 \end{pmatrix} \quad \text{with } Z_1 \in \mathbb{H}_r \text{ and } Z_2 \in \mathbb{H}_{n-r} \tag{2.3.54}$$

Since (2.3.50) converges absolutely, we can write

$$F\left(\begin{pmatrix} Z_1 & 0 \\ 0 & Z_2 \end{pmatrix}\right) = \sum_{A = \begin{pmatrix} A_1 & A_{12} \\ A_{21} & A_2 \end{pmatrix} \in \mathbb{A}_n} f(A) e\{A_1 Z_1\} e\{A_2 Z_2\}$$

$$= \sum_{A_1 \in \mathbb{A}_r} f(A_1, Z_2) e\{A_1 Z_1\},$$

where

$$f(A_1, Z_2) = \sum_{\begin{pmatrix} A_1 & A_{12} \\ A_{21} & A_2 \end{pmatrix} \in \mathbb{A}_n} f\left(\begin{pmatrix} A_1 & A_{12} \\ A_{21} & A_2 \end{pmatrix}\right) e\{A_2 Z_2\}.$$

If $M = \begin{pmatrix} A & B \\ C & D \end{pmatrix} \in \Gamma_0^{n-r}(q)$, it is clear that the matrix $M' = \begin{pmatrix} A' & B' \\ C' & D' \end{pmatrix}$ with $A' = \begin{pmatrix} E_r & 0 \\ 0 & A \end{pmatrix}$, $B' = \begin{pmatrix} 0 & 0 \\ 0 & B \end{pmatrix}$, $C' = \begin{pmatrix} 0 & 0 \\ 0 & C \end{pmatrix}$, and $D' = \begin{pmatrix} E_r & 0 \\ 0 & D \end{pmatrix}$ is contained in $\Gamma_0^n(q)$. For any Z of the form (2.3.54) we have the relations

$$M'\langle Z \rangle = \begin{pmatrix} Z_1 & 0 \\ 0 & M\langle Z_2 \rangle \end{pmatrix} \quad \text{and} \quad \det(C'Z + D') = \det(CZ_2 + D),$$

and so the functional equation (2.2.4) for the function $F \in \mathbb{M}$ and the matrix M' becomes

$$\det(CZ_2 + D)^{-k} F\left(\begin{pmatrix} Z_1 & 0 \\ 0 & M\langle Z_2 \rangle \end{pmatrix}\right) = \chi(\det D) F\left(\begin{pmatrix} Z_1 & 0 \\ 0 & Z_2 \end{pmatrix}\right).$$

Hence, for each $A_1 \in \mathbb{A}_r$, we have the relation

$$\det(CZ_2 + D)^{-k} f(A_1, M\langle Z_2 \rangle) = \chi(\det D) f(A_1, Z_2). \tag{2.3.55}$$

Let us cònsider the function $f(A_1, Z_2)$ for $A_1 = Q$. If

$$A = \begin{pmatrix} Q & A_{12} \\ A_{21} & A_2 \end{pmatrix} \in \mathbb{A}_n \quad \text{and} \quad f(A) \neq 0,$$

then, by the choice of r and Lemma 2.3.13, we can write A in the form

$$A = \begin{pmatrix} Q' & 0 \\ 0 & 0 \end{pmatrix}[V] \quad \text{with } V \in \mathsf{SL}_n(\mathbb{Z}) \text{ and } Q' \in \mathbb{A}_r.$$

Let $V = \begin{pmatrix} V_1 & * \\ * & * \end{pmatrix}$, where $V_1 = V^{(r)}$ is the block of dimension $r \times r$, then it follows from the last relation that $Q = Q'[V_1]$, hence $\det Q' \neq 0$ and $\det Q' \leq \det Q$. On the other hand, by the choice of Q, we have $\det Q \leq \det Q'$. Thus, $\det V_1 = \pm 1$ and $V_1 \in \Lambda^r$, and so we can write

$$A = \begin{pmatrix} Q & 0 \\ 0 & 0 \end{pmatrix}[W], \quad \text{where} \quad W = \begin{pmatrix} V_1^{-1} & 0 \\ 0 & E_{n-r} \end{pmatrix} V \in \Lambda^n.$$

If $\det W = -1$, we replace W by $\begin{pmatrix} E_{n-1} & 0 \\ 0 & -1 \end{pmatrix} W$, and so we can assume that $W \in \mathsf{SL}_n(\mathbb{Z})$. Then it follows from (2.3.51) that $f(A) = f\left(\begin{pmatrix} Q & 0 \\ 0 & 0 \end{pmatrix}\right)$, and hence we have

$$f(Q, Z_2) = f\left(\begin{pmatrix} Q & 0 \\ 0 & 0 \end{pmatrix}\right) \sum_{\begin{pmatrix} Q & A_{12} \\ A_{21} & A_2 \end{pmatrix} \sim_{\mathbb{Z}} \begin{pmatrix} Q & 0 \\ 0 & 0 \end{pmatrix}} e\{A_2 Z_2\}.$$

The last sum is taken over the set of all matrices of the form

$$\begin{pmatrix} Q & A_{12} \\ A_{21} & A_2 \end{pmatrix} = \begin{pmatrix} Q & 0 \\ 0 & 0 \end{pmatrix}[U] = \begin{pmatrix} Q[U_1] & {}^tU_1QU_2 \\ {}^tU_2QU_1 & Q[U_2] \end{pmatrix} \quad \text{with } U = \begin{pmatrix} U_1 & U_2 \\ U_3 & U_4 \end{pmatrix} \in \Lambda^n.$$

It follows that $Q[U_1] = Q$, hence ${}^tU_1QU_2 = QU_1^{-1}U_2$ and $Q[U_2] = Q[U_1^{-1}U_2]$. So we have

$$\begin{pmatrix} Q & A_{12} \\ A_{21} & A_2 \end{pmatrix} = \begin{pmatrix} Q & QN \\ {}^tNQ & Q[N] \end{pmatrix} \quad \text{with } N = U_1^{-1}U_2 \in \mathsf{M}_{r,n-r}.$$

Conversely, if $N \in \mathsf{M}_{r,n-r}$, then

$$\begin{pmatrix} Q & QN \\ {}^tNQ & Q[N] \end{pmatrix} = \begin{pmatrix} Q & 0 \\ 0 & 0 \end{pmatrix}[U], \quad \text{where} \quad U = \begin{pmatrix} E_r & N \\ 0 & E_{n-r} \end{pmatrix} \in \mathsf{SL}_n(\mathbb{Z}),$$

Returning to the last expression for $f(Q, Z_2)$, we obtain finally that

$$\begin{aligned} f(Q, Z_2) &= f\left(\begin{pmatrix} Q & 0 \\ 0 & 0 \end{pmatrix}\right) \sum_{N \in \mathsf{M}_{r,n-r}} e\{Q[N]Z_2\} \\ &= f\left(\begin{pmatrix} Q & 0 \\ 0 & 0 \end{pmatrix}\right) \theta^{n-r}(Z_2, Q) \end{aligned}$$

with the theta-series $\theta^{n-r}(Z_2, Q)$ defined by (2.3.48). The condition (2.3.55) shows that theta-series satisfies

$$\det(CZ_2 + D)^{-k}\theta^{n-r}(M\langle Z_2\rangle, Q) = \chi(\det D)\theta^{n-r}(Z_2, Q) \tag{2.3.56}$$

for every $M = \begin{pmatrix} A & B \\ C & D \end{pmatrix} \in \Gamma_0^{n-r}(q)$.

Lemma 2.3.17. *Let $\theta(Z) = \theta^n(Z, Q)$ be the theta-series of genus $n \geq 1$ of a matrix $Q \in \mathbb{A}_r^+$ with $r \geq 1$. Suppose that $\theta(Z)$ satisfies*

$$\det(CZ + D)^{-k}\theta(M\langle Z\rangle) = \chi(\det D)\theta(Z) \tag{2.3.57}$$

for every $M = \begin{pmatrix} A & B \\ C & D \end{pmatrix} \in \Gamma_0^n(q)$ with some $q \in \mathbb{N}$, where k is an integer and χ is a Dirichlet character modulo q. Then the following statements hold:

(1) *$k = r/2$;*

(2) *the level $q(Q)$ of Q divides q;*

(3) *$\chi(a) = \chi_Q(a)$ if* g.c.d.$(a, q) = 1$, *where χ_Q is the character of Q; in particular, χ is a real character satisfying $\chi(-1) = (-1)^k$.*

Proof of the lemma. Since, clearly $\Gamma_0^n(q)^{(n-1)} = \Gamma_0^{n-1}(q)$, it follows from (2.3.36) that the function $\theta|\Phi$, where Φ is the Siegel operator, satisfies the functional equation of the same shape as (2.3.57) for every $M' \in \Gamma_0^{n-1}(q)$. By Proposition 2.3.10, we have $\theta|\Phi = \theta^{n-1}(Z', Q)$. Thus, with the assumption of the lemma we can replace n by $n-1$ if $n>1$. Hence, we may assume without loss of generality that $n=1$. In view of Theorem 2.2.2, we have

$$\theta^1(z,Q)\in\mathfrak{M}^1_{r/2}(q(Q),\chi_Q).$$

Let $M=\begin{pmatrix}a & b\\ c & d\end{pmatrix}$ be a matrix of $\Gamma^1_0(q\cdot q(Q))$ with $c\neq 0$. It follows from the inclusion that

$$|\theta^1(M\langle z\rangle,Q)\theta^1(z,Q)^{-1}|=|cz+d|^{r/2}.$$

On the other hand, (2.3.57) with $n=1$ shows that

$$|\theta^1(M\langle Z\rangle,Q)\theta^1(z,Q)^{-1}|=|cz+d|^k.$$

Hence we conclude that $k=r/2$. By comparing the functional equations of functions of $\mathbb{M}^1_{r/2}(q(Q),\chi_Q)$ and (2.3.57), we see that $\chi(d)=\chi_Q(d)$, if g.c.d.$(d,qq(Q))=1$. Thus, it remains to prove that $q(Q)$ divides q. For this we use the inversion formula

$$\theta^1(-z^{-1},Q)=(\det Q)^{-1/2}(-iz)^{r/2}\theta^1(z,Q^{-1})$$

which is a particular case of (1.3.33) with $n=1$, $W=W'=0$ and Q^{-1} in place of Q. Take

$$M=\begin{pmatrix}1 & 0\\ -q & 1\end{pmatrix}\begin{pmatrix}0 & 1\\ -1 & 0\end{pmatrix}=\begin{pmatrix}0 & 1\\ -1 & 0\end{pmatrix}\begin{pmatrix}1 & q\\ 0 & 1\end{pmatrix}.$$

On one hand, since $\begin{pmatrix}1 & 0\\ -q & 1\end{pmatrix}\in\Gamma^1_0(q)$, we have

$$\begin{aligned}\theta^1(M\langle z\rangle,Q)&=\theta^1\left(\begin{pmatrix}1 & 0\\ -q & 1\end{pmatrix}\langle -z^{-1}\rangle,Q\right)\\ &=(qz^{-1}+1)^k\theta^1(-z^{-1},Q)\\ &=(qz^{-1}+1)^k(\det Q)^{-1/2}(-iz)^{r/2}\theta^1(z,Q^{-1}).\end{aligned}$$

On the other hand, we can write

$$\begin{aligned}\theta^1(M\langle z\rangle,Q)&=\theta^1(-(z+q)^{-1},Q)\\ &=(\det Q)^{-1/2}(-i(z+q))^{r/2}\theta^1(z+q,Q^{-1}).\end{aligned}$$

By comparing, we get

$$\theta^1(z+q,Q^{-1})=\theta^1(z,Q^{-1}),$$

since $r=2k$ is even. Then it follows from the definition of theta-series that the number $qQ^{-1}[\mathbf{m}]$ must be integral and even for every $\mathbf{m}\in\mathsf{M}_{r,1}$. One can easily check that this can happen only if the matrix qQ^{-1} is an integral matrix with even diagonal entries. It follows that q must be divisible by the level of Q. □

Let us complete the proof of the theorem. The first two statements follow from (2.3.56) and Lemma 2.3.17. Now let $Q_1, \ldots, Q_h$ be a set of representatives for the classes $\{Q\}_{\mathbb{Z}}$ of matrices Q satisfying (2.3.52) (the number of the classes is finite, by Corollary of Theorem 2.1.12). We may assume that $\det Q_1 \leq \cdots \leq \det Q_h$. Consider the system of h linear equations in $x_1, \ldots, x_h$,

$$\sum_{\beta=1}^{h} r(Q_\beta, Q_\alpha)x_\beta = f\left(\begin{pmatrix} Q_\alpha & 0 \\ 0 & 0 \end{pmatrix}\right) \qquad (\alpha = 1, \ldots, h),$$

where $f(A)$ are the Fourier coefficients of F. Since $\det Q_1 \leq \cdots \leq \det Q_h$, and the matrices $Q_1, \ldots, Q_h$, are not equivalent over $\mathbb{Z}$, we conclude that $r(Q_\beta, Q_\alpha) = 0$, if $\beta > \alpha$. It is clear that $r(Q_\alpha, Q_\alpha) \geq 1$. So we see that the matrix of the system is triangular and have non-zero diagonal entries. Hence, the system has a solution $x_1 = c_1, \ldots, x_h = c_h$. By above, since $n > 2k$, it follows that the function

$$G = F - \sum_{\beta=1}^{h} c_\beta \theta^n(Z, Q_\beta)$$

is a singular form of the space $\mathfrak{M}_k^n(q, \chi)$. Let $g(A)$ be the Fourier coefficients of G. Then for $\alpha = 1, \ldots, h$ we have

$$\begin{aligned} g\left(\begin{pmatrix} Q_\alpha & 0 \\ 0 & 0 \end{pmatrix}\right) &= f\left(\begin{pmatrix} Q_\alpha & 0 \\ 0 & 0 \end{pmatrix}\right) - \sum_{\beta=1}^{h} c_\beta r\left(Q_\beta, \begin{pmatrix} Q_\alpha & 0 \\ 0 & 0 \end{pmatrix}\right) \\ &= f\left(\begin{pmatrix} Q_\alpha & 0 \\ 0 & 0 \end{pmatrix}\right) - \sum_{\beta=1}^{h} c_\beta r(Q_\beta, Q_\alpha) = 0. \end{aligned}$$

It follows that $g\left(\begin{pmatrix} Q & 0 \\ 0 & 0 \end{pmatrix}\right) = 0$ for any Q satisfying (2.3.52). Then the above considerations show that G is a constant. Since $k > 0$, we have $G = 0$. $\square$

Exercise 2.3.18 (*E. Freitag*). *Show that every singular form* $F \in \mathfrak{M}_k(\Gamma^n, 1)$ *can be written in the form*

$$F = \sum_{\nu=1}^{h} f\left(\begin{pmatrix} Q_\nu & 0 \\ 0 & 0 \end{pmatrix}\right) r(Q_\nu, Q_\nu)^{-1} \theta^n(Z, Q_\nu),$$

where $Q_1, \ldots, Q_h$ *is a set of representatives for the* $\mathbb{Z}$*-equivalence classes of the matrices* $Q \in \mathbb{A}_{2k}^+$ *with* $\det Q = 1$.

§2.4. Spaces of Modular Forms

Proceeding from certain properties of theta-series, we have introduced the modular forms. Generally speaking, the spaces of modular forms are not spanned by the theta-series. Nevertheless, the spaces are finite dimensional.

2.4.1. Zeroes of Modular Forms for Γ^1

We say that a non-zero function F on an open subset of $\mathbb{C}$ has *the order* $n = v_p(F)$ at a point p, if the function $(z-p)^{-n}F(z)$ is regular and does not vanish at $z = p$. So $v_p(F) > 0$ if and only if $F(p) = 0$. If $F \in \mathfrak{M}_k = \mathfrak{M}_k(\Gamma^1, 1)$ is a modular form of weight k and unit character for Γ^1, then the functional equations

$$(cz+d)^{-k}F(M\langle z\rangle) = F(z)\left(M = \begin{pmatrix} a & b \\ c & d \end{pmatrix} \in \Gamma^1\right) \tag{2.4.1}$$

imply that $v_p(F) = v_{M\langle p\rangle}(F)$ for $p \in \mathbb{H}_1$ and $M \in \mathbb{C}^1$, and so the order $v_p(F)$ depends only on the orbit $\Gamma^1\langle p\rangle$ of p under Γ^1. By Theorem 2.3.1., the function F has the Fourier expansion of the form

$$F(z) = \sum_{a=0}^{\infty} f(2a)\exp(2\pi i a z) \tag{2.4.2}$$

which converges uniformly on each subset $\mathbb{H}_1(\varepsilon) \subset \mathbb{H}_1$ with $\varepsilon > 0$. It follows that F can be considered as a function in the variable $q = \exp(2\pi i z)$:

$$F(z) = F'(q) = \sum_{a=0}^{\infty} f(2a)q^a, \tag{2.4.3}$$

and the function F' is regular in the open unit disk $|q| < 1$, including its center

$$q = 0 = \lim_{z\to i\infty} \exp(2\pi i z).$$

The order of F' at $q = 0$ will be called the order of F at $i\infty$ and will be denoted by $v_{i\infty}(F)$. In other words, $v_{i\infty}(F) = n$, if

$$f(0) = f(2) = \cdots = f(2n-2) = 0, \quad \text{but} \quad f(2n) \neq 0.$$

Proposition 2.4.1. *Each non-zero modular form of integral weight k and unit character for Γ^1 vanishes only at finitely many orbits of Γ^1 on $\mathbb{H}_1$; if $p_1, \ldots, p_m$ is a set of representatives for these orbits, then the following relation holds*

$$v_{i\infty}(F) + \sum_{\alpha=1}^{m} e(p_\alpha)^{-1} v_{p_\alpha}(F) = k/12, \tag{2.4.4}$$

where $e(p) = 2$ for $p \in \Gamma^1\langle i\rangle$; $e(p) = 3$ for $p \in \Gamma^1(\rho)$ with $\rho = \exp(\pi i/3)$; and $e(p) = 1$ in other cases.

Proof. By Theorem 2.1.1, each orbit of Γ^1 on $\mathbb{H}_1$ meets the fundamental domain D_1. Thus, to prove the finiteness, we only have to show that D_1 contains only finitely many zeroes of F. Since $F'(q)$ is a non-zero function regular at $q = 0$, it follows that F' does not vanish in a domain of the shape $0 < |q| < \varepsilon$ with $\varepsilon > 0$. Hence $F(x+iy) \neq 0$ if $y > \dfrac{1}{2\pi}\ln(1/\varepsilon)$, and so all zeroes of F in D_1 lie in the set

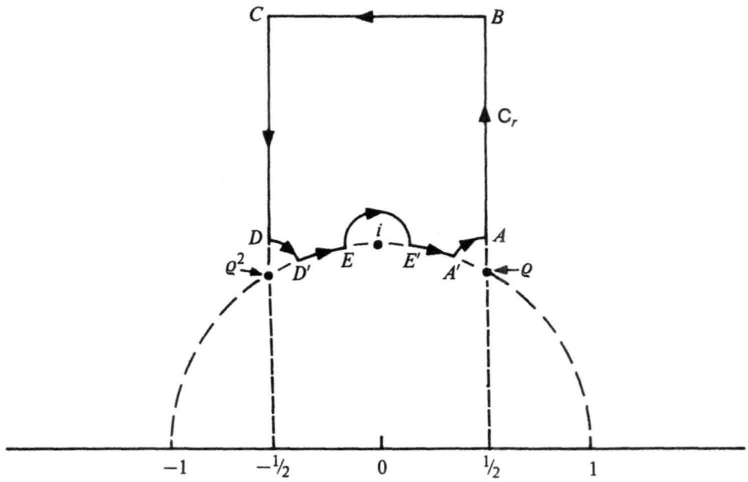

Fig. 2

$\{x + iy \in \mathsf{D}_1, y \leq (2\pi)^{-1} \ln \varepsilon^{-1}\}$, which is compact. Since F is regular, the number of the zeroes is finite.

Proving (2.4.4), we can assume that $p_1, \ldots, p_m \in \mathsf{D}_1$. We assume first that the boundary of D_1 does not contain zeroes of F, exept, possibly, at points i, ρ, ρ^2. Then there is a contour C_r of the shape pictured by Fig. 2, where DD', EE' and $A'A$ are the circular arcs of a small radius r with the centers at ρ^2, i and ρ, respectively, which encloses all the zeroes $p_1, \ldots, p_m$, exept i, ρ, ρ^2. Since the inner points of D_1 lie in different orbits of Γ^1, there are no other zeroes of F enclosed by C_r. Then we have

$$\sum_{p_\alpha \neq \rho, \rho^2, i} \nu_{p_\alpha}(F) = \frac{1}{2\pi i} \int_{\mathsf{C}_r} \frac{\mathrm{d}F}{F} = \lim_{r \to 0} \frac{1}{2\pi i} \int_{\mathsf{C}_r} \frac{\mathrm{d}F}{F}. \tag{2.4.5}$$

In order to compute the integral over C_r, we cut the contour by the points A, B, C, D, D', E, E', A' (see Fig. 2), and consider the integrals over the separate parts. First, since $F(z-1) = F(z)$ and the transformation $z \to z - 1$ carries the segment AB onto DC, we have

$$\frac{1}{2\pi i} \int_{AB} \frac{\mathrm{d}F}{F} + \frac{1}{2\pi i} \int_{CD} \frac{\mathrm{d}F}{F} = \frac{1}{2\pi i} \left(\int_{AB} \frac{\mathrm{d}F(z-1)}{F(z-1)} + \int_{CD} \frac{\mathrm{d}F}{F} \right)$$
$$= \frac{1}{2\pi i} \left(\int_{DC} \frac{\mathrm{d}F}{F} + \int_{CD} \frac{\mathrm{d}F}{F} \right) = 0.$$

Further, the mapping $z \to q = \exp(2\pi i z)$ carries the segment BC onto a circle L with clockwise direction and with the center at $q = 0$, which does not enclose any zeroes of F', exept, possibly, the zero of order $v_{i\infty}(F)$ at $q = 0$. Hence we get

$$\frac{1}{2\pi i}\int_{BC}\frac{dF}{F} = \frac{1}{2\pi i}\int_{L}\frac{dF'}{F'} = -v_{i\infty}(F).$$

If we integrate the element $(2\pi i)^{-1}F^{-1}dF$ over the whole circle of radius r with the center at ρ^2 in the clockwise direction, then, for small r, we get the value $-v_{\rho^2}(F) = -v_\rho(F)$ (note $\rho^2 = \rho - 1$). Since the angle at ρ^2 in the triangle $\rho^2 DD'$ is $2\pi/6$, we conclude that

$$\lim_{r\to 0}\frac{1}{2\pi i}\int_{DD'}\frac{dF}{F} = -\frac{1}{6}v_\rho(F).$$

In the same way, we have

$$\lim_{r\to 0}\frac{1}{2\pi i}\int_{EE'}\frac{dF}{F} = -\frac{1}{2}v_i(F), \quad \text{and} \quad \lim_{r\to 0}\frac{1}{2\pi i}\int_{AA'}\frac{dF}{F} = -\frac{1}{6}v_\rho(F).$$

Finally, since the transformation $z \to -1/z$ carries the segment $A'E'$ onto $D'E$ and the relation $F(-1/z) = z^k F(z)$ implies the relation

$$\frac{dF(-1/z)}{F(-1/z)} = \frac{dz^k F(z)}{z^k F(z)} = \frac{kdz}{z} + \frac{dF}{F},$$

it follows that

$$\begin{aligned}\frac{1}{2\pi i}\left(\int_{D'E}\frac{dF}{F} + \int_{E'A'}\frac{dF}{F}\right) &= \frac{1}{2\pi i}\left(\int_{A'E'}\frac{dF(-1/z)}{F(-1/z)} + \int_{E'A'}\frac{dF}{F}\right)\\ &= \frac{k}{2\pi i}\int_{A'E'}\frac{dz}{z} + \frac{1}{2\pi i}\left(\int_{A'E'}\frac{dF}{F} + \int_{E'A'}\frac{dF}{F}\right) = \frac{k}{2\pi i}\int_{A'E'}\frac{dz}{z},\end{aligned}$$

and so the limit of the sum as $r \to 0$ is

$$\frac{k}{2\pi i}\int_{(\rho, i)}\frac{dz}{z} = \frac{k}{12}.$$

Our computations show that

$$\lim_{r\to 0}\frac{1}{2\pi i}\int_{C_r}\frac{dF}{F} = -v_{i\infty}(F) - \frac{2}{6}v_\rho(F) - \frac{1}{2}v_i(F) + \frac{k}{12}.$$

Combining this with (2.4.5), we get (2.4.4).

Now, if the boundary of D_1 contains zeroes of F different from ρ, ρ^2, i, we can proceed in the same way with a similar contour, which avoids the zeroes and encloses only one of each pair of them belonging to the same orbit. For example, if one pair of the zeroes, say λ and $\lambda + 1$, lie in $|x| = \frac{1}{2}$ and the other pair, say β and $-1/\beta$, lie in $|z| = 1$, we can take the contour in Fig. 3. □

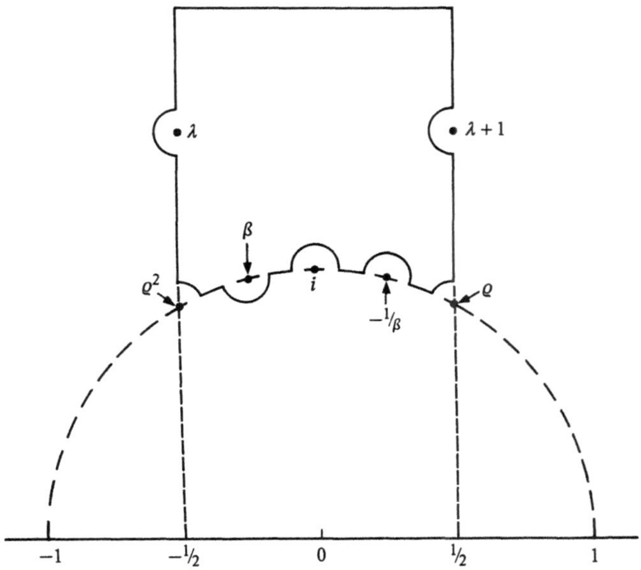

Fig. 3

Exercise 2.4.2. *Show that*

(1) $\dim \mathbb{M}_k = 0$, *if k is negative, or k is odd, or $k = 2$;*

(2) $\dim \mathbb{M}_k \leq 1$ *if $k = 0, 4, 6, 8, 10$;*

(3) $\dim \mathbb{M}_k \leq [k/12] + 1$ *if $k \geq 0$, where $[\alpha]$is the largest integer satisfying $[\alpha] \leq \alpha$.*

2.4.2. Modular Forms Whose Initial Fourier Coefficients Equal Zero

Proposition 2.4.1 shows that any non-zero form $F \in \mathbb{M}_k$ satisfies

$$\nu_{i\infty}(F) \leq k/12. \tag{2.4.6}$$

This means that F is zero, if the coefficients $f(2a)$ of the expansion (2.4.2) equal zero for $a = 0, 1, \ldots, [k/12]$. A similar property holds for modular forms for congruence subgroups of the Siegel modular group Γ^n. This allows us to prove that the spaces of modular forms of given type are finite dimensional.

Theorem 2.4.3. *Let K be a congruence subgroup of Γ^n, χ a character of K trivial on $\Gamma^n(q) \subset K$,* and *k a non-negative integer. Assume that a modular form*

$$F(Z) = \sum_{A \in \mathbb{A}_n} f(A) e\{q^{-1} AZ\} \in \mathfrak{M}_k(K, \chi)$$

satisfies

$$f(A)=0, \quad \text{if } \sigma(A) \le \frac{k}{2\pi}\sigma_n v(K) q, \tag{2.4.7}$$

where $\sigma_n = 2n^n c_n 3^{-1/2}$, c_n *being defined in Proposition* 2.1.13, *and where* $v(K) = [\Gamma^n : K]$ *is the index of* K *in* Γ^n. *Then* F *is identically zero.*

Proof. The theorem will be proved in three steps; First, using (2.4.6), we consider the case $K = \Gamma^1$ and $\chi = 1$; then, by induction on n, we prove the theorem for $K = \Gamma^n$ and $\chi = 1$; finally, we reduce the general case to the case $K = \Gamma^n$ and $\chi = 1$.

Let $K = \Gamma^1$ and $\chi = 1$. If the Fourier coefficients $f(2a)$ of a non-zero form $F \in \mathfrak{M}_k$ satisfy (2.4.7), then we have

$$v_{i\infty}(F) > k\sigma_1/4\pi \ge k/2\pi 3^{1/2} > k/12,$$

for $c_1 \ge 1$ and $\sigma_1 \ge 2, 3^{-1/2}$. This contradicts the inequality (2.4.6).

Now we use induction on n to prove the theorem for $K = \Gamma^n$ and $\chi = 1$. The case $n = 1$ has been considered. Suppose that $n > 1$ and the theorem is already proved for $K = \Gamma^{n-1}$ and $\chi = 1$. Let $F \in \mathfrak{M}_k^n = \mathfrak{M}_k(\Gamma^n, 1)$ satisfies (2.4.7). By Proposition 2.3.9, the function

$$F' = F|\Phi = \sum_{A' \in \mathbb{A}_{n-1}} f\left(\begin{pmatrix} A' & 0 \\ 0 & 0 \end{pmatrix}\right) e\{A'Z'\} \, (Z' \in \mathbb{H}_{n-1})$$

where Φ is the Siegel operator (2.3.29), is in $\mathfrak{M}_k^{n-1}$ If $\sigma(A') \le k\sigma_{n-1}/2\pi$, then

$$\sigma\left(\begin{pmatrix} A' & 0 \\ 0 & 0 \end{pmatrix}\right) = \sigma(A') \le k\sigma_n/2\pi,$$

for $c_{n-1} \le c_n$, by Proposition 2.1.13, and hence $\sigma_{n-1} < \sigma_n$. Thus, it follows from (2.4.7) that

$$f'(A') = f\left(\begin{pmatrix} A' & 0 \\ 0 & 0 \end{pmatrix}\right) = 0.$$

Hence, by the inductive hypothesis, we conclude that $F' = 0$, and therefore F is a cusp form. Then, by Theorem 2.3.12, F has the Fourier expansion of the form

$$F = \sum_{A \in \mathbb{A}_n^+} f(A) e\{AZ\}.$$

Now it follows from Theorem 2.1.21 and the second inequality of Lemma 2.3.14 for the function

$$\sum_{A \in \mathbb{A}_n^+} |f(A)| \exp(-\pi\sigma(AY)) = \sum_{A \in \mathbb{A}_n^+} |f(A) e\{AZ\}| \ge |F|$$

that, for all $Z = X + iY \in \mathsf{D}_n$, $F(Z)$ satisfies

$$|F(Z)| \le \delta \exp(-\delta' (\det Y)^{1/n})$$

with positive contains δ, δ'. This implies that

$$G(Z) = G(X + iY) = (\det Y)^{k/2}|F(Z)| \to 0$$

and $\det Y \to +\infty$ and $Z = X + iY$ remains in D_n. On the other hand, by the definition of D_n and Theorem 2.1.21, we see that any subset of D_n of the form

$$\{X + iY \in \mathsf{D}_n; \det Y \leq \varepsilon\}$$

with $\varepsilon > 0$ is bounded and closed, and therefore is compact. It follows that $G(Z)$ attains its maximum μ on D_n at some point $Z_0 = X_0 + iY_0$ of D_n. Further, it follows from Lemma 1.2.8 and the definition of $\mathbb{M}_k^n$ that G satisfies

$$\begin{aligned} G(M\langle Z\rangle) &= (|\det(CZ + D)|^{-2} \det Y)^{k/2} |\det(CZ + D)^k F(Z)| \\ &= (\det Y)^{k/2}|F(Z)| = G(Z) \end{aligned}$$

for every $M = \begin{pmatrix} A & B \\ C & D \end{pmatrix} \in \Gamma^n$, and so $G(Z)$ depends only on the orbit $\Gamma^n\langle Z\rangle$ of Z. Since D_n is a fundamental domain for Γ^n, we see that μ is the maximum of G on $\mathbb{H}_n$, i.e. $G(Z) \leq G(Z_0) = \mu$ for all $Z \in \mathbb{H}_n$. Let $t = u + iv$ be a complex parameter. Let us set $Z_t = Z_0 + tE_n$ and consider the function

$$g(t) = F(Z_t)\exp(-i\lambda\sigma(Z_t)),$$

where λ is defined by $\lambda n/\pi = 1 + [k\sigma_n/2\pi]$. Using the Fourier expansion of F, we get

$$\begin{aligned} g(t) &= \sum_{A \in \mathbb{A}_n^+} f(A)e\{AZ_t\}\exp(-i\lambda\sigma(Z_t)) \\ &= \sum_{A \in \mathbb{A}_n^+} f(A)e\{AZ_0\}\exp(-i\lambda\sigma(Z_0))q^{\sigma(A)-\lambda n/\pi} \\ &= g'(w), \end{aligned}$$

where $w = \exp(\pi i t)$. The assumption of the theorem implies that $f(A) = 0$ if $\sigma(A) - \lambda n/\pi < 0$. Hence the expansion of $g'(w)$ does not contain negative powers of w. If $\varepsilon > 0$ is so small that $Z_t \in \mathbb{H}_n$ for $v \geq -\varepsilon$, then the expansion of $g(t)$ converges absolutely and uniformly on the half-plane $v \geq -\varepsilon$, and so the function $g'(w)$ is regular in the disk $|w| \leq \exp(-\pi\varepsilon) = \tau$. Since $\tau > 1$, it follows, by the maximum-modulus principle, that there is a point $w_0 = \exp(\pi i t_0)$ with $|w_0| = \tau$ and $|g'(1)| \leq |g'(w_0)|$. Returning to the function g, we can rewrite the last inequality as

$$|F(Z_0)\exp(\lambda\sigma(Y_0))| \leq |F(Z_{t_0})|\exp(\lambda\sigma(Y_0))\exp(\lambda n v_0),$$

where $t_0 = u_0 + iv_0$, hence

$$(\det Y_0)^{-k/2}G(Z_0) \leq (\det Y_{t_0})^{-k/2}G(Z_{t_0})\exp(\lambda n v_0).$$

Since $G(Z_0) = \mu$ and $G(Z_{t_0}) \leq \mu$, the last inequality implies that

$$\mu \leq \mu(\det Y_0)^{k/2}(\det Y_{t_0})^{-k/2}\exp(\lambda n v_0) = \mu\psi(v_0), \tag{2.4.8}$$

where $\psi(v) = \det(E_n + vY_0^{-1})^{-k/2}\exp(\lambda nv)$. It is clear that $\psi(0) = 1$. Let us show that the derivative ψ' of ψ is positive at $v = 0$. We have

$$\psi(v) = \prod_{j=1}^{n}(1 + v\lambda_j)^{-k/2}\exp(\lambda nv),$$

where $\lambda_1, \ldots, \lambda_n$ are the characteristic values of Y_0^{-1}, hence

$$\psi'(0) = \lambda n - \frac{k}{2}(\lambda_1 + \cdots + \lambda_n) = \lambda n - \frac{k}{2}\sigma(Y_0^{-1})$$

$$\geq \lambda n - k\sigma_n/2 = \pi(1 + [k\sigma_n/2\pi] - k\sigma_n/2\pi) > 0,$$

for $X_0 + iY_0 \in \mathsf{D}_n$ and $\sigma(Y_0^{-1}) \leq \sigma_n$, by Theorem 2.1.21. It follows that, for small ε, we have $\psi(v_0) = \psi(-\varepsilon) < 1$. Then the inequality (2.4.8) shows that $\mu = 0$, and so F is identically zero. This completes the proof for $K = \Gamma^n$ and $\chi = 1$.

Finally, consider the general case. Since χ can be considered as a character of the finite factor group $K/\Gamma^n(q)$, there exists a positive integer d such that

$$\chi(M)^d = 1 \qquad \text{for all } M \in K. \tag{2.4.9}$$

Let $F \in \mathfrak{M}_k(K, \chi)$. Consider the functions $(F|_k M)^d$ with $M \in \Gamma^n$, where $F|_k M$ is defined by (2.3.14). It follows from (2.3.15), (2.3.16) and (2.4.9) that every such function depends only on the left coset KM of the corresponding matrix M modulo K. Hence, the function

$$F_0(Z) = \prod_{\alpha=1}^{\nu}(F|_k M_\alpha)^d,$$

where $M_1, \ldots, M_\nu$ is a set of representatives for the left cosets of Γ^n modulo K, does not depend on the choice of the representatives. Since, for any $M \in \Gamma^n$, the set $M_1M, \ldots, M_\nu M$ is again a set of representatives for $K\backslash\Gamma^n$, we have

$$\det(CZ + D)^{-kd\nu}F_0(M\langle Z\rangle) = \prod_{\alpha=1}^{\nu}(\det(CZ + D)^{-k}(F|_k M_\alpha)(M\langle Z\rangle))^d$$

$$= \prod_{\alpha=1}^{\nu}(F|_k M_\alpha|_k M)^d$$

$$= F_0 \quad \left(M = \begin{pmatrix} A & B \\ C & D \end{pmatrix} \in \Gamma^n\right).$$

On the other hand, F_0 is clearly regular in $\mathbb{H}_n$ and is bounded on any subset $\mathbb{H}_n(\varepsilon) \subset \mathbb{H}_n$ with $\varepsilon > 0$, by Theorem 2.3.4. So we have proved that $F_0 \in \mathfrak{M}^n_{kd\nu}$. Let $f_\alpha(A) = f_{M_\alpha}(A)$ be the coefficients of the Fourier expansion (2.3.18) of $F|_k M_\alpha$, then one can easily see that the Fourier coefficients $g(A)$ of F_0 are given by

$$g(A) = \sum_{A_{\alpha\beta} \in \mathbb{A}_n,\ \sum_{\alpha\beta} A_{\alpha\beta} = qA} \ \prod_{\alpha,\beta} f_\alpha(A_{\alpha\beta})$$

where $A \in \mathbb{A}_n$, $1 \le \alpha \le \nu$, and $1 \le \beta \le d$. We now assume that F satisfies (2.4.7), and let

$$A \in \mathbb{A}_n \quad \text{with } \sigma(A) \le kd\nu\sigma_n/2\pi.$$

Since the trace of any matrix of $\mathbb{A}_n$ is non-negative, the last inequality implies the inequalities

$$\sigma\left(\sum_{\beta=1}^{d} A_{\alpha\beta}\right) = \sum_{\beta=1}^{d} \sigma(A_{\alpha\beta}) \le kd\nu\sigma_n q/2\pi$$

for each decomposition $qA = \sum_{\alpha,\beta} A_{\alpha\beta}$ with $A_{\alpha\beta} \in \mathbb{A}_n$ and each $\alpha = 1, \ldots, \nu$. It follows that for each $\alpha = 1, \ldots, \nu$ there is a β with $1 \le \beta \le d$ satisfying $\sigma(A_{\alpha\beta}) \le k\nu\sigma_n q/2\pi$. If, for example, $F|_k M_1 = F$, and so $f_1 = f$, then we see that each summond of the sum for $g(A)$ has a factor of the form $f_1(A_{1\beta}) = f(A_{1\beta})$ with $\sigma(A_{1,\beta}) \le k\nu\sigma_n q/2\pi$, which is zero by (2.4.7). So and $F_0 = 0$, by the above. Thus, $F = 0$. □

Exercise 2.4.4. *Let K, k and χ be the same as in Theorem 2.4.3. Show that a function $F \in \mathfrak{M}_k(K, \chi)$ is identically zero, if, for each $M \in \Gamma^n$, the coefficients $f_M(A)$ of expansion* (2.3.18) *equal zero for all $A \in \mathbb{A}_n$ satisfying $\sigma(A) \le k\sigma_n q/2\pi$.*

2.4.3. Dimension of Spaces of Modular Forms

Theorem 2.4.5. *Let K be a congruence subgroup of Γ^n, χ a character of K trivial on $\Gamma^n(q) \subset K$, and k an integer. Then the dimension of $\mathfrak{M}_k(K, \chi)$ over $\mathbb{C}$ satisfies*

(1) $\dim \mathfrak{M}_k(K, \chi) \le d_n(k\nu(K)q)^{\langle n \rangle}$, *if $k > 0$, where d_n depends only on n, and $\nu(K)$ is the index of K in Γ^n;*

(2) $\dim \mathfrak{M}_0(K, \chi) = \begin{cases} 1, & \text{if } \chi = 1 \\ 0, & \text{if } \chi \ne 1; \end{cases}$

(3) $\dim \mathfrak{M}_k(K, \chi) = 0$, *if $k < 0$.*

Proof. Let $k > 0$. Using the estimate (2.3.13) for the number of $A \in \mathbb{A}_n$ with $\sigma(A) \le N$, we see that the number of $A \in \mathbb{A}_n$ satisfying the inequality in (2.4.7) is bounded by

$$d = d_n(k\nu(K)q)^{\langle n \rangle},$$

where d_n depends only on n. Taking d_n to be integral, we see that $d + 1$ arbitrary functions $F_1, \ldots, F_{d+1}$ of $\mathfrak{M}_k(K, \chi)$ are linearly dependent, since one can always find $c_1, \ldots, c_{d+1} \in \mathbb{C}$ with $(c_1, \ldots, c_{d+1}) \ne (0, \ldots, 0)$ so that $F = c_1 F_1 + \cdots + c_{d+1} F_{d+1}$ satisfies (2.4.7) and therefore is identically zero, by Theorem 2.4.3.

Now let $k = 0$. Then Theorem 2.4.3 shows that $F = 0$, if $f(0) = 0$. Since, clearly, $1 \in \mathfrak{M}_0(K, 1)$, we have $F - f(0) \cdot 1 \in \mathfrak{M}_0(K, 1)$ for every $F \in \mathfrak{M}_0(K, 1)$, hence $F - f(0) = 0$, and so $F = f(0)$. If $F \in \mathfrak{M}_0(K, \chi)$ and $\chi(M) \ne 1$ for some $M \in K$, then the functional equation $F|_0 M = F(M\langle Z\rangle) = \chi(M)F$ implies that $f(0) = \chi(M)f(0)$, and so $f(0) = 0$. Hence $F = 0$.

Finally consider the case $k < 0$. First, by induction on n, we show that the statement is true for $K = \Gamma^n$ and $\chi = 1$. If $n = 1$, then this follows from Proposition 2.4.1, since the left side of (2.4.4) is non-negative. Let $n > 1$ and we know that $\mathfrak{M}_k^{n-1} = \{0\}$. If $F \in \mathfrak{M}_k^n$, then $F|\Phi \in \mathfrak{M}_k^{n-1}$, by Proposition 2.3.9, and so $F|\Phi = 0$, i.e. F is a cusp form. Let G be a non-zero modular form of integral positive weight l and character ψ for a congruence subgroup K_1 of Γ^n (for example, a theta-series). Then clearly

$$F^l G^{|k|} \in \mathfrak{M}_0(K_1, \psi^{|k|}),$$

and the constant term of the Fourier expansion of the function equals zero. So, by the above, we have $F^l G^{|k|} = 0$, hence $F = 0$. The general case for $k < 0$ is reduced to the case $K = \Gamma^n$ and $\chi = 1$ in the same way as in the proof of Theorem 2.4.3: If $F \in \mathfrak{M}_k(K, \chi)$, and d satisfies (2.4.9), then

$$F_0 = \sum_{M \in K\backslash\Gamma^n} (F|_k M)^d \in \mathfrak{M}^n_{kdv(K)};$$

hence, $F_0 = 0$, and so $F = 0$. □

Exercise 2.4.6. *Show that*

(1) $\dim \mathfrak{M}_k^2 = 0$ *for* $k = 1, 2, 3$;

(2) $\dim \mathfrak{M}_4^2 \le 1$.

[Hint: Deduce from Exercise 2.1.16 and Theorem 2.4.3 that $\mathfrak{M}_k^2$ contains no non-zero cusp forms if $k = 1, 2, 3, 4$, and then use Exercise 2.4.2].

Exercise 2.4.7. *Show that* $\dim \mathfrak{M}_k^n = 0$, *if nk is odd.*

Exercise 2.4.8. *Let K be a congruence subgroup of $\Gamma_0^n(4)$, χ a character of K trivial on $\Gamma^n(q) \subset K$, and k an integer. Show that*

(1) *if* $k \ge 1$, *then*

$$\dim \mathfrak{M}_{k-1/2}(K, \chi) \le d_n (k v(K) q)^{\langle n\rangle}$$

with the same constant d_n as in Theorem 2.4.5.

[Hint: Consider the space $\theta^n(Z, (2))\, \mathfrak{M}_{k-1/2}(K, \chi)$.]

(2) *If* $k < 1$, *then* $\dim \mathfrak{M}_{k-1/2}(K, \chi) = 0$.

[Hint: Consider the square of a form of the space.]

§ 2.5. Scalar Product and Orthogonal Decomposition

A space of modular forms admits an invariant scalar product; in particular, this allows us to define a natural complement of the subspace of cusp forms.

2.5.1. Scalar Product

For two functions F and G on $\mathbb{H}_n$, we consider the differential form on $\mathbb{H}_n$ given by

$$\omega_k(F, G)(Z) = F(Z)\overline{G(Z)}h(Z)^k \mathrm{d}^*Z, \tag{2.5.1}$$

where k is an integer, $h(Z) = h(X + iY) = \det Y$ is the height of Z, and d^*Z is the invariant element of volume defined in Proposition 1.2.9.

Lemma 2.5.1. *The following transformation formulae hold for every matrix* $M = \begin{pmatrix} A & B \\ C & D \end{pmatrix} \in \mathrm{GSp}_n^+(\mathbb{R})$:
If $Z = X + iY \in \mathbb{H}_n$ *and* $M\langle Z\rangle = X' + iY'$, *then*

$$Y' = \mu(M)\,{}^t(C\bar{Z} + D)^{-1} Y (CZ + D)^{-1}, \tag{2.5.2}$$

in particular,

$$h(M\langle Z\rangle) = \det Y' = \mu(M)^n |\det(CZ + D)|^{-2} h(Z); \tag{2.5.3}$$

$$\mathrm{d}^* M\langle Z\rangle = \mathrm{d}^* Z; \tag{2.5.4}$$

$$\omega_k(F, G)(M\langle Z\rangle) = \mu(M)^{nk} \omega_k(F|_k M, G|_k M)(Z), \tag{2.5.5}$$

where $|_k M$ *is the operator* (2.3.14).

Proof. Since clearly $\mu(\lambda M) = \lambda^2 \mu(M)$ for $\lambda > 0$, it follows that $M_1 = \mu(M)^{-1/2} M \in \mathrm{Sp}_n(\mathbb{R})$. The formulae (2.5.2) and (2.5.3) then follow from Lemma 1.2.8 for the matrix M_1. (2.5.4) follows from Proposition 1.2.9, since $\mathrm{d}^* \lambda Z = \mathrm{d}^* Z$ for $\lambda > 0$. Finally, by (2.5.3) and (2.5.4), we get

$$\begin{aligned} \omega_k(F, G)(M\langle Z\rangle) &= F(M\langle Z\rangle)\overline{G(M\langle Z\rangle)}h(M\langle Z\rangle)^k \mathrm{d}^* M\langle Z\rangle \\ &= \mu(M)^{nk} \det(CZ + D)^{-k} F(M\langle Z\rangle)\overline{\det(CZ + D)^{-k} G(M\langle Z\rangle)}h(Z)^k \mathrm{d}^* Z, \end{aligned}$$

which proves (2.5.5). □

Let us assume now that F and G are modular forms of integral weight k and character χ for a subgroup K of Γ^n of finite index. It follows from (2.5.5) that the differential form $\omega_k(F, G)(Z)$ is invariant under K, since for $M \in K$ we have

$$\begin{aligned} \omega_k(F, G)(M\langle Z\rangle) &= \omega_k(F|_k M, G|_k M)(Z) \\ &= \omega_k(\chi(M)F, \chi(M)G)(Z) = \omega_k(F, G)(Z). \end{aligned}$$

It follows that the integral

$$\int_{D(K)} \omega_k(F, G)(Z), \tag{2.5.6}$$

where $D(K)$ is a fundamental domain for K on $\mathbb{H}_n$, does not depend on the choice of the fundamental domain, provided that the integral converges absolutely.

Lemma 2.5.2. *Let K be a congruence subgroup of Γ^n, χ a congruence character of K and k an integer. Assume that at least one of two forms $F, G \in \mathfrak{M}_k(K, \chi)$ is a cusp form. Then the integral* (2.5.6) *converges absolutely.*

Proof. We take as $\mathsf{D}(K)$ a fundamental domain of the form (2.1.30), which is a finite union of sets of the form $M\langle \mathsf{D}_n \rangle$ with $M \in \Gamma^n$, D_n being the fundamental domain for Γ^n defined in Theorem 2.1.20. Then it suffices to show that each integral of the form

$$\int_{M\langle \mathsf{D}_n \rangle} \omega_k(F, G)(Z) = \int_{\mathsf{D}_n} \omega_k(F|_k M, G|_k M)(Z) \qquad \text{with } M \in \Gamma^n$$

converges absolutely. Let (say) F be a cusp form. Then, by Theorem 2.3.12 and Theorem 2.3.4, respectively, we can write the expansions

$$F|_k M = \sum_{A \in \mathbb{A}_n^+} f_M(A) e\{q^{-1} A Z\},$$

$$G|_k M = \sum_{A \in \mathbb{A}_n} g_M(A) e\{q^{-1} A Z\},$$

where q is such that $\Gamma^n(q) \subset K$ and χ is trivial on $\Gamma^n(q)$. Since the two series converge absolutely in $\mathbb{H}_n$, it follows that

$$\begin{aligned} &|(F|_k M)(Z)\overline{(G|_k M)}(Z)| \\ &\quad \leq \sum_{A_1 \in \mathbb{A}_n^+, A_2 \in \mathbb{A}_n} |f_M(A_1) g_M(A_2)| \exp(-\pi q^{-1} \sigma((A_1 + A_2) Y)) \\ &\quad = \sum_{A \in \mathbb{A}_n^+} t(A) \exp(-\pi q^{-1} \sigma(A Y)), \end{aligned}$$

where

$$t(A) = \sum_{A_1 + A_2 = A} |f_M(A_1) g_M(A_2)| \qquad \text{(a finite sum)},$$

and the last series is convergent in $\mathbb{H}_n$. Then, by Theorem 2.1.21 and the first inequality of Lemma 2.3.14, we obtain the inequality

$$|(F|_k M)(Z) \cdot (G|_k M)(Z)| \leq \delta_1 \exp(-\delta_2 \sigma(Y)) \quad (Z = X + iY \in \mathsf{D}_n),$$

where δ_1 and δ_2 are positive constants. Thus, to prove the lemma, it suffices to show that the integral

$$\begin{aligned} &\int_{\mathsf{D}_n} \exp(-\delta \sigma(Y)) (\det Y)^k \mathrm{d}^* Z \\ &\quad = \int_{\mathsf{D}_n} \exp(-\delta \sigma(Y)) (\det Y)^{k-n-1} \prod_{1 \leq \alpha \leq \beta \leq n} \mathrm{d}x_{\alpha\beta} \mathrm{d}y_{\alpha\beta} \end{aligned}$$

with $\delta > 0$ is convergent. If $(x_{\alpha\beta}) + i(y_{\alpha\beta}) \in \mathsf{D}_n$, then it follows from the definition of D_n and (2.1.12) that $|x_{\alpha\beta}| \leq \frac{1}{2} (1 \leq \alpha, \beta \leq n)$ and $|y_{\alpha\beta}| \leq y_{\alpha\alpha}/2 (\alpha \neq \beta)$. It was proved in the beginning of the proof of Theorem 2.1.21, that $y_{\alpha\alpha} \geq 3^{1/2}/2$. Then the inequality (2.1.13), if $k < n + 1$, and the inequality (A.1.8), if $k \geq n + 1$, imply that the last integral is dominated by

$$c \int\limits_{\substack{|x_{\alpha\beta}|\le 1/2,\\ y_{\alpha\alpha}\ge 3^{1/2}/2,\, |y_{\alpha\beta}|\le y_{\alpha\alpha}/2}} \prod_{\alpha=1} y_{\alpha\alpha}^{k-n-1}\exp(-\delta y_{\alpha\alpha}) \prod_{1\le\alpha\le\beta\le n} dx_{\alpha\beta}dy_{\alpha\beta}$$

$$= c\prod_{\alpha=1}^{n} \int\limits_{3^{1/2}/2}^{\infty} y_{\alpha\alpha}^{k-n-1+n-\alpha}\exp(-\delta y_{\alpha\alpha})dy_{\alpha\alpha} < \infty. \qquad \square$$

Now we can prove

Theorem 2.5.3. *Let K be a congruence subgroup of Γ^n, χ a congruence character of K and k an integer. If at least one of the modular forms $F, G \in \mathfrak{M}_k(K,\chi)$ is a cusp form, we define their scalar product (F, G) by*

$$(F,G) = v(K')^{-1} \int\limits_{D(K)} \omega_k(F,G)(Z), \tag{2.5.7}$$

where $K' = K \cup (-E_{2n})K$, $v(K') = [\Gamma^n : K']$, and $\mathsf{D}(K)$ is a fundamental domain for K on $\mathbb{H}_n$. Then the scalar product satisfies the following rules:

(1) (F, G) converges absolutely and is independent of the choice of fundamental domain $\mathsf{D}(K)$;

(2) (F, G) is independent of the choice of the group K with $F, G \in \mathfrak{M}_k(K,\chi)$;

(3) (F, G) is linear in F and conjugate linear in G;

(4) $(G,F) = \overline{(F,G)}$;

(5) $(F,F) \ge 0$, and $(F,F) = 0$ only if $F = 0$;

(6) If $M \in \mathsf{GSp}_n^+(\mathbb{Q})$, then

$$(F|_k M, G|_k M) = \mu(M)^{-nk}(F,G), \tag{2.5.8}$$

the functions $F|_k M$, $G|_k M$ being considered as elements of $\mathfrak{M}_k(K_m, \chi_M)$ (see Proposition 2.3.6 and Theorem 2.3.12 (3)).

Proof. The first statement has been proved.

Let us prove (2). If we have also $F, G \in \mathfrak{M}_k(K_1, \chi_1)$, where K_1 is a congruence subgroup and χ_1 is a congruence character of K_1, then, on replacing K_1 by $K_1 \cap K$, we may assume that $K_1 \subset K$, because the intersection of two congruence subgroups is clearly a congruence subgroup. Let

$$\Gamma^n = \bigcup_\alpha K' M_\alpha \quad \text{and} \quad K' = \bigcup_\beta K_1' N_\beta \quad \text{with } K_1' = K_1 \cup (-E_{2n})K_1$$

be decompositions into different left cosets. Then we have obviously that

$$\Gamma^n = \bigcup_{\alpha,\beta} K_1' N_\beta M_\alpha$$

and the left cosets are pairwise distinct. Thus, by Theorem 2.1.22, we can take

$$D(K_1) = \bigcup_{\alpha,\beta} N_\beta M_\alpha \langle \mathsf{D}_n \rangle;$$

hence we have

$$v(K_1')^{-1} \int\limits_{\mathsf{D}(K_1)} \omega_k(F,G)(Z) = v(K_1')^{-1} \sum_\nu \sum_\beta \int\limits_{N_\beta \langle M_\alpha \langle \mathsf{D}_n \rangle\rangle} \omega_k(F,G)(Z).$$

Since $N_\beta \in K'$, it follows from (2.5.5) that

$$\int_{N_\beta\langle M_\alpha\langle \mathsf{D}_n\rangle\rangle} \omega_k(F,G)(Z) = \int_{M_\alpha\langle \mathsf{D}_n\rangle} \omega_k(F,G)(Z),$$

and so we have

$$\begin{aligned} \nu(K_1')^{-1} \int_{\mathsf{D}(K_1)} \omega_k(F,G)(Z) &= [K':K_1']\nu(K_1')^{-1} \sum_\alpha \int_{M_\alpha\langle \mathsf{D}_n\rangle} \omega_k(F,G)(Z) \\ &= \nu(K')^{-1} \int_{\mathsf{D}(K)} \omega_k(F,G)(Z), \end{aligned}$$

which proves (2).

The properties (3), (4) and (5) follow directly from the definitions.

Finally, let us prove (2.5.8). Since the groups K and K_M are both congruence subgroups of Γ^n, the same is true for their intersection

$$K_{(M)} = K \cap K_M = K \cap M^{-1}KM$$

which is therefore of finite index in Γ^n. Then it follows from (2.5.5) and the property (2) of the scalar product that

$$\begin{aligned} (F|_k M, G|_k M) &= \nu(K'_{(M)})^{-1} \int_{\mathsf{D}} \omega(F|_k M, G|_k M)(Z) \\ &= \mu(M)^{-nk} \nu(K'_{(M)})^{-1} \int_{\mathsf{D}} \omega(F,G)(M\langle Z\rangle) \\ &= \mu(M)^{-nk} \nu(K'_{(M)})^{-1} \int_{M\langle \mathsf{D}\rangle} \omega(F,G)(Z), \end{aligned}$$

where $K'_{(M)} = K_{(M)} \cup (-E)K_{(M)}$ and $\mathsf{D} = \mathsf{D}(K_{(M)})$ is a fundamental domain for $K_{(M)}$. It is easy to see that the set $M\langle \mathsf{D}\rangle$ is a fundamental domain for $MK_{(M)}M^{-1} = MKM^{-1} \cap K = K_{(M^{-1})}$. Hence, we can use the property (2) once again to write the last expression in the form

$$\begin{aligned} &= \mu(M)^{-nk} \nu(K'_{(M)})^{-1} \nu(K'_{(M^{-1})}) \nu(K'_{(M^{-1})})^{-1} \int_{D(K_{(M^{-1})})} \omega(F,G)(Z) \\ &= \mu(M)^{-nk} \nu(K'_{(M)})^{-1} \nu(K'_{(M^{-1})})(F,G) \end{aligned}$$

which proves (2.5.8) modulo the relation $\nu(K'_{(M)}) = \nu(K'_{(M^{-1})})$. Since, clearly,

$$[\Gamma_{(M^{-1})} : K'_{(M^{-1})}] = [M\Gamma_{(M)}M^{-1} : MK'_{(M)}M^{-1}] = [\Gamma_{(M)} : K'_{(M)}]$$

with $\Gamma = \Gamma^n$, it sufficies to show that $[\Gamma : \Gamma_{(M^{-1})}] = [\Gamma : \Gamma_{(M)}]$. It will be convenient to prove here a more general statement:

Lemma 2.5.4. *Let K be a congruence subgroup of Γ^n. Then, for each $M \in \mathsf{GSp}_n^+(\mathbb{Q})$, the group $K_{(M)}$ is a congruence subgroup of Γ^n, and the subgroups $K_{(M)}$, $K_{(M^{-1})}$ have the same index K:*

$$[K : K_{(M)}] = [K : K_{(M^{-1})}]. \tag{2.5.9}$$

Proof. As it was mentioned above, the first statement follows from Proposition 2.3.6, since $K_{(M)} = K \cap (M^{-1}KM \cap \Gamma^n)$. Now let D be a fundamental domain for

$K_{(M)}$ on $\mathbb{H}_n$. Since $K_{(M^{-1})} = MK_{(M)}M^{-1}$, it follows that we can take the set $M\langle \mathsf{D}\rangle$ as a fundamental domain for $K_{(M^{-1})}$. Then it follows from Proposition 2.1.23 (2) that

$$v(\mathsf{D}) = [K : K_{(M)}]v(\mathsf{D}(K)), \qquad \text{and} \qquad v(M\langle \mathsf{D}\rangle) = [K : K_{(M^{-1})}]v(\mathsf{D}(K)),$$

where $\mathsf{D}(K)$ is a fundamental domain for K and v denote the invariant volume. On the other hand, by (2.5.4), we have

$$v(\mathsf{D}) = \int_{\mathsf{D}} \mathrm{d}^*Z = \int_{\mathsf{D}} \mathrm{d}^*M\langle Z\rangle = \int_{M\langle \mathsf{D}\rangle} \mathrm{d}^*Z = v(M\langle Z\rangle).$$

The relations imply (2.5.9). □

2.5.2. Orthogonal Decomposition

Let $\mathfrak{E}_k(K,\chi)$ be the subspace of all modular forms of $\mathfrak{M}_k(K,\chi)$ which are orthogonal to the subspace of cusp forms $\mathfrak{N}_k(K,\chi)$ with respect to the scalar product (2.5.7):

$$\mathfrak{E}_k(K,\chi) = \{F \in \mathfrak{M}_k(K,\chi); \quad (F,G) = 0 \text{ for all } G \in \mathfrak{N}_k(K,\chi)\}. \tag{2.5.10}$$

Proposition 2.5.5. (1) *The space* $\mathfrak{M}_k(K,\chi)$ *is a direct sum of* $\mathfrak{E}_k(K,\chi)$ *and* $\mathfrak{N}_k(K,\chi)$:

$$\mathfrak{M}_k(K,\chi) = \mathfrak{E}_k(K,\chi) \oplus \mathfrak{N}_k(K,\chi); \tag{2.5.11}$$

(2) *If* $F \in \mathfrak{E}_k(K,\chi)$ *and* $M \in \mathsf{GSp}_n^+(\mathbb{Q})$, *then* $F|_k M \in \mathfrak{E}_k(K_M, \chi_M)$ (*see Proposition* 2.3.6).

Proof. The first statement follows from the properties (3)–(5) of the scalar product listed in Theorem 2.5.3 and linear algebra. The second statement follows from Theorem 2.3.12 (3) and relations (2.5.8). □

By Proposition 2.5.5, each modular form $F \in \mathfrak{M}_k(K,\chi)$ has a unique decomposition of the form

$$F = F_1 + F_2$$

with $F_1 \in \mathfrak{E}_k(K,\chi)$ and $F_2 \in \mathfrak{N}_k(K,\chi)$. By comparing the Fourier coefficients $f(A)$, $f_1(A)$ and $f_2(A)$ of F, F_1 and F_2, respectively, we obtain the relations

$$f(A) = f_1(A) + f_2(A) \qquad \text{for } A \in \mathbb{A}_n. \tag{2.5.12}$$

It turns out, that in some cases the Fourier coefficients $f_1(A)$ of $F_1 \in \mathfrak{E}_k(K,\chi)$ can be computed. On the other hand, the Fourier coefficients $f_2(A)$ of the cusp form F_2 satisfy (2.3.43). This allows us to show that under certain conditions the decomposition (2.5.12) gives an asymptotic formula for $f(A)$ as $\det A \to \infty$ with $f_1(A)$ as the main term. In Exercise 2.5.6 this approach is illustrated for the simplest case $K = \Gamma^1$, $\chi = 1$.

Exercise 2.5.6. *Let $k > 2$ be an integer. Show that*

(1) $$G_k(z) = \sum_{(c,d)\in \mathsf{M}_{1,2}-(0,0)} (cz+d)^{-k} \in \mathfrak{M}_k \quad (z \in \mathbb{H}_1).$$

(The function G_k are called *the Eisenstein series*).

[Hint: If S is a compact subset in $\mathbb{H}_1$, then $|\alpha z + \beta| \geq b(|\alpha| + |\beta|)$ for all α, $\beta \in \mathbb{R}$ and $z \in \mathsf{S}$ with a positive constant b. Since there are only $4r$ pairs of integers (c, d) with $|c| + |d| = r$, it follows that $G_k(z)$ with $z \in \mathsf{S}$ is dominated term-by-term by

$$b^{-k} \sum_{r=1}^{\infty} 4r \cdot r^{-k} = 4b^{-k}\zeta(k-1).];$$

(2) *The function $G_k(z)$ belongs to the subspace $\mathfrak{E}_k(\Gamma^1, 1) \subset \mathfrak{M}_k$ and spans the subspace.*

[Hint: Note that

$$G_k = (1 + (-1)^k)\zeta(k) \sum_{M \in \Gamma_0^1 \backslash \Gamma^1} j(M, z)^k;$$

hence, for a cusp form $F \in \mathfrak{N}_k(\Gamma^1, 1)$, it follows that

$$(G_k, F) = \gamma \sum_{M \in \Gamma_0^1 \backslash \Gamma^1} \int_{\mathsf{D}_1} j(M, z)^{-k} \bar{F} y^{k-2} \mathrm{d}x\mathrm{d}y = \gamma \int_{\mathsf{D}(\Gamma_0^1)} \bar{F} y^{k-2} \mathrm{d}x\mathrm{d}y,$$

where $\mathsf{D}(\Gamma_0^1) = \bigcup_{M \in \Gamma_0^1 \backslash \Gamma^1} M\langle \mathsf{D}_1 \rangle$ is a fundamental domain for Γ_0^1 on $\mathbb{H}_1$. Replacing $\mathsf{D}(\Gamma_0^1)$ by $\{z = x + iy \in \mathbb{H}_1; |x| \leq \frac{1}{2}\}$ and integrating on $|x| \leq \frac{1}{2}$, we see that the last integral equals zero.];

(3) *If k is even, the Fourier expansion of G_k is given by*

$$G_k(z) = 2\sum_{d \in \mathbb{N}} d^{-k} + 2\frac{(2\pi i)^k}{(k-1)!} \sum_{n \in \mathbb{N}} \left(\sum_{d|n} d^{k-1} \right) e^{2\pi i n z}.$$

[Hint: $\sum_{d \in \mathbb{Z}} (z+d)^{-2} = \pi^2/\sin^2 \pi z = (2\pi i)^2 t/(1-t)^2 = (2\pi i)^2 \sum_{n \in \mathbb{N}} n t^n$ with $t = \exp(2\pi i z)$, for the difference is entire, of period 1, even and $\to 0$ as $\operatorname{Im} z \to \infty$. Differentiating with respect to z, one gets

$$\sum_{d \in \mathbb{Z}} (z+d)^{-k} = \frac{(-2\pi i)^k}{(k-1)!} \sum_{n \in \mathbb{N}} n^{k-1} t^n.]$$

(4) *Let*

$$F(z) = \sum_{n=0}^{\infty} f(2n) e^{2\pi i n z} \in \mathfrak{M}_k$$

where k is even. Then the Fourier coefficients $f(2n)$ satisfy

$$f(2n) = f(0)\zeta(k)^{-1} \frac{(2\pi i)^k}{(k-1)!} \sum_{d|n} d^{k-1} + f_2(2n) \qquad \textit{with } |f_2(2n)| \leq \delta_F n^{k/2}.$$

In particular, the formula holds when $f(2n) = r(Q_{2k}, 2n)$ *is the number of in-*

tegral representations of n by quadratic form with a matrix $Q_{2k} \in \mathbb{A}_{2k}^{+}$ *satisfying* $\det Q_{2k} = 1$.

[Hint: Use (2.5.12), (2.3.43) and Theorem 2.2.2.]

Exercise 2.5.7. *Show that in the notation of Proposition* 2.4.1 *the following equalities hold:*

$$\begin{aligned} v_\rho(G_4) &= 1, \quad \text{and} \quad v_p(G_4) = 0 \quad \text{if } p \notin \Gamma^1\langle \rho \rangle; \\ v_i(G_6) &= 1, \quad \text{and} \quad v_p(G_6) = 0 \quad \text{if } p \notin \Gamma^1\langle i \rangle. \end{aligned}$$

Show that the function

$$\Delta'(z) = ((2\zeta(4))^{-1} G_4(z))^3 - ((2\zeta(6))^{-1} G_6(z))^2$$

is a non-zero cusp-form of $\mathfrak{M}_{12}$ *which has no finite zeroes in* $\mathbb{H}_1$ *and has order* 1 *at* $i\infty$.

[Hint: Use Proposition 2.4.1.]

Exercise 2.5.8. *Show that*

(1) $\dim \mathfrak{M}_k = 1$ *if* $k = 0, 4, 6, 8, 10$. *For these k,* $\mathfrak{N}_k(\Gamma^1, 1) = \{0\}$ *and the space* $\mathfrak{M}_k$ *is spanned respectively by* 1, G_4, G_6, G_8, *and* G_{10};

(2) *The multiplication by* $\Delta'(z)$ *(see Exercise* 2.5.7*) defines an isomorphism of* $\mathfrak{M}_{k-12}$ *with even* $k \geq 2$ *onto* $\mathfrak{N}_k(\Gamma^1, 1)$;

(3) *If k is even,* then

$$\dim \mathfrak{M}_k = \begin{cases} [k/12], & \text{if } k \equiv 2 \pmod{12},\ k \geq 0 \\ [k/12] + 1, & \text{if } k \not\equiv 2 \pmod{12},\ k \geq 0 \end{cases}$$

(4) *The product* $G_4^a G_6^b$ *with* $a, b \in \mathbb{Z}$, $a, b \geq 0$, $4a + 6b = k$ *form a basis of* $\mathfrak{M}_k$.

[Hint: See Exercise 2.4.2.]

Exercise 2.5.9. *Let* $Q_m \in \mathbb{A}_m^{+}$ *with* $\det Q_m = 1$. *Show that for* $n \in \mathbb{N}$ *the following formulae hold*

$$r(Q_8, 2n) = \zeta(4)^{-1}((2\pi)^4/6) \sum_{d|n} d^3,$$

$$r(Q_{16}, 2n) = \zeta(8)^{-1}((2\pi)^8/7!) \sum_{d|n} d^7.$$

Deduce from the formulae that

$$\zeta(4) = \pi^4/2 \cdot 3^2 \cdot 5, \qquad \zeta(8) = \pi^8/2 \cdot 3^3 \cdot 5^2 \cdot 7.$$

[Hint: Use Exercise 1.4.4.]

Exercise 2.5.10. *Try to develop the theory of the Eisenstein series of the form*

$$G_k(z, c_0, d_0; q) = \sum_{\substack{(c,d) \in M_{1,2} - (0,0) \\ (c,d) \equiv (c_0, d_0) \pmod q}} (cz + d)^{-k}$$

for $K = \Gamma^1(q)$, $\chi = 1$, *where* $k > 2$ *is an integer, in the same way as it was done for* $q = 1$ *in Exercise* 2.5.6. *In particular, show that the Fourier expansion of* $G_k(z, c_0, d_0; q)$ *is given by*

$$G_k(z, c_0, d_0; q) = \delta(c_0/q) \sum_{\substack{d \equiv d_0 \ (\mathrm{mod}\ q) \\ d \neq 0}} d^{-k}$$

$$+ \frac{(-2\pi i)^k}{q^k (k-1)!} \sum_{\substack{c, d \in \mathbb{Z}, cd > 0 \\ c \equiv c_0 \ (\mathrm{mod}\ q)}} d^{k-1} \operatorname{sign} d \cdot \exp(2\pi i d_0 d/q) \exp\left(\frac{2\pi i c d z}{q}\right)$$

where $\delta(\alpha) = 1$ *for* $\alpha \in \mathbb{Z}$ *and* $\delta(\alpha) = 0$, *if* $\alpha \notin \mathbb{Z}$.

Chapter 3. Hecke Rings

One of the most fruitful ideas in the theory of modular forms, the idea of Hecke operators, is based on an averaging of modular forms for a group K over certain double cosets modulo K. The theory of Hecke operators and its applications will be considered in the next two chapters. The majority of relations between Hecke operators only reflect relations between the corresponding double cosets. This chapter deals with the latter.

§ 3.1. Abstract Hecke Rings

3.1.1. Averaging over Double Cosets

As in § 1.4.1, let S be a multiplicative semigroup operating on a set H, say $g\colon h \to g(h)$ for $g \in \mathsf{S}$ and $h \in \mathsf{H}$, as a subsemigroup of the group of all one-to-one mappings of H onto itself. Let $\varphi\colon \mathsf{S} \times \mathsf{H} \to \mathsf{T}$ be a factor of automorphy of S on H with values in a group T, and let V be a left T-module. A function $F\colon \mathsf{H} \to \mathsf{V}$ is called *an automorphic function of weight φ for a subgroup $\Gamma \subset \mathsf{S}$* if, for any $\gamma \in \Gamma$, it satisfies the functional equation

$$(F|_{\varphi}\gamma)(h) = \varphi(\gamma, h)^{-1}F(\gamma(h)) = F(h). \tag{3.1.1}$$

The set $\mathfrak{M} = \mathfrak{M}_{\varphi}(\Gamma)$ of all automorphic functions of weight φ for Γ is clearly an abelian group.

For $F \in \mathfrak{M}$ and $g \in \mathsf{S}$, we consider the function $F'(h) = (F|_{\varphi}g)(h) = \varphi(g, h)^{-1}F(g(h))$. Generally speaking, F' does not belongs to $\mathfrak{M}$. It can even happen that there are infinitely many different functions of the form

$$F'|_{\varphi}\gamma = F|_{\varphi}g|_{\varphi}\gamma = F|_{\varphi}g\gamma \qquad \text{with } \gamma \in \Gamma. \tag{3.1.2}$$

Often, however, the number of different functions of the form (3.1.2) is finite. Then one may hope that the sum of these function is again a function of $\mathfrak{M}$. The typical situation of this kind occurs when the double coset $\Gamma g \Gamma$ contains only finitely many left cosets modulo Γ:

$$|\Gamma \backslash \Gamma g \Gamma| < \infty. \tag{3.1.3}$$

Really, each of the products $g\gamma$ with $\gamma \in \Gamma$ belongs to the double coset $\Gamma g \Gamma$; when $g\gamma$ belongs to a fixed left coset $\Gamma g' \subset \Gamma g \Gamma$, then $g\gamma = \gamma' g'$ with $\gamma' \in \Gamma$, and so, by Lemma 1.4.1(3), we get

$$F'|_\varphi \gamma = F|_\varphi g\gamma = F|_\varphi \gamma'|_\varphi g' = F|_\varphi g'.$$

This implies that any function of the form (3.1.2) coincides with one of the functions $F|_\varphi g_1, \ldots, F|_\varphi g_\nu$, where $g_1, \ldots, g_\nu$ is a set of representatives for the left cosets modulo Γ contained in $\Gamma g \Gamma$ Let us consider the following averaging of F over $\Gamma g \Gamma$:

$$F|(g) = F|_\varphi(g) = \sum_{g_i \in \Gamma \backslash \Gamma g \Gamma} F|_\varphi g_i. \tag{3.1.4}$$

Lemma 3.1.1. *Let $F \in \mathfrak{M} = \mathfrak{M}_\varphi(\Gamma)$. Suppose that a double coset $\Gamma g \Gamma$ with $g \in \mathrm{S}$ satisfies* (3.1.3). *Then the function $F|(g)$ is independent of the choice of representatives $g_1, \ldots, g_\nu$ for the left cosets of $\Gamma g \Gamma$ modulo Γ and again belongs to $\mathfrak{M}$.*

Proof. If $g_1' = \gamma_1 g_1, \ldots, g_\nu' = \gamma_\nu g_\nu$ with $\gamma_i \in \Gamma$ is another set of representatives, then by Lemma 1.4.1(3) and (3.1.1) we have

$$\sum_i F|_\varphi \gamma_i g_i = \sum_i F|_\varphi \gamma_i|_\varphi g_i = \sum_i F|_\varphi g_i.$$

Now let $\gamma \in \Gamma$. Since, clearly, the set $g_1\gamma, \ldots, g_\nu\gamma$ is again a set of representatives for $\Gamma \backslash \Gamma g \Gamma$, if follows that

$$(F|_\varphi(g))|_\varphi \gamma = \sum_i F|_\varphi g_i|_\varphi \gamma = \sum_i F|_\varphi g_i \gamma = F|_\varphi(g),$$

and so $F|_\varphi(g) \in \mathfrak{M}$. □

The operators of the form

$$|_\varphi(g)\colon \mathfrak{M}_\varphi(\Gamma) \to \mathfrak{M}_\varphi(\Gamma) \tag{3.1.5}$$

are called *Hecke operators on* $\mathfrak{M}_\varphi(\Gamma)$.

3.1.2. Hecke Rings

Intending to consider relations between Hecke operators, we first look at relations between the corresponding double cosets.

Lemma 3.1.2. *Let* G *be a group, Γ a subgroup of* G, *and $g \in \mathrm{G}$. Let*

$$\Gamma = \bigcup_{\gamma_i \in \Gamma_{(g)} \backslash \Gamma} \Gamma_{(g)} \gamma_i, \quad \text{where} \quad \Gamma_{(g)} = \Gamma \cap g^{-1} \Gamma g, \tag{3.1.6}$$

be the decomposition of Γ into the left cosets modulo $\Gamma_{(g)}$. Then

$$\Gamma g \Gamma = \bigcup_{\gamma_i \in \Gamma_{(g)} \backslash \Gamma} \Gamma g \gamma_i, \tag{3.1.7}$$

where the union is disjoint; in particular,

$$\nu(g) = \nu_\Gamma(g) = |\Gamma\backslash\Gamma g\Gamma| = [\Gamma : \Gamma_{(g)}]. \tag{3.1.8}$$

Proof. It is clear that the right side in (3.1.7) is contained in the left side. Now let $g' = \gamma g\delta$ with $\gamma, \delta \in \Gamma$ be an element of $\Gamma g\Gamma$. By (3.1.6), the element δ belongs to a left coset $\Gamma_{(g)}\gamma_i$, i.e. $\delta = \alpha\gamma_i$, where $\alpha \in \Gamma$ and $g\alpha g^{-1} \in \Gamma$. Then we have

$$g' = \gamma g\alpha\gamma_i = \gamma g\alpha g^{-1} g\gamma_i \in \Gamma g\gamma_i,$$

and so the left side in (3.1.7) is contained in the right side.

Finally, if $\Gamma g\gamma_i$ meets $\Gamma g\gamma_j$, then we have $\gamma g\gamma_i = \delta g\gamma_j$ with $\gamma, \delta \in \Gamma$, whence $g^{-1}\delta^{-1}\gamma g\gamma_i = \gamma_j$, and so $\Gamma_{(g)}\gamma_i = \Gamma_{(g)}\gamma_j$, for the element $g^{-1}\delta^{-1}\gamma g = \gamma_j\gamma_i^{-1}$ is clearly contained in $\Gamma_{(g)}$. □

We see that, for an invertible g, the condition (3.1.3) is equivalent with the condition $[\Gamma: \Gamma \cap g^{-1}\Gamma g] < \infty$. We say that two subgroups Γ_1 and Γ_2 of a multiplicative group G are *commensurable* if $\Gamma_1 \cap \Gamma_2$ is of finite index in Γ_1 and Γ_2. In this case we write $\Gamma_1 \sim \Gamma_2$.

Lemma 3.1.3. *If $\Gamma_1 \sim \Gamma_2$ and $\Gamma_2 \sim \Gamma_3$, then $\Gamma_1 \sim \Gamma_3$.*

Proof. The imbedding $\Gamma_1 \cap \Gamma_2 \subset \Gamma_2$ implies the imbedding

$$\Gamma_1 \cap \Gamma_2 \cap \Gamma_3\backslash\Gamma_1 \cap \Gamma_2 \subset \Gamma_2 \cap \Gamma_3\backslash\Gamma_2.$$

Hence, $[\Gamma_1 \cap \Gamma_2 : \Gamma_1 \cap \Gamma_2 \cap \Gamma_3] \le [\Gamma_2 : \Gamma_2 \cap \Gamma_3] < \infty$, and $[\Gamma_1 : \Gamma_1 \cap \Gamma_3] \le [\Gamma_1 : \Gamma_1 \cap \Gamma_2 \cap \Gamma_3] = [\Gamma_1 : \Gamma_1 \cap \Gamma_2][\Gamma_1 \cap \Gamma_2 : \Gamma_1 \cap \Gamma_2 \cap \Gamma_3] < \infty$. In the same way we see that $[\Gamma_3 : \Gamma_1 \cap \Gamma_3] < \infty$. So $\Gamma_1 \sim \Gamma_3$. □

Lemma 3.1.4. *Let* G *be a group, and Γ a subgroup of* G. *Then the set*

$$\tilde{\Gamma} = \{g \in \mathrm{G}; g^{-1}\Gamma g \sim \Gamma\}$$

is a group.

Proof. If Γ_1 and Γ_2 are two commensurable subgroups of G and $g \in \mathrm{G}$, then clearly $g^{-1}\Gamma_1 g \sim g^{-1}\Gamma_2 g$. Thus, for $g \in \tilde{\Gamma}$, we have $\Gamma \sim g^{-1}\Gamma g$, hence $g\Gamma g^{-1} \sim g(g^{-1}\Gamma g)g^{-1} = \Gamma$, and so $g^{-1} \in \tilde{\Gamma}$. Now, for $g_1, g_2 \in \tilde{\Gamma}$, we have $g_1^{-1}\Gamma g_1 \sim \Gamma$, hence $g_2^{-1}g_1^{-1}\Gamma g_1 g_2 \sim g_2^{-1}\Gamma g_2 \sim \Gamma$, and so by Lemma 3.1.3, $(g_1 g_2)^{-1}\Gamma g_1 g_2 \sim \Gamma$, i.e. $g_1 g_2 \in \tilde{\Gamma}$. □

We call $\tilde{\Gamma}$ *the commensurator of Γ in* G.

Let G be a multiplicative group, Γ a subgroup of G, and S a multiplicatively closed subset in G. We say that the pair (Γ, S) is *a Hecke pair* if it satisfies

$$\Gamma \subset \mathrm{S} \subset \tilde{\Gamma}, \tag{3.1.9}$$

where $\tilde{\Gamma}$ is the commensurator of Γ in G. For a Hecke pair (Γ, S) we denote by

$\mathsf{L} = \mathsf{L}(\Gamma, \mathsf{S})$ a free $\mathbb{Z}$-module consisting of all formal finite linear combinations with coefficients in $\mathbb{Z}$ of the symbols (Γg) with $g \in \mathsf{S}$ associated in a one-to-one manner to the left cosets Γg of S modulo Γ. For brevity, the symbols (Γg) will be called *the left cosets* (*of* S *modulo* Γ). Elements of S operate on L by

$$\mathsf{S} \supset g\colon t = \sum_i a_i(\Gamma g_i) \to tg = \sum_i a_i(\Gamma g_i g).$$

Let us consider the submodule $\mathsf{D} = \mathsf{D}(\Gamma, \mathsf{S})$ of L consisting of those $t \in \mathsf{L}$ which are Γ-invariant:

$$\mathsf{D}(\Gamma, \mathsf{S}) = \{t \in \mathsf{L}(\Gamma, \mathsf{S}); \quad t\gamma = t \quad \text{for all } \gamma \in \Gamma\}.$$

If

$$t = \sum_i a_i(\Gamma g_i) \quad \text{and} \quad t' = \sum_j b_j(\Gamma h_j)$$

are two elements of D, then the element

$$tt' = \sum_{i,j} a_i b_j (\Gamma g_i h_j) \in \mathsf{L} \tag{3.1.10}$$

is independent of the choice of representatives g_i and h_j in the corresponding left cosets and again belongs to D. To see this, we first note that tt' is independent of the choice of representatives $g_i \in \Gamma g_i$. Now let us replace the representatives h_j by $\gamma_j h_j$ with $\gamma_j \in \Gamma$. By the definition of D, we have

$$\sum_i a_i(\Gamma g_i \gamma_j) = t\gamma_j = t = \sum_i a_i(\Gamma g_i)$$

for each j. Hence,

$$\begin{aligned}\sum_{i,j} a_i b_j (\Gamma g_i \gamma_j h_j) &= \sum_j b_j \left(\sum_i a_i(\Gamma g_i \gamma_j)\right) h_j \\ &= \sum_j b_j \left(\sum_i a_i(\Gamma g_i)\right) h_j = \sum_{i,j} a_i b_j (\Gamma g_i h_j),\end{aligned}$$

and so tt' is independent of the choice of $h_j \in \Gamma h_j$. Finally, if $\gamma \in \Gamma$, we have

$$(t \cdot t')\gamma = \sum_{i,j} a_i b_j (\Gamma g_i h_j \gamma) = t \cdot (t'\gamma) = t \cdot t',$$

whence, $tt' \in \mathsf{D}$. Since the multiplication $(t, t') \to t \cdot t'$ is clearly bilinear and associative, we see that $\mathsf{D}(\Gamma, \mathsf{S})$ is an associative ring. The ring $\mathsf{D}(\Gamma, \mathsf{S})$ will be called *the Hecke ring of the pair* (Γ, S).

If (Γ, S) is a Hecke pair, then, by Lemma 3.1.2, each of the double cosets $\Gamma g \Gamma$ with $g \in S$ is a finite union of disjoint left cosets modulo Γ:

$$\Gamma g \Gamma = \bigcup_{i=1}^{\nu} \Gamma g_i.$$

Since the set $g_1\gamma, \ldots, g_\nu\gamma$ with $\gamma \in \Gamma$ is clearly still a set of representatives for the left cosets of $\Gamma g \Gamma$ modulo Γ, it follows that each element of $\mathsf{L}(\Gamma, \mathsf{S})$ of the form

$$(g) = (g)_\Gamma = \sum_{g_i \in \Gamma\backslash\Gamma g\Gamma} (\Gamma g_i) \tag{3.1.11}$$

satisfies $(g)\gamma = g$ for all $\gamma \in \Gamma$ and so belongs to the ring $\mathsf{D} = \mathsf{D}(\Gamma, \mathsf{S})$. The element

$$(e) = (e)_\Gamma = (\Gamma),$$

where e is the identity element of G, is clearly the identity element of the ring D.

Lemma 3.1.5. *The elements* (3.1.11) *associated to distinct double cosets of* S *modulo* Γ *form a basis of the* $\mathbb{Z}$*-module* $\mathsf{D}(\Gamma, \mathsf{S})$.

Products of the elements (3.1.11) *can be computed by the following formulae: Let* $g, g' \in \mathsf{S}$, *and let*

$$\Gamma g\Gamma = \bigcup_{i=1}^{\nu} \Gamma g_i, \qquad \Gamma g'\Gamma = \bigcup_{j=1}^{\nu'} \Gamma g'_j$$

be decompositions of the double cosets into left cosets. Then

$$(g)(g') = \sum_{\Gamma h\Gamma \subset \Gamma g\Gamma g'\Gamma} c(g, g'; h)(h),$$

where h *runs over a set of representatives for the double cosets modulo* Γ *contained in* $\Gamma g\Gamma g'\Gamma$, *the coefficient* $c(g, g'; h)$ *being defined as the number of pairs* g_i, g'_j *such that* $g_i g'_j \in \Gamma h$. *The coefficients* $c(g, g'; h)$ *can also be written in the form*

$$c(g, g'; h) = c'(g, g'; h)\nu(g')\nu(h)^{-1},$$

where $c'(g, g'; h)$ *is the number of the representatives* g_i *satisfying* $g_i g' \in \Gamma h\Gamma$, *and* $\nu(g')$, $\nu(h)$ *are the indices* (3.1.8).

Proof. Let D' be the submodule of D consisting of all finite linear combinations of elements (3.1.11) with coefficients in $\mathbb{Z}$. Each non-zero element $t \in \mathsf{D}$ can be written in the form

$$t = \sum_{i=1}^{\nu} a_i(\Gamma g_i), \tag{3.1.12}$$

where all a_i are non-zero, and all left cosets Γg_i are distinct. In this case we call the number $\nu = \nu(t)$ *the length of* t. We shall prove the inclusion $t \in \mathsf{D}'$ by induction on $\nu(t)$. If $\nu = 1$, then $t = a(\Gamma g)$. Since $t \in \mathsf{D}$, we have $t\gamma = t$, i.e. $a(\Gamma g\gamma) = a(\Gamma g)$ for all $\gamma \in \Gamma$, which means that $\Gamma g = \Gamma g\Gamma$, and so $t = a(g) \in D'$. We now assume that D' contains all of the elements $t' \in \mathsf{D}$ with $\nu(t') < \nu$, where $\nu > 1$. Let $t \in \mathsf{D}$ be an element of the form (3.1.12) with $\nu(t) = \nu$. Since $t\gamma = t$ for $\gamma \in \Gamma$, it follows that, for each $\gamma \in \Gamma$ and $i = 1, \ldots, \nu$, the left coset $(\Gamma g_i\gamma)$ occurs with the coefficients a_i in (3.1.12). According to Lemma 3.1.2, each coset of $\Gamma g_i\Gamma$ modulo Γ is of the form $\Gamma g_i\gamma$ with $\gamma \in \Gamma$. It follows that $\nu(t - a_i(g_i)) < \nu$, and therefore $t - a_i(g_i) \in D'$. Thus, $t \in \mathsf{D}'$. The first part is proved.

By definitions, we have

$$(g) = \sum_i (\Gamma g_i), \quad (g') = \sum_j (\Gamma g'_j), \quad \text{and} \quad (g)(g') = \sum_{i,j} (\Gamma g_i g'_j).$$

Since each of the products $g_i g'_j$ is clearly contained in $\Gamma g \Gamma g' \Gamma$, it follows, by the first part, that the product $(g)(g')$ can be rewritten in the form

$$(g)(g') = \sum_{\Gamma h \Gamma \subset \Gamma g \Gamma g' \Gamma} c(g, g'; h)(h) = \sum_{\Gamma h \Gamma \subset \Gamma g \Gamma g' \Gamma} c(g, g'; h) \sum_{\Gamma h_k \in \Gamma \backslash \Gamma h \Gamma} (\Gamma h_k)$$

with some coefficients $c(g, g'; h)$. Comparing the coefficients of (Γh) in the two decompositions of $(g)(g')$ into the left cosets, we see that $c(g, g'; h)$ is equal to the number of pairs g_i, g'_j such that $\Gamma g_i g'_j = \Gamma h$, i.e. $g_i g'_j \in \Gamma h$. It is easy to see that $c(g, g'; h)$ depends only on the double cosets of g, g', and h. Thus, the sum

$$\sum_{\Gamma h_k \in \Gamma \backslash \Gamma h \Gamma} c(g, g'; h) = \nu(h) c(g, g'; h)$$

is equal to the number of pairs g_i, g'_j with $g_i g'_j \in \Gamma h \Gamma$ By Lemma 3.1.2, we can take $g'_j = g' \gamma_j$ with $\gamma_j \in \Gamma$. Then we see that the last number equals the number of g_i with $g_i g' \in \Gamma h \Gamma$ multiplied by $\nu(g')$. □

3.1.3. The Imbedding i

Here we shall consider relations between Hecke rings of different Hecke pairs contained in the same group G. Let (Γ, S) and (Γ_0, S_0) be two Hecke pairs. We assume that the following conditions are fulfilled:

$$\Gamma_0 \subset \Gamma, \qquad \mathsf{S} \subset \Gamma \mathsf{S}_0, \qquad \text{and} \qquad \Gamma \cap \mathsf{S}_0 \mathsf{S}_0^{-1} \subset \Gamma_0. \tag{3.1.13}$$

Because $\mathsf{S} \subset \Gamma \mathsf{S}_0$, each element $t \in \mathsf{L}(\Gamma, \mathsf{S})$ can be written in the form

$$t = \sum_i a_i (\Gamma g_i) \qquad \text{with } g_i \in \mathsf{S}_0.$$

We then set

$$\mathsf{i}(t) = \sum_i a_i (\Gamma_0 g_i) \in \mathsf{L}(\Gamma_0, \mathsf{S}_0).$$

The element $\mathsf{i}(t)$ is independent of the choice of representatives $g_i \in \mathsf{S}_0$ in Γg_i, for if $g'_i \in \Gamma g_i \cap \mathsf{S}_0$, the third condition in (3.1.13) implies that $\gamma_i = g'_i g_i^{-1} \in \Gamma \cap \mathsf{S}_0 \mathsf{S}_0^{-1} \subset \Gamma_0$, hence

$$\sum_i a_i (\Gamma_0 g'_i) = \sum_i a_i (\Gamma_0 \gamma_i g_i) = \sum_i a_i (\Gamma_0 g_i).$$

The condition $\Gamma_0 \subset \Gamma$ implies that i carries different left cosets modulo Γ into different left cosets modulo Γ_0. Thus, i is an imbedding of $\mathsf{L}(\Gamma, \mathsf{S})$ into $\mathsf{L}(\Gamma_0, \mathsf{S}_0)$.

Proposition 3.1.6. *If Hecke pairs* (Γ, S) *and* (Γ_0, S_0) *satisfy* (3.1.13), *then the restriction of* i *to the Hecke ring* $\mathsf{D}(\Gamma, \mathsf{S})$ *is a ring monomorphism of* $\mathsf{D}(\Gamma, \mathsf{S})$ *into* $\mathsf{D}(\Gamma_0, \mathsf{S}_0)$:

$$\mathsf{i}\colon \mathsf{D}(\Gamma, \mathsf{S}) \to \mathsf{D}(\Gamma_0, \mathsf{S}_0). \tag{3.1.14}$$

If, moreover,

$$\mathsf{S}_0 \subset \mathsf{S} \quad \text{and} \quad \nu_\Gamma(g) = \nu_{\Gamma_0}(g) \quad \text{for all } g \in \mathsf{S}_0, \tag{3.1.15}$$

where ν *are the indices* (3.1.8), *then the mapping* (3.1.14) *is an isomorphism of* $\mathsf{D}(\Gamma, \mathsf{S})$ *onto* $\mathsf{D}(\Gamma_0, \mathsf{S}_0)$.

Proof. The first assertion directly follows from definitions and the condition $\Gamma_0 \subset \Gamma$. In order to prove the second assertion, in view of Lemma 3.1.5, we only have to show that

$$\mathsf{i}(g)_\Gamma = (g)_{\Gamma_0} \quad \text{for all } g \in \mathsf{S}_0, \tag{3.1.16}$$

if the conditions (3.1.15) are fulfilled. Let $\gamma_1, \ldots, \gamma_\nu$ with $\nu = \nu_{\Gamma_0}(g)$ be a set of representatives for the left cosets of Γ_0 modulo $\Gamma_0 \cap g^{-1}\Gamma_0 g$. By the definition of $(g)_{\Gamma_0}$ and Lemma 3.1.2, we have

$$(g)_{\Gamma_0} = \sum_{i=1}^{\nu} (\Gamma_0 g \gamma_i).$$

On the other hand, the elements $g\gamma_1, \ldots, g\gamma_\nu$ lie in $\Gamma g \Gamma$ and belong to different left cosets modulo Γ, for a relation $g\gamma_i = \delta g \gamma_j$ with $\delta \in \Gamma$ would imply that $\delta = (g\gamma_i\gamma_j^{-1})g^{-1} \subset \Gamma \cap \mathsf{S}_0\mathsf{S}_0^{-1} \subset \Gamma_0$, hence $i = j$ and $\delta = e$. The number of these elements equals the number of different left cosets of $\Gamma g \Gamma$ modulo Γ, by (3.1.15) and Lemma 3.1.2. It follows that

$$(g)_\Gamma = \sum_{i=1}^{\nu} (\Gamma g \gamma_i),$$

and so $\mathsf{i}(g)_\Gamma = (g)_{\Gamma_0}$, by the definition of i. □

3.1.4. The Anti-Isomorphism j

Proposition 3.1.7. *Let* G *be a group, and* (Γ, S) *a Hecke pair with* $\Gamma, \mathsf{S} \subset \mathsf{G}$. *Then the pair* $(\Gamma, \mathsf{S}^{-1})$ *with* $\mathsf{S}^{-1} = \{g^{-1}; g \in \mathsf{S}\}$ *is again a Hecke pair, and the linear mapping*

$$\mathsf{j}\colon \mathsf{D}(\Gamma, \mathsf{S}) \to \mathsf{D}(\Gamma, \mathsf{S}^{-1}) \tag{3.1.17}$$

given on the elements of $\mathsf{D}(\Gamma, \mathsf{S})$ *of the form* (3.1.11) *by*

$$\mathsf{j}((g)_\Gamma) = (g^{-1})_\Gamma \quad (g \in \mathsf{S})$$

is an anti-isomorphism of the rings. In particular, if S *is a group, then* j *is an anti-isomorphism of the ring* $\mathsf{D}(\Gamma, \mathsf{S})$.

First we prove a lemma.

Lemma 3.1.8. *If* Γ *is a subgroup of* G *and* $\tilde{\Gamma}$ *is the commensurator of* Γ *in* G, *then the function* $\lambda(g) = \nu(g)\nu(g^{-1})^{-1}$ *on* $\tilde{\Gamma}$, *where* $\nu(h) = [\Gamma\colon \Gamma \cap h^{-1}\Gamma h]$, *satisfies* $\lambda(g_1 g_2) = \lambda(g_1)\lambda(g_2)$ *for all* $g_1, g_2 \in \tilde{\Gamma}$.

Proof of the Lemma. Let us denote by X the set of all subgroups of G which are commensurable with Γ. If $\Gamma_1, \Gamma_2 \in \mathsf{X}$, there is a group $\Gamma' \in \mathsf{X}$ of finite index in Γ_1 and Γ_2 (for example, by Lemma 3.1.3, one can take $\Gamma' = \Gamma_1 \cap \Gamma_2$). We then set

$$\lambda(\Gamma_1/\Gamma_2) = [\Gamma_1 : \Gamma'][\Gamma_2 : \Gamma']^{-1}.$$

It is readily verified that λ is independent of the choice of Γ' and satisfies the relations

$$\begin{aligned} \lambda(\Gamma_1/\Gamma_2)\lambda(\Gamma_2/\Gamma_3) &= \lambda(\Gamma_1/\Gamma_3), \\ \lambda(g^{-1}\Gamma_1 g/g^{-1}\Gamma_2 g) &= \lambda(\Gamma_1/\Gamma_2) \qquad (g \in \mathsf{G}). \end{aligned} \tag{3.1.18}$$

Let $g \in \tilde{\Gamma}$ and $\Gamma' \in \mathsf{X}$. We set $\lambda'(g) = \lambda(\Gamma'/g^{-1}\Gamma' g)$. It is easy to see that $\lambda'(g)$ is independent of the choice of Γ'. Hence, by (3.1.18), we have

$$\begin{aligned} \lambda'(g_1 g_2) &= \lambda(\Gamma'/(g_1 g_2)^{-1}\Gamma' g_1 g_2) \\ &= \lambda(\Gamma'/g_1^{-1}\Gamma' g_1)\lambda(g_1^{-1}\Gamma' g_1/g_2^{-1}(g_1^{-1}\Gamma' g_1)g_2) = \lambda'(g_1)\lambda'(g_2) \end{aligned}$$

for any $g_1, g_2 \in \tilde{\Gamma}$. On the other hand, if $g \in \tilde{\Gamma}$, then

$$\begin{aligned} \lambda'(g) &= \lambda(\Gamma/\Gamma \cap g^{-1}\Gamma g)\lambda(\Gamma \cap g^{-1}\Gamma g/g^{-1}\Gamma g) \\ &= v(g)\lambda(g\Gamma g^{-1} \cap \Gamma/\Gamma) = v(g)v(g^{-1})^{-1} = \lambda(g). \end{aligned}$$ □

Proof of Proposition 3.1.7. Since the elements of the form (3.1.11) span the $\mathbb{Z}$-module $\mathsf{D}(\Gamma, \mathsf{S})$, it suffices to show that

$$\mathrm{j}((g_1)(g_2)) = \mathrm{j}((g_2))\mathrm{j}((g_1)) = (g_2^{-1})(g_1^{-1})$$

for any $g_1, g_2 \in \mathsf{S}$. To prove this, by Lemma 3.1.5 and the definition of j, we only have to show that

$$c'(g_1, g_2; h)v(g_2)v(h)^{-1} = c'(g_2^{-1}, g_1^{-1}; h^{-1})v(g_1^{-1})v(h^{-1})^{-1}, \tag{3.1.19}$$

if $h \in \Gamma g_1 \Gamma g_2 \Gamma$. By Lemma 3.1.8, we have

$$v(h)v(h^{-1})^{-1} = \lambda(h) = \lambda(g_1)\lambda(g_2) = v(g_1)v(g_1^{-1})^{-1}v(g_2)v(g_2^{-1})^{-1},$$

for the mapping $g \to \lambda(g)$ is trivial on Γ. It follows that (3.1.19) is equivalent with

$$c'(g_1, g_2; h)v(g_2^{-1}) = c'(g_2^{-1}, g_1^{-1}; h^{-1})v(g_1). \tag{3.1.20}$$

By the definition of $c'(g_1, g_2; h)$ and (3.1.7), we see that $c'(g_1, g_2; h)$ is equal to the number of elements of the set

$$\{\gamma \in \Gamma_{(g_1)}\backslash\Gamma; g_1\gamma g_2 \in \Gamma h\Gamma\}.$$

Let $\gamma \in \Gamma$. It is readily verified that the double coset $\Gamma g_1 \gamma g_2 \Gamma$ depends only on $\Gamma_{(g_1)}\gamma\Gamma_{(g_2^{-1})}$, and the relation $\Gamma_{(g_1)}\gamma t = \Gamma_{(g_1)}\gamma$ with $t \in \Gamma_{(g_2^{-1})}$ holds if and only if $t \in \gamma^{-1}\Gamma_{(g_1)}\gamma$. Thus, we can write $c'(g_1, g_2; h)$ in the form

$$c'(g_1, g_2; h) = \sum_{\substack{\gamma \in \Gamma_{(g_1)}\backslash\Gamma/\Gamma_{(g_2^{-1})},\\ g_1\gamma g_2 \in \Gamma h \Gamma}} [\Gamma_{(g_2^{-1})} : (\Gamma_{(g_2^{-1})} \cap \gamma^{-1}\Gamma_{(g_1)}\gamma)]. \tag{3.1.21}$$

Similarly,

$$c'(g_2^{-1}, g_1^{-1}; h^{-1}) = \sum_{\substack{\gamma \in \Gamma_{(g_1)}\backslash\Gamma/\Gamma_{(g_2^{-1})},\\ g_1\gamma g_2 \in \Gamma h \Gamma}} [\Gamma_{(g_1)} : (\Gamma_{(g_1)} \cap \gamma\Gamma_{(g_2^{-1})}\gamma^{-1})]$$

Using the symbol $\lambda(\Gamma_1/\Gamma_2)$ defined in the proof of Lemma 3.1.8 and its properties (3.1.18), we get

$$\begin{aligned}
c'(g_2^{-1}, g_1^{-1}; h^{-1}) &= \sum_\gamma \lambda(\Gamma_{(g_1)}/\Gamma_{(g_1)} \cap \gamma\Gamma_{(g_2^{-1})}\gamma^{-1}) \\
&= \sum_\gamma \lambda(\gamma^{-1}\Gamma_{(g_1)}\gamma/\gamma^{-1}\Gamma_{(g_1)}\gamma \cap \Gamma_{(g_2^{-1})}) \\
&= \sum_\gamma \lambda(\gamma^{-1}\Gamma_{(g_1)}\gamma/\Gamma_{(g_1)})\lambda(\Gamma_{(g_1)}/\Gamma)\lambda(\Gamma/\Gamma_{(g_2^{-1})}) \\
&\quad \times \lambda(\Gamma_{(g_2^{-1})}/\Gamma_{(g_2^{-1})} \cap \gamma^{-1}\Gamma_{(g_1)}\gamma) \\
&= \sum_\gamma \lambda(\gamma)\nu(g_1)^{-1}\nu(g_2^{-1})\lambda(\Gamma_{(g_2^{-1})}/\Gamma_{(g_2^{-1})} \cap \gamma^{-1}\Gamma_{(g_1)}\gamma) \\
&= \nu(g_1)^{-1}\nu(g_2^{-1})c'(g_1, g_2; h),
\end{aligned}$$

where γ runs over the same set as in (3.1.21). □

The next Lemma shows that the anti-isomorphism j is consistent with the imbedding i.

Lemma 3.1.9. *Let* (Γ, S) *and* (Γ_0, S_0) *be two Hecke pairs satisfying* (3.1.13). *Assume that Hecke pairs* $(\Gamma, \mathsf{S}^{-1})$ *and* $(\Gamma_0, \mathsf{S}_0^{-1})$ *also satisfy* (3.1.13). *Then the diagram*

$$\begin{array}{ccc}
\mathsf{D}(\Gamma, \mathsf{S}) & \xrightarrow{\ \mathsf{i}\ } & \mathsf{D}(\Gamma_0, \mathsf{S}_0) \\
\mathsf{j}\downarrow & & \downarrow\mathsf{j} \\
\mathsf{D}(\Gamma, \mathsf{S}^{-1}) & \xrightarrow{\ \mathsf{i}\ } & \mathsf{D}(\Gamma_0, \mathsf{S}_0^{-1})
\end{array}$$

is commutative.

Proof. By the definition of i we have

$$\mathsf{i}(g)_\Gamma = \sum_{g_i \in \Gamma_0\backslash\Gamma g\Gamma \cap \mathsf{S}_0/\Gamma_0} (g_i)_{\Gamma_0} \qquad (g \in \mathsf{S}),$$

hence

$$\begin{aligned}
\mathsf{ji}(g)_\Gamma &= \sum_{g_i \in \Gamma_0\backslash\Gamma g\Gamma \cap \mathsf{S}_0/\Gamma_0} (g_i^{-1})_{\Gamma_0} = \sum_{h_i \in \Gamma_0\backslash\Gamma g^{-1}\Gamma \cap \mathsf{S}_0^{-1}/\Gamma_0} (h_i)_{\Gamma_0} \\
&= \mathsf{i}(g^{-1})_\Gamma = \mathsf{ij}(g)_\Gamma.
\end{aligned}$$

Lemma follows from the relations and Lemma 3.1.5. □

3.1.5. Representations on Automorphic Functions

Let (Γ, S) be a Hecke pair with the Hecke ring $\mathsf{D} = \mathsf{D}(\Gamma, \mathsf{S})$. As in the beginning of this section, we assume that S operates on a set H. Let φ be a factor of automorphy of S on H with values in a group T, V a left T-module, and $\mathfrak{M} = \mathfrak{M}_\varphi(\mathsf{T})$ the $\mathbb{Z}$-module of all automorphic functions $F\colon \mathsf{H} \to \mathsf{V}$ of weight φ for Γ. If

$$t = \sum_i a_i(\Gamma g_i) \in \mathsf{D} \quad \text{and} \quad F \in \mathfrak{M},$$

we set

$$F|t = F|_\varphi t = \sum_i a_i F|_\varphi g_i, \tag{3.1.22}$$

where $(F|_\varphi g)(h) = \varphi(g, h)^{-1} F(g(h))$. It follows from Lemma 1.4.1(3) and the relation (3.1.1) that $F|t$ is independent of the choice of representatives $g_i \in \Gamma g_i$ and again belongs to $\mathfrak{M}$. In fact, if $\gamma_i g_i$ with $\gamma_i \in \Gamma$ are other representatives, then

$$\sum_i a_i F|\gamma_i g_i = \sum_i a_i F|\gamma_i|g_i = \sum a_i F|g_i,$$

and since $t\gamma = \sum_i a_i(\Gamma g_i \gamma) = t$ for $\gamma \in \Gamma$, it follows that

$$(F|t)|\gamma = \sum_i a_i F|g_i\gamma = F|t\gamma = F|t.$$

Thus, each element $t \in \mathsf{D} = \mathsf{D}(\Gamma, \mathsf{S})$ defines a linear operator

$$|t = |_\varphi t\colon \mathfrak{M}_\varphi(\Gamma) \to \mathfrak{M}_\varphi(\Gamma).$$

When $t = (g)_\Gamma$, the operator $|t$ clearly coincides with the operator (3.1.4). The operators $|t$ with $t \in \mathsf{D}(\Gamma, \mathsf{S})$ will be called Hecke operators on $\mathfrak{M}_\varphi(\Gamma)$. By Lemma 3.1.5, any Hecke operator on $\mathfrak{M}_\varphi(\Gamma)$ is a linear combination of the operators (3.1.4).

Proposition 3.1.10. *The mapping*

$$t \to |_\varphi t \qquad (t \in \mathsf{D}(\Gamma, \mathsf{S}))$$

is a homomorphism of $\mathsf{D}(\Gamma, \mathsf{S})$ *into the ring of endomorphism of the* $\mathbb{Z}$*-module* $\mathfrak{M}_\varphi(\Gamma)$.

Proof. The mapping is clearly linear. If $F \in \mathfrak{M}_\varphi(\Gamma)$, $t = \sum_i a_i(\Gamma g_i)$, $t' = \sum_j b_j(\Gamma h_j) \in \mathsf{D}(\Gamma, \mathsf{S})$, then we have

$$\begin{aligned} F|tt' &= \sum_{i,j} a_i b_j F|g_i h_j = \sum_j b_j \left(\sum_i a_i F|g_i\right)\Bigg| h_j \\ &= F|t|t', \end{aligned}$$

by definitions and Lemma 1.4.1(3). □

The last proposition implies that all algebraic relations for elements of a Hecke ring are also satisfied by the corresponding Hecke operators.

3.1.6. Hecke Rings over a Commutative Ring

Let (Γ, S) be a Hecke pair, and K a commutative ring with the identity element. In the same way as it was done in § 3.1.2 for $\mathsf{K} = \mathbb{Z}$, we can define a free K-module $\mathsf{L}_\mathsf{K} = \mathsf{L}_\mathsf{K}(\Gamma, \mathsf{S})$ consistiny of all formal finite linear combinations with coefficients in K of the left cosets $(\Gamma g)(g \in \mathsf{S})$, and the submodule $\mathsf{D}_\mathsf{K} = \mathsf{D}_\mathsf{K}(\Gamma, \mathsf{S})$ of L_K consisting of all Γ-invariant elements. The multiplication

$$\left(\sum_i a_i(\Gamma g_i)\right)\left(\sum_j b_j(\Gamma h_j)\right) = \sum_{i,j} a_i b_j(\Gamma g_i h_j)$$

of elements in D_K is again independent of the choice of representatives in the left cosets and turns D_K into an associative ring with the identity element, the Hecke ring of the pair (Γ, S) over K. All results of § 3.1.2–3.1.4 hold for the Hecke rings $\mathsf{D}_\mathsf{K}(\Gamma, \mathsf{S})$ without any changes. The same is true for the results of § 3.1.5, if we assume that V is also a K-module and the action of K commutes with that of T. It is clear that

$$\mathsf{D}_\mathsf{K}(\Gamma, S) = \mathsf{D}(\Gamma, \mathsf{S}) \otimes_\mathbb{Z} \mathsf{K}, \qquad \text{if } \mathbb{Z} \subset \mathsf{K}. \tag{3.1.23}$$

(the tensor product of rings)

Exercise 3.1.11. *Let Γ be a normal subgroup of finite index in a group S. Show that the pair (Γ, S) is a Hecke pair and the Hecke ring $\mathsf{D}_\mathsf{K}(\Gamma, \mathsf{S})$ of (Γ, S) over a ring K is isomorphic to the group algebra of the factor group $\Gamma \backslash \mathsf{S}$ over K.*

Exercise 3.1.12. *Let (Γ, S) be a Hecke pair. Let us define a mapping v: $\mathsf{D}_\mathsf{K}(\Gamma, \mathsf{S}) \to \mathsf{K}$ by*

$$v\left(\sum_i a_i(g_i)\right) = \sum_i a_i v_\Gamma(g_i),$$

where $(g_i) = (g_i)_\Gamma$ are elements of the form (3.1.1) *and $v_\Gamma(g_i)$ is the index* (3.1.8). *Show that v is a ring homomorphism.*

Exercise 3.1.13. *Let (Γ, S) be a Hecke pair. Let us assume that there exists an anti-automorphism $g \to g^*$ of S (i.e. a one-to-one mapping $S \to S$ with $(g_1 g_2)^* = g_2^* g_1^*$) such that $\Gamma^* = \Gamma$ and $(\Gamma g \Gamma)^* = \Gamma g \Gamma$ for any $g \in S$. Show that the ring $\mathsf{D}_\mathsf{K}(\Gamma, \mathsf{S})$ is commutative for any commutative K.*

[Hint: see proof of Theorem 3.2.3 below.]

§ 3.2. Hecke Rings of the General Linear Group

Although our main goal is to study Hecke rings of the symplectic group and their representations on modular forms, our methods are often based on reduction to the case of the general linear group.

3.2.1. Global Rings

Let us consider the groups

$$\Lambda = \Lambda^n = \mathsf{GL}_n(\mathbb{Z}) \quad \text{and} \quad \mathsf{G} = \mathsf{G}^n = \mathsf{GL}_n(\mathbb{Q}) \tag{3.2.1}$$

with $n \in \mathbb{N}$, where $\mathbb{Q}$ is the field of rational numbers.

Lemma 3.2.1. *The commensurator of Λ in G is equal to G; in particular, the pair (Λ, G) is a Hecke pair.*

Proof. Considering the entries of matrices in Λ modulo some $q \in \mathbb{N}$, we get a homomorphism of Λ into the finite group $\mathsf{GL}_n(\mathbb{Z}/q\mathbb{Z})$. The kernel of the homomorphism is *the principal congruence subgroup of level q of Λ*,

$$\Lambda(q) = \Lambda^n(q) = \{\lambda \in \Lambda; \lambda \equiv E_n \,(\mathrm{mod}\, q)\}, \tag{3.2.2}$$

which is, therefore, a normal subgroup of finite index in Λ.

Let $g \in \mathsf{G}$. We can write g in the form $g = tg_0$ with $t \in \mathbb{Q}$ and an integral matrix g_0. Then it is readily verified that $g\Lambda(d)g^{-1} \subset \Lambda$ and $g^{-1}\Lambda(d)g \subset \Lambda$, where $d = |\det g_0|$. It follows that $\Lambda \cap g^{-1}\Lambda g$ contains the subgroup $\Lambda(d)$ and therefore has finite index in Λ. Also $\Lambda \cap g\Lambda g^{-1}$ has finite index in Λ, and therefore the group $g^{-1}(\Lambda \cap g\Lambda g^{-1})g = g^{-1}\Lambda g \cap \Lambda$ has finite index in $g^{-1}\Lambda g$. We see that $g^{-1}\Lambda g$ is commensurable with Λ. □

In this section we shall consider the Hecke ring

$$\mathsf{H} = \mathsf{H}^n = \mathsf{D}_{\mathbb{Q}}(\Lambda^n, \mathsf{G}^n) = \mathsf{D}(\Lambda^n, \mathsf{G}^n) \otimes_{\mathbb{Z}} \mathbb{Q} \tag{3.2.3}$$

of the pair $(\Lambda^n, \mathsf{G}^n)$ over $\mathbb{Q}$ and its subrings. In view of Lemma 3.1.5, one can take the set of elements $(g) = (g)_\Lambda$ of the form (3.1.11) as a $\mathbb{Q}$-basis of H. These elements correspond in a one-to-one manner to the double cosets $\Lambda g \Lambda$ of G modulo Λ. The following lemma gives a parametrization of the double cosets.

Lemma 3.2.2. *Each of the double cosets $\Lambda g \Lambda$ with $\Lambda = \Lambda^n$ and $g \in \mathsf{G}^n$ contains the unique representative of the form*

$$\mathrm{ed}(g) = \mathrm{diag}(d_1, \ldots, d_n) \quad \text{with } d_i > 0 \text{ and } d_i | d_{i+1}. \tag{3.2.4}$$

Proof. Let d_1 be a positive rational number such that $g = d_1 g_1$ where g_1 is a primitive matrix, i.e. an integral matrix with coprime entries. We denote by $\delta = \delta(g_1)$ the minimum of the greatest common divisors of entries of columns of g_1 and prove by induction on $\delta(g_1)$ that the double coset $\Lambda g_1 \Lambda$ of any primitive g_1 contains a representative of the form $\begin{pmatrix} 1 & 0 \\ 0 & g_2 \end{pmatrix}$ with an integral matrix g_2 of order $n-1$. If $\delta = 1$, which means that g_1 has a column with coprime entries, we can multiply g_1 from the right by a permutation matrix to obtain a matrix g_1' in the same double coset, whose first column has coprime entries. So we can assume that $g_1' = g_1$. Then, by Lemma 1.3.8, we can multiply g_1 from the left by a matrix of Λ to obtain a matrix of the form

$$\begin{pmatrix} 1 & \mathbf{l} \\ 0 & g_2 \end{pmatrix} = \begin{pmatrix} 1 & 0 \\ 0 & g_2 \end{pmatrix} \begin{pmatrix} 1 & \mathbf{l} \\ 0 & E_{n-1} \end{pmatrix} \quad \text{with } \mathbf{l} \in \mathsf{M}_{1,n-1},$$

which proves the assertion for g_1 with $\delta(g_1) = 1$. If the assertion is already proved for all primitive g_1' with $\delta(g_1') < \delta$, where $\delta > 1$, and g_1 is a primitive matrix with $\delta(g_1) = \delta$, then again, after a permutation of columns, we may assume that the greatest common divisor of the entries of the first column of g_1 is δ. Then, by Lemma 1.3.8, we can find in Λg_1 a matrix of the form

$$\begin{pmatrix} \delta & \mathbf{l} \\ 0 & h \end{pmatrix} = \begin{pmatrix} \delta & \mathbf{l} + \delta\mathbf{l}' \\ 0 & h \end{pmatrix} \begin{pmatrix} 1 & -\mathbf{l}' \\ 0 & E_{n-1} \end{pmatrix} = g_1' \begin{pmatrix} 1 & -\mathbf{l}' \\ 0 & E_{n-1} \end{pmatrix},$$

where $\mathbf{l} \in \mathsf{M}_{1,n-1}$, h is an integral matrix of order $n-1$, and $\mathbf{l}'$ is an arbitrary integral row of order $n-1$. The row $\mathbf{l}'$ can be choosen in such a way that all entries of $\mathbf{l}'' = \mathbf{l} + \delta\mathbf{l}'$ are positive and not greater than δ. Then the matrix $g_1' \in \Lambda g_1 \Lambda$ satisfies $\delta(g_1') < \delta$ (the inequality $\delta(g_1') \le \delta$ is clear; the equality $\delta(g_1') = \delta$ would imply that δ divides all entries of g_1', which are coprime). By the inductive hypothesis, the assertion is true for g_1' and therefore for g_1.

Returning to the proof of the lemma, we see that the double coset $\Lambda g \Lambda = d_1 \Lambda g_1 \Lambda$ contains a representative of the form $g' = \begin{pmatrix} d_1 & 0 \\ 0 & d_1 g_2 \end{pmatrix}$, where d_1 is the greatest common divisor of entries of g and g_2 is an integral matrix of order $n-1$. It is readily verified that the matrix g' remains in $\Lambda g \Lambda$, when g_2 remains in $\Lambda^{n-1} g_2 \Lambda^{n-1}$. So we can apply the same arguments as above to the matrix $d_1 g_2$ and so on. Finally, we get a representative of the form (3.2.4).

Let $\mathsf{D} = \operatorname{diag}(d_1, \dots, d_n)$ and $\mathsf{D}' = \operatorname{diag}(d_1', \dots, d_n')$ be two matrices of the form (3.2.4) which satisfy $\mathsf{D}' = \lambda_1 \mathsf{D} \lambda_2$ with $\lambda_1, \lambda_2 \in \Lambda$. The Binet-Cauchy formulae imply that each minor of order r of D', where $1 \le r \le n$, is divisible by $d_1 \cdots d_r$; in particular, $d_1 \cdots d_r | d_1' \cdots d_r'$. Similarly, $d_1' \cdots d_r' | d_1 \cdots d_r$. It follows that $d_1' \cdots d_r' = d_1 \cdots d_r$, hence $\mathsf{D}' = \mathsf{D}$. □

The matrix $\operatorname{ed}(g) = \operatorname{diag}(d_1, \dots, d_n)$ is called *the matrix of elementary divisors of* g, and $d_r = d_r(g)$ for $1 \le r \le n$ are *the elementary divisors of* g. The Binet-Cauchy formulae imply that $d_1(g) \cdots d_r(g)$ is equal to the greatest common divisor of minors of order r of g; in particular,

$$d_1(g)\cdots d_n(g) = |\det g|. \tag{3.2.5}$$

Let us now turn to multiplicative properties of the ring H^n.

Theorem 3.2.3. *The ring* H^n *is commutative for any* $n \in \mathbb{N}$.

Proof. By Lemma 3.1.5, it suffices to show that

$$(g)(g') = (g')(g) \qquad \text{if } g, g' \in \mathsf{G}^n.$$

Since each double coset $\Lambda g \Lambda$ with $g \in \mathsf{G}$ contains a diagonal matrix, it satisfies ${}^t(\Lambda g \Lambda) = \Lambda\, {}^t g \Lambda = \Lambda g \Lambda$. This implies that the number of the left cosets of $\Lambda g \Lambda$ modulo Λ is equal to the number of the right cosets. Since every right coset in $\Lambda g \Lambda$ meets every left coset, it follows that there is a set $g_1, \ldots, g_\nu$ of common representatives for the left and right cosets of $\Lambda g \Lambda$ modulo Λ. Then it is easy to see that the set ${}^t g_1, \ldots, {}^t g_\nu$ is again a set of common representatives for the left and right cosets of $\Lambda g \Lambda$ modulo Λ. We let $g'_1, \ldots, g'_{\nu'}$ denote a similar set of representatives of $\Lambda g' \Lambda$ modulo Λ. Then, by Lemma 3.1.5, we have

$$(g)(g') = \sum_{\Lambda h \Lambda \in \Lambda g \Lambda g' \Lambda} c''(g, g'; h)\nu(h)^{-1}(h),$$

where $c''(g, g'; h) = c'(g, g'; h)\nu(g')$ is the number of pairs g_i, g'_j with $g_i g'_j \in \Lambda h \Lambda$. If we replace g_i by ${}^t g_i$ and g'_j by ${}^t g'_j$, we see that $c''(g, g'; h)$ is also the number of pairs g_i, g'_j with

$${}^t g_i\, {}^t g'_j = {}^t(g'_j g_i) \in \Lambda h \Lambda = {}^t(\Lambda h \Lambda), \qquad \text{i.e. } g'_j g_i \in \Lambda h \Lambda.$$

It follows that $c''(g, g'; h) = c''(g', g; h)$. The set $\Lambda g \Lambda g' \Lambda$ is a finite union of double cosets modulo Λ, and therefore it coincides with the set ${}^t(\Lambda g \Lambda g' \Lambda) = \Lambda\, {}^t g' \Lambda\, {}^t g \Lambda = \Lambda g' \Lambda g \Lambda$. Thus, $(g)(g') = (g')(g)$. □

Lemma 3.2.4. *Let* $g \in \mathsf{G}^n$, *and let* $r \in \mathsf{G}^1 = \mathbb{Q}^*$. *Then the following relation hold in* H^n:

$$(rE_n)\cdot(g) = (g)\cdot(rE_n) = (rg). \tag{3.2.6}$$

Proof. Since clearly $\Lambda_{(g)} = \Lambda \cap g^{-1}\Lambda g = \Lambda_{(rg)}$, we can write

$$\Lambda g \Lambda = \bigcup_{\gamma_i \in \Lambda_{(g)} \backslash \Lambda} \Lambda g \gamma_i, \qquad \text{and} \qquad \Lambda r g \Lambda = \bigcup_{\gamma_i \in \Lambda_{(g)} \backslash \Lambda} \Lambda r g \gamma_i,$$

by Lemma 3.1.2. On the other hand, we have $\Lambda r E_n \Lambda = \Lambda r E_n$. Then by definitions we have

$$(rE_n)(g) = \sum_i (\Lambda g \gamma_i r E_n) = \sum_i (\Lambda r g \gamma_i) = (rg). \qquad \square$$

Proposition 3.2.5. *Let* $g, g' \in \mathsf{G}^n$. *Let us assume that the quotients of elementary divisors* $d_n(g)/d_1(g)$ *and* $d_n(g')/d_1(g')$ *are coprime. Then the corresponding elements*

(3.1.11) *of* H^n *satisfy*

$$(g)(g') = (gg'). \tag{3.2.7}$$

Proof. It follows from Lemma 3.2.4 that one can assume $d_1(g) = d_1(g') = 1$. So it suffices to consider the case, when the matrices g, g' are integral, and the numbers $a = |\det g|$, $a' = |\det g'|$ are coprime. By Lemma 3.1.5 we can write

$$(g)(g') = \sum_{\Lambda h \Lambda \subset \Lambda g \Lambda g' \Lambda} c(g, g'; h)(h),$$

where $c(g, g'; h)$ are some positive integers, which depend only on the double cosets $\Lambda g \Lambda$, $\Lambda g' \Lambda$ and $\Lambda h \Lambda$. Since $\Lambda g g' \Lambda \subset \Lambda g \Lambda g' \Lambda$, it follows that $c(g, g'; gg') \geq 1$, and so we can write

$$(g)(g') = (gg') + \sum_{\Lambda h \Lambda \subset \Lambda g \Lambda g' \Lambda} c_1(g, g'; h)(h), \tag{3.2.8}$$

where $c_1(g, g'; h)$ are some non-negative integers, which depend only on the double cosets $\Lambda g \Lambda$, $\Lambda g' \Lambda$ and $\Lambda h \Lambda$. We let

$$\mathsf{ED}(a) = \mathsf{ED}_n(a) = \{\mathrm{diag}(d_1, \ldots, d_n); d_i \in \mathbb{N}, d_i | d_{i+1}, d_1 \cdots d_n = a\} \tag{3.2.9}$$

denote the set of all integral matrices of the form (3.2.4) of determinant $a \in \mathbb{N}$. By Lemma 3.2.2, we can assume that the matrices g, g' and h in (3.2.8) satisfy $g \in \mathsf{ED}(a)$, $g' \in \mathsf{ED}(a')$, and $h \in \mathsf{ED}(aa')$, for any h in (3.2.8) is integral and satisfies $|\det h| = |\det g| |\det g'|$. Since each of the sets $\mathsf{ED}(a)$ is finite, we can define elements of H^n of the form

$$t(a) = t_n(a) = \sum_{g \in \mathsf{ED}_n(a)} (g). \tag{3.2.10}$$

Summing up the relations (3.2.8) over all $g \in \mathsf{ED}(a)$ and $g' \in \mathsf{ED}(a')$, we get

$$t(a)t(a') = \sum_{\substack{g \in \mathsf{ED}(a) \\ g' \in \mathsf{ED}(a')}} (gg') + \sum_{h \in \mathsf{ED}(aa')} \left(\sum_{\substack{g \in \mathsf{ED}(a) \\ g' \in \mathsf{ED}(a')}} c_1(g, g'; h) \right)(h). \tag{3.2.11}$$

If a, a' are coprime, it is easy to see that $(g, g') \to gg'$ is a one-to-one mapping of $\mathsf{ED}(a) \times \mathsf{ED}(a')$ onto $\mathsf{ED}(aa')$. Thus, the first sum on the right side in (3.2.11) equals $t(aa')$. So if we prove that $t(a)t(a') = t(aa')$ for coprime a, a', it will follow from (3.2.11) that the double sum on the right side equals zero, and therefore each of the coefficients $c_1(g, g'; h)$ equals zero too. (Note that all of them are non-negative.) Then (3.2.8) will turn into (3.2.7), and the proposition will be proved. So it suffices to prove the following lemma.

Lemma 3.2.6. *Let* $a \in \mathbb{N}$. *Then the set*

$$\mathsf{M}_n(\pm a) = \{\mathsf{M} \in \mathsf{M}_n; \det \mathsf{M} = \pm a\} \tag{3.2.12}$$

is a finite union of left cosets modulo $\Lambda = \Lambda^n$, *and the element* $t(a) = t_n(a) \in \mathsf{H}^n$ *can be written in the form*

$$t(a) = \sum_{g \in \Lambda \backslash \mathsf{M}_n(\pm a)} (\Lambda g). \tag{3.2.13}$$

If a, a' are coprime, then

$$t(a)t(a') = t(aa'). \tag{3.2.14}$$

Proof. By Lemma 3.2.2, the set $\mathsf{M}_n(\pm a)$ is a finite union of double cosets modulo Λ, and each one of the cosets has single representative in $\mathsf{ED}_n(a)$. Then the first statement of the lemma follows from Lemma 3.2.1 and definitions.

Let g.c.d.$(a, a') = 1$ and let $g_1, \ldots, g_\nu$ and $g'_1, \ldots, g'_\mu$ be sets of representatives for the left cosets of $\mathsf{M}_n(\pm a)$ and $\mathsf{M}_n(\pm a')$ respectively modulo Λ. Each of the products $g_i g'_j$ is clearly contained in $\mathsf{M}_n(\pm aa')$. Let us assume that two of the products are in the same left coset modulo Λ, say $\lambda g_i g'_j = g_k g'_l$ with $\lambda \in \Lambda$. Set $h = g_k^{-1} \lambda g_i = g'_l (g'_j)^{-1}$. Then the matrices $ah = a g_k^{-1} \lambda g_i$ and $a'h = g'_l a' (g'_j)^{-1}$ are both integral. Hence, h is integral, for g.c.d.$(a, a') = 1$. Also $\det h = \pm 1$, and so $h \in \Lambda$. It follows that $g'_l = h g'_j \in \Lambda g'_j$, hence $l = j$. Then we have $\lambda g_i = g_k$, hence $k = i$. We have proved that the products $g_i g'_j$ belong to different left cosets of $\mathsf{M}_n(\pm aa')$ modulo Λ. Let now Λg with $g \in \mathsf{M}_n(\pm aa')$ be a left coset of $\mathsf{M}_n(\pm aa')$ modulo Λ. Then it follows from Lemma 3.2.2 that g can be written in the form $g = vw$ with $v \in \mathsf{M}_n(\pm a)$ and $w \in \mathsf{M}_n(\pm a')$. The matrix w lies in a left coset $\Lambda g'_j$, i.e. $w = \lambda_1 g'_j$ with $\lambda_1 \in \Lambda$. Now the matrix $v\lambda_1$ lies in a left coset λg_i, i.e. $v\lambda_1 = \lambda g_i$ with $\lambda \in \Lambda$. Then $g = \lambda g_i g'_j$, and therefore the left coset Λg contains the product $g_i g'_j$, (3.2.14) follows from the above considerations and definitions. □

The following lemma is useful, when one wants to fix representatives for the left classes of G^n modulo Λ. An integral matrix $C = (c_{ij}) \in \mathsf{G}^n$ will be called *reduced* if it satisfies

$$0 \le c_{ij} < c_{jj}, \quad \text{and} \quad c_{ji} = 0 \quad \text{for } 1 \le i < j \le n. \tag{3.2.15}$$

Lemma 3.2.7. *Each left coset $\Lambda^n g$ with an integral $g \in \mathsf{G}^n$ contains unique reduced representative.*

Proof. We prove the existence of a reduced representative by induction on n. If $n = 1$, it is clear. Let $n > 1$, and the statement is already proved for $n - 1$. Applying Lemma 1.3.8 to the first column of g, we conclude that Λg contains a representative g' of the form $\begin{pmatrix} c_{11} & * \\ 0 & g_1 \end{pmatrix}$ with $c_{11} > 0$ and an integral g_1 of G^{n-1}. By the inductive hypothesis we can write $g_1 = \varepsilon g_0$, where $\varepsilon \in \Lambda^{n-1}$ and g_0 is reduced. Then it is easy to see that there is a row $\mathsf{l} \in \mathsf{M}_{1,n-1}$ such that the matrix

$$\begin{pmatrix} 1 & \mathsf{l} \\ 0 & E_{n-1} \end{pmatrix} \begin{pmatrix} 1 & 0 \\ 0 & \varepsilon^{-1} \end{pmatrix} g' \in \Lambda g$$

is reduced. The uniqueness trivially follows by contradiction. □

The study of the "global" Hecke ring H^n can be reduced to that of its "local" subrings H_p^n, where p runs over all prime numbers. For a prime number p we set

$$\mathsf{G}_p = \mathsf{G}_p^n = \mathsf{GL}_n(\mathbb{Z}[p^{-1}]), \tag{3.2.16}$$

where

$$\mathbb{Z}[p^{-1}] = \{a/p^b; a, b \in \mathbb{Z}\}. \tag{3.2.17}$$

Since $\Lambda^n \subset \mathsf{G}_p^n \subset \mathsf{G}^n$, we can define the Hecke ring

$$\mathsf{H}_p = \mathsf{H}_p^n = \mathsf{D}_{\mathbb{Q}}(\Lambda^n, \mathsf{G}_p^n) = \mathsf{D}(\Lambda^n, \mathsf{G}_p^n) \otimes_{\mathbb{Z}} \mathbb{Q}, \tag{3.2.18}$$

which can be considered as a subring of H^n. The subrings $\mathsf{H}_p \subset \mathsf{H}$ will be called *local Hecke ring of* G.

Theorem 3.2.8. *The ring* H^n *is generated by the local subrings* H_p^n *for all prime* p.

Proof. For a rational $r \neq 0$ we let $p^{v_p(r)}$ denote the highest power of prime p which devides r. If $g \in \mathsf{G}^n$ and p is a prime, we define the matrix $\operatorname{ed}_p(g)$ of elementary p-divisors of g by

$$\operatorname{ed}_p(g) = \operatorname{diag}(p^{v_p(d_1(p))}, \ldots, p^{v_p(d_n(g))}), \tag{3.2.19}$$

where $d_1(g), \ldots, d_n(g)$ are the elementary divisors of g. The numbers $p^{v_p(d_i(g))}$ will be called *the elementary p-divisors of* g. It is clear that, for a fixed g, we have $\operatorname{ed}_p(g) = E_n$ for all but a finite number of p, and

$$\prod_{p \in \mathbb{P}} \operatorname{ed}_p(g) = \operatorname{ed}(g) \qquad (g \in \mathsf{G}) \tag{3.2.20}$$

where $\mathbb{P}$ is the set of all prime numbers. Since $g \in \Lambda \operatorname{ed}(g)\Lambda$, it follows that each $g \in \mathsf{G}$ can be written in the form

$$g = \prod_p g_p, \qquad \text{where } \operatorname{ed}(g_p) = \operatorname{ed}_p(g). \tag{3.2.21}$$

(Note $g_p = E_n$ for all but a finite number of p.) By Proposition 3.2.5, we then obtain the decomposition

$$(g)_\Lambda = \prod_p (g_p)_\Lambda.$$

Since H consists of finite linear combinations of elements $(g)_\Lambda$ with $g \in G$, it follows that each element of H is a finite sum of finite products of elements $(g_p)_\Lambda \in \mathsf{H}_p$. □

Exercise 3.2.9. *For* $g, g' \in \mathsf{G}^n$, *let us assume that* $\text{g.c.d.}(d_n(g)/d_1(g), d_n(g')/d_1(g')) = 1$. *Prove then the following relations*

$$d_i(gg') = d_i(g)d_i(g') \qquad \text{for } 1 \leq i \leq n.$$

Exercise 3.2.10. *Show that for* $d \in \mathbb{N}$ *one can take*

$$\Lambda^n \backslash \mathsf{M}_n(\pm a) = \{C \in \mathsf{M}_n;\ C \textit{ is reduced}, \det C = a\}.$$

Use this to show that the zeta-function of H^n *defined by*

$$Z(s, v) = \sum_{a=1}^{\infty} \frac{v(t(a))}{a^s},$$

where $v\colon \mathsf{H}^n \to \mathbb{Q}$ *is the homomorphism of Exercise* 3.1.12, *and* $\operatorname{Re} s > n$, *is equal to* $\zeta(s)\zeta(s-1)\cdots\zeta(s-n+1)$, $\zeta(s)$ *being the Riemann zeta-function.*

Exercise 3.2.11. *Show the following identities for the formal Dirichlet series with coefficients in* H^n:

$$Z(s) = \sum_{a=1}^{\infty} \frac{t(a)}{a^s} = \prod_{p \in \mathbb{P}} Z_p(p^{-s}),$$

where

$$Z_p(v) = \sum_{\delta=0}^{\infty} t(p^\delta) v^\delta;$$

$$Z(s_1, \dots, s_n) = \sum_{d_1 | d_2 \cdots | d_n, d_i \in \mathbb{N}} d_1^{-s_1} \cdots d_n^{-s_n} (\operatorname{diag}(d_1, \dots, d_n))_\Lambda = \prod_{p \in \mathbb{P}} Z_p(p^{-s_1}, \dots, p^{-s_n}),$$

where

$$Z_p(v_1, \dots, v_n) = \sum_{0 \le \delta_1 \le \cdots \le \delta_n} (\operatorname{diag}(p^{\delta_1}, \dots, p^{\delta_n}))_\Lambda v_1^{\delta_1} \cdots v_n^{\delta_n}.$$

Exercise 3.2.12. *Let* $n = 2$. *Prove the relations*

$$t(p)t(p^\delta) = t(p^{\delta+1}) + p(pE_2)_\Lambda t(p^{\delta-1}),$$

where $t(a) = t_2(a)$, $p \in \mathbb{P}$, *and* $\delta \in \mathbb{N}$. *Use the relations to prove the following formal identities of formal power series over* H_p^2:

$$Z_p(v) = \sum_{\delta=0}^{\infty} t(p^\delta) v^\delta = (1 - t(p)v + p(pE_2)v^2)^{-1};$$

$$Z_p(v_1, v_2) = \sum_{0 \le \delta_1 \le \delta_2} (\operatorname{diag}(p^{\delta_1}, p^{\delta_2})) v_1^{\delta_1} v_2^{\delta_2} = (1 - (pE_2)v_2^2)(1 - (pE_2)v_1 v_2)^{-1}(1 - t(p)v_2 + p(pE_2)v_2^2)^{-1}.$$

Exercise 3.2.13. *Show that the elements* $t(a) = t_2(a) \in \mathsf{H}^2$ *satisfy*

$$t(a)t(b) = \sum_{d | \text{g.c.d.}(a,b)} d \cdot (dE_2) \cdot t(ab/d^2). \qquad (a, b \in \mathbb{N})$$

Exercise 3.2.14. *Show that* G_p^n *coincides with the set of all* $g \in \mathsf{G}^n$ *satisfying* $\operatorname{ed}(g) = \operatorname{ed}_p(g)$.

3.2.2. Local Rings

Here we shall study the structure of local Hecke rings H_p^n defined by (3.2.18). We first note that the study of H_p^n can be reduced to that of *the integral subring*

$$\underline{\mathsf{H}}_p^n = \mathsf{D}_{\mathbb{Q}}(\Lambda^n, \mathsf{G}_p^n \cap \mathsf{M}_n) \qquad (3.2.22)$$

consisting of linear combinations of elements $(g)_\Lambda$ with integral matrices $g \in \mathsf{G}_p$.

Lemma 3.2.15. *The element*

$$(p) = (p)_n = (pE_n)_\Lambda \in \mathsf{H}_p^n \qquad (3.2.23)$$

is invertible in H_p^n, *and* $(p)^{-1} = (p^{-1}E_n)_\Lambda$.

The element $(p)^{-1}$ *generates the ring* H_p^n *over the subring* $\underline{\mathsf{H}}_p^n$: $\mathsf{H}_p^n = \underline{\mathsf{H}}_p^n[(p)^{-1}]$.

Proof. The lemma follows from Lemma 3.2.4 and definitions. □

We shall write

$$(\mathrm{diag}(p^{\delta_1}, \dots, p^{\delta_n}))_\Lambda = t(p^{\delta_1}, \dots, p^{\delta_n}).$$

Then the ring $\underline{\mathsf{H}}_p^n$ consists of linear combinations of elements $t(p^{\delta_1}, \dots, p^{\delta_n})$ with $0 \le \delta_1 \le \cdots \le \delta_n$. An element of this form will be called *primitive* if $\delta_1 = 0$, and *imprimitive* otherwise. An element of $\underline{\mathsf{H}}_p^n$ will be called *primitive* (resp. *imprimitive*) if it is a linear combination of primitive (resp. imprimitive) elements $t(p^{\delta_1}, \dots, p^{\delta_n})$. It is clear that any $t \in \underline{\mathsf{H}}_p^n$ has unique decomposition of the form $t = t^{pr} + t^{im}$, where t^{pr} is primitive, and t^{im} is imprimitive. It follows from Lemma 3.2.4 that the set I *of all imprimitive elements of* $\underline{\mathsf{H}}_p$ is the principal ideal generated by (p): $\mathsf{I} = (p)\underline{\mathsf{H}}_p$.

Lemma 3.2.16. *Let* $\psi: \underline{\mathsf{H}}_p^n \to \underline{\mathsf{H}}_p^{n-1}$ *be the* $\mathbb{Q}$*-linear mapping satisfying*

$$\psi(t(p^{\delta_1}, \dots, p^{\delta_n})) = \begin{cases} t(p^{\delta_2}, \dots, p^{\delta_n}), & \text{if } 0 = \delta_1 \le \delta_2 \le \cdots \le \delta_n \\ 0, & \text{if } 0 < \delta_1 \le \delta_2 \le \cdots \le \delta_n. \end{cases}$$

Then ψ *is a ring homomorphism of* $\underline{\mathsf{H}}_p^n$ *onto* $\underline{\mathsf{H}}_p^{n-1}$. *The kernel of* ψ *is the ideal* I *of all imprimitive elements of* $\underline{\mathsf{H}}_p^n$.

Proof. It follows from Lemmas 3.1.5 and 3.2.2 that the mapping gives an isomorphism of the linear space of all imprimitive elements of $\underline{\mathsf{H}}_p^n$ onto the space $\underline{\mathsf{H}}_p^{n-1}$. It is clear that the kernel of ψ coincides with I. Thus, we only have to show that ψ is a ring homomorphism. To prove this, it suffices to check the relations

$$\psi(t(1, p^{\delta_2}, \dots, p^{\delta_n})t(1, p^{\delta_2'}, \dots, p^{\delta_n'})) = t(p^{\delta_2}, \dots, p^{\delta_n})t(p^{\delta_2'}, \dots, p^{\delta_n'}), \qquad (3.2.24)$$

where $0 \le \delta_2 \le \cdots \le \delta_n$, $0 \le \delta_2' \le \cdots \le \delta_n'$. We set $g_0 = \mathrm{diag}(p^{\delta_2}, \dots, p^{\delta_n})$, $g_0' =$

$\operatorname{diag}(p^{\delta_2'},\ldots,p^{\delta_n'})$, $g = \begin{pmatrix} 1 & 0 \\ 0 & g_0 \end{pmatrix}$ and $g' = \begin{pmatrix} 1 & 0 \\ 0 & g_0' \end{pmatrix}$. By Lemma 3.1.5, we have

$$(g)(g') = \sum_{\Lambda h \Lambda \subset \Lambda g \Lambda g' \Lambda} c(g, g'; h) \cdot (h).$$

It follows from Lemma 3.1.5 and (3.2.5) that each of the elements (h) occurring on the right is of the form $t(p^{\gamma_1},\ldots,p^{\gamma_n})$, where $0 \le \gamma_1 \le \cdots \le \gamma_n$ and $\gamma_1 + \cdots + \gamma_n = \delta_2 + \cdots + \delta_n + \delta_2' + \cdots + \delta_n'$. Thus, by the definition of ψ, we see that

$$\psi((g)(g')) = \sum_{\substack{0 \le \gamma_2 \le \cdots \le \gamma_n, \\ \gamma_2 + \cdots + \gamma_n = \delta_2 + \cdots + \delta_n + \delta_2' + \cdots + \delta_n'.}} c(g, g'; \operatorname{diag}(1, p^{\gamma_2}, \ldots, p^{\delta_n})) t(p^{\gamma_2}, \ldots, p^{\gamma_n}).$$

Similarly, in the ring $\underline{\mathrm{H}}_p^{n-1}$ we have the relation

$$(g_0)(g_0') = \sum_{\substack{0 \le \gamma_2 \le \cdots \le \gamma_n, \\ \gamma_2 + \cdots + \gamma_n = \delta_2 + \cdots + \delta_n + \delta_2' + \cdots + \delta_n'.}} c(g_0, g_0'; \operatorname{diag}(p^{\gamma_2}, \ldots, p^{\gamma_n})) t(p^{\gamma_2}, \ldots, p^{\gamma_n}).$$

Thus, in order to prove (3.2.24), we only have to show that

$$c(g, g'; h) = c(g_0, g_0'; h_0), \tag{3.2.25}$$

if $h = \begin{pmatrix} 1 & 0 \\ 0 & h_0 \end{pmatrix}$ with $h_0 = \operatorname{diag}(p^{\gamma_2},\ldots,p^{\gamma_n})$. Since $h' = \begin{pmatrix} h_0 & 0 \\ 0 & 1 \end{pmatrix} \in \Lambda h \Lambda$, we may replace h by h'. By Lemma 3.1.5 and Lemma 3.2.7, the coefficient $c(g, g'; h')$ is equal to the number of pairs d, d' of reduced matrices satisfying $d \in \Lambda g \Lambda$, $d' \in \Lambda g' \Lambda$, and $dd' = \lambda h'$ with $\lambda \in \Lambda$. Since $\lambda = dd'(h')^{-1}$, the matrix λ must be upper triangular; in particular, it can be written in the form $\lambda = \begin{pmatrix} \lambda_0 & \mathbf{l} \\ 0 & \lambda_{nn} \end{pmatrix}$ with $\lambda_0 \in \Lambda^{n-1}$ and $\lambda_{nn} = \pm 1$. We can write $d = \begin{pmatrix} d_0 & \mathbf{v} \\ 0 & d_{nn} \end{pmatrix}$, $d' = \begin{pmatrix} d_0' & \mathbf{v}' \\ 0 & d_{nn}' \end{pmatrix}$, where d_0 and d_0' are reduced matrices of order $n-1$, $d_{nn} > 0$, $d_{nn}' > 0$, and the entries of $\mathbf{v}$ (resp. $\mathbf{v}'$) are non-negative integers smaller than d_{nn} (resp. d_{nn}'). Then the relation $dd' = \lambda h'$ implies that $d_{nn} d_{nn}' = \lambda_{nn}$ and $d_0 d_0' = \lambda_0 h_0$. By the first relation, we have $d_{nn} = d_{nn}' = 1$, hence $\mathbf{v} = \mathbf{v}' = \mathbf{0}$. Thus, $d = \begin{pmatrix} d_0 & 0 \\ 0 & 1 \end{pmatrix}$, $d' = \begin{pmatrix} d_0' & 0 \\ 0 & 1 \end{pmatrix}$, where d_0 and d_0' are reduced matrices of order $n-1$ satisfying $d_0 d_0' = \lambda_0 h_0$ with $\lambda_0 \in \Lambda^{n-1}$. It is clear that $d_0 \in \Lambda^{n-1} g_0 \Lambda^{n-1}$ and $d_0' \in \Lambda^{n-1} g_0' \Lambda^{n-1}$. Conversely, arbitrary matrices d, d' satisfying these conditions are reduced, belong to the double cosets $\Lambda g \Lambda$, $\Lambda g' \Lambda$ respectively, and satisfy $dd' = \lambda h'$ with $\lambda \in \Lambda$. Thus, $c(g, g'; h') = c(g_0, g_0'; h_0)$. □

Now we can prove

Theorem 3.2.17. *Let $n \in \mathbb{N}$, and let p be a prime number. Then the following assertions hold:*

(1) *The ring $\underline{\mathrm{H}}_p^n$ is generated over $\mathbb{Q}$ by the elements*

$$\pi_i(p) = \pi_i^n(p) = \left(\begin{pmatrix} E_{n-i} & 0 \\ 0 & pE_i \end{pmatrix}\right)_\Lambda, \quad \textit{with } 1 \le i \le n; \tag{3.2.26}$$

(2) *The ring* H_p^n *is generated over* $\mathbb{Q}$ *by the elements* $\pi_1(p), \ldots, \pi_n(p)$, *and* $\pi_n(p)^{-1} = (p)_n^{-1}$;

(3) *The elements* $\pi_1(p), \ldots, \pi_n(p)$ *are algebraically independent over* $\mathbb{Q}$.

Remark. We identify $\mathbb{Q}$ with the subring $\mathbb{Q}\cdot(E_n)_\Lambda \subset \underline{\mathsf{H}}_p^n \subset \mathsf{H}_p^n$.

Proof. To prove (1), it suffices to show that each element $t(p^{\delta_1}, \ldots, p^{\delta_n})$ with $0 \le \delta_1 \le \cdots \le \delta_n$ is a polynomial in $\pi_1(p), \ldots, \pi_n(p)$ with coefficients in $\mathbb{Q}$. We shall prove this by induction on n, and, for a fixed $n > 1$, by induction on $N = \delta_1 + \cdots + \delta_n$. For $n = 1$ the assertion is clear, since $t(p^\delta) = t(p)^\delta = \pi_1^1(p)^\delta$. Now assume that $n > 1$, and the assertion is already proved for $n - 1$. If $N = \delta_1 + \cdots + \delta_n = 1$, then $t(p^{\delta_1}, \ldots, p^{\delta_n}) = \pi_1(p)$. Let us consider an element $t = t(p^{\delta_1}, \ldots, p^{\delta_n})$ with $0 \le \delta_1 \le \cdots \le \delta_n$ and $\delta_1 + \cdots + \delta_n = N > 1$, assuming that each element $t(p^{\nu_1}, \ldots, p^{\nu_n})$ with $\nu_1 + \cdots + \nu_n < N$ is known to be a polynomial in $\pi_1(p), \ldots, \pi_n(p)$. If $\delta_1 \ge 1$, then by Lemma 3.2.4, we have

$$t = \pi_n(p)^{\delta_1} t(1, p^{\delta_2 - \delta_1}, \ldots, p^{\delta_n - \delta_1}) = \pi_n(p)^{\delta_1} t'.$$

Since $(\delta_2 - \delta_1) + \cdots + (\delta_n - \delta_1) < N$, it follows that t' is a polynomial in $\pi_1(p), \ldots, \pi_n(p)$. Hence, the same is true for t. Now let $\delta_1 = 0$. By the first inductive hypothesis, the element

$$\psi(t(1, p^{\delta_2}, \ldots, p^{\delta_n})) = t(p^{\delta_2}, \ldots, p^{\delta_n}) \in \underline{\mathsf{H}}_p^{n-1},$$

where ψ is the mapping defined in Lemma 3.2.16, is a polynomial in $\pi_1^{n-1}(p), \ldots, \pi_{n-1}^{n-1}(p)$, say

$$t(p^{\delta_2}, \ldots, p^{\delta_n}) = F(\pi_1^{n-1}(p), \ldots, \pi_{n-1}^{n-1}(p))$$

with

$$F(x_1, \ldots, x_{n-1}) = \sum_{(\alpha_1, \ldots, \alpha_{n-1})} a_{\alpha_1, \ldots, \alpha_{n-1}} x_1^{\alpha_1} \cdots x_{n-1}^{\alpha_{n-1}}.$$

Since any element (h) occuring in the decomposition of a product $\pi_1^{n-1}(p)^{\alpha_1} \cdots \pi_{n-1}^{n-1}(p)^{\alpha_{n-1}}$ satisfies $|\det h| = p^{\alpha_1 + 2\alpha_2 + \cdots + (n-1)\alpha_{n-1}}$, we can assume that $a_{\alpha_1, \ldots, \alpha_{n-1}} \ne 0$ only if $\alpha_1 + \cdots + (n-1)\alpha_{n-1} = \delta_2 + \cdots + \delta_n = N$. Since $\psi(\pi_i^n(p)) = \pi_i^{n-1}(p) (1 \le i \le n-1)$ it follows that the element

$$t_1 = t(1, p^{\delta_2}, \ldots, p^{\delta_n}) - \sum_{\alpha_1 + 2\alpha_2 + \cdots + (n-1)\alpha_{n-1} = N} a_{\alpha_1, \ldots, \alpha_{n-1}} \pi_1^n(p)^{\alpha_1} \cdots \pi_{n-1}^n(p)^{\alpha_{n-1}}$$

satisfies $\psi(t_1) = 0$. Hence, t_1 is imprimitive. On the other hand, t_1 is a linear combination of elements $t(p^{\nu_1}, \ldots, p^{\nu_n})$ with $\nu_1 + \cdots + \nu_n = N$. It follows that $t_1 = \pi_n(p) t_1'$, where t_1' is a linear combination of elements $t(p^{\nu_1}, \ldots, p^{\nu_n})$ with $\nu_1 + \cdots + \nu_n < N$. Then the second inductive hypothesis implies that t_1' is a polynomial in $\pi_1^n(p), \ldots, \pi_n^n(p)$. Hence, the t is. The first assertion is proved.

The second assertion follows from the first one and Lemma 3.2.15.

The third assertion will be proved by induction on n. If $n = 1$, it is clear, since the elements $\pi_1^1(p)^\delta = (p^\delta)$ with $\delta = 0, 1, 2, \ldots$ correspond to the pairwise distinct double cosets modulo $\Lambda^1 = \{\pm 1\}$ and therefore are linearly independent. We now assume that $n > 1$ and the assertion is already proved for $n - 1$. Suppose that $\pi_1^n(p), \ldots, \pi_n^n(p)$ are algebraically dependent over $\mathbb{Q}$ and let $F(x_1, \ldots, x_n)$ be a polynomial of minimal total degree such that $F(\pi_1^n(p), \ldots, \pi_n^n(p)) = 0$. Then we have

$$\psi(F(\pi_1^n(p), \ldots, \pi_n^n(p))) = F(\psi(\pi_1^n(p)), \ldots, \psi(\pi_n^n(p)))$$
$$= F(\pi_1^{n-1}(p), \ldots, \pi_{n-1}^{n-1}(p), 0) = 0.$$

By the inductive hypothesis, it follows that $F(x_1, \ldots, x_n) = x_n F(x_1, \ldots, x_n)$, where F_1 is again a polynomial. By Lemma 3.2.15, the element $\pi_n^n(p) = (p)_n$ is not a zero divisor. So the equality $F(\pi_1^n(p), \ldots, \pi_n^n(p)) = \pi_n^n(p) F_1(\pi_1^n(p), \ldots, \pi_n^n(p)) = 0$ implies that $F_1(\pi_1^n(p), \ldots, \pi_n^n(p)) = 0$, a contradiction. □

We close this subsection with several technical results on the generators $\pi_i(p)$ and their products, which will be used later.

Lemma 3.2.18. *For $n \in \mathbb{N}$ and $p \in \mathbb{P}$ the following assertions hold*

(1) *For $0 \le i \le n$, the set*

$$\{C = (c_{\alpha\beta}) \in \mathsf{M}_n;\ C \text{ is reduced}, \det C = p^i$$
$$c_{\alpha\alpha} = 1 \text{ or } p,\ c_{\alpha\beta} = 0, \text{ if } 1 \le \alpha < \beta \le n \text{ and } c_{\alpha\alpha} = p\} \tag{3.2.27}$$

is a set of representatives for the left cosets of $\Lambda D_i \Lambda$ modulo $\Lambda = \Lambda^n$, the number of these left cosets is given by

$$\nu_\Lambda(D_i) = \frac{\varphi_n(p)}{\varphi_i(p)\varphi_{n-i}(p)} \quad \text{with } D_i = D_i^n(p) = \begin{pmatrix} E_{n-i} & 0 \\ 0 & pE_i \end{pmatrix}, \tag{3.2.28}$$

where

$$\varphi_r(x) = (x-1)(x^2-1)\cdots(x^r-1) \quad \text{for } r \in \mathbb{N}, \quad \text{and} \quad \varphi_0(x) = 1; \tag{3.2.29}$$

(2) *For $1 \le i, j \le n$, the following formula holds*

$$\pi_i \pi_j = \sum_{\substack{0 \le a \le n-j,\, 0 \le b \le j, \\ a+b=i}} \frac{\varphi_{a+j-b}(p)}{\varphi_a(p)\varphi_{j-b}(p)} \pi_{a+j-b,b} \tag{3.2.30}$$

where $\pi_\alpha = \pi_\alpha^n(p)$, and

$$\pi_{\alpha\beta} = \pi_{\alpha\beta}^n(p) = (D_{\alpha\beta})_\Lambda \quad \text{with } D_{\alpha\beta} = D_{\alpha\beta}^n(p) = \begin{pmatrix} E_{n-\alpha-\beta} & 0 & 0 \\ 0 & pE_\alpha & 0 \\ 0 & 0 & p^2 E_\beta \end{pmatrix}. \tag{3.2.31}$$

The number of the left cosets of $\Lambda D_{\alpha\beta} \Lambda$ modulo Λ is given by

$$\nu_\Lambda(D_{\alpha\beta}) = p^{\beta(n-\alpha-\beta)} \frac{\varphi_n(p)}{\varphi_{n-\alpha-\beta}(p)\varphi_\alpha(p)\varphi_\beta(p)}. \tag{3.2.32}$$

Proof. Each matrix of the set (3.2.27) can be written in the form $\operatorname{diag}(p^{\delta_1}, \dots, p^{\delta_n})C'$, where $\delta_1 + \dots + \delta_n = i$, $\delta_\alpha = 0, 1$, and $C' \in \Lambda$. So it is contained in $\Lambda D_i \Lambda$. By Lemma 3.2.7, all of these matrices are in distinct left cosets modulo Λ. Now let $C = (c_{\alpha\beta})$ be a reduced matrix in $\Lambda D_i \Lambda$. Then the diagonal entries $c_{\alpha\alpha}$ are positive integers whose product is p^i. It follows that $c_{\alpha\alpha} = p^{\delta_\alpha}$ with $\delta_\alpha \geq 0$ and $\delta_1 + \dots + \delta_n = i$. Since the matrix pC^{-1} is integral, we conclude that $\delta_\alpha = 0$ or 1. Let us assume that $c_{\alpha\beta} \neq 0$, where $1 \leq \alpha < \beta \leq n$. Then we have $\delta_\beta = 1$ and $c_{\alpha\beta} \not\equiv 0 \pmod p$. We let $\gamma_1, \dots, \gamma_{n-i}$ denote the indexes γ satisfying $\delta_\gamma = 0$. If $\delta_\alpha = 1$, then the columns of c with indexes $\gamma_1, \dots, \gamma_{n-i}, \beta$ are linearly independent modulo p, while the rank of C modulo p is clearly $n - i$. So $\delta_\alpha = 0$ and C belongs to the set (3.2.27).

It is easy to see that the number of matrices in the set (3.2.27) with a fixed diagonal $(p^{\delta_1}, \dots, p^{\delta_n})$, where $\delta_\alpha = 1$ for $\alpha = \alpha_1, \alpha_2, \dots, \alpha_i$ with $1 \leq \alpha_1 < \alpha_2 < \dots < \alpha_i \leq n$, equals $p^{\alpha_1 - 1} \cdot p^{\alpha_2 - 2} \cdots p^{\alpha_i - i}$. Hence, the number of elements of the set (3.2.27) equals

$$\sum_{1 \leq \alpha_1 < \dots < \alpha_i \leq n} p^{\alpha_1 - 1} \cdots p^{\alpha_i - i} = p^{-\langle i \rangle} \sum_{1 \leq \alpha_1 < \dots < \alpha_i \leq n} p^{\alpha_1 + \dots + \alpha_i}.$$

Thus, to prove (3.2.28), it suffices to check the following identity in the variable x:

$$\sum_{1 \leq \alpha_1 < \dots < \alpha_i \leq n} x^{\alpha_1 + \dots + \alpha_i} = x^{\langle i \rangle} \frac{\varphi_n(x)}{\varphi_i(x)\varphi_{n-i}(x)} \qquad (1 \leq i \leq n), \tag{3.2.33}$$

where $\varphi_r(x)$ are the polynomials (3.2.29). This identity follows by comparing the coefficients of t^i in both sides of the following identity in two variables x, t:

$$\prod_{\alpha=1}^{n} (1 + tx^\alpha) = \sum_{i=0}^{n} t^i \frac{x^{\langle i \rangle} \varphi_n(x)}{\varphi_i(x)\varphi_{n-i}(x)}, \tag{3.2.34}$$

whose correctness is easily checked by induction on n in the same way as the standard proof of the binomial formula.

To prove (3.2.30), we shall use induction on n. For $n = 1$ we have to prove the relation $\pi_1^1 \cdot \pi_1^1 = \pi_{01}^1$, which is clear. Now assume that $n > 1$, and (3.2.30) is already proved for $n - 1$. By Lemma 3.1.5 and the first part of this lemma, we can write

$$\pi_i \pi_j = \sum_{g \in \Lambda \backslash \mathrm{G}_p / \Lambda} c_{ij}(g)(g)_\Lambda,$$

where

$$c_{ij}(g) = c'(D_i, D_j; g)\nu(D_j)\nu(g)^{-1},$$

$c'(D_i, D_j; g)$ being the number of matrices C in the set (3.2.27) satisfying $CD_j \in \Lambda g \Lambda$. The number c' can be easily computed. If the diagonal entries of C are $p^{\delta_1}, \dots, p^{\delta_n}$, then $C = \operatorname{diag}(p^{\delta_1}, \dots, p^{\delta_n})C'$, where C' is an upper triangular matrix in

Λ. It follows that

$$CD_j = \operatorname{diag}(p^{\delta_1}, \dots, p^{\delta_n}) D_j D_j^{-1} C' D_j \in \Lambda \operatorname{diag}(p^{\delta_1}, \dots, p^{\delta_n}) D_j \Lambda,$$

since clearly $D_j^{-1} C' D_j \subset \Lambda$. Now let $\alpha_1 < \cdots < \alpha_i$ be the indexes α with $\delta_\alpha = 1$. If a is defined by $\alpha_a \le n - j < \alpha_{a+1}$, and $b = i - a$, it is clear that

$$\operatorname{diag}(p^{\delta_1}, \dots, p^{\delta_n}) D_j = \Lambda D_{a+j-b,b} \Lambda,$$

where $D_{\alpha\beta} = D^n_{\alpha\beta}(p)$. As it was noted above, the number of C with fixed $\alpha_1, \dots, \alpha_i$ is $p^{\alpha_1 - 1} \cdots p^{\alpha_i - i}$. Hence, the number of C in (3.2.27) with $CD_j \in \Lambda D_{a+j-b,b} \Lambda$ for fixed a, b satisfying $0 \le a \le n - j$, $0 \le b \le j$, and $a + b = i$ is equal to

$$\begin{aligned}
&\sum_{1 \le \alpha_1 < \cdots < \alpha_a \le n-i < \alpha_{a+1} < \cdots < \alpha_i \le n} p^{\alpha_1 + \cdots + \alpha_i - \langle i \rangle} \\
&= p^{-\langle i \rangle} \sum_{\substack{1 \le \alpha_1 < \cdots < \alpha_a \le n-j, \\ 1 \le \beta_1 < \cdots < \beta_b \le j}} p^{\alpha_1 + \cdots + \alpha_a + (\beta_1 + (n-j)) + \cdots + (\beta_b + (n-j))} \\
&= p^{b(n-j) - \langle i \rangle} \frac{p^{\langle a \rangle} \varphi_{n-j}(p)}{\varphi_a(p) \varphi_{n-j-a}(p)} \cdot \frac{p^{\langle b \rangle} \varphi_j(b)}{\varphi_b(p) \varphi_{j-b}(p)} \\
&= p^{b(n-j-a)} \frac{\varphi_{n-j}(p) \varphi_j(p)}{\varphi_a(p) \varphi_{n-j-a}(p) \varphi_b(p) \varphi_{j-b}(p)}
\end{aligned}$$

(we have also used (3.2.33)). The above considerations and (3.2.28) imply that

$$\pi_i \pi_j = \sum_{\substack{0 \le a \le n-j, 0 \le b \le j, \\ a+b=i}} c_{ij}(D_{a+j-b,b}) \pi_{a+j-b,b}, \tag{3.2.35}$$

where

$$c_{ij}(D_{a+j-b,b}) = p^{b(n-j-a)} \frac{\varphi_n(p)}{\varphi_a(p) \varphi_{n-j-a}(p) \varphi_b(p) \varphi_{j-b}(p)} \nu(D_{a+j-b,b})^{-1}.$$

If either $i = n$ or $j = n$, then the decomposition (3.2.30) follows from definitions and Lemma 3.2.4. Thus, one can assume that $1 \le i, j < n$. Then applying the homomorphism ψ defined in Lemma 3.2.16 to the two sides of (3.2.35), we obtain

$$\pi_i^{n-1}(p) \pi_j^{n-1}(p) = \sum_{\substack{0 \le a \le n-1-j, 0 \le b \le j, \\ a+b=i}} c_{ij}(D_{a+j-b,b}) \pi^{n-1}_{a+j-b,b}(p).$$

On the other hand, by the inductive hypothesis, we have

$$\pi_i^{n-1}(p) \pi_j^{n-1}(p) = \sum_{\substack{0 \le a \le n-1-j, 0 \le b \le j, \\ a+b=i}} \frac{\varphi_{a+j-b}(p)}{\varphi_a(p) \varphi_{j-b}(p)} \pi^{n-1}_{a+j-b,b}(p).$$

Since the double cosets occurring in these decompositions are linearly independent, it follows that

$$c_{ij}(D_{a+j-b,b}) = \frac{\varphi_{a+j-b}(p)}{\varphi_a(p) \varphi_{j-b}(p)},$$

where $0 \le a \le n-1-j$, $0 \le b \le j$, and $a+b=i$. If $a=n-j$ the same formula follows from the original formula for $c_{ij}(D_{2n-i-j,i+j-n})$ and (3.2.28), for

$$\nu(D_{2n-i-j,i+j-n}) = \nu(D^n_{i+j-n}(p)) = \frac{\varphi_n(p)}{\varphi_{i+j-n}(p)\varphi_{2n-i-j}(p)}.$$

Formula (3.2.30) is proved. By comparing the expressions for $c_{ij}(D_{a+j-b,b})$ obtained above, we get

$$\nu(D_{a+j-b,b}) = p^{b(n-j-a)}\frac{\varphi_n(p)}{\varphi_{n-j-a}(p)\varphi_{a+j-b}(p)\varphi_b(p)},$$

which proves (3.2.32), if we set $a+j-b=\alpha$ and $b=\beta$. □

Exercise 3.2.19. *Let $\mathbb{Q}_p$ be the field of p-adic numbers, and $\mathbb{Z}_p$ the ring of p-adic integers. Then the group $\mathsf{G} = \mathsf{GL}_n(\mathbb{Q}_p)$ is a locally compact group in the p-adic topology, and $\Gamma = \mathsf{GL}_n(\mathbb{Z}_p)$ is a compact subgroup of G. For a subring $\mathsf{K} \subset \mathbb{C}$, we let D_K denote the K-module of all continuous functions $f\colon G \to \mathsf{K}$ with compact support, which satisfy $f(\gamma_1 g \gamma_2) = f(g)$ for any $\gamma_1, \gamma_2 \in \Gamma$ and $g \in \mathsf{G}$. We fix the Haar measure on G by $\mu(\Gamma) = 1$, and define the product $f * f_1$ of two functions $f, f_1 \in \mathsf{D}_\mathsf{K}$ by*

$$(f * f_1)(h) = \int_G f(hg^{-1}) f_1(g)\,d\mu(g) \qquad (h \in \mathsf{G}).$$

Show that the K-linear mapping of the Hecke ring $\mathsf{D}_\mathsf{K}(\Lambda^n, \mathsf{G}^n_p)$ in D_K defined by

$$(g)_\Lambda \to f_g(h) = \begin{cases} 1, & \text{if } h \in \Gamma g \Gamma \\ 0, & \text{otherwise} \end{cases}$$

is an isomorphism of the rings.

3.2.3. The Spherical Mapping

We have seen, that any element of a local Hecke ring of the general linear group can be written as a polynomial in finite number of generators. However, the problem of finding these polynomials proves to be rather difficult. To simplify it, certain mappings of the Hecke rings into rings of symmetric polynomials are defined, by analogy with the spherical functions in the theory of representations of topological groups.

As above, we assume that $n \in \mathbb{N}$ and p is a prime number. We first define a linear mapping

$$\omega = \omega^n_p\colon \mathsf{L}_\mathbb{Q}(\Lambda^n, \mathsf{G}^n_p) \to \mathbb{Q}[x_1^{\pm 1}, \dots, x_n^{\pm 1}]$$

of the module of left cosets of G^n_p modulo Λ^n over $\mathbb{Q}$ into the ring of polynomials in $x_1^{\pm 1}, \dots, x_n^{\pm 1}$ over $\mathbb{Q}$, where $x_1, \dots, x_n$ are algebraically independent over $\mathbb{Q}$.

It follows from Lemma 3.2.7 that each left coset Λg with $g \in \mathrm{G}_p$ contains a representative of the form

$$\begin{pmatrix} p^{\delta_1} & c_{12} & \cdots & c_{1n} \\ 0 & p^{\delta_2} & \cdots & c_{2n} \\ \cdots & \cdots & \cdots & \cdots \\ 0 & 0 & \cdots & p^{\delta_n} \end{pmatrix} \quad \text{with } \delta_1, \ldots, \delta_n \in \mathbb{Z}, \tag{3.2.36}$$

where the diagonal $(p^{\delta_1}, \ldots, p^{\delta_n})$ depends only on the left coset. We then set

$$\omega(\Lambda g) = \prod_{i=1}^{n} (x_i p^{-i})^{\delta_i},$$

and for a given element

$$t = \sum_j a_j(\Lambda g_j) \in \mathrm{L}_{\mathbb{Q}}(\Lambda^n, \mathrm{G}_p^n)$$

define

$$\omega(t) = \sum_j a_j \omega((\Lambda g_j)).$$

We call the mapping ω_p^n *the spherical mapping*. In order to describe the image of H_p^n under ω, we shall remind that an element of the field $\mathbb{Q}(x_1, \ldots, x_n)$ of rational functions in n variables $x_1, \ldots, x_n$ is called symmetric if it is invariant under all permutations of the variables.

Theorem 3.2.20. *The restriction of $\omega = \omega_p^n$ to the Hecke ring $\mathrm{H}_p^n = \mathrm{D}_{\mathbb{Q}}(\Lambda^n, \mathrm{G}_p^n) \subset \mathrm{L}_{\mathbb{Q}}(\Lambda^n, \mathrm{G}_p^n)$ is an isomorphism of H_p^n onto the subring $\mathbb{Q}[x_1^{\pm 1}, \ldots, x_n^{\pm 1}]_s$ of all symmetric elements in $\mathbb{Q}[x_1^{\pm 1}, \ldots, x_n^{\pm 1}]$. The image of the integral subring $\underline{\mathrm{H}}_p^n$ under ω is equal to the ring $\mathbb{Q}[x_1, \ldots, x_n]_s$ of all symmetric polynomials in $x_1, \ldots, x_n$ over $\mathbb{Q}$.*

We first prove a lemma.

Lemma 3.2.21. *The images of elements of the form (3.2.26) under $\omega = \omega_p^n$ are given by*

$$\omega(\pi_i^n(p)) = p^{-\langle i \rangle} s_i(x_1, \ldots, x_n) \qquad (1 \le i \le n),$$

where

$$s_i(x_1, \ldots, x_n) = \sum_{1 \le \alpha_1 < \cdots < \alpha_i \le n} x_{\alpha_1} \cdots x_{\alpha_i}$$

are the elementary symmetric functions.

Proof. Taking the set (3.2.27) as a set of representatives for the left cosets occurring in $\pi_i^n(p)$, we get

$$\omega(\pi_i) = \sum_{1 \le \alpha_1 < \cdots < \alpha_i \le n} p^{(\alpha_1 - 1) + (\alpha_2 - 2) + \cdots + (\alpha_i - i)} (x_{\alpha_1} p^{-\alpha_1}) \cdots (x_{\alpha_i} p^{-\alpha_i})$$
$$= p^{-\langle i \rangle} s_i(x_1, \ldots, x_n),$$

since the number of elements in (3.2.27) with $c_{\alpha_1, \alpha_1} = \cdots = c_{\alpha_i \alpha_i} = p$, where $1 \le \alpha_1 < \cdots < \alpha_i \le n$, is $p^{\alpha_1 - 1} \cdots p^{\alpha_i - i}$. □

Proof of Theorem 3.2.20. We first show that the restriction of ω to H_p is a ring homomorphism. Let

$$t = \sum_i a_i(\Lambda g_i) \quad \text{and} \quad t' = \sum_j b_j(\Lambda g_j')$$

be two elements of H_p. It follows from Lemma 3.2.7 that all representatives $g_i \in \Lambda g_i$ and $g_j' \in \Lambda g_j'$ can be taken in the form (3.2.36). Then all representatives $g_i g_j'$ in

$$tt' = \sum_{i,j} a_i b_j (\Lambda g_i g_j')$$

have the same form, and diagonal entries of matrices $g_i g_j'$ equal the products of the corresponding diagonal entries of g_i and g_j'. Hence, by the definition of ω, we have

$$\omega(tt') = \sum_{i,j} a_i b_j \omega((\Lambda g_i g_j'))$$
$$= \sum_{i,j} a_i b_j \omega((\Lambda g_i)) \omega((\Lambda g_j')) = \omega(t)\omega(t').$$

It is clear that the identity element $(E) = (E_n)_\Lambda$ of H_p satisfies $\omega((E)) = 1$. Theorem 3.2.17 then implies that the image $\omega(\underline{\mathsf{H}}_p)$ consists of all polynomials in $\omega(\pi_1(p)), \ldots, \omega(\pi_n(p))$ over $\mathbb{Q}$, and the image $\omega(\mathsf{H}_p)$ consists of all polynomials in $\omega(\pi_1(p)), \ldots, \omega(\pi_n(p))$ and $\omega(\pi_n(p)^{-1}) = \omega(\pi_n(p))^{-1}$ over $\mathbb{Q}$. Since every symmetric polynomial is a polynomial in the elementary symmetric functions, it follows from Lemma 3.2.21 that $\omega(\underline{\mathsf{H}}_p)$ coincides with the ring of all symmetric polynomials in n variables over $\mathbb{Q}$, and $\omega(\mathsf{H}_p)$ is given by

$$\omega(\mathsf{H}_p) = \omega(\underline{\mathsf{H}}_p)[(x_1 \cdots x_n)^{-1}] = \mathbb{Q}[x_1^{\pm 1}, \ldots, x_n^{\pm 1}]_s.$$

Finally, if $\omega(t) = 0$ for a $t \in \mathsf{H}_p$, then, by Lemma 3.2.15, we can write $t = \pi_n(p)^\delta t_1$ with $\delta \in \mathbb{Z}$ and $t_1 \in \underline{\mathsf{H}}_p$. Since $\omega(\pi_n(p)^{\pm 1}) \ne 0$, it follows that $\omega(t_1) = 0$. By Theorem 3.2.17, $t_1 = F(\pi_1(p), \ldots, \pi_n(p))$, where $F(x_1, \ldots, x_n)$ is a polynomial. In view of Lemma 3.2.21, the equality $\omega(t_1) = 0$ implies the identity

$$F(p^{-1}s_1(x_1, \ldots, x_n), \ldots, p^{-\langle n \rangle} s_n(x_1, \ldots, x_n)) = 0$$

which shows that $F(x_1, \ldots, x_n) = 0$, since $s_1, \ldots, s_n$ are algebraically independent. Thus, $t_1 = 0$ and $t = 0$. □

Theorem 3.2.20 reduces calculations in the ring H_p^n to calculations in the ring of symmetric polynomials. It also allows us to express an element $t \in \mathsf{H}_p^n$ in terms of the generators, if we can express the symmetric function $\omega(t)$ in terms of the elementary symmetric functions.

To illustrate the technique of passing back and forth between local Hecke rings and rings of symmetric functions, we shall find the sum of generating power series for elements of the form $t(p^\delta)$ with $p \in \mathbb{P}$. Note that there is no simple law of multiplication of these elements for $n > 2$ (the case $n = 1$ is trivial, for $n = 2$ see Exercise 3.2.12).

Proposition 3.2.22. *In the notation* (3.2.10) *and* (3.2.26) *the following identity of formal power series over* $\underline{\mathrm{H}}_p^n$ *holds:*

$$\sum_{\delta=0}^{\infty} t(p^\delta)v^\delta = \left(\sum_{i=0}^{n}(-1)^i p^{\langle i-1\rangle}\pi_i(p)v^i\right)^{-1}.$$

Proof. It follows from the definition of $t(p^\delta)$ and Lemma 3.2.7 that

$$t(p^\delta) = \sum_{\substack{\delta_i \geq 0,\, \delta_1 + \cdots + \delta_n = \delta,\\ 0 \leq c_{ij} < p^{\delta_j}}} \left(\Lambda \begin{pmatrix} p^{\delta_1} & c_{12} & \cdots & c_{1n} \\ 0 & p^{\delta_2} & \cdots & c_{2n} \\ \cdots & \cdots & \cdots & \cdots \\ 0 & 0 & \cdots & p^{\delta_n} \end{pmatrix}\right).$$

Hence, by the definition of $\omega = \omega_p^n$, we have

$$\begin{aligned} \omega(t(p^\delta)) &= \sum_{\delta_i \geq 0,\, \delta_1 + \cdots + \delta_n = \delta} p^{\delta_2} \cdot p^{2\delta_3} \cdots p^{(n-1)\delta_n}(x_1 p^{-1})^{\delta_1} \cdots (x_n p^{-n})^{\delta_n} \\ &= \sum_{\delta_i \geq 0,\, \delta_1 + \cdots + \delta_n = \delta} (x_1 p^{-1})^{\delta_1} \cdots (x_n p^{-1})^{\delta_n}. \end{aligned} \tag{3.2.37}$$

Thus,

$$\begin{aligned} \sum_{\delta=0}^{\infty} \omega(t(p^\delta))v^\delta &= \sum_{\delta_1, \ldots, \delta_n = 0}^{\infty} (x_1 p^{-1})^{\delta_1} \cdots (x_n p^{-1})^{\delta_n} v^{\delta_1} \cdots v^{\delta_n} \\ &= \prod_{i=1}^{n}(1 - p^{-1}x_i v)^{-1} = \left(\sum_{i=0}^{n}(-1)^i p^{-i} s_i(x_1, \ldots, x_n)v^i\right)^{-1} \\ &= \left(\sum_{i=0}^{n}(-1)^i p^{\langle i-1\rangle}\omega(\pi_i(p))v^i\right)^{-1} \end{aligned} \tag{3.2.38}$$

(see Lemma 3.2.21). The last identity implies that each coefficient of the formal power series

$$\left(\sum_{\delta=0}^{\infty} t(p^\delta)v^\delta\right)\left(\sum_{i=0}^{n}(-1)^i p^{\langle i-1\rangle}\pi_i(p)v^i\right) - (E)_\Lambda$$

belongs to the kernel of ω and therefore equals zero. □

One can use the proposition to express the elements $t(p^\delta)$ in terms of the generators $\pi_i(p)$.

Finally, we prove a lemma describing the anti-automorphism j of H_p^n (see §3.1.4) in terms of the corresponding symmetric functions.

Lemma 3.2.23. *The diagram*

$$\begin{array}{ccc} \mathrm{H}_p^n & \xrightarrow{\ \omega\ } & \mathbb{Q}[x_1^{\pm 1}, \ldots, x_n^{\pm 1}] \\ {\scriptstyle \mathrm{j}}\downarrow & & \downarrow{\scriptstyle \mathrm{k}} \\ \mathrm{H}_p^n & \xrightarrow{\ \omega\ } & \mathbb{Q}[x_1^{\pm 1}, \ldots, x_n^{\pm 1}] \end{array}$$

where k *is the homomorphism of rings over* $\mathbb{Q}$ *defined by* $\mathrm{k}(x_\alpha) = p^{n+1}x_\alpha^{-1}$ *for* $1 \le \alpha \le n$, *is commutative.*

Proof. Since H_p^n is commutative, it follows that all of the mappings in the diagram are $\mathbb{Q}$-linear ring homomorphisms. Thus, according to Theorem 3.2.17 and Lemma 3.2.4, we only have to show that $\omega(\mathrm{j}(\pi_\alpha(p))) = \mathrm{k}(\omega(\pi_\alpha(p)))$ for $1 \le \alpha \le n$. By definitions and Lemma 3.2.21, we have

$$\begin{aligned} \omega(\mathrm{j}(\pi_\alpha(p))) &= \omega((D_\alpha^{-1})_\Lambda) = \omega((p^{-1}D_{n-\alpha})_\Lambda) \\ &= \omega(\pi_n(p)^{-1}\pi_{n-\alpha}(p)) = p^{\langle n\rangle}(x_1 \cdots x_n)^{-1}p^{-\langle n-\alpha\rangle}s_{n-\alpha}(x_1, \ldots, x_n) \\ &= p^{-\langle\alpha\rangle}s_\alpha(p^{n+1}x_1^{-1}, \ldots, p^{n+1}x_n^{-1}) = \mathrm{k}(\omega(\pi_\alpha(p))). \end{aligned}$$ □

Exercise 3.2.24. *Show that any non-zero* $\mathbb{Q}$*-linear homomorphism of* H_p^n *into* $\mathbb{C}$ *is of the form*

$$t \to \lambda(t) = \omega(t)|_{x_1=\alpha_1, \ldots, x_n=\alpha_n} \qquad (t \in \mathrm{H}_p^n)$$

where ω *is the spherical mapping and* $\alpha_1, \ldots, \alpha_n$ *are non-zero complex numbers which are determined by* λ *up to their order. The numbers* $\alpha_1, \ldots, \alpha_n$ *are called the parameters of* λ.

Exercise 3.2.25. *Let* $\lambda\colon \mathrm{H}^n \to \mathbb{C}$ *be a non-zero* $\mathbb{Q}$*-linear homomorphism. Show that the formal zeta-function of* H^n *with the "character* λ*", i.e. the formal Dirichlet series*

$$Z(s, \lambda) = \sum_{a=1}^{\infty} \frac{\lambda(t(a))}{a^s},$$

has an Euler product expansion of the form

$$Z(s, \lambda) = \prod_{p \in \mathbb{P}} Z_p(p^{-s}, \lambda_p),$$

where λ_p *is the restriction of* λ *to* H_p^n, *and where the local zeta-functions* $Z_p(v, \lambda_p)$ *are given by*

$$Z_p(v, \lambda_p) = \prod_{i=1}^{n} (1 - \alpha_i(p)p^{-1}v)^{-1},$$

$\alpha_1(p), \ldots, \alpha_n(p)$ *being the parameters of* λ_p.

Exercise 3.2.26. *In the notation of Exercise 3.2.19, we recall that a continuous function* ω: $\mathsf{G} = \mathsf{GL}_n(\mathbb{Q}_p) \to \mathbb{C}$ *is called a (zonal) spherical function if it satisfies* $\omega(\gamma_1 g \gamma_2) = \omega(g)$ *for any* $\gamma_1, \gamma_2 \in \Gamma = \mathsf{GL}_n(\mathbb{Z}_p)$ *and* $g \in \mathsf{G}$, $\omega(E_n) = 1$, *and the mapping*

$$f \to \int_{\mathsf{G}} f(g)\omega(g)\, d\mu(g)$$

is a homomorphism of $\mathsf{D}_{\mathbb{C}}$ *into* $\mathbb{C}$. *Show that each spherical function on* G *is of the form*

$$\omega_{(\alpha_1,\ldots,\alpha_n)}(g) = \int_{\Gamma} \varphi_{(\alpha_1,\ldots,\alpha_n)}(g\gamma)\, d\mu\gamma$$

where $\alpha_1, \ldots, \alpha_n$ *are non-zero complex numbers, and the function* $\varphi = \varphi_{(\alpha_1,\ldots,\alpha_n)}$ *is determined by the conditions:* $\varphi(\gamma g) = \varphi(g)$ *for* $\gamma \in \Gamma$ *and* $g \in \mathsf{G}$, *and*

$$\varphi\left(\begin{pmatrix} p^{\delta_1} & c_{12} & \cdots & c_{1n} \\ 0 & p^{\delta_2} & \cdots & c_{2n} \\ \cdots & \cdots & \cdots & \cdots \\ 0 & 0 & \cdots & p^{\delta_n} \end{pmatrix}\right) = \prod_{i=1}^{n} (\alpha_i p^{-i})^{\delta_i}.$$

Show that $\omega_{(\alpha_1,\ldots,\alpha_n)} = \omega_{(\beta_1,\ldots,\beta_n)}$ *if and only if* $\beta_1, \ldots, \beta_n$ *can be obtained by a permutation of* $\alpha_1, \ldots, \alpha_n$.

[Hint: Use Exercises 3.2.19 and 3.2.24, Theorem 3.2.20, Lemma 3.2.21 and common sense.]

§ 3.3. Hecke Rings of the Symplectic Group

3.3.1. Global Rings

In the symplectic case the second components of all Hecke pairs under consideration will be subgroups of the group

$$\mathsf{S} = \mathsf{S}^n = \mathsf{GSp}_n^+(\mathbb{Q}) = \{M \in \mathsf{M}_{2n}(\mathbb{Q}); {}^tMJ_nM = \mu(M)J_n, \mu(M) > 0\}.$$

It follows from the definition that a rational matrix $M = \begin{pmatrix} A & B \\ C & D \end{pmatrix}$ of order $2n$ is contained in S if and only if its $n \times n$-blocks A, B, C, and D satisfy

$$ {}^tAC = {}^tCA, \quad {}^tBD = {}^tDB, \quad {}^tAD - {}^tCB = \mu(M)E_n \tag{3.3.1}$$

with $\mu(M) > 0$. Since M is contained in S if and only if ${}^tM = \mu(M)J_nM^{-1}J_n$ is, it follows that (3.3.1) is equivalent with

$$A\,{}^tB = B\,{}^tA, \quad C\,{}^tD = D\,{}^tC, \quad A\,{}^tD - B\,{}^tC = \mu(M)E_n. \tag{3.3.2}$$

Lemma 3.3.1. *Let K be a congruence subgroup of $\Gamma = \Gamma^n = \mathsf{Sp}_n(\mathbb{Z})$. Then the commensurator of K in S^n coincides with S^n; in particular, (K, S^n) is a Hecke pair.*

Proof. Let $M \subset \mathsf{S}$. By Proposition 2.3.6, the intersection $M^{-1}KM \cap \Gamma$ is a congruence subgroup of Γ. It follows that the group $K_{(M)} = K \cap M^{-1}KM = \Gamma \cap K \cap M^{-1}KM$ is also a congruence subgroup and therefore has finite index in K. Replacing M by M^{-1}, we see that $K_{(M^{-1})}$ has finite index in K and so $M^{-1}K_{(M^{-1})}M = K_{(M)}$ has finite index in $M^{-1}KM$. □

The lemma allows us to define the Hecke ring D of the pair (K, S), say, over $\mathbb{Q}$. However, at present the structure of the ring is, in general, unknown, and there is no substantial general theory of the corresponding Hecke operators. Since the theory of Hecke rings and operators is not the final goal of our consideration, but rather a tool, we may limit ourselves to those classes of congruence subgroups which occur in arithmetic; we may also replace the ring D by certain subrings.

First we prove a lemma on approximation modulo q.

Lemma 3.3.2 (1). *The natural mappings modulo q,*

$$\mathsf{SL}_n(\mathbb{Z}) \to \mathsf{SL}_n(\mathbb{Z}/q\mathbb{Z}),$$

and

$$\Gamma^n = \mathsf{Sp}_n(\mathbb{Z}) \to \mathsf{Sp}_n(\mathbb{Z}/q\mathbb{Z})$$

with $n, q \in \mathbb{N}$ are homomorphisms onto;

(2) *Let q and q_1 be coprime, then*

$$\Gamma^n(q)\Gamma^n(q_1) = \Gamma^n.$$

Proof. We shall prove (1) by induction on n. (1) is clearly true for SL_1. Let us assume that the assertion is already proved for SL_{n-1}, where $n > 1$. If g is an integral matrix of order n with $\det g \equiv 1 \pmod q$ then the greatest common divisor of entries of the first column of g and q are coprime. Hence, we can replace the entries by integers in the same residue classes modulo q which are coprime. Then, by Lemma 1.3.8, there is a $V \in \mathsf{SL}_n(\mathbb{Z})$ such that $Vg = \begin{pmatrix} 1 & l \\ 0 & g' \end{pmatrix}$. Since $\det g' = \det g \equiv 1 \pmod q$, it follows by the inductive hypothesis that there is a $V' \in \mathsf{SL}_{n-1}(\mathbb{Z})$ satisfying $V' \equiv g' \pmod q$. Then, clearly, $g \equiv V^{-1}\begin{pmatrix} 1 & l \\ 0 & V' \end{pmatrix} \pmod q$, and the last matrix belongs to $\mathsf{SL}_n(\mathbb{Z})$. We have proved (1) for the SL-case. Since $\Gamma^1 = \mathsf{SL}_2(\mathbb{Z})$ it follows that (1) is true for Γ^1. Let us assume that (1) is already proved for Sp_{n-1}, where $n > 1$. If M is an integral matrix of order $2n$ satisfying ${}^tMJ_nM \equiv J_n \pmod q$, we can again replace the entries of the first column of M by coprime integers in the same residue classes modulo q. Then, by Lemma 1.3.9, we see that there is a $g \in \Gamma^n$ such that the first column of gM equals the first column of E_{2n}. It is clear that there is a $g' \in \Gamma_0^n$ such that $gMg' = M' = \begin{pmatrix} A & B \\ C & D \end{pmatrix}$ with

$$A = \begin{pmatrix} 1 & 0 \\ 0 & A' \end{pmatrix}, \quad B = \begin{pmatrix} 0 & 0 \\ b & B' \end{pmatrix}, \quad C = \begin{pmatrix} 0 & c \\ 0 & C' \end{pmatrix}, \quad D = \begin{pmatrix} d_1 & d_2 \\ d_3 & D' \end{pmatrix},$$

A', B', C', and D' being the blocks of dimension $(n-1) \times (n-1)$. The matrix M' satisfies ${}^tM'J_nM' = J_n \pmod q$, which is equivalent with the congruences

$${}^tAC \equiv {}^tCA, \quad {}^tBD \equiv {}^tDB, \quad {}^tAD - {}^tCB \equiv E_n \pmod q.$$

The first of the congruences implies that $c \equiv 0 \pmod q$ and ${}^tA'C' \equiv {}^tC'A' \pmod q$, the third implies that $d_1 \equiv 1$, $d_2 \equiv 0$ and ${}^tA'D' - {}^tC'B' \equiv E_{n-1} \pmod q$, and the second implies that ${}^tB'D' \equiv {}^tD'B' \pmod q$. It follows, in particular, that $N = \begin{pmatrix} A' & B' \\ C' & D' \end{pmatrix}$ satisfies ${}^tNJ_{n-1}N \equiv J_{n-1} \pmod q$. Hence, by the inductive hypothesis, there is a h in Γ^{n-1} satisfying $hN \equiv E_{2n-2} \pmod q$. We have $\hat{h} \in \Gamma^n$, and it follows from the above congruences that $\hat{h}M' \equiv E_{2n} \pmod q$, since the matrix $\hat{h}M'$ is symplectic modulo q. Returning to our development of the case Sp_n, we see that $M \equiv g^{-1}(\hat{h})^{-1}(g')^{-1} \pmod q$, which proves the first assertion.

By the first part, the index $\nu(\Gamma^n(q)) = [\Gamma^n : \Gamma^n(q)]$ is equal to the order of $\mathsf{Sp}_n(\mathbb{Z}/q\mathbb{Z})$, which is the number of solutions of a system of polynomial congruences modulo q. Then the Chinese Remainder Theorem implies that the index $\nu(\Gamma^n(q))$ is multiplicative in q, i.e.

$$\nu(\Gamma^n(qq_1)) = \nu(\Gamma^n(q))\nu(\Gamma^n(q_1)), \quad \text{if g.c.d.}(q, q_1) = 1.$$

On the other hand, if q and q_1 are coprime, we have clearly that $\Gamma^n(q) \cap \Gamma^n(q_1) = \Gamma^n(qq_1)$, which implies that the imbedding $\Gamma^n(q) \subset \Gamma^n$ defines an imbedding of factor groups

$$\Gamma^n(q)/\Gamma^n(qq_1) \to \Gamma^n/\Gamma^n(q_1).$$

It follows from the multiplicativity of the index, that the orders of the factor groups are equal. Hence, the last imbedding is one-to-one. Thus, for any $\gamma \in \Gamma^n$ there is a $\gamma_1 \in \Gamma^n(q_1)$, such that $\gamma\gamma_1 \in \Gamma^n(q)$. Then $\gamma \in \Gamma^n(q)\gamma_1^{-1} \subset \Gamma^n(q)\Gamma^n(q_1)$. □

Let us return to Hecke rings. For $n, q \in \mathbb{N}$ we define the group

$$\mathsf{S}(\Gamma^n(q)) = \left\{ M \in \mathsf{S}^n \cap \mathsf{GL}_{2n}(\mathbb{Z}_{(q)}); M \equiv \begin{pmatrix} E_n & 0 \\ 0 & \mu(M)E_n \end{pmatrix} \pmod q \right\}, \quad (3.3.3)$$

where $\mathbb{Z}_{(q)}$ is the ring of rational q-integral numbers. We shall limit ourselves to the consideration of the groups K with $\Gamma^n(q) \subset K \subset \Gamma^n$ for a $q \in \mathbb{N}$, which satisfy the condition

$$K \cdot \mathsf{S}(\Gamma^n(q)) = \mathsf{S}(\Gamma^n(q)) \cdot K. \quad (3.3.4)$$

Such a group K will be called a *q-symmetric group* (*of genus n*). If K is a q-symmetric group of genus n, then the set

$$\mathsf{S}(K) = \mathsf{S}(K, q) = K\mathsf{S}(\Gamma^n(q))K = K\mathsf{S}(\Gamma^n(q)) = \mathsf{S}(\Gamma^n(q)) \cdot K \qquad (3.3.5)$$

is a subgroup of S^n. According to Lemma 3.3.1, the pair $(K, \mathsf{S}(K))$ is a Hecke pair. We let

$$\mathsf{L}(K) = \mathsf{D}_{\mathbb{Q}}(K, \mathsf{S}(K)) \qquad (3.3.6)$$

denote the Hecke ring of the pair $(K, \mathsf{S}(K))$ over $\mathbb{Q}$.

Theorem 3.3.3. *Let K and K_1 be two q-symmetric groups of genus n with $q, n \in \mathbb{N}$. Suppose that $K_1 \subset K$. Then the following assertions hold*:

(1) $\mathsf{S}(K) = K\mathsf{S}(K_1) = \mathsf{S}(K_1)K$;

(2) $\Gamma^n \cap \mathsf{S}(K) = K$;

(3) *If $M, M' \in \mathsf{S}(K_1)$ and $M' \in KMK$, then $M' \in \Gamma^n(q)MK_1$*;

(4) *For $M \in \mathsf{S}(K_1)$, a decomposition $(M)_K = \sum_i (KM_i)$ with $M_i \in \mathsf{S}(K_1)$ implies that $(M)_{K_1} = \sum_i (K_1 M_i)$ and vice versa, where $(M)_K$ and $(M)_{K_1}$ are the elements of the form* (3.1.11);

(5) *The linear mappings* $\mathsf{i}\colon \mathsf{L}(K) \to \mathsf{L}(K_1)$ *and* $\nu\colon \mathsf{L}(K_1) \to \mathsf{L}(K)$ *defined by*

$$i((M)_K) = (M')_{K_1} \qquad (M \in \mathsf{S}(K), M' \in KMK \cap \mathsf{S}(K_1))$$

and

$$\nu((M')_{K_1}) = (M')_K \qquad (M' \in \mathsf{S}(K_1))$$

are ring isomorphisms satisfying $\mathsf{i}\nu = 1$ *and* $\nu\mathsf{i} = 1$.

Proof. By (3.3.5) we have

$$\mathsf{S}(K) = K\mathsf{S}(\Gamma^n(q)) = K \cdot K_1 \cdot \mathsf{S}(\Gamma^n(q)) = K\mathsf{S}(K_1).$$

In the same way we obtain $\mathsf{S}(K) = \mathsf{S}(K_1)K$. (1) is proved.

If $g \in \Gamma^n \cap \mathsf{S}(K)$, then $g = \gamma M$ with $\gamma \in K$ and $M \in \mathsf{S}(\Gamma^n(q))$. Hence, $M = \gamma^{-1} g \in \Gamma^n \cap \mathsf{S}(\Gamma^n(q))$. The last intersection is, clearly, $\Gamma^n(q)$. Thus, $g \in K\Gamma^n(q) \subset K$, which proves (2).

To prove (3), we write $M' = \gamma M \gamma'$ with $\gamma, \gamma' \in K$. Let us choose an integer q_1 prime to q such that the matrices $q_1 M^{\pm 1}$ are both integral. By Lemma 3.3.2(2) we can write γ in the form $\gamma = \gamma_1 \gamma_2$ with $\gamma_1 \in \Gamma^n(q)$ and $\gamma_2 \in \Gamma^n(q_1^2)$. Then we can write $M' = \gamma_1 M \gamma_3$, where $\gamma_3 = M^{-1}\gamma_2 M \gamma'$. Since $\gamma_2 \in \Gamma^n(q_1^2)$, it follows that γ_3 is integral, and so $\gamma_3 \in \Gamma^n$. On the other hand, $\gamma_3 = M^{-1}\gamma_1^{-1}M' \in \mathsf{S}(K_1)$. We conclude that $\gamma_3 \in \Gamma^n \cap \mathsf{S}(K_1) = K_1$, which proves (3).

To prove (4), we note that all of the left cosets $K_1 M_i$ are pairwise distinct and their union contains the double coset $K_1 M K_1$. Thus, to prove the direct assertion, we only have to show that M_i is contained in $K_1 M K_1$ for every i, but this follows from (3). The inverse assertion follows from the direct one and (1). (4) is proved.

Finally, it follows from (1), (2) and (4) that the Hecke pairs $(K, \mathsf{S}(K))$ and $(K_1, \mathsf{S}(K_1))$ satisfy the conditions (3.1.13) and (3.1.14). Then, by Proposition 3.1.6, the mapping i: $\mathsf{D}_{\mathbb{Q}}(K, \mathsf{S}(K)) \to \mathsf{D}_{\mathbb{Q}}(K_1, \mathsf{S}(K_1))$ is defined which is a ring isomorphism onto. It follows from (4) and the definition of the mapping that this i coincides with the first mapping defined in (5). The second mapping is clearly the inverse of the first one, and therefore is also an isomorphism. □

The theorem shows that the Hecke rings $\mathsf{L}(K)$ of all q-symmetric groups K of genus n are canonically isomorphic to each other (all of them are isomorphic to $\mathsf{L}(\Gamma^n(q))$, for example). So, from the abstract point of view, any of the groups K can be taken to study the ring $\mathsf{L}(K)$. In our presentation we take as K the group $\Gamma_0^n(q)$, which has appeared in Chapter I in connection with the transformation theory for theta-series. The group has not only a clear arithmetical genesis but also certain technical advantages which simplify the consideration. The reader, who is interested in another K, may easily translate from the $\Gamma_0^n(q)$-language into the K-language, using Theorem 3.3.3 as a dictionary.

We start from technical lemmas on the left and double cosets modulo $\Gamma_0^n(q)$: Let us consider the groups

$$\mathsf{S}^n(q, q_1) = \left\{M = \begin{pmatrix} A & B \\ C & D \end{pmatrix} \in \mathsf{S}^n \cap \mathsf{GL}_{2n}(\mathbb{Z}_{(q)}); C \equiv 0 \pmod{q_1}\right\}, \quad (3.3.7)$$

where $q, q_1 \in \mathbb{N}$ and $q_1 | q$.

Lemma 3.3.4. *Each of the left cosets $\Gamma_0^n(q_1)M$ with $M \in \mathsf{S}^n(q, q_1)$ and $q_1 | q$ contains a representative of the form $M_0 = \begin{pmatrix} A & B \\ 0 & D \end{pmatrix}$.*

Proof. By Proposition 1.3.7, there is $\gamma \in \Gamma^n$ such that $\gamma M = M_0$ is of the form $\begin{pmatrix} A & B \\ 0 & D \end{pmatrix}$. But then we have $\gamma = M_0 M^{-1} \in \mathsf{S}^n(q, q_1) \cap \Gamma^n = \Gamma_0^n(q_1)$. □

Lemma 3.3.5. *Let $q_1 | q$, then*

$$\mathsf{S}^n(q, q_1) = \Gamma_0^n(q_1) \cdot \mathsf{S}(\Gamma^n(q)) = \mathsf{S}(\Gamma^n(q))\Gamma_0^n(q_1);$$

in particular, the group $\Gamma_0^n(q_1)$ is a q-symmetric group with $\mathsf{S}(\Gamma_0^n(q_1), q) = \mathsf{S}^n(q, q_1)$.

Proof. It suffices to prove the first relation, since the second relation follows from the first one by application of the anti-automorphism $M \to M^{-1}$ of the group $\mathsf{S}^n(q, q_1)$. Let $M \in \mathsf{S}^n(q, q_1)$. By Lemma 3.3.4, we can assume that $M = M_0 = \begin{pmatrix} A & B \\ 0 & D \end{pmatrix}$. The matrix $M_1 = \begin{pmatrix} A_1 & B_1 \\ 0 & D_1 \end{pmatrix}$ with $A_1 = A^{-1}$, $B_1 = -A^{-1}BD^{-1}$ and $D_1 = {}^tA$ is q-integral, and M_1 modulo q belongs to $\mathsf{Sp}_n(\mathbb{Z}/q\mathbb{Z})$. Then, by Lemma 3.3.2(1), there is $\gamma \in \Gamma^n$ satisfying $\gamma \equiv M_1 \pmod{q}$. It follows that $\gamma \in \Gamma_0^n(q) \subset \Gamma_0^n(q_1)$, and $\gamma M_0 \equiv \begin{pmatrix} E & 0 \\ 0 & \mu(M)E \end{pmatrix} \pmod{q}$. □

By Lemma 3.3.5, $\mathsf{S}(\Gamma_0^n(q), q) = \mathsf{S}^n(q, q)$, and so the Hecke ring (3.3.6) for $K = \Gamma_0^n(q)$ is

$$\mathsf{L}^n(q) = \mathsf{L}(\Gamma_0^n(q)) = \mathsf{D}_{\mathbb{Q}}(\Gamma_0^n(q), \mathsf{S}^n(q)) \quad \text{with } \mathsf{S}^n(q) = \mathsf{S}^n(q, q). \tag{3.3.8}$$

According to Lemma 3.1.5, the elements $(M)_{\Gamma_0^n(q)}$ associated to different double cosets $\Gamma_0^n(q) M \Gamma_0^n(q)$ of $\mathsf{S}^n(q)$ modulo $\Gamma_0^n(q)$ form a $\mathbb{Q}$-basis for $\mathsf{L}^n(q)$.

Lemma 3.3.6. *Each of the double cosets $\Gamma_0^n(q) M \Gamma_0^n(q)$ with $M \in \mathsf{S}^n(q)$ contains the unique representative of the form*

$$\operatorname{sd}(M) = \operatorname{diag}(d_1, \ldots, d_n, e_1, \ldots, e_n), \tag{3.3.9}$$

where $d_i, e_j > 0$; $d_i | d_{i+1}$, $d_n | e_n$, $e_{i+1} | e_i$, $d_i e_i = \mu(M)$.

Proof. We first consider the case $q = 1$ by induction on n. Since $\Gamma^1 = \mathsf{SL}_2(\mathbb{Z})$ and $\mathsf{S}^1(1)$ is the group of rational matrices of order 2 with positive determinants, the case $n = 1$ follows from Lemma 3.2.2. We assume now that the assertion is already proved for matrices of order $2(n-1)$, where $n > 1$. In our consideration we shall often divide the standard blocks A, B, C, D of a matrix $M = \begin{pmatrix} A & B \\ C & D \end{pmatrix} \in \mathsf{S}^n$ into subblocks with fixed indexing, whose size is clear from the contents. For example, we write $A = \begin{pmatrix} A_1 & A_2 \\ A_3 & A_4 \end{pmatrix}$ and call A_1 the block A_1 of M and so on. In this proof the blocks A_1, B_1, C_1, and D_1 will be of dimension 1×1. Let d_1 be the greatest common divisor of entries of a matrix $M' \in \mathsf{S}^n$. Then $M' = d_1 M$, where $M \in \mathsf{S}^n$ is an integral matrix with coprime entries. We let $\delta = \delta(M)$ denote the minimum of greatest common divisors of entries of columns of M. Using the induction on δ, we shall prove that the double coset $\Gamma^n M \Gamma^n$ contains a representative $\begin{pmatrix} A & B \\ C & D \end{pmatrix}$ with the blocks of the form

$$A = \begin{pmatrix} 1 & 0 \\ 0 & A_4 \end{pmatrix}, \quad B = \begin{pmatrix} 0 & 0 \\ 0 & B_4 \end{pmatrix}, \quad C = \begin{pmatrix} 0 & 0 \\ 0 & C_4 \end{pmatrix}, \quad D = \begin{pmatrix} \mu(M) & 0 \\ 0 & D_4 \end{pmatrix}, \tag{3.3.10}$$

where $\begin{pmatrix} A_4 & B_4 \\ C_4 & D_4 \end{pmatrix} \in \mathsf{S}^{n-1}$. Assume first that $\delta = 1$, and let the index i be minimal among the indexes of columns of M with coprime entries. Replacing M by MJ_n, if it is necessary, we may assume that $i \le n$. Replacing then M by $MU(V^*)$, with a permutation matrix $V \in \Lambda^n$, we may assume that $i = 1$. In this case Lemma 1.3.9 for the first column of M implies that the left coset $\Gamma^n M$ contains a representative whose block A_1 is 1 and the blocks A_3, C_1 and C_3 are the zero-blocks. Right multiplication of the representative by $U(W)$ with $W = \begin{pmatrix} 1 & 0 \\ {}^t\!A_2 & E_{n-1} \end{pmatrix}$, where A_2 is its block A_2, gives us a matrix whose block A_2 is also the zero-block. Right multiplication of the matrix by $T(S)$ with $S = \begin{pmatrix} -B_1 & -B_2 \\ -{}^t\!B_2 & 0 \end{pmatrix}$, where B_1 and B_2

are its blocks B_1 and B_2, leads us to a matrix whose blocks B_1 and B_2 are also the zero-blocks. Thus, the double coset contains a matrix whose blocks $A_1 = 1$, $A_2 = 0, B_1 = 0, B_2 = 0, A_3 = 0, C_1 = 0$, and $C_3 = 0$. Then it follows from (3.3.1)–(3.3.2) that the matrix is of the form (3.3.10). For example, the first relation in (3.3.1) implies that $C_2 = 0$ and ${}^tA_4C_4 = {}^tC_4A_4$, the first relation in (3.3.2) implies that $B_3 = 0$ and $A_4{}^tB_4 = B_4{}^tA_4$ and so forth. If $\delta > 1$, we can use the same arguments as above to show that the coset $\Gamma^n M \Gamma^n$ contains a representative M_0 with blocks $A_1 = \delta, A_3 = 0, C_1 = 0, C_3 = 0$, and such that all entries of the blocks A_2, B_1 and B_2 belong to the set $1, 2, \ldots, \delta$. It is clear that $\delta(M_0) \le \delta$. If $\delta(M_0) = \delta$, then all entries of M_0 are divisible by δ and so M is divisible by δ, a contradiction. Hence $\delta(M_0) < \delta$. Repeating the same arguments with respect to M_0 and so on, finally, we find a representative of the form (3.3.10). Returning to the proof of the lemma for $q = 1$, we see that the double coset $\Gamma^n M \Gamma^n$ of a given $M \in \mathrm{S}^n$ contains a representative $M_0 = \begin{pmatrix} A & B \\ C & D \end{pmatrix}$ with

$$A = \begin{pmatrix} d_1 & 0 \\ 0 & A_4 \end{pmatrix}, \quad B = \begin{pmatrix} 0 & 0 \\ 0 & B_4 \end{pmatrix}, \quad C = \begin{pmatrix} 0 & 0 \\ 0 & C_4 \end{pmatrix}, \quad \text{and} \quad D = \begin{pmatrix} e_1 & 0 \\ 0 & D_4 \end{pmatrix},$$

where $d_1, e_1 > 0$, $d_1 | e_1$, $d_1 e_1 = \mu(M)$, $M_4 = \begin{pmatrix} A_4 & B_4 \\ C_4 & D_4 \end{pmatrix} \in \mathrm{S}^{n-1}$, and all entries of M_4 are divisible by d_1. By the inductive hypothesis, there are $\xi, \xi_1 \in \Gamma^{n-1}$ such that $\xi M_4 \xi_1$ is of the form (3.3.9). Then the matrix $\hat{\xi} M_4 \hat{\xi}_1$ is of the form (3.3.9). Since the numbers $d_1, \ldots, d_n, e_n, \ldots, e_1$ are clearly the elementary divisors of the matrix (3.3.9) in the sense of §3.2.1, it follows from Lemma 3.2.2 that the matrix of the form (3.3.9) is unique in the corresponding double coset. We have proved the lemma for $q = 1$.

Now let $q \in \mathbb{N}$. If $M \in \mathrm{S}^n(q)$, then, by the above, there are $\gamma_1, \gamma_2 \in \Gamma^n$ such that $\gamma_1 M \gamma_2 = \mathrm{sd}(M)$. According to Lemma 3.3.5, the groups $K = \Gamma^n$ and $K_1 = \Gamma_0^n(q)$ satisfy the condition of Theorem 3.3.3. Since $\mathrm{sd}(M) \in S^n(q)$, it follows from the part (3) of the theorem that $\mathrm{sd}(M) \in \Gamma_0^n(q) M \Gamma_0^n(q)$. $\mathrm{sd}(M)$ is unique in the double coset, for it is unique in $\Gamma^n M \Gamma^n$. □

The matrix $\mathrm{sd}(M)$ of the form (3.3.6) is called *the matrix of symplectic divisors of* M, and $d_1 = d_1(M), \ldots, d_n = d_n(M)$, $e_1 = e_1(M), \ldots, e_n = e_n(M)$ are *the symplectic divisors of* M. It follows from Lemmas 3.1.5 and 3.3.6 that the elements of the form

$$T_q(d_1, \ldots, d_n, e_1, \ldots, e_n) = (\mathrm{diag}(d_1, \ldots, d_n, e_1, \ldots, e_n))_{\Gamma_0^n(q)} \tag{3.3.11}$$

with positive rational q-integral numbers d_i, e_j satisfying

$$d_i | d_{i+1}, \quad d_n | e_n, \quad e_{i+1} | e_i, \quad d_1 e_1 = \cdots = d_n e_n \tag{3.3.12}$$

form a basis of $\mathrm{L}^n(q)$ over $\mathbb{Q}$.

Let us now turn to the multiplicative properties of $\mathrm{L}^n(q)$.

Theorem 3.3.7. *The ring* $\mathsf{L}^n(q)$ *is commutative.*

Proof. The mapping $M \to \mu(M^{-1})M$ is clearly an automorphism of $\mathsf{S}^n(q)$ which does not change any element of $\Gamma_0^n(q)$. Hence, the linear mapping of $\mathsf{L}^n(q)$ into itself given by

$$(M) \to (\mu(M^{-1})M) \qquad (M \in \mathsf{S}^n(q))$$

is an automorphism of $\mathsf{L}^n(q)$. Then Proposition 3.1.7 implies that the mapping

$$X = \sum_i a_i \cdot (M_i) \to X^* = \sum_i a_i \cdot (\mu(M_i)M_i^{-1}) \tag{3.3.13}$$

is an anti-automorphism of $\mathsf{L}^n(q)$; in particular, $(XY)^* = Y^* \cdot X^*$ for $X, Y \in \mathsf{L}^n(q)$. On the other hand, if $M \in \mathsf{S}^n(q)$ and $M_0 = \operatorname{sd}(M) = \operatorname{diag}(d_1, \ldots, d_n, e_1, \ldots, e_n)$, then Lemma 3.3.6 implies that $(M) = (M_0)$, hence

$$\begin{aligned}(M)^* = (M_0)^* &= (\mu(M)\operatorname{diag}(d_1^{-1}, \ldots, e_n^{-1}))\\ &= (\operatorname{diag}(e_1, \ldots, e_n, d_1, \ldots, d_n) = (J_n M_0 J_n^{-1}).\end{aligned}$$

Since the double coset of $J_n M_0 J_n^{-1}$ modulo Γ^n is $\Gamma^n M_0 \Gamma^n$, it follows that $\operatorname{sd}(J_n M_0 J_n^{-1}) = \operatorname{sd} M_0 \doteq M_0$. Hence, since the matrix of symplectic divisors does not depend on q, it follows from Lemma 3.3.6 that, for $\Gamma = \Gamma_0^n(q)$,

$$(J_n M_0 J_n^{-1})_\Gamma = (\operatorname{sd}(J_n M_0 J_n^{-1})) = (M_0) = (M).$$

Thus, we have

$$X^* = X \qquad \text{for } X \in \mathsf{L}^n(q), \tag{3.3.14}$$

and so $XY = (XY)^* = Y^*X^* = \mathrm{Y\,X}$. □

Let us consider now the products of specific elements of $\mathsf{L}^n(q)$.

Lemma 3.3.8. *Let* $M \in \mathsf{S}^n(q)$ *and* $r \in \mathsf{GL}_1(\mathbb{Z}_{(q)})$. *Then*

$$(rE)_\Gamma (M)_\Gamma = (rE)_\Gamma (M)_\Gamma = (rM)_\Gamma, \tag{3.3.15}$$

where $E = E_{2n}$ *and* $\Gamma = \Gamma_0^n(q)$.

Proof. In this case the proof is the same as that of Lemma 3.2.4, we only have to replace Λ by Γ and g by M. □

Proposition 3.3.9. *Let* $M, M' \in \mathsf{S}^n(q)$. *Let us suppose that the quotients* $e_1(M)/d_1(M)$ *and* $e_1(M')/d_1(M')$ *of the symplectic divisors are coprime. Then in* $\mathsf{L}^n(q)$ *the following relation holds*

$$(M)_\Gamma (M')_\Gamma = (MM')_\Gamma, \tag{3.3.16}$$

where $\Gamma = \Gamma_0^n(q)$.

Proof. The proof of (3.3.16) is completely analogous to that of Proposition 3.2.5 with clear modifications. So we shall cut a long story short. It follows from Lemma 3.3.8 and the definition of symplectic divisors that we only have to consider the case when the matrices M, M' are integral and the numbers $a = \mu(M)$, $a' = \mu(M')$ are coprime. In the same way as (3.2.8) we get

$$(M)(M') = (MM') + \sum_{\Gamma H\Gamma \subset \Gamma M\Gamma M'\Gamma} c_1(M, M'; H)(H) \tag{3.3.17}$$

with some non-negative integers $c_1(M, M'; H)$ depending only on the double cosets of M, M' and H modulo Γ. Let

$$\begin{aligned} \mathsf{SD}(a) &\\ &= \mathsf{SD}_n(a) \\ &= \{\operatorname{diag}(d_1, \ldots, d_n, e_1, \ldots, e_n); d_i, e_j \in \mathbb{N}, d_i | d_{i+1}, d_n | e_n, e_{i+1} | e_i, d_i e_i = a\} \end{aligned} \tag{3.3.18}$$

be the set of all integral matrices of the form (3.3.9) with $\mu(M) = a \in \mathbb{N}$. By Lemma 3.3.6 we can assume that $M \in \mathsf{SD}(a)$, $M' \in \mathsf{SD}(a')$, and $H \in \mathsf{SD}(aa')$ for all H in (3.3.17). If a is prime to q, we can define the element

$$T(a) = T^n(a) = \sum_{M \in \mathsf{SD}_n(a)} (M) \in \mathsf{L}^n(q). \tag{3.3.19}$$

Summing the relations (3.3.17) over all $M \in \mathsf{SD}(a)$ and $M' \in \mathsf{SD}(a')$, we get

$$T(a)T(a') = T(aa') + \sum_{H \in \mathsf{SD}(aa')} \left(\sum_{\substack{M \in \mathsf{SD}(a), \\ M' \in \mathsf{SD}(a')}} c_1(M, M'; H) \right) (H).$$

Thus, as above, in order to prove (3.3.16), we only have to check that

$$T(a)T(a') = T(aa'), \quad \text{if g.c.d.}(a, a') = 1. \tag{3.3.20}$$

Lemma 3.3.10. *Suppose a belongs to the set $\mathbb{N}_{(q)}$ of all numbers of $\mathbb{N}$ prime to q. Then the set*

$$\mathsf{SM}(a) = \{M \in \mathsf{S}^n(q) \cap \mathsf{M}_{2n}; \mu(M) = a\} \tag{3.3.21}$$

is a finite union of left cosets modulo $\Gamma = \Gamma_0^n(q)$, and the element (3.3.19) *has the decomposition of the form*

$$T(a) = \sum_{M \in \Gamma \backslash \mathsf{SM}(a)} (\Gamma M). \tag{3.3.22}$$

If a and a' are coprime (and prime to q) and M (resp. M') runs over a system of representatives for the left cosets of $\mathsf{SM}(a)$ (resp. $\mathsf{SM}(a')$) modulo Γ), then the product MM' runs a system of representatives for the left cosets of $\mathsf{SM}(aa')$ modulo Γ; in particular, the relation (3.3.20) *holds.*

Proof. The first assertion follows from Lemmas 3.3.6 and 3.3.1.

Let $M_1, \ldots, M_\nu$ and $M'_1, \ldots, M'_{\nu'}$ be systems of representatives for $\Gamma \backslash \mathsf{SM}(a)$ and $\Gamma \backslash \mathsf{SM}(a')$ respectively. It is clear that any product $M_i M'_j$ is in $\mathsf{SM}(aa')$.

Suppose now, that two of the products are in the same left coset modulo Γ, say

$$\gamma M_i M'_j = M_{i_1} M'_{j_1} \qquad \text{with } \gamma \in \Gamma.$$

We then set $H = M_{i_1}^{-1} \gamma M_i = M'_{j_1}(M'_j)^{-1}$. The matrices aH and $a'H$ are integral. Since g.c.d.$(a, a') = 1$, it follows that H is integral. On the other hand, $H \in \mathsf{S}^n(q)$ and $\mu(H) = 1$. Thus, $H \in \Gamma^n \cap \mathsf{S}^n(q) = \Gamma_0^n(q) = \Gamma$, whence $M_{i_1} \in \Gamma M_i$ and $M'_{j_1} \in \Gamma M'_j$. It follows that $i_1 = i$ and $j_1 = j$. We have proved that all of the products $M_i M'_j$ belong to different left cosets of $\mathsf{SM}(aa')$ modulo Γ. Now, if ΓM_0 with $M_0 \in SM(aa')$ is an arbitrary left coset, it follows from Lemma 3.3.6 that M_0 can be written in the form $M_0 = MM'$, where $M \in \mathsf{SM}(a)$ and $M' \in \mathsf{SM}(a')$. Then $M' = \gamma' M'_j$ with $\gamma' \in \Gamma$, and $M\gamma' = \gamma M_i$ with $\gamma \in \Gamma$, hence $M_0 = \gamma M_i M'_j$, and so the coset ΓM_0 contains the product $M_i M'_j$. Lemma 3.3.10 and Proposition 3.3.9 are proved. □

The following lemma in useful when one wants to write an element of $\mathsf{L}^n(q)$ in terms of left cosets.

Lemma 3.3.11. *Each of the left coset ΓM with $\Gamma = \Gamma_0^n(q)$ and $M \in \mathsf{S}^n(q)$ contains the unique representative of the form*

$$\begin{pmatrix} A & B \\ 0 & D \end{pmatrix}, \tag{3.3.23}$$

where D belongs to a fixed left coset of $\mathsf{GL}_n(\mathbb{Z}_{(q)})$ modulo Λ^n, B belongs to a fixed class of

$$\mathsf{B}(D) = \mathsf{B}_{\mathbb{Q}}(D) = \{B \in \mathsf{M}_n(\mathbb{Q}); {}^tBD = {}^tDB\} \tag{3.3.24}$$

modulo the equivalence relation

$$B \equiv B_1 \pmod{D} \Leftrightarrow (B - B_1)D^{-1} \in \mathsf{M}_n, \tag{3.3.25}$$

and $A = \mu(M)D^*$.

Proof. The lemma follows from Lemma 3.3.4, (3.3.1) and definitions. □

In the same way as it was done in the case of the general linear group, the consideration of the "global" Hecke rings $\mathsf{L}^n(q)$ can be reduced to the consideration of "local" subrings. Let p belongs to the set $\mathbb{P}_{(q)}$ of all prime numbers prime to q. Consider the group

$$\mathsf{S}_p(q) = \mathsf{S}_p^n(q) = \mathsf{S}^n(q) \cap \mathsf{GL}_{2n}(\mathbb{Z}[p^{-1}]), \tag{3.3.26}$$

where $\mathbb{Z}[p^{-1}]$ is the ring (3.2.17). Since $\Gamma_0^n(q) \subset \mathsf{S}_p^n(q) \subset \mathsf{S}^n(q)$, it follows that the pair $(\Gamma_0^n(q), \mathsf{S}_p^n(q))$ is a Hecke pair, and we can consider the Hecke ring

$$\mathsf{L}_p(q) = \mathsf{L}_p^n(q) = \mathsf{D}_{\mathbb{Q}}(\Gamma_0^n(q), \mathsf{S}_p^n(q)) \tag{3.3.27}$$

as a subring of $\mathsf{L}^n(q)$. The ring $\mathsf{L}_p(q)$ will be called *a local subring of* $\mathsf{L}(q)$.

Theorem 3.3.12. *The ring* $\mathsf{L}^n(q)$ *is generated by the local subrings* $\mathsf{L}^n_p(q)$ *with* $p \in \mathbb{P}_{(q)}$.

Proof. For $M \in \mathsf{S}(q)$ and $p \in \mathbb{P}_{(q)}$ we define *the matrix of symplectic p-divisors of M by*

$$\operatorname{sd}_p(M) = \operatorname{diag}(p^{v_p(d_1)}, \dots, p^{v_p(d_n)}, p^{v_p(e_1)}, \dots, p^{v_p(e_n)}), \tag{3.3.28}$$

where $d_i = d_i(M)$ and $e_j = e_j(M)$ are the symplectic divisors of M and where $v_p(r)$ denotes te p-order of a non-zero rational number r. It is clear that $\operatorname{sd}_p(M) \in \mathsf{S}_p(q)$. If M is fixed, then $\operatorname{sd}_p(M) = E_{2n}$ for all p but a finite number, and

$$\prod_{p \in \mathbb{P}_{(q)}} \operatorname{sd}_p(M) = \operatorname{sd}(M) \qquad (M \in \mathsf{S}(q)). \tag{3.3.29}$$

Since $M \in \Gamma^n_0(q) \operatorname{sd}(M) \Gamma^n_0(q)$, it follows from (3.3.29) that M can be written in the form

$$M = \prod_{p \in \mathbb{P}_{(q)}} M_p \qquad \text{with } M_p \in \mathsf{S}_p(q),\ \operatorname{sd}_p(M_p) = \operatorname{sd}_p(M) \tag{3.3.30}$$

(in particular, $M_p = E$ for all p but a finite number). It then follows from Proposition 3.3.9 that

$$(M)_\Gamma = \prod_{p \in \mathbb{P}_{(q)}} (M_p)_\Gamma \qquad \text{with } \Gamma = \Gamma^n_0(q).$$

Since $(M_p) \subset \mathsf{L}_p(q)$ and $\mathsf{L}(q)$ consists of finite linear combinations of elements (M) with $M \in \mathsf{S}(q)$, the theorem follows. □

Exercise 3.3.13. *Let* $M, M' \in \mathsf{S}^n(q)$. *Suppose the quotients* $e_n(M)/d_1(M)$, $e_n(M')/d_1(M')$ *are coprime. Show that* $\operatorname{sd}(MM') = \operatorname{sd}(M) \cdot \operatorname{sd}(M')$.

Exercise 3.3.14. *Show that, for a prime to q, the set*

$$\left\{\begin{pmatrix} A & B \\ 0 & D \end{pmatrix}; D \in \Lambda^n \backslash \mathsf{GL}_n(\mathbb{Q}) \cap \mathsf{M}_n, d_n(D) | a, B \in \mathsf{B}(D) \cap \mathsf{M}_n/\operatorname{mod} D, A = aD^*\right\}$$

is a set of representatives for the left cosets of $\mathsf{SM}_n(a, q)$ *modulo* $\Gamma^n_0(q)$.

Exercise 3.3.15. *Let D be an integral matrix of order n with* $\det D \neq 0$. *Show that the number* $b(D) = |\mathsf{B}(D) \cap \mathsf{M}_n/\operatorname{mod} D|$ *of different right residue classes of* $\mathsf{B}(D) \cap \mathsf{M}_n$ *modulo D is finite and satisfies the rules*

$$b(\lambda_1 D \lambda_2) = b(D), \qquad \text{if } \lambda_1, \lambda_2 \in \Lambda^n;$$
$$b(D) = d_1^n \cdot d_2^{n-1} \cdots d_n, \qquad \text{if } \operatorname{ed}(D) = \operatorname{diag}(d_1, \dots, d_n).$$

Exercise 3.3.16. *Show that the zeta-function of* $\mathsf{L}^n(q)$ *defined by*

$$Z(s, v) = \sum_{a \in \mathbb{N}_{(q)}} v(T(a))/a^s,$$

where $v\colon \mathsf{L}^n(q) \to \mathbb{Q}$ *is the mapping of Exercise* 3.1.12, *and where the real part of s is sufficiently large, converges and has an Euler product expansion of the form*

$$Z(s, v) = \prod_{p \in \mathbb{P}_{(q)}} \left(\sum_{\delta=0}^{\infty} v(T(p^{\delta}))p^{-\delta s} \right).$$

Show that

$$v(T(a)) = \sum_{d_i \in \mathbb{N}, d_1 | d_2 | \cdots | d_n | a} v(t(d_1, \ldots, d_n)) d_1^n d_2^{n-1} \cdots d_n,$$

where $t(d_1, \ldots, d_n) = (\operatorname{diag}(d_1, \ldots, d_n)) \in \mathsf{H}^n$ *and on the right* v *means the corresponding mapping of* H^n.

[Hint: To prove the last relation, use two previous exercises.]

Exercise 3.3.17. *Show that, for* $L^2(q)$,

$$\begin{aligned} Z(s, v) &= \prod_{p \in \mathbb{P}_{(q)}} (1 + p^{1-s}) \{(1 - p^{-s})(1 - p^{2-s})(1 - p^{3-s})\}^{-1} \\ &= \zeta_q(2s - 2)^{-1} \zeta_q(s) \delta_q(s - 1) \zeta_q(s - 2) \zeta_q(s - 3), \end{aligned}$$

where

$$\zeta_q(s) = \sum_{a \in \mathbb{N}_{(q)}} a^{-s}.$$

[Hint: Use the previous exercise and the identity

$$\sum_{0 \le \delta_1 \le \delta_2} v(t(p^{\delta_1}, p^{\delta_2})) v_1^{\delta_1} v_2^{\delta_2} = (1 - v_2^2)(1 - v_1 v_2)^{-1} (1 - (p + 1)v_2 + p v_2^2)^{-1},$$

which follows from Exercise 3.2.12.]

Exercise 3.3.18. *Show that the Hecke pairs* $(\Lambda^2, \mathsf{G}^2)$, (Γ^1, S^1) *satisfy* (3.1.13) *and* (3.1.15), *so that the mapping* (3.1.14) *gives a natural isomorphism of* H^2 *onto* L^1. *Use the isomorphism and Exercise* 3.2.13 *to show that, for* $n = 1$, *the elements* (3.3.19) *satisfy*

$$T(a)T(a') = \sum_{d | a, a'} d \cdot (dE_2)_{\Gamma} \cdot T(aa'/d^2),$$

for $a, a' \in \mathbb{N}_{(q)}$, *with* $\Gamma = \Gamma_0^1(q)$.

Exercise 3.3.19. *Show that, for* $n = 1$, *the following formal identity holds*

$$\sum_{a \in \mathbb{N}_{(q)}} T(a)/a^s = \prod_{p \in \mathbb{P}_{(q)}} (1 - T(p)p^{-s} + (pE_2) \cdot p^{1-2s})^{-1}.$$

3.3.2. Local Rings

Here we shall study the structure of the local Hecke rings $\mathsf{L}_p^n(q)$ with $p \in \mathbb{P}_{(q)}$. We note first that the structure does not depend on q:

Lemma 3.3.20. *Let p be a prime number prime to q. Then the restriction* i *of the mapping* $\mathsf{i}\colon \mathsf{D}_{\mathbb{Q}}(\Gamma^n, \mathsf{S}^n(q,1)) \to \mathsf{L}^n(q)$ *to*

$$\mathsf{L}_p^n = \mathsf{L}_p^n(1) = \mathsf{D}_{\mathbb{Q}}(\Gamma^n, \mathsf{S}_p^n(1)) \subset \mathsf{D}_{\mathbb{Q}}(\Gamma^n, \mathsf{S}^n(q,1))$$

is an isomorphism of L_p^n *onto* $\mathsf{L}_p^n(q)$ *satisfying*

$$\mathsf{i}((M)_{\Gamma^n}) = (M)_{\Gamma_0^n(q)} \quad \textit{for } M \in \mathsf{S}_p^n(q). \tag{3.3.31}$$

Proof. The lemma follows from Theorem 3.3.3(5) and Lemma 3.3.5. □

In the same way as in § 3.2.2, we define *the integral subring* $\underline{\mathsf{L}}_p(q)$ of $\mathsf{L}_p(q)$ by

$$\underline{\mathsf{L}}_p^n(q) = \mathsf{D}_{\mathbb{Q}}(\Gamma_0^n(q), \mathsf{S}_p^n(q) \cap \mathsf{M}_{2n}). \tag{3.3.32}$$

It follows from Lemma 3.3.20 that the ring $\underline{\mathsf{L}}_p^n(q)$ is naturally isomorphic to $\underline{\mathsf{L}}_p^n = \underline{\mathsf{L}}_p^n(1)$.

Lemma 3.3.21. *The element*

$$[p] = [p]_n = (pE_{2n})_{\Gamma_0^n(q)} \subset \mathsf{L}_p^n(q) \tag{3.3.33}$$

has the inverse $[p]^{-1} = (p^{-1}E_{2n})$. *The ring* $\mathsf{L}_p^n(q)$ *is generated by the subring* $\underline{\mathsf{L}}_p^n(q)$ *and the element* $[p]^{-1}$.

Proof. The lemma follows from Lemma 3.3.8 and definitions. □

Set

$$T(p^{\delta_1}, \ldots, p^{\delta_n}, p^{\varepsilon_1}, \ldots, p^{\varepsilon_n}) = (\operatorname{diag}(p^{\delta_1}, \ldots, p^{\delta_n}, p^{\varepsilon_1}, \ldots, p^{\varepsilon_n}))_{\Gamma_0^n(q)}, \tag{3.3.34}$$

where $\delta_i, \varepsilon_j \in \mathbb{Z}$, $\delta_1 + \varepsilon_1 = \cdots = \delta_n + \varepsilon_n$. Then, by Lemmas 3.1.5 and 3.3.6, the ring $\underline{\mathsf{L}}_p^n(q)$ is spanned by the elements of the form (3.3.34) with

$$0 \le \delta_1 \le \cdots \le \delta_n \le \varepsilon_n \le \cdots \le \varepsilon_1. \tag{3.3.35}$$

Such an element will be called *primitive* if $\delta_1 = 0$, and *imprimitive* if $\delta_1 \ge 1$. An arbitrary element $T \in \underline{\mathsf{L}}_p^n(q)$ will be called *primitive* (resp. *imprimitive*) if it is a linear combination of primitive (resp. imprimitive) elements of the form (3.3.34)–(3.3.35). It is clear that any $T \in \underline{\mathsf{L}}_p^n(q)$ has the unique decomposition of the form

$$T = T^{pr} + T^{im}, \tag{3.3.36}$$

where T^{pr} is primitive and T^{im} is imprimitive. It follows from Lemma 3.3.21 that the subset I of all imprimitive elements of $\underline{\mathsf{L}}_p^n(q)$ is the principal ideal generated by $[p]$:

$$\mathsf{I} = [p]\underline{\mathsf{L}}_p^n(q). \tag{3.3.37}$$

Lemma 3.3.22. *Let*

$$\psi\colon \underline{\mathsf{L}}_p^n(q) \to \underline{\mathsf{L}}_p^{n-1}(q) \tag{3.3.38}$$

be the $\mathbb{Q}$*-linear mapping given for the elements of the form* (3.3.34)–(3.3.35) *by*

$$\psi(T(p^{\delta_1},\dots,p^{\varepsilon_n})) = \begin{cases} T(p^{\delta_2},\dots,p^{\delta_n};p^{\varepsilon_2},\dots,p^{\varepsilon_n}), & \text{if } \delta_1 = 0 \\ 0, & \text{if } \delta_1 > 0. \end{cases}$$

Then ψ *is an epimorphism of the rings whose kernel coincides with the ideal* I *of imprimitive elements.*

Proof. It follows from Lemmas 3.1.5 and 3.3.6 and definitions that ψ is an epimorphism of the linear spaces with the kernel I. So it suffices to show that ψ is a ring homomorphism. To prove this, we only have to check that ψ maps product of two primitive elements of the form (3.3.34)–(3.3.35) into product of their images. Let

$$M = \operatorname{diag}(p^{\delta_1},\dots,p^{\varepsilon_n}), \qquad M' = \operatorname{diag}(p^{\delta'_1},\dots,p^{\varepsilon'_n}),$$

where the exponents satisfy (3.3.35) and $\delta_i + \varepsilon_i = \delta$, $\delta'_i + \varepsilon'_i = \delta'$. Let us assume that $\delta_1 = \delta'_1 = 0$ and set

$$M_0 = \operatorname{diag}(p^{\delta_2},\dots,p^{\delta_n},p^{\varepsilon_2},\dots,p^{\varepsilon_n}), \qquad M'_0 = \operatorname{diag}(p^{\delta'_2},\dots,p^{\delta'_n},p^{\varepsilon'_2},\dots,p^{\varepsilon'_n}).$$

By Lemmas 3.1.5 and 3.3.6 and the definition of ψ, we have

$$\psi((M)_\Gamma\cdot(M')_\Gamma) = \sum_{H=\operatorname{sd}(H) \text{ is primitive},\ \mu(H)=p^{\delta+\delta'}} c(M,M';H)(H_0)_{\Gamma'},$$

where $\Gamma = \Gamma_0^n(q)$, $\Gamma' = \Gamma_0^{n-1}(q)$, and where for $H = \operatorname{diag}(p^{\alpha_1},\dots,p^{\beta_n})$, we set $H_0 = \operatorname{diag}(p^{\alpha_2},\dots,p^{\alpha_n},p^{\beta_2},\dots,p^{\beta_n})$, $c(M,M';H)$ being the number of pairs $(M_i,M'_j)\in(\Gamma\backslash\Gamma M\Gamma,\Gamma\backslash\Gamma M'\Gamma)$ (fixed sets of representatives) satisfying

$$M_iM'_j = \gamma H \qquad \text{with } \gamma\in\Gamma. \tag{3.3.39}$$

Similarly,

$$(M_0)_{\Gamma'}(M'_0)_{\Gamma'} = \sum_{H_0=\operatorname{sd} H_0 \text{ is integral},\ \mu(H_0)=p^{\delta+\delta'}} c(M_0,M'_0;H_0)(H_0)_{\Gamma'},$$

where $c(M_0,M'_0;H_0)$ is the number of pairs $(N_k,N'_l)\in(\Gamma'\backslash\Gamma'M_0\Gamma',\Gamma'\backslash\Gamma'M'_0\Gamma')$ satisfying

$$N_kN'_l = \gamma' H_0 \qquad \text{with } \gamma'\in\Gamma'. \tag{3.3.40}$$

Since H_0 runs over the same set in the two decompositions, it follows that in order to prove the relation

$$\psi((M)_\Gamma(M')_\Gamma) = (M_0)_{\Gamma'}(M'_0)_{\Gamma'} = \psi((M)_\Gamma)\cdot\psi((M_0)_\Gamma),$$

we only have to check that

$$c(M,M';H) = c(M_0,M'_0;H_0) \tag{3.3.41}$$

for any primitive $H = \operatorname{sd}(H)$ of order n with $\mu(H) = p^{\delta+\delta'}$. Since the coefficient $c(M,M';H)$ depends only on $\Gamma H\Gamma$, it follows from Lemma 3.3.6 that we may assume without loss of generality that

$$H = \operatorname{diag}(p^{\alpha_1},\dots,p^{\alpha_n},p^{\beta_1},\dots,p^{\beta_n}) \qquad \text{with } \alpha_i + \beta_i = \delta + \delta' \text{ and } \beta_n = 0.$$

Then, by the same reason, we can take

$$H_0 = \operatorname{diag}(p^{\alpha_1}, \ldots, p^{\alpha_{n-1}}; p^{\beta_1}, \ldots, p^{\beta_{n-1}}).$$

According to Lemmas 3.3.11 and 3.2.7, we may assume that

$$M_i = \begin{pmatrix} A_i & B_i \\ 0 & D_i \end{pmatrix}, \quad M_j' = \begin{pmatrix} A_j' & B_j' \\ 0 & D_j' \end{pmatrix},$$

where D_i, D_j' are of the form (3.2.15) and B_i, B_j' are fixed modulo D_i, D_j' respectively. Then it follows from (3.3.39) that

$$\gamma = \begin{pmatrix} \gamma_{11} & \gamma_{12} \\ 0 & \gamma_{22} \end{pmatrix} \quad \text{and} \quad D_i D_j' = \gamma_{22} \operatorname{diag}(p^{\beta_1}, \ldots, p^{\beta_n}).$$

The last relation implies that the matrix $\gamma_{22} \in \Lambda^n$ is upper triangular. Hence diagonal entries of γ_{22} are equal to ± 1. Since the matrices D_i, D_j' are reduced and $\beta_n = 0$, it follows that

$$D_i = \begin{pmatrix} d_i & 0 \\ 0 & 1 \end{pmatrix}, \quad D_j' = \begin{pmatrix} d_j' & 0 \\ 0 & 1 \end{pmatrix}, \quad \gamma_{22} = \begin{pmatrix} \gamma_{22}' & 0 \\ 0 & 1 \end{pmatrix},$$

and

$$d_i d_j' = \gamma_{22}' \cdot \operatorname{diag}(p^{\beta_1}, \ldots, p^{\beta_{n-1}}).$$

Hence we have

$$A_i = \begin{pmatrix} a_i & 0 \\ 0 & p^{\delta} \end{pmatrix}, \quad A_j' = \begin{pmatrix} a_j' & 0 \\ 0 & p^{\delta'} \end{pmatrix}, \quad \gamma_{11} = \begin{pmatrix} \gamma_{11}' & 0 \\ 0 & 1 \end{pmatrix},$$

and $a_i a_j' = \gamma_{11}' \operatorname{diag}(p^{\alpha_1}, \ldots, p^{\alpha_{n-1}})$. Replacing B_i by $B_i + SD_i$ with a suitable $S \in \mathrm{S}_n(\mathbb{Z})$, we may assume that $B_i = \begin{pmatrix} b_i & 0 \\ 0 & 0 \end{pmatrix}$. In the same way we may assume that $B_j' = \begin{pmatrix} b_j' & 0 \\ 0 & 0 \end{pmatrix}$. Then

$$\gamma_{12} = \begin{pmatrix} \gamma_{12}' & 0 \\ 0 & 0 \end{pmatrix} \quad \text{and} \quad a_i b_j' + b_i d_j' = \gamma_{12}' \operatorname{diag}(p^{\beta_1}, \ldots, p^{\beta_{n-1}}).$$

It follows that the matrices

$$m_i = \begin{pmatrix} a_i & b_i \\ 0 & d_i \end{pmatrix}, \quad m_j' = \begin{pmatrix} a_j' & b_j' \\ 0 & d_j' \end{pmatrix}$$

satisfy

$$m_i m_j' = \gamma' H_0 \quad \text{with } \gamma' = \begin{pmatrix} \gamma_{11}' & \gamma_{12}' \\ 0 & \gamma_{22}' \end{pmatrix} \in \Gamma'.$$

The same arguments in reverse order show that the last relation implies (3.3.29).

The relation (3.3.41) follows, since the matrices of the form m_i and m'_j (m is defined if and only if the block D of M is of the form $\begin{pmatrix} d & 0 \\ 0 & 1 \end{pmatrix}$) make up a set of representatives for the left cosets of $\Gamma' M_0 \Gamma'$ and $\Gamma' M'_0 \Gamma'$, respectively, modulo Γ'. □

Now we can describe the structure of $\underline{\mathsf{L}}_p^n(q)$ and $\mathsf{L}_p^n(q)$.

Theorem 3.3.23. *Let* $p \in \mathbb{P}_{(q)}$. *Then the following assertions hold:*

(1) *The ring* $\underline{\mathsf{L}}_p^n(q)$ *is generated over* $\mathbb{Q}$ *by the elements*

$$T(p) = T^n(p) = T(\underbrace{1, \ldots, 1}_{n}; \underbrace{p, \ldots, p}_{n}) \tag{3.3.42}$$

and

$$T_i(p^2) = T_i^n(p^2) = T(\underbrace{1, \ldots, 1}_{n-i}, \underbrace{p, \ldots, p}_{i}; \underbrace{p^2, \ldots, p^2}_{n-i}, \underbrace{p, \ldots, p}_{i})$$

for $i = 1, 2, \ldots, n$;

(2) *The ring* $\mathsf{L}_p^n(q)$ *is generated over* $\mathbb{Q}$ *by the elements* (3.3.42) *and the element* $T_n(p^2)^{-1} = [p]^{-1}$;

(3) *The elements* (3.3.42) *are algebraically independent over* $\mathbb{Q}$.

Proof. Let first $n = 1$. Since

$$\Gamma^1 = \mathsf{SL}_2(\mathbb{Z}) = \{M \in \Lambda^2; \det M > 0\} \quad \text{and} \quad \mathsf{S}_p^1(1) = \{M \in \mathsf{G}_p^2; \det M > 0\}$$

it follows from Proposition 3.1.6 that there exists an isomorphism i of H_p^2 onto L_p^1 which satisfies $\mathsf{i}(t(p^{\delta_1}, p^{\delta_2})) = T(p^{\delta_1}, p^{\delta_2})$ for $\delta_1, \delta_2 \in \mathbb{Z}$ (see (3.1.16)). This shows that Theorem 3.3.23 for $n = 1$ and $q = 1$ follows from Theorem 3.2.17. The case $n = 1$ and $q \in \mathbb{N}$ then follows from Lemma 3.3.20.

Now let $n \in \mathbb{N}$. To prove the first assertion, it suffices to show that each element of the form (3.3.34), where the exponents satisfy (3.3.35), is a polynomial in the elements (3.3.42). We shall prove it by induction on n and, for a fixed $n > 1$, by induction on $\delta = \delta_i + \varepsilon_i$. The case $n = 1$ have been considered. The case $\delta = 1$ is trivial for every n, since in the case $T(p^{\delta_1}, \ldots, p^{\varepsilon_n}) = T(p)$. Suppose the statement is already proved for all $n' < n$ and for all elements with $\delta' < \delta$, where n, $\delta > 1$. Let $T = T(p^{\delta_1}, \ldots, p^{\varepsilon_n})$ be an element with $\delta_i + \varepsilon_i = \delta$. If $\delta_1 \geq 1$, by Lemma 3.3.8, we have $T = [p]^{\delta_1} T(p^{\delta_1 - \delta_1}, \ldots, p^{\varepsilon_n - \delta_1})$. Since $(\delta_i - \delta_1) + (\varepsilon_i - \delta_1) < \delta$ and $[p] = T_n^n(p^2)$, it follows from the inductive hypothesis that T is a polynomial in the elements (3.3.42). Now let $\delta_1 = 0$. By the inductive hypothesis on n, the element

$$\psi(T) = T(p^{\delta_2}, \ldots, p^{\delta_n}, p^{\varepsilon_2}, \ldots, p^{\varepsilon_n}),$$

where ψ is the homomorphism (3.3.38), can be written in the form

$$T(p^{\delta_2}, \ldots, p^{\delta_n}; p^{\varepsilon_2}, \ldots, p^{\varepsilon_n}) = F(\psi(T(p)), \psi(T_1(p^2)), \ldots, \psi(T_{n-1}(p^2))),$$

where

$$F(x_1,\dots,x_n) = \sum_{(\alpha_1,\dots,\alpha_n)} a_{\alpha_1,\dots,\alpha_n} x_1^{\alpha_1}\cdots x_n^{\alpha_n}.$$

Since the product $\psi(T(p))^{\alpha_1}\cdots\psi(T_{n-1}(p^2))^{\alpha_n}$ is a linear combination of elements $(H)_{\Gamma_0^{n-1}(q)}$ with $\mu(H) = p^{\alpha_1+2\alpha_2+\cdots+2\alpha_n}$, we may assume that $a_{\alpha_1,\dots,\alpha_n} = 0$ unless $\alpha_1 + 2\alpha_2 + \cdots + 2\alpha_n = \delta$. Then the element

$$T_1 = T - \sum_{\alpha_1+2\alpha_2+\cdots+2\alpha_n=\delta} a_{\alpha_1,\dots,\alpha_n}(T(p))^{\alpha_1}(T_1(p^2))^{\alpha_2}\cdots(T_{n-1}(p^2))^{\alpha_n}$$

satisfies $\psi(T_1) = 0$. So T_1 is imprimitive. On the other hand T_1 is a linear combination of elements $T(p^{\delta_1'},\dots,p^{\varepsilon_n'})$ with $\delta_i' + \varepsilon_i' = \delta$. It follows that $T_1 = [p]T_1'$, where T_1' is a linear combination of $T(p^{\beta_1},\dots,p^{\gamma_n})$ with $\beta_i + \gamma_i = \delta - 2 < \delta$. By the inductive hypothesis on δ, T_1' is a polynomial in the elements (3.3.42); hence the same is true for T. The assertion (1) is proved.

The assertion (2) follows from (1) and Lema 3.3.21.

The assertion (3) we shall prove by induction on n. For $n = 1$ it was proved above. Suppose that the assertion is true for all $n' < n$ where $n > 1$. Assume that the elements (3.3.42) are algebraically dependent over $\mathbb{Q}$ and let $F(x_0, x_1,\dots,x_n)$ be a polynomial of minimal total degree such that $F(T(p),\dots,T_n(p^2)) = 0$. Applying ψ, we get

$$\begin{aligned}\psi(F(T(p),\dots,T_n(p^2))) &= F(\psi(T(p)),\dots,\psi(T_n(p^2)))\\ &= F(T^{n-1}(p), T_1^{n-1}(p^2),\dots,T_{n-1}^{n-1}(p^2),0) = 0.\end{aligned}$$

Then the inductive hypothesis implies that $F(x_0,x_1,\dots,x_n) = x_nF_1(x_0,\dots,x_n)$, where F_1 is again a polynomial (of smaller degree). Since $T_n^n(p^2) = [p]_n$ is not a zero divisor in $\mathsf{L}_p^n(q)$, the relation

$$F(T(p), T_1(p^2),\dots,T_n(p^2)) = T_n(p^2)F_1(T(p),\dots,T_n(p^2)) = 0$$

implies that $F_1(T(p),\dots,T_n(p^2)) = 0$, a contradiction. □

Exercise 3.3.24. *Reformulate and prove the assertion of Exercise* 3.2.19 *in the case when*

$$\mathsf{G} = \{M \in \mathsf{M}_{2n}(\mathbb{Q}_p); {}^tMJ_nM = \mu(M)J_n, \mu(M) \neq 0\}$$

and

$$\Gamma = \mathsf{G} \cap \mathsf{GL}_{2n}(\mathbb{Z}_p),$$

where again $\mathbb{Q}_p$ is the field of p-adic numbers and $\mathbb{Z}_p$ is the ring of p-adic integers.

3.3.3. The Spherical Mapping

The proof of Theorem 3.3.23 gives us an effective way to express elements of a local Hecke ring in terms of generators. However, the way is practically inapplicable. Calculations in local Hecke rings can be considerably simplified, if one uses another realization of the rings given by the spherical mapping.

For further use, we define the corresponding mappings for some extensions of the local Hecke rings. The extensions appear as Hecke rings of the triangular subgroup Γ_0^n of Γ^n (see (2.3.1)), where the second components of the corresponding Hecke pairs are subgroups of the group

$$\begin{aligned} \mathcal{S} = \mathcal{S}^n &= \left\{\begin{pmatrix} A & B \\ 0 & D \end{pmatrix} \in \mathsf{S}^n\right\} \\ &= \left\{M = \begin{pmatrix} A & B \\ 0 & D \end{pmatrix} \in \mathsf{M}_{2n}(\mathbb{Q}); {}^tAD = \mu(M)E_n, \mu(M) > 0, {}^tBD = {}^tDB\right\}, \end{aligned} \tag{3.3.43}$$

the "triangular" subgroup of S^n.

Lemma 3.3.25. (1) *The left coset* $\Gamma_0^n M$ *of a matrix* $M = \begin{pmatrix} A & B \\ 0 & D \end{pmatrix}$ *of* $\mathcal{S}^n$ *is uniquely determined by the number* $\mu(M)$, *the left coset* $\Lambda^n D$, *and, for a fixed* D, *the right residue class* $B + \mathsf{S}_n(\mathbb{Z})D$ *of* B *modulo* D, *and vice versa;*

(2) *If* $M = \begin{pmatrix} A & B \\ 0 & D \end{pmatrix} \in \mathcal{S}^n$, *then the double coset* $\Gamma_0^n M \Gamma_0^n$ *has the following decomposition into the left cosets:*

$$\Gamma_0 M \Gamma_0 = \bigcup_{\substack{L \in \Lambda\backslash \Lambda D \Lambda, \\ N \in \mathsf{B}(L, M)/\mathrm{mod}\, L}} \Gamma_0 \begin{pmatrix} \mu L^* & N \\ 0 & L \end{pmatrix}, \tag{3.3.44}$$

where $\mu = \mu(M)$, $\Lambda = \Lambda^n$, *and where*

$$\mathsf{B}(L, M) = \left\{N; \begin{pmatrix} \mu L^* & N \\ 0 & L \end{pmatrix} \in \Gamma_0 M \Gamma_0\right\};$$

(3) *The pair* $(\Gamma_0^n, \mathcal{S}^n)$ *is a Hecke pair.*

Proof. The assertion (1) follows from definitions. Since the inclusion $\begin{pmatrix} L' & N \\ 0 & L \end{pmatrix} \in \Gamma_0 M \Gamma_0$ implies that $L' = \mu L^*$ and $L \in \Lambda D \Lambda$, the assertion (2) follows from (1). To prove (3), is suffices to check that each of the double cosets $\Gamma_0 M \Gamma_0$ with $M \in \mathcal{S}$ is a finite union of the left cosets. We can assume without loss of generality that M is integral. Then, for each $L \in \Lambda D \Lambda$, the set $\mathsf{B}(L, M)$ is a subset of $\{B \in \mathsf{M}_n; {}^tBD = {}^tDB\}$, which is clearly a finite union of the right residue classes modulo D. Hence, by Lemma 3.2.1, the union (3.3.44) is finite. □

We shall return to the group $\mathcal{S}^n$ in the next section. Here, in order to extend the local Hecke rings, we shall use the subgroups of $\mathcal{S}^n$ of the form

$$\mathcal{S}_p = \mathcal{S}_p^n = \mathcal{S}^n \cap \mathsf{GL}_{2n}(\mathbb{Z}[p^{-1}]),$$

where p is a prime number. Lemma 3.3.25(3) implies that $(\Gamma_0^n, \mathcal{S}_p^n)$ is a Hecke pair.

Lemma 3.3.26. *The Hecke pairs* $(\Gamma_0^n(q), \mathsf{S}_p^n(q))$ *and* $(\Gamma_0^n, \mathcal{S}_p^n)$, *where* $p \in \mathbb{P}_{(q)}$, *satisfy the conditions* (3.1.13). *The diagram*

$$\begin{array}{ccc} \mathsf{L}_p^n & \xrightarrow{\ \mathsf{i}\ } & \mathcal{L}_p^n = \mathsf{D}_{\mathbb{Q}}(\Gamma_0^n \mathcal{S}_p^n), \\ & {\scriptstyle \mathsf{i}_{1,q}} \searrow \quad \nearrow {\scriptstyle \mathsf{i}_q} & \\ & \mathsf{L}_p^n(q) & \end{array}$$

where $\mathsf{i} = \mathsf{i}_1$ *and* i_q *are imbeddings* (3.1.14), *and* $\mathsf{i}_{1,q}$ *is the isomorphism of Lemma* 3.3.20, *is commutative.*

Proof. For the Hecke pairs, the first and third conditions in (3.1.13) are obvious; the second condition follows from Lemma 3.3.4. The commutativity of the diagram follows from the definition of the mappings i. □

The lemma allows us to replace the ring $\mathsf{L}_p^n(q)$ by the isomorphic subring

$$\mathbf{L}_p^n = \mathsf{i}_q(\mathsf{L}_p^n(q)) = \mathsf{i}(\mathsf{L}_p^n) \tag{3.3.45}$$

of the local Hecke ring of Γ_0^n,

$$\mathcal{L}_p^n = \mathsf{D}_{\mathbb{Q}}(\Gamma_0^n, \mathcal{S}_p^n). \tag{3.3.46}$$

It also shows that $\mathbf{L}_p^n$ is independent of q.

In complete analogy with Lemma 3.3.21, the study of rings $\mathcal{L}_p^n$ and $\mathbf{L}_p^n$ can be reduced to that of their *integral subrings*

$$\underline{\mathcal{L}}_p^n = \mathsf{D}_{\mathbb{Q}}(\Gamma_0^n, \mathcal{S}_p^n \cap \mathsf{M}_{2n})$$

and (3.3.47)

$$\underline{\mathbf{L}}_p^n = \mathbf{L}_p^n \cap \underline{\mathcal{L}}_p^n = \mathsf{i}_q(\underline{\mathsf{L}}_p^n(q)),$$

which may be helpful.

Lemma 3.3.27. *The element*

$$[\mathbf{p}] = [\mathbf{p}]_n = (pE_{2n})_{\Gamma_0^n} \in \mathbf{L}_p^n \tag{3.3.48}$$

belong to the center of $\mathcal{L}_p^n$ *and has the inverse* $[\mathbf{p}]^{-1} = (p^{-1}E_{2n})_{\Gamma_0^n}$. *The ring* $\mathbf{L}_p^n$ *(resp.* $\mathcal{L}_p^n$*) is generated over* $\underline{\mathbf{L}}_p^n$ *(resp.* $\underline{\mathcal{L}}_p^n$*) by* $[\mathbf{p}]^{-1}$.

Proof. The lemma follows from definition, since $\Gamma p^{\pm 1} E_{2n} \Gamma = \Gamma p^{\pm 1} E_{2n} = p^{\pm 1} E_{2n} \Gamma$ for every subgroup Γ of Γ^n. □

Remark. The element (3.3.48) is clearly the image of the element (3.3.33) under the mapping i_q. In order to simplify the notation, we generally denote the elements of $\mathsf{i}_q(\mathsf{L}_p^n(q)) = \mathbf{L}_p^n \subset \mathcal{L}_p^n$ by the same letters as their inverse images, but in **boldface**.

The spherical mapping of $\mathscr{L}_p^n$ will be defined in two steps. First we map $\mathscr{L}_p^n$ into certain extension of the local Hecke ring H_p^n of general linear group; then we use the spherical mapping of H_p^n defined in § 3.2.3. We start from the left cosets. Let

$$\Gamma_0 M \quad \text{with } M = \begin{pmatrix} p^\delta D^* & B \\ 0 & D \end{pmatrix} \in \mathscr{S}_p$$

be a left coset of $\mathscr{S}_p$ modulo Γ_0. By Lemma 3.3.25, the left coset ΛD of $D \in \mathsf{G}_p$ as well as the exponent δ is uniquely determined by $\Gamma_0 M$. It follows that the element

$$\Phi((\Gamma_0 M)) = x_0^\delta(\Lambda D) \in \mathsf{L}_{\mathbb{Q}[x_0^{\pm 1}]}(\Lambda, \mathsf{G}_p),$$

where x_0 is supposed to be transcendental over $\mathbb{Q}$, depends only on the left coset $\Gamma_0 M$. Extending Φ by linearity, we get a mapping

$$\Phi = \Phi_p^n \colon \mathsf{L}_{\mathbb{Q}}(\Gamma_0^n, \mathscr{S}_p^n) \to \mathsf{L}_{\mathbb{Q}[x_0^{\pm 1}]}(\Lambda^n, G_p^n).$$

Proposition 3.3.28. *The restriction of Φ to the ring $\mathscr{L}_p^n$ is a ring homomorphism of $\mathscr{L}_p^n$ onto the ring*

$$\mathsf{H}_p^n[x_0^{\pm 1}] = \mathsf{D}_{\mathbb{Q}[x_0^{\pm 1}]}(\Lambda^n, \mathsf{G}_p^n).$$

Proof. Let $X \in \mathscr{L}_p^n$. By the definition, X is invariant under right multiplication by any matrix of the form $U(\gamma)$ with $\gamma \in \Lambda$. Hence, $\Phi(X)$ is invariant under right multiplication by any $\gamma \in \Lambda$, and so $\Phi(X) \in \mathsf{H}_p^n[x_0^{\pm 1}]$. It follows from definition that Φ is a ring homomorphism. Finally, if $D \in \mathsf{G}_p$ and $\delta \in \mathbb{Z}$, then $M = \begin{pmatrix} p^\delta D^* & 0 \\ 0 & D \end{pmatrix} \in \mathscr{S}_p^n$, and the decomposition (3.3.44) shows that

$$\Phi((M)_{\Gamma_0}) = \alpha(M) x_0^\delta (D)_\Lambda$$

with $\alpha(M) \in \mathbb{N}$, which implies that Φ is onto. □

Now let $x_0, x_1, \ldots, x_n$ be algebraically independent (commuting) variables over $\mathbb{Q}$, and let $\omega = \omega_p^n$ be the $\mathbb{Q}$-linear mapping of $\mathsf{H}_p^n[x_0^{\pm 1}]$ into $\mathbb{Q}[x_0^{\pm 1}, \ldots, x_n^{\pm 1}]$ which satisfies $\omega(x_0) = x_0$, and whose restriction to H_p^n is equal to the spherical mapping ω defined in § 3.2.3. By Theorem 3.2.20 and definitions, we then obtain the following lemma.

Lemma 3.3.29. *The mapping $\omega = \omega_p^n$ is an isomorphism of the ring $\mathsf{H}_p^n[x_0^{\pm 1}]$ onto the subring $\mathbb{Q}[x_0^{\pm 1}, x_1^{\pm 1}, \ldots, x_n^{\pm 1}]_s$ of all symmetric in $x_1, \ldots, x_n$ elements of $\mathbb{Q}[x_0^{\pm 1}, \ldots, x_n^{\pm 1}]$.*

Finally, we define *the spherical mapping $\Omega = \Omega_p^n$* of $\mathscr{L}_p^n$ into the ring $\mathbb{Q}[x_0^{\pm 1}, \ldots, x_n^{\pm 1}]_s$ by setting

$$\Omega(X) = \omega(\Phi(X)) \qquad (X \in \mathscr{L}_p^n). \tag{3.3.49}$$

Thus, we have the commutative diagram

$$\begin{array}{ccc} \mathscr{L}_p^n & \xrightarrow{\ \Omega\ } & \mathbb{Q}[x_0^{\pm 1}, \ldots, x_n^{\pm 1}]_s \\ & {\scriptstyle \Phi} \searrow \quad \nearrow {\scriptstyle \omega} & \\ & \mathrm{H}_p^n[x_0^{\pm 1}] & \end{array} \quad . \tag{3.3.50}$$

Since Φ and ω are $\mathbb{Q}$-linear ring homomorphisms onto, the same is true for Ω.

It turns out that the restriction of Ω to the subring $\mathbf{L}_p^n \subset \mathscr{L}_p^n$ is an isomorphic imbedding of $\mathbf{L}_p^n$. To prove this and describe the image of $\mathbf{L}_p^n$, we first introduce some definitions and notation. Let $\mathrm{W} = \mathrm{W}_n$ be the group of $\mathbb{Q}$-automorphisms of the rational function field $\mathbb{Q}(x_0, x_1, \ldots, x_n)$ generated by all permutations in variables $x_1, \ldots, x_n$ and by the automorphisms $\tau_1, \ldots, \tau_n$ satisfying

$$\tau_i(x_0) = x_0 x_i, \qquad \tau_i(x_i) = x_i^{-1}, \qquad \tau_i(x_j) = x_j (j \neq 0, i). \tag{3.3.51}$$

It is easy to see that the coefficients $r_a = r_a^n(x_1, \ldots, x_n)$ of the polynomial

$$r(x_1, \ldots, x_n; v) = \prod_{i=1}^{n} (1 - x_i v)(1 - x_i^{-1} v) = \sum_{a=0}^{2n} (-1)^a r_a^n(x_1, \ldots, x_n) v^a, \tag{3.3.52}$$

as well as the polynomials

$$\rho_a = \rho_a^n(x_0, x_1, \ldots, x_n) = x_0^2 x_1 \cdots x_n r_a^n(x_1, \ldots, x_n), \tag{3.3.53}$$

and the polynomial

$$t = t^n(x_0, x_1, \ldots, x_n) = x_0 \prod_{i=1}^{n} (1 + x_i) \tag{3.3.54}$$

are invariant relative to all of the transformations in W_n. The polynomials t, ρ_0, $\ldots$, ρ_{n-1} play the same part for the group W_n as the elementary symmetric functions play for the permutation group.

Theorem 3.3.30. *For any $n \in \mathbb{N}$ and $p \in \mathbb{P}$ the following assertions hold:*

(1) *The restriction of $\Omega = \Omega_p^n$ to the integral subring $\underline{\mathbf{L}}_p^n$ of $\mathbf{L}_p^n$ is an isomorphism of $\underline{\mathbf{L}}_p^n$ onto the ring $\mathbb{Q}[x_0, x_1, \ldots, x_n]_W$ of all W_n-invariant polynomials in x_0, x_1, $\ldots$, x_n over $\mathbb{Q}$;*

(2) *Each element of $\mathbb{Q}[x_0, x_1, \ldots, x_n]_W$ is a polynomial in the $n + 1$ elements*

$$t = t(x_0, x_1, \ldots, x_n), \rho_0 = \rho_0(x_0, x_1, \ldots, x_n), \ldots, \rho_{n-1} = \rho_{n-1}(x_0, x_1, \ldots, x_n) \tag{3.3.55}$$

with rational coefficients:

$$\mathbb{Q}[x_0, x_1, \ldots, x_n]_W = \mathbb{Q}[t, \rho_0, \rho_1, \ldots, \rho_{n-1}] \tag{3.3.56}$$

The elements (3.3.55) *are algebraically independent over $\mathbb{Q}$;*

(3) *The restriction of $\Omega = \Omega_p^n$ to the whole subring $\mathbf{L}_p^n$ of $\mathscr{L}_p^n$ is an isomorphism of $\mathbf{L}_p^n$ onto the ring $\mathbb{Q}[x_0^{\pm 1}, x_1^{\pm 1}, \ldots, x_n^{\pm 1}]_W$ of all W_n-invariant polynomials in $x_0^{\pm 1}$,*

$x_1^{\pm 1}, \ldots, x_n^{\pm 1}$ *over* $\mathbb{Q}$. *The last ring is generated over* $\mathbb{Q}[x_0, x_1, \ldots, x_n]_W$ *by* ρ_0^{-1}:

$$\mathbb{Q}[x_0^{\pm 1}, \ldots, x_n^{\pm 1}]_W = \mathbb{Q}[x_0, \ldots, x_n]_W[(x_0^2 x_1 \cdots x_n)^{-1}]. \tag{3.3.57}$$

Corollary 3.3.31. *The restriction of* $\Phi = \Phi_p^n$ *to* $\mathbf{L}_p^n$ *is an isomorphic imbedding.*

The scheme of the proof of Theorem 3.3.30 is analogous to that of Theorem 3.2.20: computing the images of the generators of $\underline{\mathbf{L}}_p$ under Ω, we get the generators of $\Omega(\underline{\mathbf{L}}_p)$, which allows us to investigate the structure of $\Omega(\underline{\mathbf{L}}_p)$ and show that the restriction $\Omega|_{\underline{\mathbf{L}}_p}$ is an imbedding; the assertions on $\mathbf{L}_p$ then follow from Lemma 3.3.27. However, in the symplectic case the calculation of the images of generators of $\underline{\mathbf{L}}_p$ requires some preparation. This is done in Lemmas 3.3.32–3.3.34.

Lemma 3.3.32. *The images of elements* (3.3.42) *under* $\mathrm{i}_q \colon \mathsf{L}_p^n(q) \to \mathbf{L}_p^n$ *have the following decompositions into the left cosets modulo* $\Gamma_0 = \Gamma_0^n$:

$$\mathbf{T}(p) = \mathbf{T}^n(p) = \sum_{a=0}^{n} \Pi_a \tag{3.3.58}$$

with

$$\Pi_a = \Pi_a^n(p) = \sum_{\substack{D \in \Lambda \backslash \Lambda D_a \Lambda, \\ B \in \underline{\mathrm{B}}(D)/\mathrm{mod}\, D}} \left(\Gamma_0 \begin{pmatrix} pD^* & B \\ 0 & D \end{pmatrix} \right) \tag{3.3.59}$$

where $D_a = D_a^n(p)$ *are the matrices* (3.2.28), $\Lambda = \Lambda^n$,

$$\underline{\mathrm{B}}(D) = \{B \in \mathrm{M}_n; {}^tBD = {}^tDB\}, \tag{3.3.60}$$

and where the equivalence modulo D is understood in the sense (3.3.25);

$$\mathbf{T}_i(p^2) = \mathbf{T}_i^n(p^2) = \sum_{a+b \le n,\, a \ge i} \Pi_{a,b}^{(a-i)} \quad (0 \le i \le n) \tag{3.3.61}$$

with

$$\Pi_{a,b}^{(r)} = \Pi_{a,b}^{(r)}(p) = \sum_{\substack{D \in \Lambda \backslash \Lambda D_{a,b} \Lambda,\, B \in \underline{\mathrm{B}}(D)/\mathrm{mod}\, D, \\ r_p\left(\begin{pmatrix} p^2 D^* & B \\ 0 & D \end{pmatrix}\right) = n-a+r}} \left(\Gamma_0 \begin{pmatrix} p^2 D^* & B \\ 0 & D \end{pmatrix} \right), \tag{3.3.62}$$

where $D_{a,b} = D_{a,b}^n(p)$ *are the matrices* (3.2.31), *and* $r_p(M)$ *denotes the rank of* M *over the field* $\mathbb{Z}/p\mathbb{Z}$.

The elements Π_a *with* $0 \le a \le n$, *and* $\Pi_{a,b}^{(r)}$ *with* $a + b \le n$, $r \le a$ *belong to the ring* $\mathscr{L}_p^n$, *besides,*

$$\Pi_a = (M_a)_{\Gamma_0} \quad \text{with } M_a = M_a^n(p) = \begin{pmatrix} pD_a^{-1} & 0 \\ 0 & D_a \end{pmatrix}. \tag{3.3.63}$$

Proof. Without loss of generality we may consider the elements (3.3.42) for $q = 1$. It follows from Lemma 3.3.6 that the double coset $\Gamma M_n \Gamma$ with $\Gamma = \Gamma^n$ is equal to the set $\mathsf{SM}(p) = \mathsf{SM}_n(p, 1)$ defined in (3.3.21). The definition of $\mathsf{SM}(p)$ implies that it contains a matrix of the form $\begin{pmatrix} A & B \\ 0 & D \end{pmatrix}$ if and only if $A, D \in \mathsf{M}_n$, ${}^tAD = pE_n$, and $B \in \mathsf{B}(D)$. According to Lemma 3.2.2, the matrix ${}^tA = pD^{-1}$ with an integral D is integral if an only if D belongs to one of the double cosets $\Gamma D_a \Gamma$ with $a = 0, 1, \ldots, n$. Thus, by Lemma 3.3.11, we get the decomposition

$$\Gamma \begin{pmatrix} E & 0 \\ 0 & pE \end{pmatrix} \Gamma = \bigcup_{a=0}^{n} \bigcup_{\substack{D \in \Lambda \backslash \Lambda D_a \Lambda \\ B \in \underline{\mathsf{B}}(D)/\mathrm{mod}\, D}} \Gamma \begin{pmatrix} pD^* & B \\ 0 & D \end{pmatrix}.$$

Hence, by the definition of the imbedding i, we obtain (3.3.58).

Applying Lemma 3.3.6 to the set $\mathsf{SM}(p^2) = \mathsf{SM}_n(p^2, 1)$, we obtain the decomposition

$$\mathsf{SM}(p^2) = \bigcup_{i=0}^{n} \mathsf{SM}^i(p^2), \qquad \text{where} \qquad \mathsf{SM}^i(p^2) = \Gamma \begin{pmatrix} D_i & 0 \\ 0 & p^2 D_i^{-1} \end{pmatrix} \Gamma.$$

On the other hand, in the same way as the above decomposition of $\mathsf{SM}(p)$, by Lemmas 3.2.2 and 3.3.11, we can write that

$$\mathsf{SM}(p^2) = \bigcup_{\substack{a+b \le n, \\ D \in \Lambda \backslash \Lambda D_{a,b} \Lambda,\, B \in \underline{\mathsf{B}}(D)/\mathrm{mod}\, D}} \Gamma \begin{pmatrix} p^2 D^* & B \\ 0 & D \end{pmatrix}. \tag{3.3.64}$$

Since each of the sets $\mathsf{SM}^i(p^2)$ is a single double coset modulo $\Gamma \subset \Lambda^{2n}$, it consists of matrices of the same rank over $\mathbb{Z}/p\mathbb{Z}$, which is clearly equal to $n - i$. Hence, $\mathsf{SM}^i(p^2)$ is the union of those left cosets ΓM in (3.3.64) which satisfy $r_p(M) = n - i$ The decomposition (3.3.61) follows from these considerations and definitions.

Set

$$\mathsf{U}_a = \left\{ M = \begin{pmatrix} pD^* & B \\ 0 & D \end{pmatrix}; D \in \Lambda D_a \Lambda, B \in \underline{\mathsf{B}}(D) \right\},$$

and

$$\mathsf{U}_{a,b}^r = \left\{ M = \begin{pmatrix} p^2 D^* & B \\ 0 & D \end{pmatrix}; D \in \Lambda D_{a,b} \Lambda, B \in \underline{\mathsf{B}}(D), r_p(M) = n - a + r \right\}.$$

Then, according to Lemma 3.3.25, we can write

$$\Pi_a = \sum_{M \in \Gamma_0 \backslash \mathsf{U}_a} (\Gamma_0 M), \tag{3.3.65}$$

$$\Pi_{a,b}^{(r)} = \sum_{M \in \Gamma_0 \backslash \mathsf{U}_{a,b}^r} (\Gamma_0 M). \tag{3.3.66}$$

Since clearly $\Gamma_0 U_a \Gamma_0 = \mathsf{U}_a$ and $\Gamma_0 \mathsf{U}_{a,b}^r \Gamma_0 = \mathsf{U}_{a,b}^r$, it follows that Π_a and $\Pi_{a,b}^r$ are invariant under right multiplication by any element of Γ_0 and so belong to $\mathscr{L}_p^n$. Finally, if $M = \begin{pmatrix} pD^* & B \\ 0 & D \end{pmatrix}$ is an arbitrary element of U_a, then replacing M by

$\gamma M \gamma_1$ with some $\gamma, \gamma_1 \in \Gamma_0$, we may assume that $D = D_a$. Then B is an integral matrix of the form $\begin{pmatrix} B_1 & B_2 \\ B_3 & B_4 \end{pmatrix}$ with $B_1 \in \mathsf{S}_{n-a}(\mathbb{Z})$, $B_4 \in \mathsf{S}_a(\mathbb{Z})$, and $B_2 = p\,{}^tB_3$. Hence we have

$$M = \begin{pmatrix} pD_a^{-1} & B \\ 0 & D_a \end{pmatrix} = \begin{pmatrix} E & S \\ 0 & E \end{pmatrix} \begin{pmatrix} pD_a^{-1} & 0 \\ 0 & D_a \end{pmatrix} \begin{pmatrix} E & S_1 \\ 0 & E \end{pmatrix} \in \Gamma_0 M_a \Gamma_0,$$

where $S = \begin{pmatrix} B_1 & {}^tB_3 \\ B_3 & 0 \end{pmatrix}$, $S_1 = \begin{pmatrix} 0 & 0 \\ 0 & B_4 \end{pmatrix}$. Thus, $\mathsf{U}_a = \Gamma_0 M_a \Gamma_0$, and (3.3.63) follows from (3.3.65). □

We now describe the sets $\underline{\mathsf{B}}(D)/\mathrm{mod}\, D$.

Lemma 3.3.33. *Let $D \in \mathsf{M}_n$ and* $\det D \neq 0$. *Then the following assertions hold:*

(1) *If $\alpha, \beta \in \Lambda^n$, then*

$$\underline{\mathsf{B}}(\alpha D \beta) = \alpha^* \underline{\mathsf{B}}(D) \beta,$$

and one can take

$$\underline{\mathsf{B}}(\alpha D \beta)/\mathrm{mod}\, \alpha D \beta = \alpha^* \{\underline{\mathsf{B}}(D)/\mathrm{mod}\, D\} \beta;$$

in particular, the number $b(D) = |\underline{\mathsf{B}}(D)/\mathrm{mod}\, D|$ depends only on the double coset $\Lambda D \Lambda$.

(2) *Let* $D = \mathrm{ed}(D) = \mathrm{diag}(d_1, \ldots, d_n)$ *be a matrix of elementary divisors (see* (3.2.4)). *Then one can take*

$$\underline{\mathsf{B}}(D)/\mathrm{mod}\, D$$

$$= \{B = (b_{ij}); b_{ij} = d_j d_i^{-1} b_{ji} (1 \le i < j \le n), 0 \le b_{ji} < d_i (1 \le i \le j \le n)\};$$

in particular, $b(D) = d_1^n d_2^{n-1} \cdots d_n$.

Proof. The relation ${}^tBD = {}^tDB$ is obviously equivalent with ${}^t(\alpha^* B \beta)(\alpha D \beta) = {}^t(\alpha D \beta)(\alpha^* B \beta)$, and the congruence $B \equiv B_1 \pmod{D}$ for $B, B_1 \in \underline{\mathsf{B}}(D)$, is equivalent with $\alpha^* B \beta \equiv \alpha^* B_1 \beta \pmod{\alpha D \beta}$. This implies the assertion (1). The assertion (2) follows from definitions. □

Now we can compute the images of the elements Π_a and $\Pi_{a,b}^{(r)}$ under the mappings $\Phi = \Phi_p^n$ and $\Omega = \Omega_p^n$.

Lemma 3.3.34. *The following formulae hold:*

$$\Phi(\Pi_a^n(p)) = p^{\langle a \rangle} x_0 \pi_a^n(p),$$

$$\Omega(\Pi_a^n(p)) = x_0 s_a(x_1, \ldots, x_n),$$

where $\pi_a^n(p) \in \mathsf{H}_p^n$ are the elements (3.2.26) *and s_a is the elementary symmetric function of degree a;*

$$\Phi(\Pi_{a,b}^{(r)}) = p^{b(a+b+1)}\,\mathrm{sm}_p(r,a)x_0^2\pi_{a,b}^n(p),$$
$$\Omega(\Pi_{a,b}^{(r)}) = p^{b(a+b+1)}\,\mathrm{sm}_p(r,a)x_0^2\omega(\pi_{a,b}^n(p)),$$

where $\mathrm{sm}_p(r,a)$ *is the number of symmetric matrices of order* a *and of rank* r *over the field* $\mathbb{Z}/p\mathbb{Z}$, $\pi_{a,b}^n(p) \in \mathsf{H}_p^n$ *are the elements* (3.2.31), *and* ω *is the spherical mapping of* H_p^n; *in particular, for the element* (3.3.48),

$$\Phi([\mathbf{p}]_n) = x_0^2(p)_n, \qquad \Omega([\mathbf{p}]_n) = p^{-\langle n\rangle}x_0^2x_1\cdots x_n.$$

Proof. It follows from (3.3.59), Lemma 3.3.33 and definitions that

$$\Phi(\Pi_a) = \sum_{D\in\Lambda\backslash\Lambda D_a\Lambda} x_0 b(D)(\Lambda D)$$
$$= x_0 b(D_a) \sum_{D\in\Lambda\backslash\Lambda D_a\Lambda} (\Lambda D) = x_0 p^{\langle a\rangle}\pi_a,$$

which proves the first formula. The second formula follows from the first formula and Lemma 3.2.21.

Now let $D \in \Lambda D_{a,b}\Lambda$, where $a + b \le n$. Then $D = \alpha D_{a,b}\beta$ with $\alpha, \beta \in \Lambda$ and, by Lemma 3.3.33, we can take

$$\underline{\mathsf{B}}(D)/\mathrm{mod}\, D = \alpha^*\{\underline{\mathsf{B}}(D_{a,b})/\mathrm{mod}\, D_{a,b}\}\beta$$
$$= \alpha^*\left\{B = \begin{pmatrix} 0 & 0 & 0 \\ 0 & B_{22} & B_{23} \\ 0 & B_{32} & B_{33}\end{pmatrix}; B_{22} \in \mathsf{S}_a(\mathbb{Z})/\mathrm{mod}\, p,\right. \tag{3.3.67}$$
$$\left. B_{33} \in \mathsf{S}_b(\mathbb{Z})/\mathrm{mod}\, p^2, B_{23} = p\,{}^tB_{32}, B_{32} \in \mathsf{M}_{b,a}/\mathrm{mod}\, p\right\}\beta.$$

It is easy to see that, for a fixed matrix $B' = \alpha^* B\beta$ in the last set, we have the inclusion

$$M = \begin{pmatrix} p^2D^* & B' \\ 0 & D\end{pmatrix} \in \Gamma_0 \begin{pmatrix} p^2D_{a,b}^{-1} & B_0 \\ 0 & D_{a,b}\end{pmatrix}\Gamma_0 \quad \text{with } B_0 = \begin{pmatrix} 0 & 0 & 0 \\ 0 & B_{22} & 0 \\ 0 & 0 & 0\end{pmatrix}. \tag{3.3.68}$$

This clearly implies that $r_p(M) = b + r_p(B_{22}) + n - a - b$. Thus, we have

$$\Phi(\Pi_{a,b}^{(r)}) = \sum_{D\in\Lambda\backslash\Lambda D_{a,b}\Lambda} x_0^2 p^{ba+b(b+1)}\,\mathrm{sm}_p(r,a)(\Lambda D)$$
$$= p^{b(a+b+1)}\,\mathrm{sm}_p(r,a)x_0\cdot\pi_{a,b}^n(p),$$

which proves the third formula. The forth formula follows from the third formula and the definition of Ω. Since clearly $[\mathbf{p}]_n = \Pi_{n,0}^{(0)}$ and $\pi_{n,0}^n(p) = (p)_n$, the last formulae follow from the previous formulae and Lemma 3.2.21. □

Proof of Theorem 3.3.30. We first show that

$$\Omega(\mathbf{L}_p^n) = \mathbb{Q}[t, \rho_0, \rho_1, \ldots, \rho_{n-1}]. \tag{3.3.69}$$

By Theorem 3.3.23, the ring $\mathbf{L}_p^n$ is generated over $\mathbb{Q}$ by the elements $\mathbf{T}(p)$, $\mathbf{T}_1(p^2)$, $\ldots, \mathbf{T}_n(p^2)$. Thus, the ring $\Omega(L_p^n)$ is generated by $\Omega(\mathbf{T}(p)), \Omega(\mathbf{T}_1(p^2)), \ldots, \Omega(\mathbf{T}_n(p^2))$. By (3.3.58) and Lemma 3.3.34, we have

$$\begin{aligned}\Omega(\mathbf{T}(p)) &= \sum_{a=0}^{n} \Omega(\Pi_a) = \sum_{a=0}^{n} x_0 s_a(x_1, \ldots, x_n)\\ &= x_0 \prod_{i=1}^{n} (1 + x_i) = t.\end{aligned} \tag{3.3.70}$$

Hence, in order to prove (3.3.69), we only have to check that the space

$$\mathsf{V}_1 = \left\{\sum_{i=1}^{n} \alpha_i \Omega(\mathbf{T}_i(p^2)); \alpha_i \in \mathbb{Q}\right\}$$

is equal to

$$\mathsf{V}_2 = \left\{\sum_{j=0}^{n-1} \beta_j \rho_j; \beta_j \in \mathbb{Q}\right\}.$$

By (3.3.61) and Lemma 3.3.34, we have

$$\begin{aligned}\Omega(\mathbf{T}_i(p^2)) &= \sum_{a+b \le n, a \ge i} p^{b(a+b+1)} \operatorname{sm}_p(a-i, a) x_0^2 \omega(\pi_{a,b})\\ &= \sum_{a=i}^{n} \operatorname{sm}_p(a-i, a) x_0^2 \cdot \psi_a,\end{aligned}$$

where

$$\psi_a = \sum_{b=0}^{n-a} p^{b(a+b+1)} \omega(\pi_{a,b}).$$

We set

$$V_3 = \left\{\sum_{a=1}^{n} \gamma_a x_0^2 \psi_a; \gamma_a \in \mathbb{Q}\right\}.$$

The formulae for $\Omega(\mathbf{T}_i(p^2))$ show that $\mathsf{V}_1 \subset \mathsf{V}_3$. The same formulae imply that the matrix of coefficients of expansion of $\Omega(\mathbf{T}_1(p^2)), \ldots, \Omega(\mathbf{T}_n(p^2))$ in $x_0^2\psi_1, \ldots, x_0^2\psi_n$ is an integral triangular matrix with diagonal entries $\operatorname{sm}_p(0, a) = 1$ $(a = 1, 2, \ldots, n)$. Hence, the matrix has the inverse of the same form. It follows that each ψ_a for $a = 1, \ldots, n$ is an integral linear combination of $\Omega(\mathbf{T}_a(p^2)), \ldots, \Omega(\mathbf{T}_n(p^2))$; in particular, $\mathsf{V}_3 \subset \mathsf{V}_1$. Thus, $\mathsf{V}_1 = \mathsf{V}_3$. Now let us look at the polynomials ρ_a. By (3.3.52) we have

$$\begin{aligned}\sum_{a=0}^{2n} (-1)^a r_a v^a &= \prod_{i=1}^{n} (1 - x_i v)(1 - x_i^{-1} v)\\ &= \left(\sum_{i=0}^{n} (-1)^i s_i(x_1, \ldots, x_n) v^i\right)\left(\sum_{j=0}^{n} (-1)^j s_j(x_1^{-1}, \ldots, x_n^{-1}) v^j\right).\end{aligned}$$

Since $s_j(x_1^{-1},\ldots,x_n^{-1}) = (x_1\cdots x_n)^{-1}s_{n-j}(x_1,\ldots,x_n)$, it follows that

$$\rho_a = x_0^2 x_1\cdots x_n\cdot r_a = x_0^2\sum_{i+j=a} s_i(x_1,\ldots,x_n)s_{n-j}(x_1,\ldots,x_n).$$

Using the spherical mapping $\omega = \omega_p^n$ and Lemma 3.2.21, we can write the last formula in the form

$$\rho_a = x_0^2\omega\left(\sum_{i+j=a} p^{\langle i\rangle+\langle n-j\rangle}\pi_i\pi_{n-j}\right),$$

where $\pi_\alpha = \pi_\alpha^n(p)$. Computing the products $\pi_i\pi_{n-j}$ by formulae (3.2.30), we have

$$\rho_a = x_0^2\omega\left(\sum_{i+j=a} p^{\langle i\rangle+\langle n-j\rangle}\sum_{\substack{0\le\alpha\le j,\,0\le\beta\le n-j,\\ \alpha+\beta=i}}\frac{\varphi_{\alpha+n-j-\beta}(p)}{\varphi_\alpha(p)\varphi_{n-j-\beta}(p)}\pi_{\alpha+n-j-\beta,\beta}\right).$$

Now set $i=\alpha+\beta, j=a-\alpha-\beta$. Then the conditions $0\le\alpha\le j, 0\le\beta\le n-j$ are equivalent with $\alpha,\beta\ge 0$, $2\alpha+\beta\le a$. Hence we can write

$$\begin{aligned}\rho_a &= x_0^2\sum_{\alpha,\beta\ge 0,\,2\alpha+\beta\le a} p^{\langle\alpha+\beta\rangle+\langle n+\alpha+\beta-a\rangle}\frac{\varphi_{2\alpha+n-a}(p)}{\varphi_\alpha(p)\varphi_{\alpha+n-a}(p)}\omega(\pi_{2\alpha+n-a,\beta})\\ &= x_0^2\sum_{0\le\alpha\le a/2} p^{\langle\alpha\rangle+\langle n+\alpha-a\rangle}\frac{\varphi_{2\alpha+n-a}(p)}{\varphi_\alpha(p)\varphi_{\alpha+n-a}(p)}\psi_{2\alpha+n-a},\end{aligned}$$

where, according to the above notation, we write

$$\psi_{2\alpha+n-a} = \sum_{\beta=0}^{n-(2\alpha+n-a)} p^{\beta(2\alpha+n-a+\beta+1)}\omega(\pi_{2\alpha+n-a,\beta}).$$

Setting here $2\alpha+n-a=b$ and replacing a by $n-a$, we have

$$\rho_{n-a} = \sum_{\substack{a\le b\le n\\ b\equiv a\,(\mathrm{mod}\ 2)}} p^{(b-a)(b-a+2)/8+(b+a)(b+a+2)/8}\frac{\varphi_b(p)}{\varphi_{(b-a)/2}(p)\varphi_{(b+a)/2}(p)}x_0^2\psi_b.$$

The formulae imply that the polynomials $\rho_{n-1},\rho_{n-2},\ldots,\rho_0$ are linear combinations of $x_0^2\psi_1$, $x_0^2\psi_2$, $\ldots$, $x_0^2\psi_n$, and the matrix of coefficients of the linear combinations is a rational triangular matrix with non-zero diagonal entries. It follows, that $\mathsf{V}_2=\mathsf{V}_3$, and therefore $\mathsf{V}_1=\mathsf{V}_2$, which proves (3.3.69).

Now we shall prove (3.3.56). It easily follows from definitions that the elements (3.3.55) are invariant under all of the transformations in W_n. Hence, the right side in (3.3.56) is contained in the left side. The inverse inclusion will be proved by induction on n and, for a fixed n, by induction on the degree of W-invariant polynomials with respect to the variable x_0. If $n=1$, then the substitution $x_0=z_1$, $x_0x_1=z_2$ transforms $\mathbb{Q}[t,\rho_0]$ into $\mathbb{Q}[z_1+z_2,z_1z_2]$ and $\mathbb{Q}[x_0,x_1]_W$ into the ring $\mathbb{Q}[z_1,z_2]_s$ of all symmetric polynomials in z_1,z_2 over $\mathbb{Q}$ (if $F(x_0,x_1)=\sum_{i,j\ge 0}a_{ij}x_0^ix_1^j$ is W-invariant, i.e. $F(x_0x_1,x_1^{-1})=F(x_0,x_1)$, then $a_{ij}\ne 0$ implies that $i\ge j$, which shows that $F(z_1,z_2/z_1)$ is a polynomial in z_1, z_2). The last ring is equal to $\mathbb{Q}[z_1+z_2,z_1z_2]$, since every symmetric polynomial in z_1, z_2 is a polynomial in $z_1 + z_2$, $z_1 z_2$. Suppose now that (3.3.56) is already proved for $n-1$, where $n > 1$. We shall prove by induction on m that each W-invariant polynomial $F(x_0, x_1, \ldots, x_n)$ of degree m relative to x_0 is a polynomial in t, ρ_0, $\ldots$, ρ_{n-1}. If $m = 0$, then $F = F(x_1, \ldots, x_n)$ is a symmetric polynomial which satisfies $F(x_1^{-1}, x_2, \ldots, x_n) = F(x_1, x_2, \ldots, x_n)$. It easily follows that F is a constant. Suppose that the assertion has been proved for the polynomials whose degree relative to x_0 is less that m, where $m \geq 1$. Let

$$F(x_0, x_1, \ldots, x_n) = \sum_{0 \leq i \leq m} x_0^i \varphi_i(x_1, \ldots, x_n)$$

be a W-invariant polynomial of degree m relative to x_0. Since the elements x_0, x_1, $\ldots$, x_n are algebraically independent, it follows from the definition of W that each of the polynomials $x_0^i \varphi_i(x_1, \ldots, x_n)$ is also W-invariant. Thus, we may assume without loss of generality that $F = x_0^m \varphi(x_1, \ldots, x_n)$. Since F is W_n-invariant, it follows that F is W_{n-1}-invariant, where W_{n-1} acts only on x_0, x_1, $\ldots$, x_{n-1} In particular, $F(x_0, \ldots, x_{n-1}, 0)$ is W_{n-1}-invariant. Then the first inductive hypothesis implies that

$$F(x_0, x_1, \ldots, x_{n-1}, 0) = P(t^{n-1}, \rho_0^{n-1}, \ldots, \rho_{n-2}^{n-1}),$$

where

$$P(y_1, \ldots, y_n) = \sum_{i_1, \ldots, i_n} a_{i_1, \ldots, i_n} y_1^{i_1} \cdots y_n^{i_n}$$

is a polynomial over $\mathbb{Q}$. Since F is of the form $x_0^m \varphi(x_1, \ldots, x_n)$ and x_0, x_1, $\ldots$, x_{n-1} are algebraically independent, we may assume that $a_{i_1, \ldots, i_n} = 0$ if $i_1 + 2(i_2 + \cdots + i_n) \neq m$ (the polynomials t and ρ_a are homogeneous in x_0 of degree 1 and 2 respectively). Let us show that the substitution $x_n = 0$ turns t^n, $\rho_1{}^n$, $\ldots$, ρ_{n-1}^n into t^{n-1}, ρ_0^{n-1}, $\ldots$, ρ_{n-2}^{n-1} respectively. It is obvious for t^n. As to ρ_a^n, by the definition we have

$$\sum_{a=0}^{2n} (-1)^a \rho_a^n v^a = x_0^2 x_1 \cdots x_n \prod_{i=1}^{n} (1 - x_i v)(1 - x_i^{-1} v)$$

$$= \left(\sum_{b=0}^{2n-2} (-1)^b \rho_b^{n-1} v^b \right) (1 - x_n v)(x_n - v).$$

Setting here $x_n = 0$, we get

$$\sum_{a=0}^{2n} (-1)^a \rho_a^n|_{\alpha_n = 0} v^a = \sum_{a=1}^{2n-1} (-1)^a \rho_{a-1}^{n-1} v^a,$$

which proves the result. This implies that the polynomial

$$F_1(x_0, x_1, \ldots, x_n) = F(x_0, x_1, \ldots, x_n) - P(t^n, \rho_1^n, \ldots, \rho_{n-1}^n)$$

is identically zero, when $x_n = 0$. Thus, F_1 is divisible by x_n. Since F_1 is W_n-invariant, it is symmetric in x_1, $\ldots$, x_n. Hence, F_1 is divisible by $x_1 \cdots x_n$:

$$F_1(x_0, x_1, \ldots, x_n) = x_1 \cdots x_n F_2(x_0, x_1, \ldots, x_n).$$

Now let

$$F_1(x_0, x_1, \dots, x_n) = \sum_{s \geq 0} x_0^s x_1 \cdots x_n f_s(x_1, \dots, x_n)$$

be expansion of F_1 in powers of x_0. Since F_1 is W-invariant, it is invariant under τ_1 (see (3.3.51)), which gives us the relation $F_1(x_0 x_1, x_1^{-1}, x_2, \dots, x_n) = F_1(x_0, x_1, \dots, x_n)$. This implies that the polynomials f_s satisfy

$$x_1^{s-2} f_s(x_1^{-1}, x_2, \dots, x_n) = f_s(x_1, \dots, x_n).$$

If $s = 0$ or $s = 1$ and $f_s \neq 0$, then the left side is not a polynomial in x_1 and therefore $f_0(x_1, \dots, x_n) = f_1(x_1, \dots, x_n) = 0$. Thus,

$$F_1(x_0, x_1, \dots, x_n) = x_0^2 x_1 \cdots x_n F'(x_0, x_1, \dots, x_n)$$

where F' is a polynomial. F' is clearly W-invariant. It follows from our assumptions that F' is homogeneous in x_0 of degree $m - 2$. Thus, F' either equals zero or is a polynomial in t, $\rho_0, \dots, \rho_{n-1}$. The equality (3.3.56) is proved.

To prove the polynomials (3.3.55) are algebraically independent, we again use induction on n. If $n = 1$, the polynomials are $x_0(1 + x_1) = z_1 + z_2$ and $x_0^2 x_1 = z_1 z_2$, where $z_1 = x_0$, $z_2 = x_0 x_1$, and the assertion is obvious. Suppose that $n > 1$ and the assertion is already proved for $n' < n$. Let us assume that the polynomials (3.3.55) are algebraically dependent, and let $G(y, y_0, \dots, y_{n-1})$ be the polynomial of minimal degree such that $G(t^n, \rho_0^n, \dots, \rho_{n-1}^n) = 0$. Expand G in powers of y_0:

$$G = \sum_i g_i(y, y_1, \dots, y_{n-1}) y_0^i.$$

If $g_0 = 0$, then G is divisible by y_0, and $y_0^{-1} G$ is a polynomial of smaller degree with the same property as G. Hence, $g_0 \neq 0$. By the assumption, we have

$$\sum_i g_i(t^n, \rho_1^n, \dots, \rho_{n-1}^n)(\rho_0^n)^i = 0.$$

Since it is an identity in $x_0, x_1, \dots, x_n$, we can set here $x_n = 0$. As we have seen above, this turns the polynomials t^n, ρ_1^n, ..., ρ_{n-1}^n into t^{n-1}, ρ_0^{n-1}, ..., ρ_{n-2}^{n-1} respectively. Also ρ_0^n turns into zero. Thus, we obtain the identity $g_0(t^{n-1}, \rho_0^{n-1}, \dots, \rho_{n-2}^{n-1}) = 0$ which implies, by the inductive hypothesis, that $g_0 = 0$, a contradiction.

In the beginning of the proof we have seen that $\Omega(\mathbf{T}(p)) = t^n$, and $\Omega(\mathbf{T}_1(p^2))$, ..., $\Omega(\mathbf{T}_n(p^2))$ are linear combinations of $\rho_0^n, \dots, \rho_{n-1}^n$ and vice versa. This implies that the polynomials $\Omega(\mathbf{T}(p))$, $\Omega(\mathbf{T}_1(p^2))$, ..., $\Omega(\mathbf{T}_n(p^2))$ are algebraically independent over $\mathbb{Q}$. Since $\mathrm{T}(p)$, $\mathbf{T}_1(p^2)$, ..., $\mathbf{T}_n(p^2)$ generate the ring $\underline{\mathbf{L}}_p^n$, it follows that the restriction $\Omega|_{\underline{\mathbf{L}}_p^n}$ is a monomorphism. The assertion (1) and (2) of the theorem are proved.

The assertion (3) follows from (1) and (2), Lemma 3.3.27, and the formula for $\Omega([\mathbf{p}]_n)$ of Lemma 3.3.34. □

The theorem allows us to reduce calculations in local Hecke rings of symplectic group to calculations in polynomial rings. To illustrate it, we find the sums of certain formal power series over the local Hecke rings. Let us consider the formal power series

$$\sum_{\delta=0}^{\infty} T^n(p^\delta)v^\delta, \tag{3.3.71}$$

where $T^n(p^\delta) \in \mathrm{L}_p^n(q)$ are the element of the form (3.3.19) and $p \in \mathbb{P}_{(q)}$. It follows from (3.3.22), Lemma 3.3.11 and definitions that

$$\mathbf{T}^n(p^\delta) = \sum_{\substack{0 \le \delta_1 \le \cdots \le \delta_n \le \delta,\, D \in \Lambda \backslash \Lambda \operatorname{diag}(p^{\delta_1}, \ldots, p^{\delta_n})\Lambda, \\ B \in \underline{\mathrm{B}}(D)/\bmod D}} \left(\Gamma_0^n \begin{pmatrix} p^\delta D^* & B \\ 0 & D \end{pmatrix}\right), \tag{3.3.72}$$

since a matrix $\begin{pmatrix} p^\delta D^* & B \\ 0 & D \end{pmatrix}$ is integral if and only if B and D are integral, and all of the elementary divisors of D divide p^δ. Then, by the definition of Ω and Lemma 3.3.33, we obtain the formal identity

$$\sum_{\delta=0}^{\infty} \Omega(\mathbf{T}^n(p^\delta))v^\delta = \sum_{0 \le \delta_1 \le \cdots \le \delta_n \le \delta} p^{n\delta_1 + (n-1)\delta_2 + \cdots + \delta_n} \omega(t(p^{\delta_1}, \ldots, p^{\delta_n}))(x_0 v)^\delta$$

(see § 3.2.2). If $n = 1$, then we have

$$\begin{aligned} \sum_{\delta=0}^{\infty} \Omega(\mathbf{T}^1(p^\delta))v^\delta &= \sum_{\delta_1, \alpha=0}^{\infty} p^{\delta_1}(x_1 p^{-1})^{\delta_1}(x_0 v)^{\delta_1+\alpha} \\ &= [(1 - x_0 v)(1 - x_0 x_1 v)]^{-1}. \end{aligned} \tag{3.3.73}$$

If $n = 2$, we set $\delta_2 = \delta_1 + \alpha$, $\delta = \delta_2 + \beta$. Then, using Lemmas 3.2.4 and 3.2.21, we obtain

$$\begin{aligned} &\sum_{0 \le \delta_1 \le \delta_2 \le \delta} p^{2\delta_1 + \delta_2} \omega(t(p^{\delta_1}, p^{\delta_2}))(x_0 v)^\delta \\ &\quad = \sum_{\delta_1, \alpha, \beta=0}^{\infty} p^{2\delta_1 + \delta_1 + \alpha}(p^{-3}x_1 x_2)^{\delta_1} \omega(t(1, p^\alpha))(x_0 v)^{\delta_1 + \alpha + \beta} \\ &\quad = [(1 - x_0 v)(1 - x_0 x_1 x_2 v)]^{-1} \sum_{\alpha=0}^{\infty} \omega(t(1, p^\alpha))(p x_0 v)^\alpha. \end{aligned}$$

The calculation of the last series we shall reduce to that of the generating series for the polynomials $\omega(t_2(p^\delta))$, where $t(a)$ are the elements (3.2.10). First, by definitions and Lemmas 3.2.4, 3.2.21, we have

$$\begin{aligned} \sum_{\delta=0}^{\infty} \omega(t_2(p^\delta))v_1^\delta &= \sum_{\gamma, \alpha=0}^{\infty} \omega(t(p^\gamma, p^{\gamma+\alpha}))v_1^{2\gamma+\alpha} \\ &= \sum_{\gamma=0}^{\infty} \omega((p)_2^\gamma)v_1^{2\gamma} \cdot \sum_{\alpha=0}^{\infty} \omega(t(1, p^\alpha))v_1^\alpha \\ &= (1 - p^{-3}x_1 x_2 v_1^2)^{-1} \sum_{\alpha=0}^{\infty} \omega(t(1, p^\alpha))v_1^\alpha. \end{aligned}$$

Then, by this formula and formula (3.2.38) with $n = 2$, we obtain

$$\sum_{\alpha=0}^{\infty} \omega(t(1,p^\alpha))v_1^\alpha = [(1-p^{-1}x_1v_1)(1-p^{-1}x_2v_1)]^{-1}(1-p^{-3}x_1x_2v_1^2). \quad (3.3.74)$$

Finally, we have

$$\sum_{\delta=0}^{\infty} \Omega(\mathbf{T}^2(p^\delta))v^\delta$$
$$= [(1-x_0v)(1-x_0x_1v)(1-x_0x_2v)(1-x_0x_1x_2v)]^{-1}\left(1-\frac{x_0^2x_1x_2}{p}v^2\right). \quad (3.3.75)$$

The denominators of the rational fractions on the right side in (3.3.73) and (3.3.75) are special cases of the polynomial

$$\begin{aligned} q(x_0,x_1,\ldots,x_n;v) &= (1-x_0v)\prod_{r=1}^{n}\prod_{1\le i_1<\cdots<i_r\le n}(1-x_0x_{i_1}\cdots x_{i_r}v) \\ &= \sum_{i=0}^{m}(-1)^i q_i^n(x_0,x_1,\ldots,x_n)v^i \qquad (m=2^n). \end{aligned} \quad (3.3.76)$$

It is easy to see that transformations in the group W_n relative to the variables $x_0, x_1, \ldots, x_n$ only permute the factors of q. Thus, the coefficients in the expansion of q in powers of v are invariant relative to all of the transformations. Consequently, by Theorem 3.3.30, these coefficients are images under the action of Ω of uniquely determined elements of $\underline{\mathbf{L}}_p^n$:

$$q_i(x_0,x_1,\ldots,x_n) = \Omega_p^n(\mathbf{q}_i^n(p)), \qquad \mathbf{q}_i^n(p)\in\underline{\mathbf{L}}_p^n. \quad (3.3.77)$$

We let $\mathbf{Q}(v)$ denote the corresponding polynomial over $\underline{\mathbf{L}}_p^n$:

$$\mathbf{Q}(v) = \mathbf{Q}_p^n(v) = \sum_{i=0}^{m}(-1)^i\mathbf{q}_i^n(p)v^i \in \underline{\mathbf{L}}_p^n[v] \qquad (m=2^n). \quad (3.3.78)$$

It follows from the obvious relation

$$q(x_0,\ldots,x_n;v) = v^m(x_0^2x_1\cdots x_n)^{m/2}q(x_0,\ldots,x_n;(x_0^2x_1\cdots x_nv)^{-1})$$

with $m=2^n$ and the formula for $\Omega([\mathbf{p}]_n)$ of Lemma 3.3.34 that

$$\mathbf{Q}_p^n(v) = v^m(p^{\langle n\rangle}[\mathbf{p}])^{m/2}\cdot\mathbf{Q}_p^n((p^{\langle n\rangle}[\mathbf{p}]v)^{-1}),$$

which implies the following relations for the coefficients of $\mathbf{Q}$:

$$\mathbf{q}_{m-i}^n(p) = (p^{\langle n\rangle}[\mathbf{p}])^{(m/2)-i}\mathbf{q}_i^n(p) \qquad (0\le i\le m); \quad (3.3.79)$$

in particular,

$$\mathbf{q}_m^n(p) = (p^{\langle n\rangle}[\mathbf{p}])^{m/2}. \quad (3.3.80)$$

Finally, it follows from (3.3.70) that

$$\mathbf{q}_1^n(p) = \mathbf{T}^n(p). \quad (3.3.81)$$

Returning to the series (3.3.71), we have the following

Proposition 3.3.35. *For every prime p, the following formal identities hold*

$$\sum_{\delta=0}^{\infty} \mathbf{T}^1(p^\delta)v^\delta = \mathbf{Q}_p^1(v)^{-1},$$

$$\sum_{\delta=0}^{\infty} \mathbf{T}^2(p^\delta)v^\delta = \mathbf{Q}_p^2(v)^{-1}(1 - p^2[\mathbf{p}]_2 v^2),$$

where $\mathbf{T}^n(p^\delta)$ *are the* i*-images in* $\mathbf{L}_p^n$ *of the elements* (3.3.19), *and* $\mathbf{Q}_p^n(v)$ *are the polynomials* (3.3.78). *For* $n = 1, 2$, *the polynomials* $\mathbf{Q}_p^n(v)$ *can be written in the form*

$$\mathbf{Q}_p^1(v) = 1 - \mathbf{T}^1(p)v + p[\mathbf{p}]_1 v^2;$$

$$\mathbf{Q}_p^2(v) = 1 - \mathbf{T}^2(p)v + \mathbf{q}_2^2(p)v^2 - p^3[\mathbf{p}]_2\mathbf{T}^2(p)v^3 + p^6[\mathbf{p}]_2^2 v^4$$

where

$$\mathbf{q}_2^2(p) = \Omega^{-1}(x_0^2 x_1 x_2(x_1 + x_2 + x_1^{-1} + x_2^{-1} + 2)) \in \mathbf{L}_p^2.$$

Proof. It follows from (3.3.73) and definitions that the isomorphism Ω carries the free coefficient of the power series

$$\left(\sum_{\delta=0}^{\infty} \mathbf{T}^1(p^\delta)v^\delta\right)\cdot \mathbf{Q}_p^1(v)$$

on 1, and all other coefficients belong to the kernel of Ω. Hence, the free coefficient is equal to the identity element of $\mathbf{L}_p^1$ and all other coefficients are equal to zero. In the same way the second identity follows from (3.3.75). The formulae for the coefficients of $\mathbf{Q}_p^1$ and $\mathbf{Q}_p^2$ follow from (3.3.81), (3.3.79) and definitions. □

It is clear that the similar identities hold over every ring which is isomorphic to $\mathbf{L}_p^n$, for example, over $\mathsf{L}_p^n(q)$, where $p \nmid q$.

Using Theorem 3.3.30, we can construct a parametric representation of the set of all non-zero $\mathbb{Q}$-linear homomorphisms of $\mathbf{L}_p^n$ into $\mathbb{C}$.

Proposition 3.3.36. *Each non-zero* $\mathbb{Q}$*-linear homomorphism* λ *of* $\mathbf{L}_p^n$ *into* $\mathbb{C}$ *has the form*

$$\mathbf{T} \to \lambda_A(\mathbf{T}) = \Omega_p^n(\mathbf{T})|_{(x_0,\ldots,x_n)=(\alpha_0,\ldots,\alpha_n)}, \quad (\mathbf{T} \in \mathbf{L}_p^n) \tag{3.3.82}$$

where $A = (\alpha_0, \alpha_1, \ldots, \alpha_n)$ *is a set of non-zero complex numbers depending only on* λ.

Proof. According to Theorem 3.3.30, it suffices to show that each non-zero $\mathbb{Q}$-linear homomorphism

$$\eta\colon \mathbb{Q}[x_0^{\pm 1}, \ldots, x_n^{\pm 1}]_W \to \mathbb{C}$$

can be obtained by setting $x_0 = \alpha_0, \ldots, x_n = \alpha_n$ with non-zero $\alpha_0, \ldots, \alpha_n \in \mathbb{C}$. Let $r_a = r_a^n$ be the coefficients of the polynomial (3.3.52), $\rho_0 = x_0^2 x_1 \cdots x_n$, and t the polynomial (3.3.54). We set $\eta(r_a) = \beta_a$, $\eta(\rho_0) = \delta$ and $\eta(t) = \gamma$. Since $\eta \neq 0$, it follows that η carries 1 into 1, in particular, $\eta(\rho_0)\eta(\rho_0)^{-1} = 1$, hence, $\delta \neq 0$. By Theorem 3.3.30, every element of $\mathbb{Q}[x_0^{\pm 1}, \ldots, x_n^{\pm 1}]_w$ is a polynomial in $r_1, \ldots, r_{n-1}$, $\rho_0^{\pm 1}$, and t over $\mathbb{Q}$. Thus, it suffices to show that the system of equations

$$\begin{cases} r_a(x_1, \ldots, x_n) = \beta_a & (1 \leq a \leq n-1) \\ \rho_0(x_0, \ldots, x_n) = \delta \\ t(x_0, \ldots, x_n) = \gamma \end{cases} \tag{3.3.83}$$

with $\delta \neq 0$ has a solution in non-zero complex numbers.

It follows from the obvious relation $r(x_1, \ldots, x_n; v) = v^{2n} r(x_1, \ldots, x_n; v^{-1})$ for the polynomial (3.3.52) that its coefficients satisfy

$$r_a^n(x_1, \ldots, x_n) = r_{2n-a}^n(x_1, \ldots, x_n) \qquad (0 \leq a \leq 2n) \tag{3.3.84}$$

hence, $\beta_a = \eta(r_a) = \eta(r_{2n-a}) = \beta_{2n-a}$, for $a = 0, 1, \ldots, 2n$. The last relations imply that the polynomial

$$(\eta r)(v) = \sum_{a=0}^{2n} (-1)^a \beta_a v^a \in \mathbb{C}[v]$$

satisfies the relation $(\eta r)(v) = v^{2n}(\eta r)(v^{-1})$. Since clearly $\beta_0 = \beta_{2n} = 1$, it follows that the polynomial ηr has a product expansion of the form

$$(\eta r)(v) = \prod_{i=1}^{2n} (1 - \gamma_i v) \qquad \text{with } \gamma_1 \cdots \gamma_{2n} = 1.$$

Then the above relation implies that

$$\prod_{i=1}^{2n} (1 - \gamma_i v) = v^{2n} \prod_{i=1}^{2n} (1 - \gamma_i v^{-1}) = \prod_{i=1}^{2n} (1 - \gamma_i^{-1} v),$$

which shows that the numbers $\gamma_1^{-1}, \ldots, \gamma_{2n}^{-1}$ coincide with $\gamma_1, \ldots, \gamma_{2n}$ up to a permutation: $\gamma_i^{-1} = \gamma_{\sigma(i)}$, where σ is a permutation of 1, 2, ..., $2n$. If $\sigma(i) = i$ for some i, then $\gamma_i^2 = 1$, hence $\gamma_i = \pm 1$. We let $i_1, \ldots, i_k$ (resp. $j_1, \ldots, j_s$) denote all of the indexes i with $\sigma(i) = i$, and $\gamma_i = 1$ (resp. $\gamma_i = -1$). All other indexes are divided into the pairs of different indices of the form i, $\sigma(i)$. We let $l_1, \ldots, l_t$ denote the first components of the pairs. Then $k + s + 2t = 2n$, and the relation $\gamma_1 \cdots \gamma_{2n} = 1$ implies that $\gamma_{j_1} \cdots \gamma_{j_s} = 1$. Thus, the numbers k and s are even. We let $\alpha_1, \ldots, \alpha_n$ denote the numbers $\gamma_{i_1}, \ldots, \gamma_{i_{k/2}}, \gamma_{j_1}, \ldots, \gamma_{j_{s/2}}, \gamma_{l_1}, \ldots, \gamma_{l_t}$ respectively. Then we have

$$(\eta r)(v) = \sum_{a=0}^{2n} (-1)^a \beta_a v^a = \prod_{i=1}^{n} (1 - \alpha_i v)(1 - \alpha_i^{-1} v),$$

whence

$$\beta_a = r_a(\alpha_1, \ldots, \alpha_n) \qquad (0 \leq a \leq n); \tag{3.3.85}$$

in particular the numbers $\alpha_1, \ldots, \alpha_n$ form a non-zero solution of the first $n - 1$ equations in (3.3.83). Substituting the numbers into the last two equations, we obtain the system

$$\begin{cases} x_0^2\alpha_1\cdots\alpha_n = \delta \\ x_0(1+\alpha_1)\cdots(1+\alpha_n) = \gamma. \end{cases}$$

It is clear that the system has a solution in x_0 if and only if

$$\delta^{-1}\gamma^2 = (\alpha_1\cdots\alpha_n)^{-1}\prod_{i=1}^{n}(1+\alpha_i)^2 = \prod_{i=1}^{n}(1+\alpha_i)(1+\alpha_i^{-1}).$$

It follows from definitions that

$$t^n(x_0,\ldots,x_n)^2\rho_0(x_1,\ldots,x_n)^{-1} = (x_1\cdots x_n)^{-1}\prod_{i=1}^{n}(1+x_i)^2 = \prod_{i=1}^{n}(1+x_i)(1+x_i^{-1})$$

$$= r(x_1,\ldots,x_n;-1) = \sum_{a=0}^{2n} r_a(x_1\cdots x_n).$$

Hence, applying η and using (3.3.85), we obtain

$$\delta^{-1}\gamma^2 = \sum_{a=0}^{2n}\beta_a = \sum_{a=0}^{2n} r_a(\alpha_1,\ldots,\alpha_n) = \prod_{i=1}^{n}(1+\alpha_i)(1+\alpha_i^{-1}). \qquad \square$$

The numbers $\alpha_0, \alpha_1, \ldots, \alpha_n$ will be called *the parameters of* $\lambda = \lambda_{(\alpha_0,\ldots,\alpha_n)}$. If we let the group W_n operate on rows $(\alpha_0,\ldots,\alpha_n)\in \mathsf{GL}_1(\mathbb{C})^{n+1}$ in the same way as it operates on $(x_0,x_1,\ldots,x_n)$, then it is easy to see that the rows in the same orbit determine the same homomorphism λ.

Exercise 3.3.37. *Show that the order of* W_n *is* $2^n n!$.

Exercise 3.3.38. *Prove the following formulae for the middle coefficient* $\mathbf{q}_2^2(p)$ *of* $\mathbf{Q}_p^2(v)$:

$$\mathbf{q}_2^2(p) = p\mathbf{T}(1,p,p^2,p) + p(p^2+1)[\mathbf{p}]$$
$$= \mathbf{T}(p)^2 - \mathbf{T}(p^2) - p^2[\mathbf{p}],$$

where $\mathbf{T}(1,p,p^2,p) = \mathbf{T}_1(p^2)$.

[Hint: Compute the Ω-images of the right sides, using (3.3.61) and Lemma 3.3.34; then apply Theorem 3.3.30.]

Exercise 3.3.39. *Let* $\lambda\colon \mathsf{L}^2(q)\to\mathbb{C}$ *be a non-zero* $\mathbb{Q}$*-linear homomorphism satisfying* $\lambda(T^2(a)) = 0(a^\sigma)$, *for some* $\sigma\in\mathbb{R}$. *Show that the "zeta-function of* $\mathsf{L}^2(q)$ *with character* λ", *i.e. the Dirichlet series*

$$Z(s,\lambda) = \sum_{m\in\mathbb{N}_{(q)}} \lambda(T^2(a))a^{-s}$$

converges absolutely and uniformly in any domain $\{s\in\mathbb{C}; \operatorname{Re} s > 1+\sigma+\varepsilon\}$ *with* $\varepsilon>0$, *and has an Euler product of the form*

$$Z(s,\lambda) = \prod_{p\in\mathbb{P}_{(q)}} (1-\alpha_0(p)^2\alpha_1(p)\alpha_2(p)p^{-2s-1})q(\alpha_0(p),\alpha_1(p),\alpha_2(p);p^{-s})^{-1},$$

where $\alpha_0(p)$, $\alpha_1(p)$, $\alpha_2(p)$ *are the parameters of the homomorphism*

$$\mathbf{L}_p^2 \xrightarrow{i_q^{-2}} \mathsf{L}_p^2(q) \xrightarrow{\lambda} \mathbb{C},$$

and q *are the polynomials* (3.3.76) *for* $n = 2$. *Show that the parameters satisfy*

$$\max(|\alpha_0(p)|, |\alpha_0(p)\alpha_1(p)|, |\alpha_0(p)\alpha_2(p)|, |\alpha_0(p)\alpha_1(p)\alpha_2(p)|) \leq p^\sigma$$

[Hint: Use (3.3.20) and (3.3.75).]

Exercise 3.3.40. *Let* $v\colon \mathbf{L}_p^n \to \mathbb{C}$ *be the homomorphism of Exercise* 3.1.12. *Show that the parameters of* v *are* $(\alpha_0, \ldots, \alpha_n) = (1, p, \ldots, p^n)$

Exercise 3.3.41. *In the notation of Proposition* 3.3.36, *show that* $\lambda_A = \lambda_{A'}$, *where* $A, A' \in \mathsf{GL}_1(\mathbb{C})^{n+1}$, *if and only if* $A' = wA$ *with* $w \in \mathsf{W}_n$.

§ 3.4. Hecke Rings of the Triangular Subgroup of the Symplectic Group

It is often convenient to express elements of a Hecke ring of the symplectic group in terms of certain simple components. However, the components do not, in general, belong to the initial ring. According to our approach, they live in Hecke rings of the triangular subgroup Γ_0^n of Γ^n.

3.4.1. Global Rings

According to Lemma 3.3.25 (3), we can define the Hecke ring

$$\mathscr{L}^n = \mathsf{D}_{\mathbb{Q}}(\Gamma_0^n, \mathscr{S}^n) \tag{3.4.1}$$

and, for every $q \in \mathbb{N}$, the subring

$$\mathscr{L}^n(q) = \mathsf{D}_{\mathbb{Q}}(\Gamma_0^n, \mathscr{S}^n(q)) \tag{3.4.2}$$

with

$$\mathscr{S}^n(q) = \mathscr{S}^n \cap \mathsf{GL}_{2n}(\mathbb{Z}_{(q)}). \tag{3.4.3}$$

The rings $\mathscr{L}^n$ and $\mathscr{L}^n(q)$ will be called *global Hecke rings of* Γ_0^n over $\mathbb{Q}$. It is clear that a local ring $\mathscr{L}_p^n$ defined in § 3.3.3 is contained in $\mathscr{L}^n(q)$ if $p \nmid q$.

In the same way as in the local case, it follows from Lemma 3.3.5 that the Hecke pairs $(\Gamma_0^n(q), \mathscr{S}^n(q))$ and $(\Gamma_0^n, \mathscr{S}^n(q))$ satisfy the conditions (3.1.13). Thus, we can define the imbedding (3.1.14) of the corresponding Hecke rings,

$$\mathsf{i}\colon \mathsf{L}^n(q) \to \mathscr{L}^n(q) \tag{3.4.4}$$

which enables us to replace the ring $\mathsf{L}^n(q)$ by the isomorphic subring

$$\mathbf{L}^n(q) = \mathsf{i}(\mathsf{L}^n(q)) \tag{3.4.5}$$

of $\mathscr{L}^n(q)$.

We now consider multiplicative properties of the rings $\mathscr{L}^n(q)$. Unlike the Hecke rings of the symplectic group, the Hecke rings of Γ_0^n are not commutative and contain zero divisors. For example, if $n = 1$ and p is an odd prime number, then one can easily see that the double cosets

$$\left(\begin{pmatrix} p & i \\ 0 & p \end{pmatrix}\right)_{\Gamma_0^1} \quad \text{and} \quad \left(\begin{pmatrix} p^2 & i \\ 0 & p \end{pmatrix}\right)_{\Gamma_0^1} \quad \text{with } i = 1, 2, \ldots, p-1$$

are pairwise distinct and each of the cosets consists of a single left coset modulo Γ_0^1. Thus, in the ring $\mathscr{L}_p^1 \subset \mathscr{L}^1$ the following relations hold

$$\begin{aligned}\left(\begin{pmatrix} p & 0 \\ 0 & 1 \end{pmatrix}\right)\left(\sum_{i=1}^{p-1}(-1)^i\left(\begin{pmatrix} p & i \\ 0 & p \end{pmatrix}\right)\right) &= \sum_{i=0}^{p-1}(-1)^i\left(\begin{pmatrix} p^2 & pi \\ 0 & p \end{pmatrix}\right) \\ &= \sum_{i=1}^{p-1}(-1)^i\left(\begin{pmatrix} p^2 & 0 \\ 0 & p \end{pmatrix}\right) = 0,\end{aligned}$$

and

$$\left(\sum_{i=1}^{p-1}(-1)^i\left(\begin{pmatrix} p & i \\ 0 & p \end{pmatrix}\right)\right)\left(\begin{pmatrix} p & 0 \\ 0 & 1 \end{pmatrix}\right) = \sum_{i=1}^{p-1}(-1)^i\left(\begin{pmatrix} p^2 & i \\ 0 & p \end{pmatrix}\right) \neq 0.$$

However, some important properties of $\mathsf{L}^n(q)$ remain valid for $\mathscr{L}^n(q)$ Besides, as a matter of fact, we shall use only certain subrings and submoduls of $\mathscr{L}^n(q)$ and $\mathscr{L}_p^n$.

As we have seen above (see Lemmas 3.2.4 and 3.3.8), consideration of arbitrary elements of Hecke rings of the general linear and symplectic groups is easily reduced to consideration of linear combinations of double cosets consisting of integral matrices. The same is true in the triangular case by the same reasons:

Lemma 3.4.1. *Let $M \in \mathscr{S}^n(q)$ and $r \in \mathsf{GL}_1(\mathbb{Z}_{(q)})$. Then*

$$[\mathbf{r}](M)_{\Gamma_0} = (M)_{\Gamma_0}[\mathbf{r}] = (rM)_{\Gamma_0},$$

where

$$[\mathbf{r}] = [\mathbf{r}]_n = (rE_{2n})_{\Gamma_0} = (\Gamma_0 r E_{2n}) \in \mathbf{L}^n(q), \tag{3.4.6}$$

and $\Gamma_0 = \Gamma_0^n$; in particular, the elements $[r]$ are invertible in $\mathscr{L}^n(q)$ with the inverse $[\mathbf{r}^{-1}]$ and belong to the center of $\mathscr{L}^n(q)$.

Next we use the mapping j of § 3.1.4 to define an important anti-automorphism of $\mathscr{L}^n(q)$

Proposition 3.4.2. *The mapping*

$$X = \sum_i a_i(M_i)_{\Gamma_0} \to X^* = \sum_i a_i(\mu(M_i)M_i^{-1})_{\Gamma_0}, \tag{3.4.7}$$

where $a_i \in \mathbb{Q}$ and $M_i \in \mathscr{S}^n(q)$, is an anti-automorphism of order 2 of the ring $\mathscr{L}^n(q)$, i.e. it satisfies

$$(XY)^* = Y^*X^* \quad \text{and} \quad (X^*)^* = X \tag{3.4.8}$$

for every $X, Y \in \mathscr{L}^n(q)$.

Every element $\mathbf{T}$ *of the subring* $\mathbf{L}^n(q)$ *of* $\mathscr{L}^n(q)$ *is invariant under* $*$:

$$\mathbf{T}^* = \mathbf{T} \quad (\mathbf{T} \in \mathbf{L}^n(q)). \tag{3.4.9}$$

Proof. The mapping $M \to \mu(M^{-1})M$ is obviously an automorphism of order 2 of the group $\mathscr{S}^n(q)$, which is identical on the subgroup Γ_0^n. Hence, a $\mathbb{Q}$-linear mapping α of $\mathscr{L}^n(q)$ into itself given by

$$\alpha\colon (M)_{\Gamma_0} \to (\mu(M^{-1})M)$$

is an automorphism of order 2. Since the mapping $*$ is equal to $\alpha\mathrm{j} = \mathrm{j}\alpha$, where j is a $\mathbb{Q}$-linear mapping defined by

$$\mathrm{j}\colon (M)_{\Gamma_0} \to (M^{-1})_{\Gamma_0} \quad (M \in \mathscr{S}^n(q))$$

the first assertion of the proposition follows from Proposition 3.1.7.

It follows from Lemma 3.1.9 that the diagram

$$\begin{array}{ccc} \mathsf{L}^n(q) & \xrightarrow{\ \mathrm{i}\ } & \mathscr{L}^n(q) \\ {\scriptstyle *}\downarrow & & \downarrow{\scriptstyle *} \\ \mathsf{L}^n(q) & \xrightarrow{\ \mathrm{i}\ } & \mathscr{L}^n(q) \end{array}$$

where i is the imbedding (3.4.4) and the vertical arrows are the mappings (3.3.13) and (3.4.7) respectively, is commutative. Then (3.4.9) follow from (3.3.14). □

Let us now turn to subrings of $\mathscr{L}^n(q)$. It turns out that the ring $\mathscr{L}^n(q)$ contains not only Hecke rings of the symplectic group but also some commutative subrings which are naturally isomorphic to certain extensions of Hecke rings of the general linear group. This will allow us to reduce certain problems of the theory of Hecke rings and Hecke operators for the symplectic group to the similar problems for the general linear group.

Let us consider the elements of $\mathscr{L}^n(q)$ defined by

$$\Pi_-(a) = \Pi_-^n(a) = \left(\begin{pmatrix} aE_n & 0 \\ 0 & E_n \end{pmatrix}\right)_{\Gamma_0}, \quad \Pi_+(a) = \Pi_+^n(a) = \left(\begin{pmatrix} E_n & 0 \\ 0 & aE_n \end{pmatrix}\right)_{\Gamma_0} \tag{3.4.10}$$

with $a \in \mathbb{N}_{(q)}$.

Lemma 3.4.3. *The elements* (3.4.10) *have the following decompositions into the left cosets modulo* Γ_0^n:

$$\Pi_-(a) = \left(\Gamma_0\begin{pmatrix} aE & 0 \\ 0 & E \end{pmatrix}\right), \quad \Pi_+(a) = \sum_{B \in \mathrm{S}_n(\mathbb{Z})/\mathrm{mod}\, a} \left(\Gamma_0\begin{pmatrix} E & B \\ 0 & aE \end{pmatrix}\right). \tag{3.4.11}$$

The elements satisfy the rules:

$$\Pi_-(a)\Pi_-(a_1) = \Pi_-(aa_1), \qquad \Pi_+(a)\Pi_+(a_1) = \Pi_+(aa_1), \tag{3.4.12}$$

$$\Pi_-(a)\Pi_+(a) = a^{\langle n\rangle}[\mathbf{a}], \tag{3.4.13}$$

$$\Pi_-(a)\Pi_+(a_1) = \Pi_+(a_1)\Pi_-(a), \qquad \text{if g.c.d.}(a, a_1) = 1, \tag{3.4.14}$$

and

$$\Pi_-(a)^* = \Pi_+(a), \qquad \Pi_+(a)^* = \Pi_-(a), \tag{3.4.15}$$

where $*$ *stands for the anti-automorphism* (3.4.7).

Proof. The decompositions (3.4.11) follow from (3.3.44), since in this case every double coset $\Lambda D \Lambda$ consists of a single left coset and every set $\underline{\mathrm{B}}(L, M)$ coincides with $\mathrm{S}_n(\mathbb{Z})$. The relations (3.4.12)–(3.4.15) follow directly from definitions and decompositions (3.4.11). For example, we have

$$\Pi_-(a)\Pi_+(a) = \sum_{B \in \mathrm{S}_n(\mathbb{Z})/\mathrm{mod}\, a} \left(\Gamma_0\begin{pmatrix} aE & aB \\ 0 & aE \end{pmatrix}\right) = \sum_{B \in \mathrm{S}_n(\mathbb{Z})/\mathrm{mod}\, a} (\Gamma_0 aE_{2n})$$
$$= a^{\langle n\rangle}(\Gamma_0 aE_{2n}) = a^{\langle n\rangle}[\mathbf{a}]. \qquad \square$$

We now consider the subsets $\mathscr{C}_-(q)$ and $\mathscr{C}_+(q)$ of $\mathscr{L}^n(q)$ consisting of those elements which commute with each of the elements $\Pi_-(a)$ and $\Pi_+(a)$ respectively:

$$\mathscr{C}_-(q) = \mathscr{C}_-^n(q) = \{X \in \mathscr{L}^n(q); \Pi_-(a)X = X\Pi_-(a), a \in \mathbb{N}_{(q)}\}, \tag{3.4.16}$$

$$\mathscr{C}_+(q) = \mathscr{C}_+^n(q) = \{X \in \mathscr{L}^n(q); \Pi_+(a)X = X\Pi_+(a), a \in \mathbb{N}_{(q)}\}. \tag{3.4.17}$$

The subsets are clearly subrings of $\mathscr{L}^n(q)$. It follows from (3.4.15) that the subrings are dual to each other with respect to the anti-automorphism $*$:

$$\mathscr{C}_-(q)^* = \mathscr{C}_+(q) \quad \text{and} \quad \mathscr{C}_+(q)^* = \mathscr{C}_-(q). \tag{3.4.18}$$

Proposition 3.4.4. *The ring* $\mathscr{C}_-^n(q)$ *(resp.* $\mathscr{C}_+^n(q)$*) is spanned by the double cosets modulo* Γ_0^n *of matrices of the form*

$$M = U(\mu, D) \in \mathscr{S}^n(q) \quad \text{with } d_n(D)^2 | \mu \tag{3.4.19}$$

$$(\text{resp. } M = U(\mu, D) \in \mathscr{S}^n(q) \quad \text{with } \mu | d_1(D)^2) \tag{3.4.20}$$

where we set

$$U(\mu, D) = \begin{pmatrix} \mu D^* & 0 \\ 0 & D \end{pmatrix},$$

and where $d_n(D)$ (resp. $d_1(D)$) is the biggest (resp. the least) elementary divisor of D.

We first describe decomposition of the double cosets $(M)_{\Gamma_0}$ of matrices M of the shape (3.4.19) and (3.4.20) into the left cosets modulo Γ_0.

Lemma 3.4.5. *If M is of the form* (3.4.19), *then*

$$(M)_{\Gamma_0} = \sum_{D_i \in \Lambda \backslash \Lambda D \Lambda} (\Gamma_0 U(\mu, D_i)); \tag{3.4.21}$$

but if M is of the form (3.4.20), *then*

$$(M)_{\Gamma_0} = \sum_{\substack{D_i \in \Lambda\backslash \Lambda D \Lambda, \\ s \in \mathsf{S}_n(\mathbb{Z})/\mu^{-1}\,{}^tD_i\mathsf{S}_n(\mathbb{Z})D_i}} (\Gamma_0 U(\mu, D_i) T(s)) \tag{3.4.22}$$

with $\Lambda = \Lambda^n$, where in the last summation the set $\mu^{-1}\,{}^tD_i\mathsf{S}_n(\mathbb{Z})D_i$ is contained in the additive group $\mathsf{S}_n(\mathbb{Z})$ and is considered as a subgroup of the group.

Proof of the lemma. Clearly

$$\Gamma_0 U(\mu, D)\Gamma_0 = \left\{\begin{pmatrix} \mu D_i^* & B_{ij} \\ 0 & D_i \end{pmatrix}; D_i \in \Lambda D \Lambda, B_{ij} \in \mu D_i^* \mathsf{S}_n(\mathbb{Z}) + \mathsf{S}_n(\mathbb{Z}) D_i \right\}.$$

It is easy to see that matrices of this set belong to the same left coset modulo Γ_0 if and only if the corresponding matrices D_i belong to the same left coset modulo Λ and the corresponding matrices B_{ij} are congruent modulo D_i from the right (see (3.3.25)). If $d_n(D)^2 | \mu$ then the matrix $\mu D_i^* S\,{}^tD_i^*$ is integral for every $D_i \in \Lambda D \Lambda$ and $S \in \mathsf{S}_n(\mathbb{Z})$, for the matrix $d_n(D) D_i^{-1}$ is integral. In this case we see that every matrix of the set $\mu D_i^* \mathsf{S}_n(\mathbb{Z}) + \mathsf{S}_n(\mathbb{Z}) D_i$ is divisible by D_i from the right, and so one can assume that all $B_{ij} = 0$. The decomposition (3.4.21) follows. In the general case, choosing appropriate representatives in the left cosets modulo Γ_0, one can always assume that $B_{ij} = \mu D_i^* S_j$, where $S_j \in \mathsf{S}_n(\mathbb{Z})$ are such that the matrices B_{ij} are not congruent modulo D_i. A congruence $B_{ik} \equiv B_{ij} \pmod{D_i}$ means that $B_{ij} - B_{ik} = S' D_i$ with $S' \in \mathsf{S}_n(\mathbb{Z})$, that is $S_j - S_k = \mu^{-1} S'[D_i] \in \mu^{-1}\,{}^tD_i \mathsf{S}_n(\mathbb{Z}) D_i$. If $\mu | d_1(D)^2$, then the last set is contained in $\mathsf{S}_n(\mathbb{Z})$. □

Proof of Proposition 3.4.4. It follows from uniqueness of the elementary divisors that the elementary divisors of matrices D in $GL_n(\mathbb{Q})$ satisfy

$$d_i(D^{-1}) = d_{n+1-i}(D)^{-1} \qquad \text{for } i = 1, \ldots, n; \tag{3.4.23}$$

in particular, $d_1(\mu D^{-1}) = \mu d_1(D^{-1}) = \mu d_n(D)^{-1}$. This implies that a matrix M is of the form (3.4.19) if and only if the matrix

$$\mu M^{-1} = U(\mu, \mu D^{-1})$$

is of the form (3.4.20). Hence, the anti-automorphism $*$ carries linear combinations of double cosets of matrices of the form (3.4.19) to linear combinations of double cosets of matrices of the form (3.4.20) and vice versa. By (3.4.18), we then conclude that it suffices to prove the proposition, for example, for the ring $\mathscr{C}_-(q)$.

If M is of the form (3.4.19), then it follows from (3.4.11) and (3.4.21) that

$$\Pi_-(a)(M)_{\Gamma_0} = \left(\Gamma_0\begin{pmatrix} aE & 0 \\ 0 & E\end{pmatrix}\right) \sum_{D_i \in \Lambda\backslash\Lambda D\Lambda} (\Gamma_0 U(\mu, D_i))$$
$$= \sum_{D_i \in \Lambda\backslash\Lambda D\Lambda} (\Gamma_0 U(a\mu, D_i)) = (M)_{\Gamma_0}\Pi_-(a)$$

Conversely, let

$$X = \sum_\alpha a_\alpha\left(\Gamma_0\begin{pmatrix} \mu_\alpha D_\alpha^* & B_\alpha \\ 0 & D_\alpha\end{pmatrix}\right) \in \mathscr{C}_-(q),$$

where we assume that all $a_\alpha \neq 0$ and all of the left cosets are pairwise distinct. Choosing a in $\mathbb{N}_{(q)}$ so that all of the matrices $aB_\alpha D_\alpha^{-1}$ are integral, we have

$$\Pi_-(a)X = \sum_\alpha a_\alpha\left(\Gamma_0\begin{pmatrix} a\mu_\alpha D_\alpha^* & aB_\alpha \\ 0 & D_\alpha\end{pmatrix}\right)$$
$$= \sum_\alpha a_\alpha\left(\Gamma_0 T(-aB_\alpha D_\alpha^{-1})\begin{pmatrix} a\mu_\alpha D_\alpha^* & aB_\alpha \\ 0 & D_\alpha\end{pmatrix}\right)$$
$$= \sum_\alpha a_\alpha(\Gamma_0 U(a\mu_\alpha, D_\alpha))$$

(see (3.4.11)). On the other hand, we have

$$\Pi_-(a)X = X\Pi_-(a) = \sum_\alpha a_\alpha\left(\Gamma_0\begin{pmatrix} a\mu_\alpha D_\alpha^* & B_\alpha \\ 0 & D_\alpha\end{pmatrix}\right).$$

Comparing the decompositions, we conclude that each of B_α is divisible by D_α from the right. Hence, one can assume that all $B_\alpha = 0$ Since X is an element of a Hecke ring of Γ_0, it is invariant under right multiplications by matrices of Γ_0 of the form $T(S)$ with $S \in \mathrm{S}_n(\mathbb{Z})$. Hence, each of the matrices $\mu_\alpha D_\alpha^* S$ with $S \in \mathrm{S}_n(\mathbb{Z})$ is also divisible by D_α from the right, i.e.

$$\mu_\alpha D_\alpha^* \mathrm{S}_n(\mathbb{Z})\, {}^t D_\alpha^* \in S_n(\mathbb{Z}). \tag{3.4.24}$$

Since $D_\alpha = \gamma\,\mathrm{ed}(D_\alpha)\gamma_1$ with $\gamma, \gamma_1 \in \Lambda$, where $\mathrm{ed}(D_\alpha) = \mathrm{diag}(d_1, \ldots, d_n)$ is the matrix of elementary divisors of D_α, it follows that the matrix D_α in (3.4.24) can be replaced by $\mathrm{ed}(D_\alpha)$. Then (3.4.24) means obviously that all of the numbers $\mu_\alpha/d_i d_j$ are integral, which is equivalent to the condition

$$d_n(D_\alpha)^2 \mid \mu_\alpha. \tag{3.4.25}$$

Finally, since X is invariant under right multiplications by matrices of Γ_0 of the form $U(\gamma)$ with $\gamma \in \Lambda$, it follows that the decomposition of X can be written in the form

$$X = \sum_\beta a'_\beta \left\{ \sum_{D_\alpha \in \Lambda D'_\beta \Lambda} (\Gamma_0 U(\mu'_\beta, D_\alpha)) \right\},$$

where D'_β are those of D_α which belong to pairwise distinct double cosets modulo

Λ. Since each of the left cosets modulo Λ contained in a double coset $\Lambda D'_\beta \Lambda$ is equal to one of ΛD_α with $D_\alpha \in \Lambda D'_\beta \Lambda$, it follows from (3.4.25) and (3.4.21) that each of the sums in curly brackets is a decomposition of a double coset $(M_\beta)_{\Gamma_0}$, where M_β is of the form (3.4.19). □

We now introduce the global form of the mapping Φ defined in § 3.3.3, an important tool for the study of the ring $\mathscr{L}^n$ and its subrings. Let y_p with $p \in \mathbb{P}$ be the commuting independent variables indexed by prime numbers and let $\mathscr{K} = \mathbb{Q}[\ldots, y_p^{\pm 1}, \ldots]$ be the ring of all finite polynomials in the variables $y_p^{\pm 1}$ over $\mathbb{Q}$. Define a $\mathbb{Q}$-linear mapping

$$\Phi = \Phi^n \colon \mathsf{L}_{\mathbb{Q}}(\Gamma_0^n, \mathscr{S}^n) \to \mathsf{L}_{\mathscr{K}}(\Lambda^n, \mathsf{G}^n)$$

of the modules of the left cosets by setting

$$\Phi\left(\Gamma_0\begin{pmatrix} \mu D^* & B \\ 0 & D \end{pmatrix}\right) = y_{p_1}^{\nu_1} \cdots y_{p_r}^{\nu_r}(\Lambda^n D), \tag{3.4.26}$$

if $\mu = p_1^{\nu_1} \cdots p_r^{\nu_r}$ with $p_i \in \mathbb{P}$. The mapping is clearly independent of the choice of representatives of the left cosets.

Proposition 3.4.6. *The restriction of the mapping Φ to the ring $\mathscr{L}^n \subset \mathsf{L}_{\mathbb{Q}}(\Gamma_0^n, \mathscr{S}^n)$ is an homomorphism of the ring onto the Hecke ring $\mathsf{D}_{\mathscr{K}}(\Lambda^n, \mathsf{G}^n)$:*

$$\Phi \colon \mathscr{L}^n \to \mathsf{D}_K(\Lambda^n, \mathsf{G}^n) = \mathsf{H}^n[\ldots, y_p^{\pm 1}, \ldots]. \tag{3.4.27}$$

Proof. Since an element X of $\mathscr{L}$ is invariant under right multiplication by the matrices $U(\gamma)$ with $\gamma \in \Lambda$, it follows that the image $\Phi(X)$ is invariant under right multiplication by the matrices of Λ. Hence $\Phi(\mathscr{L}) \in \mathsf{D}_{\mathscr{K}}(\Lambda, \mathsf{G})$. Then the definition of multiplication in Hecke rings implies that the restriction $\Phi|_{\mathscr{L}}$ is a homomorphism. Finally, by (3.4.11), we have

$$\Phi(\Pi_-(p_1^{\nu_1} \cdots p_r^{\nu_r})) = y_{p_1}^{\nu_1} \cdots y_{p_r}^{\nu_r},$$

and by (3.3.44) we get, for $D \in \mathsf{G}$,

$$\Phi(U(D))_{\Gamma_0} = \alpha(D)(D)_\Gamma \quad \text{with } \alpha(D) \neq 0,$$

which imply that the homomorphism Φ is onto. □

Theorem 3.4.7. *The restriction $\Phi|_{\mathscr{C}_-^n(q)}$ and $\Phi|_{\mathscr{C}_+^n(q)}$ of Φ to the subring $\mathscr{C}_-^n(q)$ and $\mathscr{C}_+^n(q)$ of $\mathscr{L}^n$ are monomorphisms; in particular, the rings $\mathscr{C}_-^n(q)$ and $\mathscr{C}_+^n(q)$ are commutative rings without zero divisors.*

Proof. For example, let us consider the ring $\mathscr{C}_+(q)$. The ring $\mathscr{C}_-(q)$ is considered in the same way. According to Proposition 3.4.4 each element X of $\mathscr{C}_+(q)$ can be written in the form

$$X = \sum_i a_i (U(\mu_i, D_i))_{\Gamma_0},$$

where all D_i satisfy $\mu_i | d_1(D_i)^2$, all double cosets $(U(\mu_i, D_i))_{\Gamma_0}$ are pairwise distinct, and all $a_i \neq 0$. Then it follows from (3.4.22) that

$$\Phi(X) = \sum_i a_i \cdot \alpha_i \cdot \left(\prod_p y_p^{\nu_p(\mu_i)} \right) \cdot (D_i)_\Lambda,$$

where $\alpha_i \in \mathbb{N}$. The terms of this sum with distinct indices i have either distinct μ_i or distinct $(D_i)_\Lambda$, so that there cannot be any cancellation. Hence, $\Phi(X) \neq 0$. □

The last theorem shows that the rings $\mathscr{C}_-(q)$ and $\mathscr{C}_+(q)$ can be identified with certain extensions of the global Hecke ring of the general linear group. On the other hand, the ring $\mathscr{L}$ contains the global Hecke rings of the symplectic group. This will allow us to consider the relations between the Hecke rings of symplectic and general linear groups.

Exercise 3.4.8. (1) *Show that an element $X = \sum_i a_i(\Gamma_0 M_i)$ of $\mathscr{L}^n(q)$, where all $a_i \neq 0$ and all of the left cosets $(\Gamma_0 M_i)$ are pairwise distinct, is contained in $\mathscr{C}^n_-(q)$ if and only if each of the left cosets $\Gamma_0 M_i$ contains a representative of the form $\begin{pmatrix} A_i & 0 \\ 0 & D_i \end{pmatrix}$.*

[Hint: To prove sufficiency use that (3.4.24) is equivalent to (3.4.25).]

(2) *Let $X \in \mathscr{L}^n(q)$. Show that there are $d_- = d_-(X)$ and $d_+ = d_+(X)$ of $\mathbb{N}_{(q)}$ such that*

$$\Pi_-(a)X \in \mathscr{C}^n_-(q) \qquad \textit{for all } a \textit{ of } d_- \cdot \mathbb{N}(q),$$

and

$$X\Pi_+(a) \in \mathscr{C}^n_+(q) \qquad \textit{for all } a \textit{ of } d_+ \cdot \mathbb{N}(q)$$

[Hint: For the first inclusion, use (1) to show that, for a double coset $(M)_{\Gamma_0}$, with the decomposition (3.3.44) one can take as d_+ the least integer d of $\mathbb{N}$ such that all of the matrices $dL\mathbb{N}^{-1}$ are integral.]

3.4.2. Local Rings

In § 3.3.3, for each prime p, we have introduced the local Hecke ring $\mathscr{L}^n_p = \mathrm{D}_\mathbb{Q}(\Gamma^n_0, \mathscr{S}^n_p)$ with the subring $\mathbf{L}^n_p$ isomorphic to the local Hecke ring of the symplectic group. It is clear that

$$\mathbf{L}^n_p \subset \mathscr{L}^n_p \subset \mathscr{L}^n(q) \quad \text{if } p \nmid q. \tag{3.4.28}$$

Let us define the local subrings of the rings $\mathscr{C}^n_\pm(q)$ by setting

$$\mathscr{C}^n_{-p} = \mathscr{C}^n_-(q) \cap \mathscr{L}^n_p \quad \text{and} \quad \mathscr{C}^n_p = \mathscr{C}^n_+(q) \cap \mathscr{L}^n_p. \tag{3.4.29}$$

It follows from Proposition 3.4.4 that the rings are independent of q with $p \nmid q$. Theorem 3.4.7 implies the following local theorem:

Theorem 3.4.9. *The rings $\mathscr{C}^n_{-p}$ and $\mathscr{C}^n_p$ are commutative and contain no zero divisors. Moreover, the restrictions of the mappings Φ^n_p and Ω^n_p defined in § 3.3.3 to each of the subrings $\mathscr{C}^n_{-p}$ and $\mathscr{C}^n_p$ are monomorphisms.*

Proof. Note that if we restrict the mapping Φ to $\mathscr{L}_p$ and then set $y_p = x_0$, we get the mapping Φ_p. The theorem then follows from Theorem 3.4.7 and Lemma 3.3.29. □

The properties of local rings for a fixed p will be considered in more details in the next section. Here we shall only look at some relations between the local rings for distinct primes.

Theorem 3.4.10. *The ring $\mathscr{C}^n_-(q)$ (resp. $\mathscr{C}^n_+(q)$) is generated by the subrings $\mathscr{C}^n_{-p}$ (resp. $\mathscr{C}^n_p$) where p runs over $\mathbb{P}_{(q)}$.*

Proof. By (3.4.18), it suffices to consider, for example, the ring $\mathscr{C}_-(q)$. By Proposition 3.4.4, in this case we only have to check that the double coset $(M)_{\Gamma_0}$ of a matrix M of the form (3.4.19) is a finite product of double cosets $(M_p)_{\Gamma_0} \in \mathscr{C}_{-p}$, where p runs over a finite subset of $\mathbb{P}_{(q)}$. Let $M = U(\mu, D)$. Taking another representative in the same double coset modulo Γ_0, we may assume that $D = \mathrm{ed}(D) = \mathrm{diag}(d_1, \ldots, d_n)$. For a prime p, we set

$$M_p = (\mu_p, D_p), \qquad \text{where } \mu_p = p^{v_p(\mu)} \text{ and } D_p = \mathrm{diag}(p^{v_p(d_1)}, \ldots, p^{v_p(d_n)}).$$

It is clear that M_p is not equal to the unit matrix only for a finite number of primes and each of the primes does not divide q. Each of the matrices M_p is contained in $\mathscr{S}^n_p$. Since $d_n^2 | \mu$, it follows that $d_n(D_p)^2 = p^{2v_p(d_n)} | p^{v_p(\mu)} = \mu_p$ for every p, and so $(M_p)_{\Gamma_0} \in \mathscr{C}_{-p}$. Clearly $\Pi_p \mu_p = \mu$. By Proposition 3.2.5, we have $\Pi_p (D_p)_\Lambda = (\Pi_p D_p)_\Lambda = (D)_\Lambda$. It then follows from Lemma 3.4.5 and definitions that

$$\prod_p (M_p)_{\Gamma_0} = (M)_{\Gamma_0}. \qquad \square$$

Proposition 3.4.11. *Let p and p_1 be two different prime numbers. Then the following assertions hold:*

(1) *Each element of $\mathbf{L}^n_p$ commutes with each element of $\mathscr{C}^n_{-p_1}$ and each element of $\mathscr{C}^n_{p_1}$;*

(2) *Each element of $\mathscr{C}^n_{-p}$ commutes with each element of $\mathscr{C}^n_{p_1}$.*

Proof. By Proposition 3.4.2 and (3.4.18), it suffices to prove the first assertion, for example, for the ring $\mathscr{C}^n_{-p_1}$. According to Proposition 3.4.4, in this case we only have to check that any double coset $(M)_{\Gamma_0}$ with $M = U(\mu, A_0) \in \mathscr{S}^n_{p_1}$, where $d_n(A_0)^2 | \mu$, commutes with each of the generators $\mathbf{T}(p) = \mathrm{i}(T(p))$ and $\mathbf{T}_j(p^2) = \mathrm{i}(T_j(p^2))$ $(1 \le j \le n)$ of $\mathbf{L}^n_p$ (see Theorem 3.3.23 and Lemma 3.3.26). By (3.3.58) and (3.3.61), to show this, it suffices to prove that $(M)_{\Gamma_0}$ commutes with each of the elements of $\mathscr{L}_p$ of the form Π_a with $0 \le a \le n$ and $\Pi^{(r)}_{a,b}$ with $0 \le r \le a$, $a + b \le n$. It follows from (3.3.59) and (3.4.21) that

$$(M)_{\Gamma_0}\Pi_a = \sum_{A \in \Lambda\backslash\Lambda A_0\Lambda} \sum_{\substack{D \in \Lambda\backslash\Lambda D_a\Lambda,\\ B \in \underline{B}(D)/\mathrm{mod}\, D}} \left(\Gamma_0\begin{pmatrix}\mu p(AD)^* & \mu A^*B\\ 0 & AD\end{pmatrix}\right). \tag{3.4.30}$$

In order to transform the last expression we need some technique. For a subring K of $\mathbb{Q}$ and a matrix D of G^n we set

$$\mathsf{B}_{\mathsf{K}}(D) = \{B \in \mathsf{M}_n(\mathsf{K}); {}^tBD = {}^tDB\}.$$

In the same way as the first part of Lemma 3.3.33, one can easily check, that

$$\mathsf{B}_{\mathsf{K}}(\alpha D\beta) = \alpha^* B_K(D)\beta, \quad \text{if } \alpha, \beta \in \mathsf{GL}_n(\mathsf{K}),$$

and one can take

$$\mathsf{B}_{\mathsf{K}}(\alpha D\beta)/\mathsf{S}_n(\mathsf{K})\alpha D\beta = \alpha^*\{\mathsf{B}_{\mathsf{K}}(D)/\mathsf{S}_n(\mathsf{K})D\}\beta.$$

Let now $\mathsf{K} = \mathbb{Z}[p_1^{-1}]$, and let D be an integral matrix with $\det D$ prime to p_1. Then each of the cosets of $\mathsf{B}_{\mathsf{K}}(D)$ modulo $\mathsf{S}_n(\mathsf{K})D$ contains an integral matrix. In fact, any matrix B of $\mathsf{B}_{\mathsf{K}}(D)$ can be written in the form $q^{-1}B_0$ with an integral matrix B_0 and $q = p_1^{\alpha}$, choosing then a S_0 in $\mathsf{S}_n(\mathbb{Z})$ satisfying $B_0 + S_0D \equiv 0 \pmod q$ (note that D is invertible modulo q), we get $B + q^{-1}S_0D \in \mathsf{B}(D)$. It follows that in this case one can take

$$\mathsf{B}_K(D)/\mathsf{S}_n(\mathsf{K})D = \underline{\mathsf{B}}(D)/\mathrm{mod}\, D.$$

Returning to the expression (3.4.30), by Proposition 3.2.5, we can write it in the form

$$(M)\cdot\Pi_a = \sum_{\substack{C \in \Lambda\backslash\Lambda A_0 D_a\Lambda,\\ B \in \mu\cdot\{\mathsf{B}_{\mathsf{K}}(C)/\mathsf{S}_n(\mathsf{K})C\}}} \left(\Gamma_0\begin{pmatrix}\mu pC^* & B\\ 0 & C\end{pmatrix}\right),$$

where $\mathsf{K} = \mathbb{Z}[p_1^{-1}]$. Since μ is invertible in K, it follows that the factor μ in the last condition can be omitted. Since the Hecke ring of Λ is commutative, it follows, by Proposition 3.2.5, that $(A_0D_a)_\Lambda = (A_0)_\Lambda(D_a)_\Lambda = (D_a)_\Lambda(A_0)_\Lambda = (D_aA_0)_\Lambda$. Hence, $\Lambda A_0 D_a\Lambda = \Lambda D_a A_0\Lambda$, and so we get

$$(M)\cdot\Pi_a = \sum_{\substack{C \in \Lambda\backslash\Lambda D_a A_0\Lambda,\\ B \in \{\mathsf{B}_{\mathsf{K}}(C)/\mathsf{S}_n(\mathsf{K})C\}}} \left(\Gamma_0\begin{pmatrix}\mu pC^* & B\\ 0 & C\end{pmatrix}\right).$$

A similar consideration shows that the last expression can be written in the form

$$= \sum_{\substack{D \in \Lambda\backslash\Lambda D_a\Lambda,\, A \in \Lambda\backslash\Lambda A_0\Lambda,\\ B \in \underline{B}(D)/\mathrm{mod}\, D}} \left(\left(\Gamma_0\begin{pmatrix}\mu p(DA)^* & BA\\ 0 & DA\end{pmatrix}\right)\right) = \Pi_a\cdot(M).$$

The relations $(M)\Pi_{a,b}^{(r)} = \Pi_{a,b}^{(r)}(M)$ can be proved in the same way as above, if one uses the decompositions (3.3.62) and notices that multiplication of a matrix in (3.3.62) by any matrix of $\Gamma_0 M\Gamma_0$ does not change the rank of the first matrix over $\mathbb{Z}/p\mathbb{Z}$.

According to Proposition 3.4.4 and Lemma 3.4.2, in order to prove the second assertion, we only have to check that the double cosets $(M_0)_{\Gamma_0}$ and $(N_0)_{\Gamma_0}$, where

$M_0 = U(\mu, A_0)$ is an integral matrix of $\mathcal{S}_p^n$ satisfying (3.4.19) and $N_0 = U(\nu, D_0)$ is an integral matrix of $\mathcal{S}_{p_1}^n$ satisfying (3.4.20), commute. By Lemma 3.4.5, the representatives for left cosets in the double cosets can be taken respectively in the form

$$M = U(\mu, A) \quad \text{and} \quad N = U(\nu, D)T(S),$$

where $A \in \Lambda\backslash\Lambda A_0\Lambda$, $D \in \Lambda\backslash\Lambda D_0\Lambda$ and $S \in \mathrm{S}_n(\mathbb{Z})/\nu^{-1}\,{}^tD\mathrm{S}_n(\mathbb{Z})D$. Since $p \neq p_1$, we can assume that each of the matrices S satisfies $S \equiv 0 \pmod{\mu}$. We have

$$MN = \begin{pmatrix} \mu\nu(AD)^* & \mu\nu(AD)^*S \\ 0 & AD \end{pmatrix} \in \Gamma_0 U(\mu\nu, AD)\Gamma_0 \subset \Gamma_0 N_0 M_0 \Gamma_0,$$

for $\Lambda AD\Lambda = \Lambda A_0\Lambda \cdot \Lambda D_0\Lambda = \Lambda D_0\Lambda \cdot \Lambda A_0\Lambda = \Lambda D_0 A_0\Lambda$, by Proposition 3.2.5 and Theorem 3.2.3. Similarly,

$$NM = \begin{pmatrix} \mu\nu(DA)^* & \mu\nu(DA)^*\mu^{-1}S[A] \\ 0 & DA \end{pmatrix} \in \Gamma_0 U(\mu\nu, DA)\Gamma_0 \subset \Gamma_0 N_0 M_0 \Gamma_0.$$

It follows that

$$(M_0)(N_0) = \alpha(N_0 M_0) \quad \text{and} \quad (N_0)(M_0) = \beta(N_0 M_0)$$

with some constants α and β. Comparing the numbers of left cosets in the two sides each of the relations, we conclude that $\alpha = \beta$. □

Exercise 3.4.12. *Show that the subring $\mathcal{C}_{-p}^n$ (resp. $\mathcal{C}_p^n$) is equal to the centralizer of $\Pi_-(p)$ (resp. $\Pi_+(p)$) in $\mathcal{L}_p^n$.*

3.4.3. Decomposition of Elements $\mathbf{T}^n(a)$ for $n = 1, 2$

It often happens that elements of Hecke rings of the symplectic group can be decomposed in more simple elements belonging to certain Hecke rings of the triangular subgroup. Such decompositions for local rings will be considered in the next section. Here, remaining in the global situation, we shall obtain the decompositions of the images $\mathbf{T}^n(a) \in \mathcal{L}^n(q)$ of the elements (3.3.19) under the mapping (3.4.4) for $n = 1$ and 2.

Proposition 3.4.13. *Let $q \in \mathbb{N}$, and let $a \in \mathbb{N}_{(q)}$. Then the following relations hold:*

$$\mathbf{T}^1(a) = \sum_{d_1, d_2 \in \mathbb{N},\, d_1 d_2 = a} \Pi_+^1(d_1)\Pi_-^1(d_2), \tag{3.4.31}$$

$$\mathbf{T}^2(a) = \sum_{d_i \in \mathbb{N},\, d_1 d_2 d_3 = a} \Pi_+^2(d_1)\Pi(d_2)\Pi_-^2(d_3), \tag{3.4.32}$$

where $\Pi_\pm^n(d)$ are the elements (3.4.10), *and*

$$\Pi(d) = \Pi_1^2(d) = (U(d, D(d)))_{\Gamma_0^2} \quad \text{with } D(d) = \begin{pmatrix} 1 & 0 \\ 0 & d \end{pmatrix}. \tag{3.4.33}$$

Proof. It follows from (3.3.22), Lemma 3.3.11 and definitions that

$$\mathbf{T}^1(a) = \sum_{\substack{d_1|a,\\ b \bmod d_1}} \left(\Gamma_0\begin{pmatrix} a/d_1 & b \\ 0 & d_1\end{pmatrix}\right) = \sum_{\substack{d_1 d_2 = a,\\ b \bmod d_1}} \left(\Gamma_0\begin{pmatrix} d_2 & b \\ 0 & d_1\end{pmatrix}\right),$$

and

$$\mathbf{T}^2(a) = \sum_{\substack{d_1, d_2 \in \mathbb{N},\\ d_1 d_2 | a}} \sum_{\substack{D \in \Lambda\backslash \Lambda D(d_2)\Lambda,\\ B \in \underline{\mathrm{B}}(d_1 D)/\bmod d_1 D}} \left(\Gamma_0\begin{pmatrix} (a/d_1)D^* & B \\ 0 & d_1 D\end{pmatrix}\right)$$

$$= \sum_{d_1 d_2 d_3 = a} \sum_{\substack{D \in \Lambda\backslash \Lambda D(d_2)\Lambda,\\ B \in \underline{\mathrm{B}}(D)/\bmod d_1 D}} \left(\Gamma_0\begin{pmatrix} d_3 d_2 D^* & B \\ 0 & d_1 D\end{pmatrix}\right).$$

On the other hand, by (3.4.11), we have

$$\sum_{d_1 d_2 = a} \Pi_+^1(d_1)\Pi_-^1(d_2) = \sum_{d_1 d_2 = a} \left(\sum_{b \bmod d_1} \left(\Gamma_0\begin{pmatrix} 1 & b \\ 0 & d_1\end{pmatrix}\right)\right)\left(\Gamma_0\begin{pmatrix} d_2 & 0 \\ 0 & 1\end{pmatrix}\right)$$

$$= \sum_{\substack{d_1 d_2 = a,\\ b \bmod d_1}} \left(\Gamma_0\begin{pmatrix} d_2 & b \\ 0 & d_1\end{pmatrix}\right),$$

which proves (3.4.31). Further, using (3.4.11) and the decomposition

$$\Pi(d) = \sum_{\substack{D \in \Lambda\backslash \Lambda D(d)\Lambda,\\ B \in \underline{\mathrm{B}}(D)/\bmod D}} \left(\Gamma_0\begin{pmatrix} dD^* & B \\ 0 & D\end{pmatrix}\right) \tag{3.4.34}$$

which easily follows from Lemma 3.3.33, we see that the sum on the right side of (3.4.32) is equal to

$$\sum_{d_1 d_2 d_3 = a} \sum_{\substack{D \in \Lambda\backslash \Lambda D(d_2)\Lambda,\\ B \in \underline{\mathrm{B}}(D)/\bmod D;\\ S \in \mathrm{S}_2(\mathbb{Z})/\bmod d_1}} \left(\Gamma_0\begin{pmatrix} d_3 d_2 D^* & B + SD \\ 0 & d_1 D\end{pmatrix}\right).$$

Comparing this expression and the decomposition of $\mathbf{T}^2(a)$ obtained above, we see that in order to prove (3.4.32) it suffices to show that $B + SD$ runs over the set $\underline{\mathrm{B}}(D)/\bmod d_1 D$ as B and S run over the sets $\underline{\mathrm{B}}(D)/\bmod D$ and $\mathrm{S}_2(\mathbb{Z})/\bmod d_1$ respectively. The reader will easily check it, using Lemma 3.3.33. □

Exercise 3.4.14. *Prove the following formal identities*

$$\sum_{a \in \mathbb{N}_{(q)}} \mathbf{T}^1(a)a^{-s} = \left(\sum_{d \in \mathbb{N}_{(q)}} \Pi_+^1(d)d^{-s}\right)\left(\sum_{d \in \mathbb{N}_{(q)}} \Pi_-^1(d)d^{-s}\right)$$

$$= \left\{\prod_{p \in \mathbb{P}_{(q)}} (1 - \Pi_+^1(p)p^{-s})^{-1}\right\} \cdot \left\{\prod_{p \in \mathbb{P}_{(q)}} (1 - \Pi_-^1(p)p^{-s})^{-1}\right\}$$

$$= \prod_{p \in \mathbb{P}_{(q)}} \{(1 - \Pi_-^1(p)p^{-s})(1 - \Pi_+^1(p)p^{-s})\}^{-1}.$$

Exercise 3.4.15. *Show that*

(1) $\Pi(d)\Pi(d_1) = \Pi(dd_1)$ *if* g.c.d.$(d, d_1) = 1$;

(2) *if p is a prime, then the following formal identity holds*

$$\sum_{\delta=0}^{\infty} \Pi(p^\delta)v^\delta = (1 - \Pi_+^2(p)v)\mathbf{Q}_p^2(v)^{-1}(1 - \Pi_-^2(p)v)(1 - p^2[\mathbf{p}]v^2),$$

where $\mathbf{Q}_p^2(v)$ is the polynomial defined in Proposition 3.3.35.

[Hint: Use Proposition 3.3.35.]

§ 3.5. Factorization of Symplectic Polynomials

In this section we develop some machinery which allows us to reduce certain problems of the theory of Hecke rings and Hecke operators for the symplectic group to the analogous problems for the general linear group. Since the theories are "multiplicative", we can limit ourselves to the consideration of local rings for a fixed prime p. The Hecke ring $\mathscr{L}_p^n = \mathrm{D}_{\mathbb{Q}}(\Gamma_0^n, \mathscr{S}_p^n)$ will serve as a stage for our play. The ring $\mathbf{L}_p^n = \mathrm{i}_q(\mathrm{L}_p^n(q))$ will take the part of the local Hecke ring of $\Gamma_0^n(q)$ with $p \nmid q$; the local Hecke ring of Λ^n appears as one of the two dual extensions $\mathscr{C}_{-p}^n$ or $\mathscr{C}_p^n$. In this section we fix a prime p and a $n \in \mathbb{N}$; the indices p and n are usually omitted.

3.5.1. "Negative Powers" of Frobenius Elements

By analogy with algebraic geometry we call

$$\Pi_- = \Pi_-^n(p) \quad \text{and} \quad \Pi_+ = \Pi_+^n(p)$$

the Frobenius elements of $\mathscr{L} = \mathscr{L}_p^n$. In the previous section we have defined the commutative integral domains $\mathscr{C}_- = \mathscr{C}_{-p}^n$ and $\mathscr{C}_+ = \mathscr{C}_p^n$. The rings $\mathscr{C}_-$ and $\mathscr{C}_+$ can also be defined by

$$\mathscr{C}_- = \{X \in \mathscr{L}; \Pi_- X = X\Pi_-\} \quad \text{and} \quad \mathscr{C}_+ = \{X \in \mathscr{L}; \Pi_+ X = X\Pi_+\}, \tag{3.5.1}$$

for the inclusions $\subset$ follow from the definitions, and the inverse inclusions follow from the local variant of Proposition 3.4.4, which we leave to the reader. It follows from (3.4.18) or from (3.4.15) and (3.5.1) that the rings $\mathscr{C}_-$ and $\mathscr{C}_+$ are dual to each other with respect to the anti-automorphism $*$:

$$\mathscr{C}_-^* = \mathscr{C}_+, \qquad \mathscr{C}_+^* = \mathscr{C}_-. \tag{3.5.2}$$

We now show that every element of $\mathscr{L}$ can be "projected" to the rings $\mathscr{C}_-$ and $\mathscr{C}_+$. Let

$$X = \sum_i a_i\left(\Gamma_0\begin{pmatrix} A_i & B_i \\ 0 & D_i \end{pmatrix}\right) \tag{3.5.3}$$

be a non-zero element of $\mathscr{L}$. We assume that all $a_i \neq 0$ and all of the left cosets in (3.5.3) are pairwise distinct (we then say that the decomposition is irreducible). Then all elementary divisors of the matrices D_i as well as denominators of all entries of the matrices B_i are powers of p. Hence, there is a non-negative integer δ such that each of the matrices $p^\delta B_i D_i^{-1}$ is integral. The smallest such δ will be called *the left exponent of* X. We let $\delta_-(X)$ denote the left exponent of X. The number $\delta_+(X) = \delta_-(X^*)$ will be called *the right exponent of* X. We set $\delta_-(0) = \delta_+(0) = 0$.

Proposition 3.5.1. *Let X be a non-zero element of $\mathscr{L}$. Then for any integer d with $d \geq \delta_-(X)$ (resp. $d \geq \delta_+(X)$). The inclusion $\Pi_-^d X \in \mathscr{C}_-$ (resp. $X\Pi_+^d \in \mathscr{C}_+$) holds.*

Proof. It follows from the definition and the assumption that $\delta_-(\Pi_-^d X) = 0$ (resp. $\delta_+(X\Pi_+^d) = 0$). Thus, to prove the proposition it is sufficient to prove the following lemma.

Lemma 3.5.2. *The subsets $\mathscr{C}_-, \mathscr{C}_+ \subset \mathscr{L}$ can be described by*

$$\mathscr{C}_- = \{X \in \mathscr{L}; \delta_-(X) = 0\},$$
$$\mathscr{C}_+ = \{X \in \mathscr{L}; \delta_+(X) = 0\}.$$

Proof of the lemma. By definitions and (3.5.2), we only have to consider the subset $\mathscr{C}_-$. It follows from Proposition 3.4.4 and decomposition (3.4.21) that $\delta_-(X) = 0$, for every $X \in \mathscr{C}_-$. Conversely, let X be a non-zero element of $\mathscr{L}$ with the irreducible decomposition (3.5.3). Let us assume that $\delta_-(X) = 0$. Then each of the matrices $V_i = B_i D_i^{-1}$ is integral and symmetric. Hence, we can write

$$X = \sum_i a_i \left(\Gamma_0 \begin{pmatrix} E & -V_i \\ 0 & E \end{pmatrix} \begin{pmatrix} A_i & B_i \\ 0 & D_i \end{pmatrix}\right) = \sum_i a_i \left(\Gamma_0 \begin{pmatrix} A_i & 0 \\ 0 & D_i \end{pmatrix}\right)$$
$$= \sum_i a_i \left(\Gamma_0 \begin{pmatrix} A_i & 0 \\ 0 & D_i \end{pmatrix} \begin{pmatrix} E & S \\ 0 & E \end{pmatrix}\right) = \sum_i a_i \left(\Gamma_0 \begin{pmatrix} A_i & A_i S \\ 0 & D_i \end{pmatrix}\right)$$

with an arbitrary $S \in \mathrm{S}_n(\mathbb{Z})$. Again by $\delta_-(X) = 0$, we see that each of the matrices $A_i S D_i^{-1} = \mu_i D_i^* S D_i^{-1}$ is integral. Thus, the element X is a linear combination of the double cosets of matrices of the form $M_i = U(\mu_i, D_i)$ which satisfy (3.4.24) and therefore satisfy (3.4.25). Then $X \in \mathscr{C}_-$ by Proposition 3.4.4. □

The following lemma gives us an easy way to find integers d satisfying the conditions of Proposition 3.5.1 for an element X of the subring $\mathbf{L} = \mathbf{L}_p^n \subset \mathscr{L}$, if the image $\Omega(X)$ of X under the action of $\Omega = \Omega_p^n$ is known.

Lemma 3.5.3. *For any $X \in \mathscr{L}$ and $b \in \mathbb{Z}$ the following relations hold*

$$\delta_-([\mathbf{p}]^b X) = \delta_-(X), \qquad \delta_+([\mathbf{p}]^b X) = \delta_+(X),$$

where $[\mathbf{p}] = [\mathbf{p}]_n$ *is the element* (3.3.48). *If* $X \in \mathbf{L}$ *is a linear combination of the double cosets of integral matrices, then*

$$\delta_-(X) = \delta_+(X) \le \deg_{x_0} \Omega(X),$$

where $\deg_{x_0} \Omega(X)$ *denotes the degree of the polynomial* $\Omega_p^n(X)$ *in the variable* x_0.

Proof. The first statement follows at once from definitions. Since $X \in \mathbf{L}$, it follows that $X^* = X$, by Proposition 3.4.2. Thus, $\delta_+(X) = \delta_-(X^*) = \delta_-(X)$. We can write $X = \mathrm{i}(\sum_\alpha a_\alpha (M_\alpha)_{\Gamma^n})$ where all $a_\alpha \neq 0$, each of M_α is integral and belongs to $\mathbf{S}_p^n$, and all of the double cosets are pairwise distinct. By the definition of Ω and the first part of Theorem 3.3.30, the degree of a polynomial $\Omega(\mathrm{i}((M_\alpha)_\Gamma))$ in the variable x_0 is equal to $\delta_\alpha = v_p(\mu(M_\alpha))$, and the polynomials are linearly independent over $\mathbb{Q}$. Hence, the degree of $\Omega(X)$ in x_0 is equal to the maximum of δ_α. On the other hand, each of the left cosets in the decomposition of $\mathrm{i}((M_\alpha)_\Gamma)$ consists of integral matrices of the form $\begin{pmatrix} p^{\delta_\alpha} D^* & B \\ 0 & D \end{pmatrix}$ for which the matrix $p^{\delta_\alpha} B D^{-1}$ is clearly integral. Hence, $\delta_-(X) \le \max_\alpha \delta_\alpha$. □

The elements of the form $\Pi_-^d X \in \mathscr{C}_-$ ($X \Pi_+^d \in \mathscr{C}_+$) will be called the *left* (*right*) *projection* of X. By Lemma 3.4.5, elements of the subrings $\mathscr{C}_-$ and $\mathscr{C}_+$ have simple decomposition into left cosets, and so calculations in these subrings are considerably simpler than calculations in the whole $\mathscr{L}$. Therefore one can try to reduce operations on some elements of $\mathscr{L}$ to those on their projections. This leads to the problem of reconstructing an element $X \in \mathscr{L}$ from its projections. In general, this problem is uncertain, since Π_- (resp. Π_+) is a left (resp. right) zero divisor in $\mathscr{L}$. However, the problem has the unique solution in some important cases. For example, any element X of $\mathbf{L} = \mathbf{L}_p^n$ can be uniquely reconstructed from its projection $\Pi_-^d X \in \mathscr{C}_-$ (resp. $X \Pi_+^d \in \mathscr{C}_+$) in the form $X = \Pi_-^d X \Pi_-^{-d}$ (resp. $X = \Pi_+^{-d} X \Pi_+^d$), where Π_-^{-d} and Π_+^{-d} are some canonical elements of $\mathscr{L}$. This is based on the remarkable fact that the elements Π_- and Π_+ satisfy certain algebraic equations with coefficients in $\mathbf{L}$. More precisely:

Proposition 3.5.4. *The following equalities hold*:

$$\sum_{i=0}^{m} (-1)^i \Pi_-^i \mathbf{q}_{m-i} = 0, \quad \sum_{i=0}^{m} (-1)^i \mathbf{q}_{m-i} \Pi_+^i = 0,$$

where $m = 2^n$ *and* $\mathbf{q}_j = \mathbf{q}_j^n(p)$ *are the elements* (3.3.77) *of the ring* $\mathbf{L}$.

Proof. It follows from (3.4.9) and (3.4.15) that the anti-automorphism $*$ transform the first of the equalities to the second and vice versa. So it is sufficient to verify the first equality. Take

$$Y = \sum_{i=0}^{m} (-1)^i \Pi_-^i \mathbf{q}_{m-i}.$$

By (3.3.79), we can rewrite Y as

$$Y = \sum_{i=0}^{m} (-1)^i (p^{\langle n \rangle}[\mathbf{p}])^{(m/2)-i} \Pi_-^i \mathbf{q}_i.$$

By the definition, $\Omega(\mathbf{q}_i)$ is a polynomial in $x_0, x_1, \ldots, x_n$ whose degree in the variable x_0 is i. Then $\mathbf{q}_i$ is a linear combination of double cosets of integral matrices and $\delta_-(\mathbf{q}_i) \leq i$, by Theorem 3.3.30(1) and Lemma 3.5.3. Hence each of the products $\Pi_-^i \mathbf{q}_i$ is contained in $\mathscr{C}_-$; by Proposition 3.5.1. Since clearly $[\mathbf{p}] \in \mathscr{C}_-$, it follows that $Y \in \mathscr{C}_-$. On the other hand, by Lemma 3.3.34, we have $\Omega(\Pi_-) = \Omega(\Pi_0^n(p)) = x_0$. Hence, using the definition of the elements $\mathbf{q}_i$, we get

$$\begin{aligned}\Omega(Y) &= \sum_{i=0}^{m} (-1)^i x_0^i q_{m-i}(x_0, \ldots, x_n)\\ &= x_0^m q(x_0, \ldots, x_n; x_0^{-1}) = 0,\end{aligned}$$

where $q(x_0, \ldots, x_n; v)$ is the polynomial (3.3.76). Thus, $Y = 0$, by Theorem 3.4.7. □

Multiplying the relations of Proposition 3.5.4 on the left by Π_-^d and on the right by Π_+^d, respectively, with $d \geq 0$, we obtain the relations

$$\sum_{i=0}^{m} (-1)^i \Pi_-^{d+i} \mathbf{q}_{m-i} = 0, \quad \sum_{i=0}^{m} (-1)^i \mathbf{q}_{m-i} \Pi_+^{d+i} = 0, \tag{3.5.4}$$

which can be considered as recurrence relations for the sequences of non-negative powers of the Frobenius elements. Since $\mathbf{q}_0 = \mathbf{1}$, the relations determine the powers with exponents greater than $2^n - 1$ as linear combination with (right or left) coefficients in $\mathbf{L}$ of the powers with smaller exponents. On the other hand, by (3.3.80) and Lemma 3.3.27, the coefficient q_m is invertible in $\mathscr{L}$, and

$$\mathbf{q}_m^{-1} = (p^{\langle n \rangle}[\mathbf{p}])^{-m/2}.$$

Thus, the relations (3.5.4) also can be used to determine powers of the Frobenius elements with small exponents in terms of powers with greater exponents. It follows that the sequences $\mathbf{1} = \Pi_-^0, \Pi_-, \Pi_-^2, \ldots$ and $\mathbf{1} = \Pi_+^0, \Pi_+, \Pi_+^2, \ldots$ can be uniquely extended to the left, by defining for each negative d certain elements Π_-^d and Π_+^d of $\mathscr{L}$ in such a way that the relations (3.5.4) are valid for all $d \in \mathbb{Z}$. More precisely: if the elements Π_-^δ and Π_+^δ with $\delta > d$ are already defined, then we set

$$\Pi_-^d = -\mathbf{q}_m^{-1} \left(\sum_{i=1}^{m} (-1)^i \Pi_-^{i+d} \mathbf{q}_{m-i} \right)$$

and (3.5.5)

$$\Pi_+^d = -\mathbf{q}_m^{-1} \left(\sum_{i=1}^{m} (-1)^i \mathbf{q}_{m-i} \Pi_+^{i+d} \right).$$

For brevity we call the elements Π_-^d and Π_+^d with $d < 0$ *the "negative powers"* of

Π_- and Π_+ respectively, although, in fact, Π_- and Π_+ have no real negative powers in $\mathcal{L}$ or its extensions. The "negative powers" are not even powers of Π_-^{-1} and Π_+^{-1}. For example, $\Pi_-^{-2} \neq (\Pi_-^{-1})^2$ and $\Pi_+^{-2} \neq (\Pi_+^{-1})^2$. By induction on $|d|$, using the relations (3.4.9), (3.4.15) and (3.5.5), one can easily check that

$$(\Pi_-^d)^* = \Pi_+^d \quad \text{and} \quad (\Pi_+^d)^* = \Pi_-^d \quad \text{for all } d \in \mathbb{Z}. \tag{3.5.6}$$

Let us now turn to the problem of reconstructing of elements of $\mathcal{L}$ from their projections. For this we consider the following $\mathbb{Q}$-linear subspaces of $\mathcal{L}$:

$$\begin{aligned} \mathcal{O}_- &= \mathcal{O}_{-p}^n = \mathscr{C}_{-p}^n \mathbf{L}_p^n = \left\{ \sum_\alpha X_\alpha \mathbf{T}_\alpha ; X_\alpha \in \mathscr{C}_-, \mathbf{T}_\alpha \in \mathbf{L} \right\} \\ \mathcal{O}_+ &= \mathcal{O}_p^n = \mathbf{L}_p^n \mathscr{C}_p^n = \left\{ \sum_\alpha \mathbf{T}_\alpha Y_\alpha ; \mathbf{T}_\alpha \in \mathbf{L}, Y_\alpha \in \mathscr{C}_+ \right\}. \end{aligned} \tag{3.5.7}$$

It follows from (3.4.9) and (3.5.2) that the subspaces are dual with respect to $*$:

$$\mathcal{O}_-^* = \mathcal{O}_+, \qquad \mathcal{O}_+^* = \mathcal{O}_-. \tag{3.5.8}$$

From the definition of $\Pi_\pm^d$, by an induction, we get the inclusions

$$\Pi_-^d \in \mathcal{O}_- \quad \text{and} \quad \Pi_+^d \in \mathcal{O}_+ \quad \text{for } d \in \mathbb{Z}. \tag{3.5.9}$$

Theorem 3.5.5. *The following assertions hold:*

(1) *Each element X of $\mathcal{O}_-$ satisfies*

$$X = \Pi_-^\delta X \Pi_-^{-\delta} \quad \textit{for every } \delta \geq \delta_-(X). \tag{3.5.10}$$

Each element X of $\mathcal{O}_+$ satisfies

$$X = \Pi_+^{-\delta} X \Pi_+^\delta \quad \textit{for every } \delta \geq \delta_+(X). \tag{3.5.11}$$

Conversely, each element X of $\mathcal{L}$ satisfying one of the relations (3.5.10) (*resp.* (3.5.11)) *is contained in the subspace $\mathcal{O}_-$ (resp. $\mathcal{O}_+$).*

(2) *The restrictions of the mappings Φ and Ω to each of the subspaces $\mathcal{O}_-$, $\mathcal{O}_+$ are monomorphisms.*

We first prove a lemma.

Lemma 3.5.6. *Each element $\mathbf{T}$ of $\mathbf{L}$ satisfies the relations*

$$\Pi_-^\delta \mathbf{T} \Pi_-^d = \Pi_-^{\delta+d} \mathbf{T} \quad \textit{for every } \delta \geq \delta_-(\mathbf{T}) \textit{ and } d \in \mathbb{Z}, \tag{3.5.12}$$

$$\Pi_+^d \mathbf{T} \Pi_-^\delta = \mathbf{T} \Pi_+^{d+\delta} \quad \textit{for every } \delta \geq \delta_+(\mathbf{T}) \textit{ and } d \in \mathbb{Z}. \tag{3.5.13}$$

Proof of the lemma. The mapping $*$ carries (3.5.12) to (3.5.13) and vice versa. Thus, it suffices to prove, for example, (3.5.12). We shall apply an induction. If $d \geq 0$, then $\Pi_-^d \in \mathscr{C}_-$ and (3.5.12) follows from Proposition 3.5.1 and the fact that $\mathscr{C}_-$ is commutative. Assume that (3.5.12) is already proved for all $\delta \geq \delta_-(\mathbf{T})$ and all $d' > d$. Since the ring $\mathbf{L}$ is commutative, it follows from (3.5.5) that

$$\Pi_-^\delta \mathbf{T} \Pi_-^d = -\mathbf{q}_m^{-1}\left(\sum_{i=1}^{m}(-1)^i \Pi_-^\delta T \Pi_-^{i+d}\mathbf{q}_{m-i}\right)$$
$$= \mathbf{q}_m^{-1}\left(\sum_{i=1}^{m}(-1)^i \Pi_-^{\delta+d+i}\mathbf{q}_{m-i}\mathbf{T}\right) = \Pi_-^{\delta+d}\mathbf{T}. \qquad \square$$

Proof of the theorem. By the duality (see (3.5.6) and (3.5.8)), it suffices to prove the first assertion only for the ring $\mathcal{O}_-$. Let $X \in \mathcal{O}_-$. Then by the definition we have

$$X = \sum_\alpha X_\alpha \mathbf{T}_\alpha \qquad \text{with } X_\alpha \in \mathscr{C}_- \text{ and } \mathbf{T}_\alpha \in \mathbf{L}$$

(a finite sum). If δ' is not less than each of the exponents $\delta_-(\mathbf{T}_\alpha)$ then, by (3.5.12), we get

$$\Pi_-^{\delta'} X \Pi_-^{-\delta'} = \sum_\alpha X_\alpha \Pi_-^{\delta'} \mathbf{T}_\alpha \Pi_-^{-\delta'} = \sum_\alpha X_\alpha \mathbf{T}_\alpha = X.$$

Suppose, now, that $\delta \geq \delta_-(X)$. Then, by the above and Proposition 3.5.1, we have

$$X = \Pi_-^{\delta+\delta'} X \Pi_-^{-(\delta+\delta')} = \Pi_-^\delta X \Pi_-^{\delta'} \Pi_-^{-(\delta+\delta')} = \Pi_-^\delta X \Pi_-^{-\delta},$$

where we have also used (3.5.12) with $\mathbf{T} = \mathbf{1}$. Conversely, if an element X of $\mathscr{L}$ has the form $X = \Pi_-^\delta X \Pi_-^{-\delta}$ with some $\delta \geq \delta_-(X)$, then it follows from Proposition 3.5.1 and the first inclusion in (3.5.9) that $X \in \mathcal{O}_-$. The first part of the theorem is proved.

We prove the second part, for example, for the mapping $\Omega = \Omega_p^n$. First of all, we note that

$$\Omega(\Pi_-^d) = x_0^d \quad \text{and} \quad \Omega(\Pi_+^d) = (x_0 x_1 \cdots x_n)^d \quad \text{for every } d \in \mathbb{Z}. \tag{3.5.14}$$

In fact, if $d \geq 0$, it was proved in Lemma 3.3.34, but if $d = -\delta < 0$, then, by Lemma 3.5.6, we have $\Pi_-^\delta \Pi_-^{-\delta} = \mathbf{1}$ and $\Pi_+^{-\delta}\Pi_+^\delta = \mathbf{1}$, whence $\Omega(\Pi_-^d) = \Omega(\Pi_-^\delta)^{-1} = (x_0^\delta)^{-1} = x_0^d$, and similarly for Π_+^d. Now assume that $X \in \mathcal{O}_-$ and $\Omega(X) = 0$. Take a δ with $\delta \geq \delta_-(X)$ Then

$$0 = \Omega(X) = \Omega(\Pi_-^\delta X \Pi_-^{-\delta}) = \Omega(\Pi_-^\delta X)\cdot x_0^{-\delta},$$

by (3.5.10), hence $\Omega(\Pi_-^\delta X) = 0$. This implies the relation $\Pi_-^\delta X = 0$, by Theorem 3.4.9. Thus, $X = (\Pi_-^\delta X)\cdot \Pi_-^{-\delta} = 0$. A similar consideration holds for the space $\mathcal{O}_+$. $\square$

Exercise 3.5.7. (1) *Show that a mapping*

$$X \to (p^{\langle n\rangle}[\mathbf{p}])^{-d}\Pi_-^d X \Pi_+^d \qquad \textit{with } d \geq \min(\delta_-(X), \delta_+(X))$$

does not depend on d, commutes with $*$, and maps the space $\mathscr{L}$ onto the subspace

$$\mathscr{C}_-\mathscr{C}_+ = \left\{\sum_\alpha X_\alpha \cdot Y_\alpha; X_\alpha \in \mathscr{C}_-, Y_\alpha \in \mathscr{C}_+\right\}.$$

Show that subspace is equal to the set of all elements in $\mathscr{L}$ which are invariant *under the mapping. Deduce that the restrictions of Φ and Ω to $\mathscr{C}_-\mathscr{C}_+$ are monomorphisms;*

(2) *Prove the inclusions*

$$\mathbf{T}(p), \quad \Pi_-\mathbf{T}_i(p^2), \quad \mathbf{T}_i(p^2)\cdot\Pi_+ \in \mathscr{C}_-\mathscr{C}_+.$$

Using the inclusions, show that if $\mathbf{T}\in\mathbf{L}$ and $\Omega(\mathbf{T})$ is a polynomial in $x_0, x_1, \ldots, x_n$ of degree δ in x_0, then for any $a, b \geq 0$ with $a + b \geq \delta - 1$, the following inclusion holds

$$\Pi_-^a\mathbf{T}\Pi_+^b \in \mathscr{C}_-\mathscr{C}_+.$$

[Hint: Use (1) and decompositions (3.3.58) and (3.3.61). For the last inclusion use the fact that $\mathbf{T}$ is a polynomial in $\mathbf{T}(p)$, $\mathbf{T}_i(p^2)$];

(3) *Show that*

$$\Pi_-^{-1}, \quad \Pi_+^{-1} \in \mathscr{C}_-\mathscr{C}_+.$$

Using this, prove the formulae

$$\Pi_-^{-1} = (p^{\langle n\rangle}[\mathbf{p}])^{-1}\Pi_+, \quad \Pi_+^{-1} = (p^{\langle n\rangle}[\mathbf{p}])^{-1}\Pi_-.$$

[Hint: Use (2), the definition of negative powers, and (1).]

Exercise 3.5.8. *Let*

$$\mathscr{B}_- = \{X\in\mathscr{L}; \Pi_-^\delta X = 0 \text{ for some } \delta \geq 0\},$$

$$\mathscr{B}_+ = \{X\in\mathscr{L}; X\Pi_+^\delta = 0 \text{ for some } \delta \geq 0\}.$$

$$\mathscr{B} = \{X\in\mathscr{L}; \Pi_-^\delta X\Pi_+^\delta = 0 \text{ for some } \delta \geq 0\}.$$

Prove the following assertions

(1) $\mathscr{L} = \mathscr{O}_- \oplus \mathscr{B}_- = \mathscr{O}_+ \oplus \mathscr{B}_+ = \mathscr{C}_-\mathscr{C}_+ \oplus \mathscr{B}$ *(decompositions in direct sums of the subspaces)*;

(2) $\mathscr{B}_- = \mathscr{B}_+ = \mathscr{B} = \operatorname{Ker}\Phi = \operatorname{Ker}\Omega$; *in particular, $\mathscr{B}$ is a bilateral ideal in $\mathscr{L}$, and each of the subspaces $\mathscr{O}_-$, $\mathscr{O}_+$ and $\mathscr{C}_-\mathscr{C}_+$ is maximal one among subspaces $V\in\mathscr{L}$ for which the restrictions $\Phi|_V$ and $\Omega|_V$ are monomorphic*;

(3) *The natural isomorphism of the space $\mathscr{O}_-$ onto the ring $\mathscr{L}/\mathscr{B}_-$ carries Π_- to an invertible element, and so it carries the "negative powers" of Π_- to the ordinary negative powers. The same for $\mathscr{O}_+$.*

3.5.2. Factorization of Symplectic Polynomials

Our consideration of elements of a local Hecke ring $\mathsf{L}_p^n(q)$ or $\mathbf{L}_p^n$ is based on the following scheme: the elements appear as coefficients of a generating series, a formal local zeta-function, which is formally a rational fraction over the base ring; the denominator of the fraction usually have a simple factorization, when it is considered over the ring $\mathbb{Q}[x_0^{\pm 1}, \ldots, x_n^{\pm 1}]$ by means of the mapping Ω_p^n, then

Theorem 3.5.9 gives a simple sufficient condition to factor such a polynomial over $\mathscr{L}_p^n$; the factorization allows us to express the coefficients of the polynomial and the coefficients of the corresponding generating series in terms of more simple elements. This approach is applied both in the theory of Hecke ring and the theory of Hecke operators to simplify some calculations, or even make them possible. In this subsection we shall start to realize the scheme and give some important examples. We again fix a $n \in \mathbb{N}$ and a prime p.

Theorem 3.5.9. *Let*

$$\mathbf{P}(v) = \sum_{i=0}^{N} \mathbf{p}_i v^i$$

be a polynomial with coefficients in the subring $\mathbf{L} = \mathbf{L}_p^n$ *of* $\mathscr{L} = \mathscr{L}_p^n$. *Suppose that the polynomial*

$$\Omega(\mathbf{P})(v) = \sum_{i=0}^{N} \Omega(\mathbf{p}_i) v^i,$$

where $\Omega = \Omega_p^n$ *is the spherical mapping* (3.3.49), *is a product of two polynomials with coefficients in* $\mathbb{Q}[x_0^{\pm 1}, \dots, x_n^{\pm 1}]$:

$$\Omega(\mathbf{P})(v) = F(v)G(v),$$

where

$$F(v) = \sum_{i=0}^{N_1} f_i v^i, \quad G(v) = \sum_{j=0}^{N_2} g_j v^j \quad \text{with } f_i, g_j \in \mathbb{Q}[x_0^{\pm 1}, \dots, x_n^{\pm 1}].$$

If $f_0 = 1$ *and all of the coefficients of* F *belong to the image* $\Omega(\mathscr{C}_-)$ *of* $\mathscr{C}_- = \mathscr{C}_{-p}^n$:

$$f_i = \Omega(f_i') \quad \text{with } f_i' \in \mathscr{C}_-,$$

then all of the coefficients of G *belong to the image* $\Omega(\mathscr{O}_-)$ *of the space* $\mathscr{O}_- = \mathscr{O}_{-p}^n$:

$$g_j = \Omega(g_j') \quad \text{with } g_j' \in \mathscr{O}_-,$$

and the polynomial $\mathbf{P}(v)$ *has the following factorization over* $\mathscr{L}$:

$$\mathbf{P}(v) = \left(\sum_{i=0}^{N_1} f_i' v^i\right)\left(\sum_{j=0}^{N_2} g_j' v^j\right).$$

But if $g_0 = 1$ *and all of the coefficients of* G *belong to the image* $\Omega(\mathscr{C}_+)$ *of* $\mathscr{C}_+ = \mathscr{C}_p^n$:

$$g_j = \Omega(g_j') \quad \text{with } g_j' \in \mathscr{C}_+,$$

then all of the coefficients of F *belong to the image* $\Omega(\mathscr{O}_+)$ *of* $\mathscr{O}_+ = \mathscr{O}_p^n$:

$$f_i = \Omega(f_i') \quad \text{with } f_i' \in \mathscr{O}_+,$$

and $\mathbf{P}(v)$ *has the factorization*

$$\mathbf{P}(v) = \left(\sum_{i=0}^{N_1} f_i' v^i\right)\left(\sum_{j=0}^{N_2} g_j' v^j\right).$$

Proof. It is sufficient to consider only one of the two dual assertions of the theorem, since one of them follows from another by application of $*$. For example, we consider the first assertion. Since the restriction $\Omega|_{\mathscr{C}_-}$ is a monomorphism, and $f_0 = 1$, it follows that $f_0' = \mathbf{1}$. Since $f_0' = \mathbf{1}$ and all of the f_i' belong to the commutative ring $\mathscr{C}_-$, it follows that there exists a formal power series with coefficients $a_0, a_1, \ldots$ in $\mathscr{C}_-$ such that

$$\left(\sum_{\delta=0}^{\infty} a_\delta v^\delta\right)\left(\sum_{i=0}^{N_1} f_i' v^i\right) = \mathbf{1}.$$

Let us consider the formal power series over $\mathscr{L}$ given by

$$\left(\sum_{\delta=0}^{\infty} a_\delta v^\delta\right)\mathbf{P}(v) = \sum_{j=0}^{\infty}\left(\sum_{\delta+i=j} a_\delta \mathbf{p}_i\right)v^i = \sum_{j=0}^{\infty} g_j' v^j.$$

By the definition, all of the coefficients g_j' belong to the space $\mathcal{O}_- = \mathscr{C}_-\mathbf{L}$. On the other hand, on replacing of the coefficients by their images under Ω, we get the series

$$\left(\sum_{\delta=0}^{\infty} \Omega(a_\delta)v^\delta\right)\left(\sum_{i=0}^{N_1} f_i v^i\right)\left(\sum_{j=0}^{N_2} g_j v^j\right) = \sum_{j=0}^{N_2} g_j v^j,$$

hence, $\Omega(g_j') = g_j$ for $j = 0, 1, \ldots, N_2$, and $\Omega(g_j') = 0$ for $j > N_2$. By Theorem 3.5.5, it follows that $g_j' = 0$ for $j > N_2$ and so

$$\mathbf{P}(v) = \left(\sum_{\delta=0}^{\infty} a_\delta v^\delta\right)^{-1}\left(\sum_{j=0}^{N_2} g_j' v^j\right) = \left(\sum_{i=0}^{N_1} f_i' v^i\right)\left(\sum_{j=0}^{N_2} g_j' v^j\right).$$

Since the elements $g_j' \in \mathcal{O}_-$ are uniquely determined by their images $\Omega(g_j') = g_j$, the theorem follows. □

Remark. For practical calculation of an $g' \in \mathcal{O}_-$ with given image $\Omega(g') = g \in \Omega(\mathcal{O}_-)$, one can find an arbitrary inverse image g_1 of g in $\mathscr{L}$ and then take

$$g' = \Pi_-^d g_1 \Pi_-^{-d} \qquad \text{with } d \geq \delta_-(g_1).$$

For it follows from (3.5.9) and Proposition 3.5.1 that $g' \in \mathcal{O}_-$, and by (3.5.14), we have $\Omega(g') = x_0^d \Omega(g_1) x_0^{-d} = \Omega(g)$. A similar consideration holds for $\mathcal{O}_+$. It is not usually difficult to find an inverse images in $\mathscr{L}$. Thus, the problem of finding of inverse images in $\mathcal{O}_-$ or $\mathcal{O}_+$ is reduced to that of calculation of the "negative powers" of Frobenius elements.

Now we consider some concrete examples which illustrate the application of Theorem 3.5.9. In § 3.3 (see (3.3.52)) we have introduced the polynomial $r(x_1, \ldots, x_n; v)$ over the ring $\mathbb{Q}[x_0^{\pm 1}, \ldots, x_n^{\pm 1}]$, whose coefficients are invariant to all of the transformations in the group W_n. By the third part of Theorem 3.3.30, there is a uniquely determined polynomial

$$\mathbf{R}_p^n(v) = \sum_{a=0}^{2n} \mathbf{r}_a(p)v^a \qquad \text{with } \mathbf{r}_a(p) = \mathbf{r}_a^n(p) \in \mathbf{L}_p^n \tag{3.5.15}$$

such that

$$\Omega(\mathbf{R}_p^n)(v) = \sum_{a=0}^{2n} \Omega(\mathbf{r}_a(p))v^a = r(x_1, \ldots, x_n; v). \tag{3.5.16}$$

Proposition 3.5.10. *The polynomial $\mathbf{R}(v) = \mathbf{R}_p^n(v)$ has the following factorizations over the ring $\mathscr{L} = \mathscr{L}_p^n$:*

$$\mathbf{R}(v) = \left(\sum_{i=0}^{n} (-1)^i (p^{\langle n\rangle}[\mathbf{p}])^{-1} \Pi_- \Pi_{n-i} v^i\right)\left(\sum_{i=0}^{n} (-1)^i \Pi_- \Pi_i \Pi_-^{-2} v^i\right), \tag{3.5.17}$$

$$\mathbf{R}(v) = \left(\sum_{i=0}^{n} (-1)^i \Pi_+^{-2} \Pi_{n-i} \Pi_+ v^i\right)\left(\sum_{i=0}^{n} (-1)^i (p^{\langle n\rangle}[\mathbf{p}])^{-1} \Pi_i \Pi_+ v^i\right), \tag{3.5.18}$$

where $[\mathbf{p}] = [\mathbf{p}]_n$ is the element (3.3.48), *$\Pi_- = \Pi_-^n(p) = \Pi_0^n(p)$, $\Pi_+ = \Pi_+^n(p) = \Pi_n^n(p)$, $\Pi_a = \Pi_a^n(p)$ are the elements* (3.3.59), *and Π_-^{-2}, Π_+^{-2} are the "negative squares" of Π_- and Π_+, respectively.*

Proof. We first prove the relations

$$\Pi_a^n(p)^* = \Pi_{n-a}^n(p) \qquad \text{for } a = 0, 1, \ldots, n. \tag{3.5.19}$$

If $D_a = \begin{pmatrix} E_{n-a} & 0 \\ 0 & pE_a \end{pmatrix}$, then clearly $\Lambda^n p D_a^{-1} \Lambda^n = \Lambda^n D_{n-a} \Lambda^n$, hence, by the definition of $*$ and (3.3.63), we get

$$\Pi_a^n(p)^* = (pM_a^{-1})_{\Gamma_0} = \left(\begin{pmatrix} D_a & 0 \\ 0 & p^{-1}D_a^{-1} \end{pmatrix}\right)_{\Gamma_0} = \left(\begin{pmatrix} pD_{n-a}^{-1} & 0 \\ 0 & D_{n-a} \end{pmatrix}\right)_{\Gamma_0} = \Pi_{n-a}^n(p).$$

It follows from the relations (3.5.19) and (3.5.6) that the anti-automorphism $*$ carries (2.5.17) to (2.5.18) and vice versa. Thus, if suffices to prove, for example, the factorization (2.5.17). By the definition, we have

$$\begin{aligned}\Omega(\mathbf{R}(v)) &= \left(\prod_{i=1}^{n} (1 - x_i^{-1}v)\right)\left(\prod_{i=1}^{n} (1 - x_i v)\right) \\ &= \left(\sum_{i=1}^{n} (-1)^i s_i(x_1^{-1}, \ldots, x_n^{-1}) v^i\right)\left(\sum_{i=1}^{n} (-1)^i s_i(x_1, \ldots, x_n) v^i\right).\end{aligned}$$

Lemma 3.3.34 implies the relations

$$\begin{aligned}\Omega((p^{\langle n\rangle}[\mathbf{p}])^{-1}\Pi_-\Pi_{n-i}) &= (x_0^2 x_1 \cdots x_n)^{-1} x_0 \cdot x_0 s_{n-i}(x_1, \ldots, x_n) \\ &= s_i(x_1^{-1}, \ldots, x_n^{-1}).\end{aligned}$$

It follows from definitions that $\delta_-(\Pi_i) = 1$ and so, by Proposition 3.5.1, we have the inclusions

$$\Pi_-\Pi_i, \quad (p^{\langle n\rangle}[\mathbf{p}])^{-1}\Pi_-\Pi_{n-i} \in \mathscr{C}_-. \tag{3.5.20}$$

Further, by Lemma 3.3.34, and (3.5.14), we get

$$\Omega(\Pi_-\Pi_i\Pi_-^{-2}) = x_0 \cdot x_0 \cdot s_i(x_1, \ldots, x_n) \cdot x_0^{-2} = s_i(x_1, \ldots, x_n). \qquad (3.5.21)$$

Finally, the inclusions (3.5.9) and (3.5.20) imply that

$$\Pi_-\Pi_i\Pi_-^{-2} \in \mathcal{O}_-. \qquad (3.5.22)$$

The factorization (3.5.17) follows from the relations and inclusions, by Theorem 3.5.9. □

Let us now turn to the polynomial $\mathbf{Q}_p^n(v)$ defined by (3.3.76)–(3.3.78). Since $\Omega(\Pi_-) = x_0$ and $\Omega(\Pi_+) = x_0 x_1 \cdots x_n$, Theorem 3.5.9 immediately implies the following

Proposition 3.5.11. *The polynomial* $\mathbf{Q}(v) = \mathbf{Q}_p^n(v)$ *has the following factorizations over the ring* $\mathcal{L} = \mathcal{L}_p^n$:

$$\mathbf{Q}(v) = (1 - \Pi_- v)Q_-(v),$$

$$\mathbf{Q}(v) = Q_+(v)(1 - \Pi_+ v),$$

where Q_- *and* Q_+ *are polynomials of degree* $2^n - 1$ *with coefficients in the space* $\mathcal{O}_{-p}^n$ *and* $\mathcal{O}_p^n$ *respectively.*

In order to use the factorizations, one needs first to express the coefficients of the factors as linear combinations of left or double cosets modulo Γ_0. The rest of this section deals with this problem for the polynomials $\mathbf{Q}_p^n$ with $n = 1, 2$ and $\mathbf{R}_p^n$ with $n \in \mathbb{N}$.

Exercise 3.5.12. *Let* $F(v)$ *be a polynomial of degree* N *with coefficients in* $\mathscr{C}_{-p}^n$, *and let* $F(0) = 1$. *Show that there exists a polynomial* $G(v)$ *of degree* $\leq N(2^n - 1)$ *with coefficients in the space* $\mathcal{O}_{-p}^n$ *and* $G(0) = 1$ *such that all of the coefficients of the polynomial* $F(v)G(v)$ *are contained in* $\mathbf{L}_p^n$. *Use this to prove that each element* X *of* $\mathscr{C}_{-p}^n$ *satisfies an equation of the form* $\sum_{i=0}^N X^i \mathbf{T}_i = 0$ *with* $\mathbf{T}_i \in \mathbf{L}_p^n$, $\mathbf{T}_N = \mathbf{1}$, *and* $N \leq 2^n$. *Formulate and prove the similar assertions for the ring* $\mathscr{C}_p^n$.

[Hint: On replacing of coefficients of F by their images under Ω, we get a polynomial $f(v) = \Omega(F)(v)$ whose coefficients are symmetric with respect to $x_1, \ldots, x_n$. Hence, there is a polynomial $g(v)$ of degree $\leq (2^n - 1)N$ over $\mathbb{Q}[x_0^{\pm 1}, \ldots, x_n^{\pm 1}]$ such that all of the coefficients of $f(v)g(v)$ are invariant with respect to all of the transformations in W_n. Thus $fg = \Omega(P)(v)$, where P is a polynomial of degree $\leq N2^n$ over $\mathbf{L}_p^n$. Then apply Theorem 3.5.9 to P. For the second assertion consider the polynomial $F(v) = 1 - Xv$.]

3.5.3. A Symmetric Factorization of Polynomials $\mathbf{Q}_p^n(v)$ for $n = 1, 2$

Here we obtain a factorization of the polynomials which is invariant under $*$ and compute the coefficients of the factors.

Proposition 3.5.13. *The polynomial* $\mathbf{Q}_p^1(v)$ *has the following factorization over* $\mathscr{L}_p^1$:

$$\mathbf{Q}_p^1(v) = (1 - \Pi_- v)(1 - \Pi_+ v),$$

where $\Pi_- = \Pi_-^1(p) = \Pi_0^1(p)$ *and* $\Pi_+ = \Pi_+^1(p) = \Pi_1^1(p)$.

Proof. By (3.3.58) and (3.4.13) with $n = 1$, we have

$$\begin{aligned}(1 - \Pi_- v)(1 - \Pi_+ v) &= 1 - (\Pi_- + \Pi_+)v + \Pi_- \Pi_+ v^2 \\ &= 1 - \mathbf{T}^1(p)v + p[\mathbf{p}]_1 v^2.\end{aligned}$$

The last polynomial is equal to $\mathbf{Q}_p^1$ by Proposition 3.3.35. □

Proposition 3.5.14. *The polynomial* $\mathbf{Q}_p^2(v)$ *has the following factorization over* $\mathscr{L}_p^2$:

$$\mathbf{Q}_p^2(v) = (1 - \Pi_- v)(1 - \Pi v + p(\Pi_{2,0}^{(1)} + \Pi_{2,0}^{(2)})v^2)(1 - \Pi_+ v),$$

where $\Pi_- = \Pi_-^2(p) = \Pi_0^2(p)$, $\Pi_+ = \Pi_+^2(p) = \Pi_2^2(p)$, $\Pi = \Pi_1^2(p)$ *are the elements* (3.3.59), *and* $\Pi_{a,b}^{(r)} = \Pi_{a,b}^{(r)}(p)$ *are the elements* (3.3.62) *for* $n = 2$.

Proof. We first prove the following relations in $\mathscr{L}_p^2$:

$$\Pi_- \Pi = p\Pi_{1,0}^{(0)}, \qquad \Pi\Pi_+ = p\Pi_{1,1}^{(0)}, \tag{3.5.23}$$

$$\Pi_- \Pi \Pi_+ = p^3[\mathbf{p}]\Pi, \tag{3.5.24}$$

$$\Pi_- \Pi_{2,0}^{(1)} = (p^2 - 1)[\mathbf{p}]\Pi_-, \qquad \Pi_{2,0}^{(1)}\Pi_+ = (p^2 - 1)[\mathbf{p}]\Pi_+. \tag{3.2.25}$$

The element on the right side in each of these relations is an integral multiple of a single double coset modulo $\Gamma_0 = \Gamma_0^2$. Thus, to prove a relation, we only have to check that each of the left cosets of the product on the left side is contained in the double coset on the right side, and then to verify the coefficient on the right by comparing the numbers of the left cosets in the two sides or computing the images of the two sides under the action of the mapping $\Omega = \Omega_p^2$. For example, in the first relation in (3.5.25) on the right side we have the element $(p^2 - 1)\left(\begin{pmatrix} p^2E & 0 \\ 0 & pE \end{pmatrix}\right)_{\Gamma_0}$. By definitions, on the left side we have

$$\Pi_- \Pi_{2,0}^{(1)} = \sum_{\substack{B \in \mathrm{B}(pE_2)/\mathrm{mod}\, p, \\ r_p(B)=1}} \left(\Gamma_0 \begin{pmatrix} p^2E & pB \\ 0 & pE \end{pmatrix}\right).$$

Since $pB \equiv 0 \pmod{pE}$ it follows that each of the left cosets in the last sum is contained in $\Gamma_0 \begin{pmatrix} p^2E & 0 \\ 0 & pE \end{pmatrix} \Gamma_0$. Since $\Pi_- \Pi_{2,0}^{(1)}$ is an element of a Hecke ring, we conclude that

$$\Pi_- \Pi_{2,0}^{(1)} = \alpha \cdot \left(\begin{pmatrix} p^2E & 0 \\ 0 & pE \end{pmatrix}\right)_{\Gamma_0} = \alpha \cdot [\mathbf{p}]\Pi_-$$

with some α. Comparing the Ω-images of the two sides, by Lemma 3.3.34, we find

$$x_0 \,\mathrm{sm}_p(1, 2) x_0^2 p^{-3} x_1 x_2 = \alpha p^{-3} x_0^2 \cdot x_1 \cdot x_2 x_0,$$

hence, $\alpha = \mathrm{sm}_p(1,2)$ is the number of symmetric matrices M of order 2 over $\mathbb{Z}/p\mathbb{Z}$ with $r_p(M) = 1$, this is clearly $p^2 - 1$. We leave to the reader the verification of other relations, as an easy exercise preceding far more complicated calculations of the next subsection.

Multiplying the polynomials on the right side of the factorization which we are going to prove, we get the polynomial

$$1 - (\Pi_- + \Pi + \Pi_+)v + (\Pi_-\Pi + \Pi_-\Pi_+ + \Pi\Pi_+ + pN)v^2$$
$$- (p\Pi_-N + pN\Pi_+ + \Pi_-\Pi\Pi_+)v^3 + p\Pi_-N\Pi_+v^4,$$

where $N = \Pi_{2,0}^{(1)} + \Pi_{2,0}^{(0)}$. By definitions and the relation (3.5.23)–(3.5.25) and (3.4.13), the polynomial is equal to

$$= 1 - (\Pi_0 + \Pi_1 + \Pi_2)v + (p(\Pi_{1,0}^{(0)} + \Pi_{1,1}^{(0)} + \Pi_{2,0}^{(1)}) + (p^3 + p)\Pi_{2,0}^{(0)})v^2$$
$$- p^3[\mathbf{p}](\Pi_0 + \Pi_1 + \Pi_2)v^3 + p^6[\mathbf{p}]^2v^4$$
$$= 1 - \mathbf{T}(p)v + (p\mathbf{T}_1(p^2) + (p^3 + p)[\mathbf{p}])v^2$$
$$- p^3[\mathbf{p}]\mathbf{T}(p)v^3 + p^6[\mathbf{p}]^2v^4,$$

where we have also used the decompositions (3.3.58) and (3.3.61) of $\mathbf{T}(p) = \mathbf{T}^2(p)$ and $\mathbf{T}_1(p^2) = \mathbf{T}_1^2(p^2)$ respectively. According to the formula for $\mathbf{Q}_p^2(v)$ of Proposition 3.3.35, in order to complete our proof, it suffices to verify the relation

$$\mathbf{q}_2^2(p) = p\mathbf{T}_1^2(p^2) + (p^3 + p)[p], \tag{3.5.26}$$

where $\mathbf{q}_2^2(p)$ is the element of $\mathbf{L}_p^2$ with $\Omega(\mathbf{q}_2^2(p)) = x_0^2x_1x_2(x_1 + x_2 + x_1^{-1} + x_2^{-1} + 2)$. Since the restriction $\Omega|_{\mathbf{L}_p^2}$ is a monomorphism, it is sufficient to check that the two sides of (3.5.26) have the same image under Ω. The image of the right side can be found, if one uses the decomposition (3.3.61) of $T_1(p^2)$, Lemma 3.3.34, and, finally, Lemma 3.2.21 to calculate the polynomials $\omega(\pi_{1,0}^2(p)) = \omega(\pi_1^2(p))$, $\omega(\pi_{1,1}^2(p)) = \omega(\pi_2^2(p)\pi_1^2(p))$, and $\omega(\pi_{2,0}^2(p)) = \omega(\pi_2^2(p))$. The reader will easily verify that the images of the two side are equal. □

Exercise 3.5.15. *Prove the following factorization over $\mathscr{L}_p^n$ for every $n \in \mathbb{N}$:*

$$\mathbf{Q}_p^n(v) = (1 - \Pi_-v)Q'(v)(1 - \Pi_+v),$$

where $\Pi_- = \Pi_-^n(p)$, $\Pi_+ = \Pi_+^n(p)$, and Q' is a polynomial of degree $2^n - 2$.

[Hint: Use (3.3.79) and Exercise 3.5.7 (2) to show that all of the coefficients of the formal power series $(1 - \Pi_-v)^{-1}\mathbf{Q}_p^n(v)\cdot(1 - \Pi_+v)^{-1}$ are contained in $\mathscr{C}_-^n{}_p\mathscr{C}_p^n$, and then use the last assertion of Exercise 3.5.7 (1).]

3.5.4. Coefficients of Factors of Polynomials $\mathbf{R}_p^n(v)$

Here we shall express the coefficients of the factors of factorizations (3.5.17) and (3.5.18) as linear combinations of left (double) cosets modulo Γ_0^n. For this we shall have to compute certain products of elements in $\mathscr{L}_p^n$ of the form $\Pi_a = \Pi_a^n(p)$ and

$\Pi^{(r)}_{a,b} = \Pi^{(r)}_{a,b}(p)$ (see (3.3.59) and (3.3.62)). We first express the elements $\Pi^{(r)}_{a,b}$ as linear combinations of double cosets modulo $\Gamma_0 = \Gamma_0^n$. It follows from (3.3.67) and (3.3.68) that each of the left cosets in (3.3.62) is contained in a double coset of the form $\Gamma_0 M_{a,b}(B_0)\Gamma_0$ with $B_0 \in \mathrm{S}_a(\mathbb{Z})$ and $r_p(B_0) = r$, where we set

$$M_{a,b}(B_0) = \begin{pmatrix} p^2 D_{a,b}^{-1} & B \\ 0 & D_{a,b} \end{pmatrix} \quad \text{with } B = \begin{pmatrix} 0 & 0 & 0 \\ 0 & B_0 & 0 \\ 0 & 0 & 0 \end{pmatrix} \begin{matrix} \\ a. \\ b \end{matrix} \qquad \begin{matrix} \\ \\ \quad a \quad b \end{matrix} \tag{3.5.27}$$

Each of the left cosets contained in this double coset has a representative of the form $M_{a,b}(B_0)\gamma$ with $\gamma \in \Gamma_0$. If $\gamma = U(\varepsilon)$ with $\varepsilon \in \Lambda_{(D)} = D^{-1}\Lambda D \cap \Lambda$, where $\Lambda = \Lambda^n$ and $D = D_{a,b}$ then clearly

$$M_{a,b}(B_0)\gamma = \begin{pmatrix} p^2 D^{-1}\varepsilon^* & B\varepsilon \\ 0 & D\varepsilon \end{pmatrix} = \gamma' \begin{pmatrix} p^2 D^{-1} & {}^t\varepsilon_1 B\varepsilon \\ 0 & D \end{pmatrix}, \tag{3.5.28}$$

where $\varepsilon_1 = D\varepsilon D^{-1} \in \Lambda$ and $\gamma' = U(\varepsilon_1)$. Let $\xi_1, \ldots, \xi_k$ be a fixed set of representatives for the left cosets of Λ modulo $\Lambda_{(D)}$. Then for every $\gamma = \begin{pmatrix} \xi^* & \beta \\ 0 & \xi \end{pmatrix}$ there is an i with $1 \le i \le k$ such that $\xi = \varepsilon\xi_i$ with a $\varepsilon \in \Lambda_{(D)}$. Hence,

$$\gamma\gamma_i^{-1} = \begin{pmatrix} \varepsilon^* & \beta_1 \\ 0 & \varepsilon \end{pmatrix} = U(\varepsilon)T(S), \qquad \text{where } \gamma_i = U(\xi_i) \text{ and } S \in \mathrm{S}_n(\mathbb{Z}),$$

and so $\gamma = U(\varepsilon)T(S)\gamma_i$. Then, by (3.5.28), we have

$$M_{a,b}(B_0)\gamma = \gamma' \begin{pmatrix} p^2 D^{-1} & {}^t\varepsilon_1 B\varepsilon + p^2 D^{-1} S \\ 0 & D \end{pmatrix} \gamma_i. \tag{3.5.29}$$

Let ε_{ij} for $1 \le i, j \le 3$ be the blocks of ε of the same size as the corresponding blocks of B; for example, ε_{22} is the block of dimension $a \times a$, $\varepsilon_{3,3}$ is the block of dimension $b \times b$ and so on. Then the condition $\varepsilon \in \Lambda_{(D)}$ for a matrix $\varepsilon \in \Lambda$ is clearly equivalent to the congruences $\varepsilon_{12} \equiv 0 \pmod p$, $\varepsilon_{23} \equiv 0 \pmod p$, and $\varepsilon_{13} \equiv 0 \pmod{p^2}$. In particular, this implies that, for such a ε, $\det \varepsilon_{22}$ is prime to p. Further, by a direct computation, one can easily see that any matrix of the form ${}^t\varepsilon_1 B\varepsilon + p^2 D^{-1} S$ is congruent modulo D (see (3.3.25)) to a matrix of the form

$$\begin{pmatrix} 0 & 0 & 0 \\ 0 & B_0[\varepsilon_{22}] & B_{23} \\ 0 & B_{32} & B_{33} \end{pmatrix} \quad \text{with } B_{23} = p\,{}^tB_{32},\ B_{32} \in M_{b,a},\ B_{33} \in \mathrm{S}_b(\mathbb{Z}),$$

and the residue class of the matrix modulo D is uniquely determined if the matrices $B_0[\varepsilon_{22}]$ and B_{32} are fixed modulo p and B_{33} is fixed modulo p^2. It follows from Lemma 3.3.2 that, for $a < n$ there is a matrix $\varepsilon \in \Lambda_{(D)}$ satisfying $\varepsilon_{22} \equiv U \pmod p$ for a given $U \in \mathrm{GL}_a(\mathbb{Z}/p\mathbb{Z})$, but if $a = n$, then $\varepsilon_{22} = \varepsilon$, hence $\det \varepsilon_{22} = \det \varepsilon = \pm 1$, and the matrix U must satisfy the condition $\det U \equiv \pm 1 \pmod p$. Thus, the matrix $B_0[\varepsilon_{22}]/\mathrm{mod}\, p$ runs over the class

$$\{B_0\}_p^n = \begin{cases} \{B_0[U]/\text{mod}\, p;\ U \in \mathrm{GL}_a(\mathbb{Z}/p\mathbb{Z})\}, & \text{if } a < n \\ \{B_0[U]/\text{mod}\, p;\ U \in \mathrm{GL}_a(\mathbb{Z}/p\mathbb{Z}),\ \det U \equiv \pm 1 \ (\text{mod}\, p)\}, & \text{if } a = n. \end{cases} \tag{3.5.30}$$

It is also clear that the matrices $B_{32} \in \mathrm{M}_{b,a}$ and $B_{33} \in \mathrm{S}_b(\mathbb{Z})$ can be arbitrary as S runs over $\mathrm{S}_n(\mathbb{Z})$. Finally, by Lemma 3.1.2, the left cosets $\Lambda D\xi_1, \ldots, \Lambda D\xi_k$ are pairwise distinct, which implies that the left cosets modulo Γ_0 of matrices of the form (3.5.29) with distinct γ_i are distinct. So we have proved the decomposition

$$(M_{a,b}(B_0))_{\Gamma_0} = \sum_{\xi \in \Lambda_{(D)}\backslash\Lambda;\, B} \left(\Gamma_0 \begin{pmatrix} p^2D^{-1} & B \\ 0 & D \end{pmatrix} U(\xi)\right) \tag{3.5.31}$$

with $D = D_{a,b} = D^n_{a,b}(p)$, where B runs over the set of matrices of the form

$$\begin{pmatrix} 0 & 0 & 0 \\ 0 & B_{22} & p\,{}^tB_{32} \\ 0 & B_{32} & B_{33} \end{pmatrix} \quad \text{with } B_{22} \in \{B_0\}_p^n,\ B_{32} \in \mathrm{M}_{b,a}(\mathbb{Z}/p\mathbb{Z}),\ B_{33} \in \mathrm{S}_b(\mathbb{Z}/p\mathbb{Z}).$$

It follows from the decomposition that a equality

$$(M_{a,b}(B_0))_{\Gamma_0} = (M_{a_1,b_1}(B_0'))_{\Gamma_0}$$

is equivalent to the conditions $a = a_1$, $b = b_1$ and $\{B_0\}_p^n = \{B_0'\}_p^n$. Returning to the elements $\Pi_{a,b}^{(r)}$, we obtain:

Lemma 3.5.16. *The element $\Pi_{a,b}^{(r)}(p) \in \mathcal{L}_p^n$ with $0 \le r \le a$ and $a + b \le n$ given by* (3.3.62) *has the following decomposition into double cosets modulo Γ_0:*

$$\Pi_{a,b}^{(r)} = \sum_{\{B_0\}_p^n,\, r_p(B_0) = r} (M_{a,b}(B_0))_{\Gamma_0}, \tag{3.5.32}$$

where B_0 runs over a set of representatives of the subsets $\{B_0\}_p^n \subset \mathrm{S}_a(\mathbb{Z}/p\mathbb{Z})$ consisting of matrices of rank r over $\mathbb{Z}/p\mathbb{Z}$.

The action of $*$ on $\Pi_{a,b}^{(r)}$ is given by

$$(\Pi_{a,b}^{(r)})^* = \Pi_{a,n-a-b}^{(r)}. \tag{3.5.33}$$

Proof. We have already seen that each of the left cosets in (3.3.62) is contained in a double cosets on the right side in (3.5.32). Conversely, since clearly $r_p(M_{a,b}(B_0)) = n - a - b + b + r_p(B_0) = n - a + r_p(B_0)$, it follows that a left coset $\Gamma_0 M_{a,b}(B_0)$ with $r_p(B_0) = r$ is equal to one of the left cosets in (3.3.62), and so the corresponding double coset occurs in the decomposition of $\Pi_{a,b}^{(r)}$. Since all of the left cosets in (3.3.62), as well as all of the double cosets on the right side in (3.5.32), are distinct, it follows that each of the double cosets occurs in the decomposition of $\Pi_{a,b}^{(r)}$ with the coefficient 1. The decomposition (3.5.32) is proved. By the definition of $*$ (see (3.4.7)) we have

$$(M_{a,b}(B_0))_{\Gamma_0}^* = (p^2 M_{a,b}(B_0)^{-1})_{\Gamma_0} = \left(\begin{pmatrix} D_{a,b} & -B \\ 0 & p^2 D_{a,b}^{-1} \end{pmatrix}\right)_{\Gamma_0}$$

where $B = D_{a,b} B D_{a,b}^{-1}$ is defined in (3.5.27). Let $I = I_a = (i_{\alpha\beta})$ be the matrix of order a with the entries $i_{\alpha\beta} = 1$, if $\alpha + \beta = a + 1$ and $i_{\alpha\beta} = 0$, otherwise. Then the transformation $A \to IAI$ of a matrix A reverses the order of its rows as well as that of its columns. This implies the relation

$$\gamma \begin{pmatrix} D_{a,b} & -B \\ 0 & p^2 D_{a,b}^{-1} \end{pmatrix} \gamma = M_{a,n-a-b}(-I_a B_0 I_a) \quad \text{with } \gamma = \begin{pmatrix} I_n & 0 \\ 0 & I_n \end{pmatrix}.$$

Since $\gamma \in \Gamma_0$, it follows that

$$(M_{a,b}(B_0))^*_{\Gamma_0} = (M_{a,n-a-b}(-IB_0 I))_{\Gamma_0}.$$

The relation (3.5.33) follows from these relations and the decomposition (3.5.32), since the mapping $B_0 \to -IB_0 I$ only permutes the classes $\{B_0\}_p^n$ with $r_p(B_0) = r$. □

Proposition 3.5.17. *The following relations hold in $\mathscr{L} = \mathscr{L}_p^n$:*

$$\Pi_b \Pi_+ = p^{\langle n-b \rangle} \Pi_{n-b,b}^{(0)} \quad \text{with } 0 \le b \le n, \tag{3.5.34}$$

and

$$\Pi_{a,0}^{(r)} \Pi_{n-b,b}^{(0)} = \sum_{\substack{\max(0,a+b-n) \le t \le b, \\ 0 \le s \le r}} c(a,b,r,t,s)[p] \Pi_{a+b-2t,t}^{(r-s)} \tag{3.5.35}$$

with $0 \le r \le a \le n$, $0 \le b \le n$, and $r + b \le a$, where the coefficients are given by the formula

$$c(a,b,r,t,s) = p^{t(r+t-a-b-s-1)+b(n+1)} \operatorname{sm}_p(0_{a+s-r-t}; s, a+s-r) \frac{\varphi_{a+b+s-r-2t}(p)}{\varphi_{a+s-r-t}(p)\varphi_{b-t}(p)}$$

in which φ are the polynomials (3.2.29) and $\operatorname{sm}_p(0_{c_1}; c, d)$ for $0 \le c_1$, $c \le d$, is the number of matrices $M \in \mathrm{S}_d(\mathbb{Z}/p\mathbb{Z})$ with $r_p(M) = c$ of the form $M = \begin{pmatrix} 0_{c_1} & * \\ * & * \end{pmatrix}$, 0_{c_1} being the zero matrix of order c_1.

Proof. It follows from (3.3.59), (3.5.19) and definitions that the right and left exponents of Π_b are both not greater than 1. Then, by Proposition 3.5.1, the left side of (3.5.34) is contained in $\mathscr{C}_+ = \mathscr{C}_p^n$. The decomposition (3.5.32) shows that $\Pi_{n-b,b}^{(0)} = (M_{n-b,b}(0))_{\Gamma_0}$, and so, by Proposition 3.4.4, this element is also contained in $\mathscr{C}_+$. Thus we have

$$\Pi_b \Pi_+, \quad \Pi_{n-b,b}^{(0)} \in \mathscr{C}_+ \quad \text{for } 0 \le b \le n. \tag{3.5.36}$$

Then, by Theorem 3.4.9, to prove (3.5.34), we have to check that the two sides have the same image under Φ or Ω, which is an immediate consequence of Lemma 3.3.34. The relations (3.5.34) are proved.

The relations (3.5.35) are not so easy, and the reader must arm himself with patience. We first note that it suffices to compute the products of the form

$$(M_{a,0}(B_0))_{\Gamma_0}\Pi^{(0)}_{n-b,b}, \tag{3.5.37}$$

where $(M_{a,0}(B_0))$ with $B_0 \in \mathsf{S}_a(\mathbb{Z})$ and $r_p(B_0) = r$ is one of the double cosets in the decomposition (3.5.32) of $\Pi^{(r)}_{a,0}$. By (3.5.32), we have $\Pi^{(0)}_{n-b,b} = (M_{n-b,b}(0))_{\Gamma_0}$. Thus, by the second formula of Lemma 3.1.5 and the decomposition (3.5.31) of the double coset $(M_{a,0}(B_0))$, to compute the product (3.5.37), one has to find all of the double cosets containing at least one of the products

$$M_{a,0}(B_h)U(\xi)M_{n-b,b}(0) \quad \text{with } B_h \in \{B_0\}^n_p \text{ and } \xi \in \Lambda_{(D_a)}\backslash\Lambda, \tag{3.5.38}$$

where $D_a = D^n_a(p) = D_{a,0}$, and then calculate how many of the products is contained in each of the double cosets. For this we shall need a special system of representatives for the left cosets of Λ modulo $\Lambda_{(D_a)}$.

Let us introduce certain notation. We set

$$\mathsf{I}_{d,n} = \{(i_1, \dots, i_d) \in \mathbb{N}^d; 1 \le i_1 < \cdots < id \le n\}. \tag{3.5.39}$$

For $\mathbf{i} = (i_1, \dots, i_d) \in \mathsf{I}_{d,n}$ we denote by $\hat{\mathbf{i}}$ the system $(j_1, \dots, j_{n-d}) \in \mathsf{I}_{n-d,n}$ which is complementary to $i_1, \dots, i_d$ in $1, 2, \dots, n$. We shall associate to each permutation σ of $1, 2, \dots, n$ the matrix of order n,

$$M(\sigma) = (e_{\sigma^{-1}(i),j}),$$

where $e_{\alpha\beta}$ are entries of the unit matrix E_n. It is clear that

$$M(\sigma\tau) = M(\sigma)M(\tau) \quad \text{and} \quad M(\sigma^{-1}) = M(\sigma)^{-1} = {}^tM(\sigma).$$

Further, to each $\mathbf{i} = (i_1, \dots, i_d) \in \mathsf{I}_{d,n}$ we associate the permutation $\sigma(\mathbf{i})$ and the matrix $M(\mathbf{i})$ by

$$\sigma(\mathbf{i}) = \begin{pmatrix} i_1, \dots, i_d, & j_1, \dots, j_{n-d} \\ 1, \dots, d, & d+1, \dots, n \end{pmatrix} \quad \text{and} \quad M(\mathbf{i}) = M(\sigma(i)), \tag{3.5.40}$$

where $(j_1, \dots, j_{n-d}) = \hat{\mathbf{i}}$. We note that if $T = (\mathbf{t}_1, \dots, \mathbf{t}_n)$ is a matrix with columns $\mathbf{t}_1, \dots, \mathbf{t}_n$, then $T \cdot M(\sigma)$ is the matrix with columns $\mathbf{t}_{\sigma(1)}, \dots, \mathbf{t}_{\sigma(n)}$; in particular,

$$(\mathbf{t}_1, \dots, \mathbf{t}_n)M(\mathbf{i})^{-1} = (\mathbf{t}_{i_1}, \dots, \mathbf{t}_{i_d}, \mathbf{t}_{j_1}, \dots, \mathbf{t}_{j_{n-d}}), \tag{3.5.41}$$

where $(i_1, \dots, i_d) = \mathbf{i}$ and $(j_1, \dots, j_{n-d}) = \hat{\mathbf{i}}$. Finally, to each $\mathbf{i} = (i_1, \dots, i_d) \in \mathsf{I}_{d,n}$ with $\hat{\mathbf{i}} = (j_1, \dots, j_{n-d}) \in \mathsf{I}_{n-d,n}$ we associate the sets

$$\begin{gathered} \mathsf{V}(\mathbf{i}) = \{V = (v_{\alpha\beta}) \in \mathsf{M}_{d,n}; 0 \le v_{\alpha\beta} < p, \text{ if } i_\alpha < j_\beta, \text{ and } v_{\alpha\beta} = 0, \text{ if } i_\alpha > j_\beta\}, \\ \mathsf{W}(\mathbf{i}) = \left\{\varepsilon = \begin{pmatrix} E_d & V \\ 0 & E_{n-d} \end{pmatrix} M(\mathbf{i}); V \in \mathsf{V}(\mathbf{i})\right\}, \end{gathered} \tag{3.5.42}$$

and set

$$\mathsf{W}^n_a(p) = \bigcup_{\mathbf{i} \in \mathsf{I}_{n-a,n}} \mathsf{W}(\mathbf{i}).$$

We now show that the set $\mathsf{W}^n_a(p)$ is a complete system of representatives for the left cosets of $\Lambda = \Lambda^n$ modulo $\Lambda_{(D_a)}$:

$$\mathsf{W}^n_a(p) = \Lambda^n_{(D_a)} \backslash \Lambda. \tag{3.5.43}$$

To show this, we first note that the number of elements of $\mathsf{W}(\mathbf{i})$ with $\mathbf{i} = (i_1, \ldots, i_{n-a}) \in \mathsf{I}_{n-a,n}$ is given by

$$|\mathsf{W}(\mathbf{i})| = |\mathsf{V}(\mathbf{i})| = p^{j_1 + \cdots + j_a - \langle a \rangle}, \tag{3.5.44}$$

where $(j_1, \ldots, j_a) = \hat{\mathbf{i}}$, for the number of i_α satisfying $i_\alpha < j_\beta$ with a fixed β is equal to $j_\beta - \beta$. Hence, by the identity (3.2.33), we see that the number of elements of $\mathsf{W}^n_a(p)$ equals

$$p^{-\langle a \rangle} \sum_{1 \le j_1 < \cdots < j_a \le n} p^{j_1 + \cdots + j_a} = \frac{\varphi_n(p)}{\varphi_a(p)\varphi_{n-a}(p)}.$$

On the other hand, by Lemma 3.1.2 and the formula (3.2.28), this number is also equal to the index $\nu_\Lambda(D_a)$ of $\Lambda_{(D_a)}$ in Λ. Thus, to prove (3.5.43) we only have to check that the matrices of $\mathsf{W}^n_a(p)$ belong to pairwise distinct left cosets modulo $\Lambda_{(D_a)}$. It follows from the definition that

$$\Lambda_{(D_a)} = \left\{ \lambda = \begin{pmatrix} \lambda_1 & \lambda_2 \\ \lambda_3 & \lambda_4 \end{pmatrix} \in \Lambda;\ \lambda_2 \equiv 0 \ (\mathrm{mod}\, p) \right\},$$

λ_2 being the block of dimension $(n-a) \times a$. Hence we conclude that the block λ_1 (of dimension $(n-a) \times (n-a)$) of any matrix λ in $\Lambda_{(D_a)}$ is not singular modulo p, and if two matrices ε and ε' of Λ are contained in the same left coset modulo $\Lambda_{(D_a)}$, then the matrices $\varepsilon_{(n-a)}$ and $\varepsilon'_{(n-a)}$ composed of the first $n-a$ rows of ε and ε' respectively satisfy the relation

$$\lambda_1 \varepsilon_{(n-a)} \equiv \varepsilon'_{(n-a)} \quad (\mathrm{mod}\, p) \qquad \text{with } \lambda_1 \in \mathsf{GL}_{n-a}(\mathbb{Z}/p\mathbb{Z}). \tag{3.5.45}$$

If $\varepsilon \in \mathsf{W}(\mathbf{i})$ with $\mathbf{i} = (i_1, \ldots, i_{n-a})$, then it follows from (3.5.41) and the definition of the set $\mathsf{V}(\mathbf{i})$ that the system $(i_1, \ldots, i_{n-a})$ is, in the sense of lexicographic order, the least system of indices of columns of $\varepsilon_{(n-a)}$ which are linearly independent modulo p. Thus, if (3.5.45) holds for two matrices $\varepsilon, \varepsilon' \in \mathsf{W}^n_a(p)$, then they both belong to the same subset $\mathsf{W}(\mathbf{i})$, but then it follows from (3.5.45) that $\lambda_1 \equiv E_{n-a} \ (\mathrm{mod}\, p)$, which implies that the matrices V and V' in $\mathsf{V}(\mathbf{i})$ corresponding to ε and ε' are congruent modulo p, i.e. $\varepsilon = \varepsilon'$. The equality (3.5.43) follows.

Using the set $\mathsf{W}^n_a(p)$ as a set of representatives for the left cosets of Λ modulo $\Lambda_{(D_a)}$, we return to the consideration of the product (3.5.38). Let $\xi = \varepsilon \in \mathsf{W}^n_a(p)$. Since $D_{a,0} = D_a$ and $D_{n-b,b} = pD_b$, it follows that the product (3.5.38) is equal to

$$p \begin{pmatrix} p^2 (D_a \varepsilon D_b)^* & B\varepsilon D_b \\ 0 & D_a \varepsilon D_\beta \end{pmatrix} \quad \text{with } B = \begin{pmatrix} 0 & 0 \\ 0 & B_h \end{pmatrix}. \tag{3.5.46}$$

Let us assume that $\varepsilon \in \mathsf{W}(i)$ with $\mathbf{i} = (i_1, \ldots, i_{n-a}) \in \mathsf{I}_{n-a,n}$, that is

$$\varepsilon = \begin{pmatrix} E_{n-a} & V \\ 0 & E_a \end{pmatrix} M, \qquad \text{where } M = M(\mathbf{i}) \text{ and } V \in \mathsf{V}(\mathbf{i}).$$

We define the numbers $d_1, \ldots, d_n = 0$ or 1 by $D_b = \mathrm{diag}(p^{d_1}, \ldots, p^{d_n})$. Then from

(3.5.41) and its transpose we have

$$MD_bM^{-1} = C = \begin{pmatrix} C_1 & 0 \\ 0 & C_2 \end{pmatrix},$$

where $C_1 = \operatorname{diag}(p^{d_{i_1}}, \ldots, p^{d_{i_{n-a}}})$ and $C_2 = \operatorname{diag}(p^{d_{j_1}, \ldots, d_{j_a}})$ with $\mathbf{j} = (j_1, \ldots, j_a) = \hat{\mathbf{i}}$. It follows from the formulae that

$$D_a \varepsilon D_b = D_a \begin{pmatrix} E_{n-a} & V \\ 0 & E_a \end{pmatrix} MD_bM^{-1}M = D_a C\eta$$

with

$$\eta = C^{-1} \begin{pmatrix} E_{n-a} & V \\ 0 & E_a \end{pmatrix} CM = \begin{pmatrix} E_{n-a} & C_1^{-1}VC_2 \\ 0 & E_a \end{pmatrix} M.$$

The matrix $C_1^{-1}VC_2 = (p^{-d_{i_\alpha}+d_{j_\beta}}v_{\alpha\beta})$ is integral: if $-d_{i_\alpha} + d_{j_\beta} < 0$, then $d_{j_\beta} = 0$ and $d_{i_\alpha} = 1$, hence $i_\alpha > n - b \geq j_\beta$, and $v_{\alpha\beta} = 0$, by the definition of $\mathsf{V}(\mathbf{i})$. Thus, $\eta \in \Lambda$, and the matrix (3.5.46) belongs to the same right coset modulo Γ_0 as the matrix

$$p\begin{pmatrix} p^2(D_aC)^* & B\varepsilon D_b\eta^{-1} \\ 0 & D_aC \end{pmatrix} = p\begin{pmatrix} p^2(D_aC)^{-1} & BC \\ 0 & D_aC \end{pmatrix} \tag{3.5.47}$$

(note $\varepsilon D_b \eta^{-1} = C$). We shall denote by $t = t(\mathbf{i})$ the number of j_β in $(j_1, \ldots, j_a) = \hat{\mathbf{i}}$ which satisfy the inequality $j_\beta > n - b$. It follows from the definition of the matrices C_1 and C_2 that they have the form

$$C_1 = \begin{pmatrix} E_{n-a-b+t} & 0 \\ 0 & pE_{b-t} \end{pmatrix}, \quad C_2 = \begin{pmatrix} E_{a-t} & 0 \\ 0 & pE_t \end{pmatrix},$$

whence

$$D_aC = D_{a+b-2t,t} \quad \text{and} \quad BC = \begin{pmatrix} 0 & 0 \\ 0 & B_h \end{pmatrix}\begin{pmatrix} C_1 & 0 \\ 0 & C_2 \end{pmatrix} = \begin{pmatrix} 0 & 0 \\ 0 & B_hC_2 \end{pmatrix}.$$

For a matrix A we denote by $A^{(s)}$ the square matrix of order s located in the upper left corner of A. It follows easily from the shape of the matrix under consideration and the decomposition (3.5.31) that the matrix (3.5.47) is contained in the double coset modulo Γ_0 of the matrix

$$pM_{a+b-2t,t}(\tilde{B}_h^{(a-t)}), \quad \text{where } \tilde{B}_h^{(a-t)} = \begin{pmatrix} 0 & 0 \\ 0 & B_h^{(a-t)} \end{pmatrix}$$

(the order of last matrix is $a + b - 2t$). We see that each of the products of the form (3.5.46) is contained in the double coset of a matrix of the form $p\mathsf{M}_{a+b-2t,t}(K)$ with $K \in \mathsf{S}_{a+b-2t}(\mathbb{Z})$; and such a inclusion holds if and only if $\varepsilon \in \mathsf{W}(\mathbf{i})$ with $\mathbf{i} \in \mathsf{I}_{n-a,n}$ and $t(\mathbf{i}) = t$ and the matrix $\tilde{B}_h^{(a-t)}/\text{mod}\, p$ is contained in $\{K\}_p^n$ (see (3.5.30)). Using the formula (3.5.44) for the number of elements in $\mathsf{W}(\mathbf{i})$, we see that the number of the products (3.5.46) contained in the double coset $\Gamma_0 p\mathsf{M}_{a+b-2t,t}(K)\Gamma_0$ is equal to

$$\left(\sum_{1 \leq j_1 < \cdots < j_{a-t} \leq n-b < j_{a-t+1} < \cdots < j_a \leq n} p^{j_1 + \cdots + j_a - \langle a \rangle}\right) \nu(\{B_0\}, \{K\}, a - t),$$

where $v(\{B_0\}, \{K\}, s)$ stands for the number of $B_h \in \{B_0\}_p^n$ such that the matrix $\tilde{B}_h^{(s)} = \begin{pmatrix} 0 & 0 \\ 0 & B_h^{(s)} \end{pmatrix} \Big/ \bmod p$ (of the same order as K) is contained in $\{K\}_p^n$. Since clearly t must satisfy $t \leq a, b$; $t \geq 0, a + b - n$, it follows from Lemma 3.1.5, that

$$(M_{a,0}(B_0))_{\Gamma_0} \Pi_{n-b,b}^{(0)} = \sum_{\substack{\max(0,a+b-n) \leq t \leq \min(a,b) \\ \{K\}_p^n \in \mathrm{S}_{a+b-2t}(\mathbb{Z}/p\mathbb{Z})}} c(a,b,t;B_0,K)[p](M_{a+b-2t,t}(K))_{\Gamma_0} \tag{3.5.48}$$

with

$$c(a,b,t;B_0,K) = \left(\sum_{1 \leq j_1 < \cdots < j_{a-t} \leq n-b < j_{a-t+1} < \cdots < j_a \leq n} p^{j_1 + \cdots + j_a - \langle a \rangle} \right) v(\{B_0\}, \{K\}, a-t) \cdot v(M_{n-b,b}(0)) v(M_{a+b-2t,t}(K))^{-1},$$

where $v(M)$ is the number of left cosets modulo Γ_0 in $\Gamma_0 M \Gamma_0$.

Some factors of the last expression can be easily computed. First of all, by (3.2.33), we have

$$\begin{aligned}
&\sum_{1 \leq j_1 < \cdots < j_{a-t} \leq n-b < j_{a-t+1} < \cdots < j_a \leq n} p^{j_1 + \cdots + j_a - \langle a \rangle} \\
&= \sum_{\substack{1 \leq j_1 < \cdots < j_{a-t} \leq n-b, \\ 1 \leq \beta_1 < \cdots < \beta_t \leq b}} p^{j_1 + \cdots + j_{a-t} + (\beta_1 + (n-b)) + \cdots + (\beta_t + (n-b)) - \langle a \rangle} \\
&= p^{\langle a-t \rangle + \langle t \rangle - \langle a \rangle + t(n-b)} \frac{\varphi_{n-b}(p)}{\varphi_{a-t}(p)\varphi_{n-a-b+t}(p)} \cdot \frac{\varphi_b(p)}{\varphi_t(p)\varphi_{b-t}(p)} \\
&= p^{t(n+t-a-b)} \frac{\varphi_{n-b}(p)\varphi_b(p)}{\varphi_{a-t}(p)\varphi_{n+t-a-b}(p)\varphi_t(p)\varphi_{b-t}(p)}.
\end{aligned}$$

Further, by (3.5.31) and (3.2.28), we have

$$\begin{aligned}
v(M_{n-b,b}(0)) &= p^{b(n-b)+b(b+1)} v_\Lambda(D_{n-b,b}) \\
&= p^{b(n+1)} v_\Lambda(D_b) = p^{b(n+1)} \frac{\varphi_n(p)}{\varphi_b(p)\varphi_{n-b}(p)}.
\end{aligned}$$

Finally, from the decomposition (3.5.31) and formulae (3.2.32) we get

$$\begin{aligned}
v(M_{a+b-2t,t}(K)) &= |\{K_p^n\}| \, p^{t(a+b-2t)+t(t+1)+t(n+t-a-b)} \frac{\varphi_n(p)}{\varphi_{n+t-a-b}(p)\varphi_{a+b-2t}(p)\varphi_t(p)} \\
&= |\{K\}_p^n| \, p^{t(n+1)} \frac{\varphi_n(p)}{\varphi_{n+t-a-b}(p)\varphi_{a+b-2t}(p)\varphi_t(p)}.
\end{aligned}$$

Now, if we sum up the equalities (3.5.48) over all distinct classes $\{B_0\}_p^n \in \mathrm{S}_a(\mathbb{Z}/p\mathbb{Z})$ satisfying $r_p(B_0) = r$ and use the formulae obtained above, we get

$$\Pi_{a,0}^{(r)}\Pi_{n-b,b}^{(0)} = \sum_{\substack{\max(0,a+b-n)\le t\le b,\\ \{K\}_p^n\in \mathrm{S}_{a+b-2t}(\mathbb{Z}/p\mathbb{Z})}} c(a,b,r,t,K)[\mathbf{p}](M_{a+b-2t,t}(K))_{\Gamma_0},$$

where

$$c(a,b,r,t;K)$$
$$= \left(|\{K\}_p^n|^{-1} \sum_{\substack{\{B_0\}_p^n\in \mathrm{S}_a(\mathbb{Z}/p\mathbb{Z}),\\ r_p(B_0)=r}} \nu(\{B_0\},\{K\},a-t)\right) p^{t(t-a-b-1)+b(n+1)} \frac{\varphi_{a+b-2t}(p)}{\varphi_{a-t}(p)\varphi_{b-t}(p)} \tag{3.5.49}$$

(by the assumption, $r+b\le a$, whence $b\le a$).

We now compute the expression in parentheses in the right side of (3.5.49), which we denote by $s\{K\}$. By the definition of $\nu(\{B_0\},\{K\},a-t)$, we have

$$\begin{aligned} s\{K\} &= |\{K\}|^{-1} \sum_{\substack{T\in \mathrm{S}_a(\mathbb{Z}/p\mathbb{Z}),\, r_p(T)=r,\\ \begin{pmatrix}0 & 0\\ 0 & T^{(a-t)}\end{pmatrix}/\bmod p\in\{K\}}} 1 \\ &= |\{K\}|^{-1} \sum_{V\in \mathrm{S}_{a-t}(\mathbb{Z}/p\mathbb{Z}),\, \begin{pmatrix}0 & 0\\ 0 & V\end{pmatrix}/\bmod p\in\{K\}} \mathrm{sm}_p(V,r,a), \end{aligned} \tag{3.5.50}$$

where $\mathrm{sm}_p(V,r,a)$ stands for the number of matrices $T\in \mathrm{S}_a(\mathbb{Z}/p\mathbb{Z})$ satisfying

$$T\equiv \begin{pmatrix} V & * \\ * & * \end{pmatrix} \pmod p, \qquad r_p(T)=r. \tag{3.5.51}$$

We show that $\mathrm{sm}_p(V,r,a)$ depends only on the rank of V over $\mathbb{Z}/p\mathbb{Z}$. More precisely: let us assume that the order of V is $a-t$ and $r_p(V)=r-s$ with $t,s\ge 0$. Then the following formula holds

$$\mathrm{sm}_p(V,r,a) = p^{(r-s)t}\mathrm{sm}_p(0_{a+s-r-t},s,a+s-r). \tag{3.5.52}$$

To prove this, we note first that $\mathrm{sm}_p(V,r,a)$ depends only on the class of V over the field $\mathbb{Z}/p\mathbb{Z}$. Then, by Theorem A.1.3, one can assume that $V=\begin{pmatrix} V_1 & 0\\ 0 & 0\end{pmatrix}$, where V_1 is a matrix of order $r-s$ satisfying $\det V_1\not\equiv 0 \pmod p$. In this case, each matrix T of $\mathrm{S}_a(\mathbb{Z})$ satisfying $T\equiv\begin{pmatrix} V & *\\ * & *\end{pmatrix} \pmod p$ is congruent modulo p to a matrix of the form

$$\begin{pmatrix} V_1 & 0 & T_{13}\\ 0 & 0 & T_{23}\\ {}^tT_{13} & {}^tT_{23} & T_{33}\end{pmatrix}$$
$$= \begin{pmatrix} E_{r-s} & 0 & 0\\ 0 & E_{a+s-r-t} & 0\\ {}^tT_{12}V_1^{-1} & 0 & E_t\end{pmatrix} \begin{pmatrix} V_1 & 0 & 0\\ 0 & 0 & T_{23}\\ 0 & {}^tT_{23} & T_{33}\end{pmatrix} \begin{pmatrix} E_{r-s} & 0 & V_1^{-1}T_{12}\\ 0 & E_{a+s-r-t} & 0\\ 0 & 0 & E_t\end{pmatrix}$$

whose rank over $\mathbb{Z}/p\mathbb{Z}$ is clearly equal to

$$r_p(V_1) + r_p\left(\begin{pmatrix} 0 & T_{23} \\ {}^tT_{23} & T_{33} \end{pmatrix}\right) = r - s + r_p\left(\begin{pmatrix} 0 & T_{23} \\ {}^tT_{23} & T_{33} \end{pmatrix}\right).$$

Conversely, each of the matrices of this form with $T_{13} \in \mathsf{M}_{r-s,t}(\mathbb{Z}/p\mathbb{Z})$, $T_{23} \in \mathsf{M}_{a+s-r-t}(\mathbb{Z}/p\mathbb{Z})$, $T_{33} \in \mathsf{S}_t(\mathbb{Z}/p\mathbb{Z})$, and $r_p\left(\begin{pmatrix} 0 & T_{23} \\ {}^tT_{23} & T_{33} \end{pmatrix}\right) = s$ satisfies (3.5.51). The relation (3.5.52) follows.

Returning to our calculation of the sum $s\{K\}$ with $K \in \mathsf{S}_{a+b-2t}(\mathbb{Z}/p\mathbb{Z})$, we note that $r_p(K) \le r$ if the sum is not empty. We set $r_p(K) = r - s$. Then each of the matrices V on the right side in (3.5.50) must satisfy $r_p(V) = r_p(K) = r - s$. Hence, using (3.5.52), we can rewrite $s\{K\}$ in the form

$$s\{K\} = p^{(r-s)t}\,\mathrm{sm}_p(0_{a+s-r-t}, s, a + s - r)|\{K\}|^{-1}s_1\{K\}, \tag{3.5.53}$$

where $s_1\{K\}$ is the number of matrices in $\{K\}$ which are congruent modulo p to a matrix of the form $\begin{pmatrix} 0 & 0 \\ 0 & V \end{pmatrix}$ with a block V of dimension $(a - t) \times (a - t)$. By Theorem A.1.3, we may assume that $K = \begin{pmatrix} 0 & 0 \\ 0 & K_1 \end{pmatrix}$ with $K_1 \in \mathsf{S}_{r-s}(\mathbb{Z})$ and $r_p(K_1) = r - s$. Let us denote by G the group $\mathsf{GL}_{a+b-2t}(\mathbb{Z}/p\mathbb{Z})$, if $a + b - 2t < n$, and the subgroup of the group consisting of matrices of determinant ± 1 modulo p, if $a + b - 2t = n$. Then, by (3.5.30), the class $\{K\} = \{K\}_p^n$ consists of matrices of the form $K[U]$ with $U \in \mathsf{G}$. We set $U = (U_{ij})$ with $i, j = 1, 2, 3$, where U_{11}, U_{22}, U_{33} are the blocks of dimension $(b - t) \times (b - t)$, $(a + s - t - r) \times (a + s - t - r)$, and $(r - s) \times (r - s)$ respectively. Since U is non-singular modulo p, it easily follows by a direct calculation that the matrix $K[U]$ with $K = \begin{pmatrix} 0 & 0 \\ 0 & K_1 \end{pmatrix}$, where $K_1 \in \mathsf{GL}_{r-s}(\mathbb{Z}/p\mathbb{Z})$, is congruent modulo p to a matrix of the form $\begin{pmatrix} 0 & 0 \\ 0 & V \end{pmatrix}$ with a square block V of order $a - t$ if and only if U belongs to the set

$$\mathsf{G}_1 = \{U = (U_{ij}) \in \mathsf{G};\, U_{31} \equiv 0 \pmod{p}\}.$$

Thus, if we denote by

$$\mathsf{G}_0 = \left\{U \in \mathsf{G}; \begin{pmatrix} 0 & 0 \\ 0 & K_1 \end{pmatrix}[U] \equiv \begin{pmatrix} 0 & 0 \\ 0 & K_1 \end{pmatrix} \pmod{p}\right\},$$

the stability group of K in G, then the number $s_1\{K\}$ can be written in the form $s_1\{K\} = |\mathsf{G}_1||\mathsf{G}_0|^{-1}$. On the other hand, we have obviously that $|\{K\}| = |\mathsf{G}| \cdot |\mathsf{G}_0|^{-1}$. Substituting these expressions in (3.5.53), we see that

$$s\{K\} = p^{(r-s)t}\,\mathrm{sm}_p(0_{a+s-r-t}, s, a + s - r)|\mathsf{G}|^{-1}|\mathsf{G}_1|. \tag{3.5.54}$$

In order to calculate the quotient $|\mathsf{G}|^{-1}|\mathsf{G}_1|$ we shall use the following

Lemma 3.5.18. *The number of matrices $V \in \mathsf{M}_{d,c}(\mathbb{Z}/p\mathbb{Z})$ satisfying $r_p(V) = c$, where $1 \le c \le d$ and p is a prime, is equal to*

$$p^{\langle c-1\rangle}(p^d - 1)\cdots(p^{d-c+1} - 1) = p^{\langle c-1\rangle}\varphi_d(p)/\varphi_{d-c}(p).$$

Proof. An induction on c. If $c = 1$, then the assertion is clear. Suppose it is already proved for some c with $1 \le c < d$. It is clear that a matrix $V' = (V, \mathbf{v}) \in \mathsf{M}_{d,c+1}(\mathbb{Z}/p\mathbb{Z})$, where $\mathbf{v}$ stands for the last column of V', satisfies $r_p(V') = c + 1$ if and only $r_p(V) = c$ and $\mathbf{v}$ is not a linear combination modulo p of the columns of V. The number of these $\mathbf{v}$ modulo p, for a fixed $V \in \mathsf{M}_{d,c}(\mathbb{Z}/p\mathbb{Z})$ with $r_p(V) = c$, is obviously $p^d - p^c = p^c(p^{d-c} - 1)$. Hence, using the inductive hypothesis, we conclude that the number of $V' \in \mathsf{M}_{d,c+1}(\mathbb{Z}/p\mathbb{Z})$ with $r_p(V') = c + 1$ is equal to

$$(p^{\langle c-1\rangle}\varphi_d(p)/\varphi_{d-c}(p))\cdot p^c(p^{d-c} - 1) = p^{\langle c\rangle}\varphi_d(p)/\varphi_{d-c-1}(p). \qquad \square$$

Let us complete the calculation of $s\{K\}$. The same arguments as in the proof of Lemma 3.5.18 show that, for any matrix $M \in \mathsf{M}_{a+b-2t,b-t}(\mathbb{Z}/p\mathbb{Z})$ with $1 \le b - t < a + b - 2t$ and $r_p(M) = b - t$, there are matrices $M' \in \mathsf{M}_{a+b-2t,a-t}(\mathbb{Z}/p\mathbb{Z})$ with $(M, M') \in \mathsf{G}$ and the number of these M' is independent of M. Hence, using Lemma 3.5.18, we conclude that the number is equal to

$$|\mathsf{G}|p^{-\langle b-t-1\rangle}\varphi_{a-t}(p)\varphi_{a+b-2t}(p)^{-1}.$$

On the other hand, the number of matrices M of the form indicated above whose last $r - s$ rows consist of zeroes, by the same lemma, is equal to

$$p^{\langle b-t-1\rangle}\varphi_{a+b+s-2t-r}(p)\cdot\varphi_{a+s-t-r}(p)^{-1}.$$

Multiplying the two number, we get the number of elements of G_1:

$$|\mathsf{G}_1| = |\mathsf{G}|\,\varphi_{a-t}(p)\varphi_{a+b+s-2t-r}(p)\varphi_{a+b-2t}(p)^{-1}\varphi_{a+s-t-r}(p)^{-1}.$$

The formula is also true if $b - t = 0$ or $b - t = a + b - 2t$, for in the first case the two sides are equal to $|G|$, and the second case is reduced to the first case, because $t \le b \le a$. Using the last formula, formulae (3.5.54) and (3.5.49), after obvious simplifications we obtain the formula

$$\begin{aligned} &c(a, b, r, t; K) \\ &\quad = p^{t(r+t-a-b-s-1)+b(n+1)}\,\mathrm{sm}_p(0_{a+s-r-t}, S; a + s - r)\frac{\varphi_{a+b+s-r-2t}(p)}{\varphi_{a+s-r-t}(p)\varphi_{b-t}(p)} \\ &\quad = c(a, b, r, t, s), \end{aligned}$$

where s is defined by $r_p(K) = r - s$. Using this formula for the coefficients of the product $\Pi_{a,0}^{(r)}\cdot\Pi_{n-b,b}^{(0)}$, we get the formula (3.5.35), for the sum of the double cosets $(\mathsf{M}_{a+b-2t,t}(K))_{\Gamma_0}$ over the classes $\{K\}_p^n \subset \mathsf{S}_{a+b-2t}(\mathbb{Z}/p\mathbb{Z})$ with $r_p(K) = r - s$ equals $\Pi_{a+b-2t,t}^{(r-s)}$, by Lemma 3.5.16. $\square$

Now we can calculate coefficients of factorizations (3.5.17) and (3.5.18) of $\mathbf{R}_p^n(v)$.

Proposition 3.5.19. *The following decompositions hold in the ring* $\mathscr{L}_p^n$:

$$\Pi_- \Pi_i \Pi_-^{-2} = p^{-\langle i\rangle - i(n-i)}[\mathbf{p}]^{-1} \sum_{j=0}^{i} \alpha_{ij} \sum_{a=n-i+j}^{n} \Pi_{a,n-a}^{(i-j+a-n)}, \tag{3.5.55}$$

$$\Pi_+^{-2} \Pi_{n-i} \Pi_+ = p^{-\langle i\rangle - i(n-i)}[\mathbf{p}]^{-1} \sum_{j=0}^{i} \alpha_{ij} \sum_{a=n-i+j}^{n} \Pi_{a,0}^{(i-j+a-n)} \tag{3.5.56}$$

for $0 \le i \le n$, *where*

$$\alpha_{ij} = \alpha_{ij}^n(p) = \frac{\varphi_{n-i+j}(p)}{\varphi_{n-i}(p)} \sum_{t=0}^{j} \frac{(-p)^t}{\varphi_t(p)\varphi_{j-t}(p)}, \tag{3.5.57}$$

$\varphi_r(v)$ *being the polynomials* (3.2.29), *and where* $\Pi_{a,b}^{(r)} \in \mathscr{L}_p^n$ *are the elements* (3.3.62).

Proof. The formulae (3.5.6), (3.5.19), and (3.5.33) show that the anti-automorphism $*$ carries the decomposition (3.5.55) to (3.5.56) and vice versa. Hence, it suffices to prove only one of them. We shall prove (3.5.56). For this we first prove that the two sides in (3.5.56) are contained in the subspace $\mathcal{O}_+ = \mathcal{O}_p^n = \mathbf{L}_p^n \mathscr{C}_p^n$ of $\mathscr{L}_p^n$. Next we show that they have the same image under Ω_p^n. Then the equality follows from Theorem 3.5.5(2).

From the inclusions (3.5.9) and (3.5.36) we see that

$$\Pi_+^{-2} \Pi_{n-i} \Pi_+ \in \mathcal{O}_+ \qquad \text{for } 0 \le i \le n. \tag{3.5.58}$$

To consider the right side of (3.5.56), we introduce the sums

$$Y_{i,c} = \sum_{a=n-i}^{n-c} \Pi_{a,c}^{(i+a-n)} \tag{3.5.59}$$

with $0 \le c \le i \le n$. The right side of (3.5.56) is a linear combination of the elements $[\mathbf{p}]^{-1} Y_{i-j,0}$ with $j = 0, 1, \ldots, i$ and so is contained in $\mathcal{O}_+$, if each of the elements $Y_{0,0}, Y_{1,0}, \ldots, Y_{n,0}$ is. To show the latter, we shall prove the inclusions

$$Y_{i,c} = Y_{c+d,c} \in \mathcal{O}_+ \qquad \text{for } 0 \le c \le c+d \le n \tag{3.5.60}$$

by induction on $d = i - c$. Since $Y_{c,c} = \Pi_{n-c,c}^{(0)}$, it follows from (3.5.36) that $Y_{c,c} \in \mathscr{C}_+ \subset \mathcal{O}_+$, which proves (3.5.60) for $d = 0$. Suppose that (3.5.60) is already proved for all c, d satisfying $0 \le d < h$ and $0 \le c \le n - d$, where $0 < h \le n$. Using the decompositions (3.3.61) and (3.5.59), we get

$$\begin{aligned} \mathbf{T}_{n-h}(p^2) &= \sum_{c=0}^{h} \sum_{a=n-h}^{n-c} \Pi_{a,c}^{(h+a-n)} = \sum_{c=0}^{h} Y_{h,c} \\ &= Y_{h,0} + \sum_{c+d=h, c\ge 1} Y_{c+d,c}. \end{aligned}$$

By the inductive hypothesis, each of the sums $Y_{c+d,c}$ with $c + d = h$ and $c \ge 1$ is contained in $\mathcal{O}_+$. By the definition of $\mathcal{O}_+$, we have $\mathbf{T}_{n-h}(p^2) \in \mathbf{L}_p^n \subset \mathcal{O}_+$. Thus, $Y_{h,0} \in \mathcal{O}_+$. Now, by induction on c, we shall prove the inclusions

$$Y_{c+h,c} \in \mathcal{O}_+ \qquad \text{for } 0 \le c \le n - h. \tag{3.5.61}$$

For $c = 0$, (3.5.61) is already proved. Assume (3.5.61) is true for $0 \le c < b \le n - h$, where $0 < b \le n - h$. Let us consider the product

$$Y_{h,0} \cdot \Pi^{(0)}_{n-b,b} = \sum_{a=n-h}^{n} \Pi^{(h+a-n)}_{a,0} \Pi^{(0)}_{n-b,b}.$$

Since $b \le n - h$, it follows that $(h + a - n) + b \le a$, and we can apply the formula (3.5.35) to compute the products in the last sum (this is the only place where we use the formula!) Note that the coefficients $c(a, b, r, t, s)$ in (3.5.35) depend relative to a and r only on their difference $a - r$, and so we may set $c(a, b, r, t, s) = \gamma(a - r, b, t, s)$. In this notation, we have the relation

$$\begin{aligned} Y_{h,0} \cdot \Pi^{(0)}_{n-b,b} &= \sum_{a=n-h}^{n} \sum_{\substack{\max(0,a+b-n) \le t \le b,\\ 0 \le s \le h+a-n}} \gamma(n-h, b, t, s)\,[\mathbf{p}]\,\Pi^{(h+a-n-s)}_{a+b-2t,t} \\ &= \sum_{0 \le t \le b} \sum_{0 \le s \le h} \gamma(n-h, b, t, s)\,[\mathbf{p}] \sum_{s+n-h \le a \le t+n-b} \Pi^{(h+a-n-s)}_{a+b-2t,t} \\ &= \sum_{0 \le t \le b} \sum_{0 \le s \le h} \gamma(n-h, b, t, s)\,[\mathbf{p}]\, Y_{h+2t-b-s,t}. \end{aligned}$$

It follows from the inclusion $Y_{h,0} \in \mathcal{O}_+$ and the inclusions (3.5.36) that the left side of this relation is contained in $\mathcal{O}_+$. By the inductive hypothesis and $[\mathbf{p}] \in \mathbf{L}^n_p$, we have the inclusion $[\mathbf{p}]\, Y_{h+2t-b-s,t} \in \mathcal{O}_+$ if $h + t - b - s < h$ or $h + t - b - s = h$ and $t < b$. Hence, all terms on the right side, except, possibly, for the term corresponding to $t = b$ and $s = 0$, are contained in $\mathcal{O}_+$. Thus, the term with $t = b$ and $s = 0$ is also contained in $\mathcal{O}_+$:

$$\gamma(n-h, b, b, 0)\,[\mathbf{p}]\, Y_{h+b,b} = p^{bh}[\mathbf{p}]\, Y_{h+b,b} \in \mathcal{O}_+,$$

whence, by $[\mathbf{p}]^{-1} \in \mathbf{L}^n_p$, we have $Y_{h+b,b} \in \mathcal{O}_+$. We have proved (3.5.61) and (3.5.60). Hence, the two sides of (3.5.56) are contained in $\mathcal{O}_+$. Thus, to prove (3.5.56), we only have to verify that the two sides have equal Ω-images.

By (3.5.14) and Lemma 3.3.34, we have

$$\begin{aligned} \Omega(\Pi_+^{-2}\Pi_{n-i}\Pi_+) &= \Omega(\Pi_+^{-2})\Omega(\Pi_{n-i})\Omega(\Pi_+) \\ &= (x_1 \cdots x_n)^{-1} s_{n-i}(x_1, \ldots, x_n). \end{aligned} \tag{3.5.62}$$

On the other hand, by Lemma 3.3.34 and Lemma 3.2.21, we get

$$\begin{aligned} &\Omega\left(p^{-\langle i\rangle - i(n-i)}[\mathbf{p}]^{-1} \sum_{j=0}^{i} \alpha_{ij} \sum_{a=n-i+j}^{n} \Pi^{(i-j+a-n)}_{a,0}\right) \\ &= p^{-\langle i\rangle - i(n-i) + \langle n\rangle}(x_0^2 x_1 \cdots x_n)^{-1} \\ &\quad \times \sum_{j=0}^{i} \alpha_{ij} \sum_{a=n-i+j}^{n} \mathrm{sm}_p(i-j+a-n, a) \cdot x_0^2 p^{-\langle a\rangle} s_a(x_1, \ldots, x_n) \\ &= (x_1 \cdots x_n)^{-1} p^{\langle n-i\rangle} \sum_{a=n-i}^{n} p^{-\langle a\rangle} s_a(x_1, \ldots, x_n) \sum_{j=0}^{a+i-n} \alpha_{ij}\, \mathrm{sm}_p(i-j+a-n, a). \end{aligned}$$

To prove that the last expression is equal to (3.5.62), which will complete the proof, we only have to verify the relations

$$\sum_{j=0}^{a+i-n} \alpha_{ij}\,\mathrm{sm}_p(i-j+a-n, a) = \begin{cases} 1, & \text{if } a = n-i \\ 0, & \text{if } n-i < a \le n. \end{cases}$$

Setting $n - a = k$, we can write the relations in the form

$$\sum_{j=0}^{i-k} \alpha_{ij}\,\mathrm{sm}_p(i-j-k, n-k) = \begin{cases} 1, & \text{if } k = i \\ 0, & \text{if } 0 \le k < i. \end{cases} \tag{3.5.63}$$

To prove these relations, several properties of the numbers $\mathrm{sm}_p(r, a)$ are required. The following two lemmas are dealing with the properties.

Lemma 3.5.20. *For $0 \le r \le a$ and a prime p, set*

$$\mathsf{SM}_p(r, a) = \{A \in \mathsf{S}_a(\mathbb{Z}/p\mathbb{Z}); r_p(A) = r\},$$

and, for $\mathbf{i} \in \mathsf{I}_{r,a}$ (see (3.5.39)), denote by $\mathsf{SM}_p(r, a, \mathbf{i})$ the subset of all matrices of $\mathsf{SM}_p(r, a)$ for which $\mathbf{i}$ is the least system, in the sense of lexicographic ordering, of indices of r linearly independent modulo p rows (and columns). Then one can take

$$\mathsf{SM}_p(r, a, \mathbf{i}) = \left\{\begin{pmatrix} A & AV \\ {}^tVA & A[V] \end{pmatrix}[M(\mathbf{i})]; A \in \mathsf{SM}_p(r, r); V \in \mathsf{V}(\mathbf{i})\right\}, \tag{3.5.64}$$

where $M(\mathbf{i})$ is the matrix (3.5.40), and the set $\mathsf{V}(\mathbf{i})$ is defined in (3.5.42). In particular, the number of elements of the set $\mathsf{SM}_p(r, a, \mathbf{i})$ is given by

$$|\mathsf{SM}_p(r, a, \mathbf{i})| = p^{j_1 + \cdots + j_{a-r} - \langle a-r \rangle}\,\mathrm{sm}_p(r, r), \tag{3.5.65}$$

where $(j_1, \ldots, j_{a-r}) = \hat{\mathbf{i}} \in \mathsf{I}_{a-r,a}$. For the number of elements of the whole set $\mathsf{SM}_p(r, a)$ the following formula holds:

$$\mathrm{sm}_p(r, a) = \mathrm{sm}_p(r, r)\frac{\varphi_a(p)}{\varphi_r(p)\varphi_{a-r}(p)}, \tag{3.5.66}$$

where φ_r are the polynomials (3.2.29).

Proof of the lemma. Let $A' = (a_{st}) \in \mathsf{SM}_p(r, a, \mathbf{i})$. Then, by (3.5.41) and the similar relation for rows, we obtain

$$A'[M(\mathbf{i})^{-1}] = A'' = \begin{pmatrix} A & B \\ {}^tB & C \end{pmatrix},$$

where $A = (a_{i_\alpha, i_\beta})$, $B = (a_{i_\alpha, j_\beta})$ and $C = (a_{j_\alpha, j_\beta})$ with $(i_1, \ldots, i_r) = \mathbf{i}$ and $(j_1, \ldots, j_{a-r}) = \hat{\mathbf{i}}$. We first show that $r_p(A) = r$. We have $r_p(A'') = r_p(A') = r$ and, by the definition, the first r columns of A'' are linearly independent modulo p. Hence, each of the columns of A'' is a linear combination modulo p of the first r columns. In particular, this is true for the matrix (A, B), whose rank over $\mathbb{Z}/p\mathbb{Z}$ is again r. Hence, the first r columns of (A, B) can not be linearly dependent modulo p. Next we note that for each $\beta = 1, \dots, a-r$, the β-th column of B is a linear combination modulo p of the first s columns of A, where s is the largest index with $i_s < j_\beta$, for if the columns of (A, B) with indices $i_1, \dots, i_s, j_\beta$ were linearly independent modulo p, then the same would be true for the corresponding columns of A'', and so the index i_{s+1} could be replaced by $j_\beta < i_{s+1}$. Thus, we have

$$^t(a_{i_1,j_\beta}, \dots, a_{i_r,j_\beta}) \equiv \sum_{\alpha=1}^{s} v_{\alpha\beta}\, ^t(a_{i_1,i_\alpha}, \dots, a_{i_r,i_\alpha}) \equiv \sum_{\alpha=1}^{r} v_{\alpha\beta}\, ^t(a_{i_1,i_\alpha}, \dots, a_{i_r,i_\alpha}),$$

where $0 \le v_{\alpha\beta} < p$ for $1 \le \alpha \le r$, $1 \le \beta \le a-r$, and $v_{\alpha\beta} = 0$, if $i_\alpha > j_\beta$. Hence, $B \equiv AV \pmod p$ with $V = (v_{\alpha\beta}) \in \mathrm{V}(\mathrm{i})$. Further, since

$$r = r_p\left(\begin{pmatrix} A & AV \\ {}^tVA & C \end{pmatrix}\right) = r_p\left(\begin{pmatrix} A & AV \\ {}^tVA & C \end{pmatrix}\left[\begin{pmatrix} E_r & -V \\ 0 & E_{a-r} \end{pmatrix}\right]\right)$$
$$= r_p\begin{pmatrix} A & 0 \\ 0 & C - A[V] \end{pmatrix},$$

and $r_p(A) = r$, it follows that $r_p(C - A[V]) = 0$, whence $C \equiv A[V] \pmod p$. Since the arguments are clearly reversible, we find that each of the matrices on the right side in (3.5.64) belongs (modulo p) to the left side, and the matrices with different V or different modulo p matrices A are different modulo p. Formula (3.5.65) follows from (3.5.64) and formula (3.5.44) for the number of elements of $\mathrm{V}(\mathrm{i})$. (3.5.66) follows from (3.5.65) and (3.2.33). □

The following lemma contains, in particular, explicit formulae for the numbers $\mathrm{sm}_p(r, r)$.

Lemma 3.5.21. *For each prime p the following identities hold in formal power series over* $\mathbb{Q}$:

$$\sum_{r=0}^{\infty} \frac{\mathrm{sm}_p(r,r)}{\varphi_r(p)} v^r = \left(\sum_{a=0}^{\infty} \frac{p^{\langle a \rangle}}{\varphi_a(p)} v^a\right)\left(\sum_{b=0}^{\infty} \frac{(-1)^b p^{\langle b-1 \rangle}}{\varphi_b(p)} v^b\right),$$

$$\left(\sum_{r=0}^{\infty} \frac{\mathrm{sm}_p(r,r)}{\varphi_r(p)} v^r\right)^{-1} = \left(\sum_{a=0}^{\infty} \frac{(-p)^a}{\varphi_a(p)} v^a\right)\left(\sum_{b=0}^{\infty} \frac{1}{\varphi_b(p)} v^b\right).$$

Proof of the lemma. Since the set $\mathrm{S}_a(\mathbb{Z}/p\mathbb{Z})$ consists of $p^{\langle a \rangle}$ elements, it follows from (3.5.66) that

$$p^{\langle a \rangle} = \sum_{r=0}^{a} \mathrm{sm}_p(r, a) = \varphi_a(p) \sum_{r=0}^{a} \frac{\mathrm{sm}_p(r,r)}{\varphi_r(p)} \cdot \frac{1}{\varphi_{a-r}(p)}.$$

The set of these relations for $a = 0, 1, \dots$ is obviously equivalent to the following identity between formal power series

$$\left(\sum_{r=0}^{\infty} \frac{\mathrm{sm}_p(r,r)}{\varphi_r(p)} v^r\right)\left(\sum_{c=0}^{\infty} \frac{1}{\varphi_c(p)} v^c\right) = \sum_{a=0}^{\infty} \frac{p^{\langle a \rangle}}{\varphi_a(p)} v^a.$$

Thus, to prove the lemma, we only have to verify the following two identities:

$$\left(\sum_{c=0}^{\infty} \frac{1}{\varphi_c(p)} v^c\right)\left(\sum_{b=0}^{\infty} \frac{(-1)^b p^{\langle b-1\rangle}}{\varphi_b(p)} v^b\right) = 1, \tag{3.5.67}$$

$$\left(\sum_{c=0}^{\infty} \frac{(-p)^c}{\varphi_c(p)} v^c\right)\left(\sum_{b=0}^{\infty} \frac{p^{\langle b\rangle}}{\varphi_b(p)} v^b\right) = 1. \tag{3.5.68}$$

If we put $x = p$ and $t = -p^{-1}$ in (3.2.34), we obtain, for $n = 0, 1, \ldots$ the relations

$$\varphi_n(p) \sum_{b,c\geq 0, b+c=n} \frac{(-1)^b p^{\langle b-1\rangle}}{\varphi_b(p)} \cdot \frac{1}{\varphi_c(p)} = \begin{cases} 1, & \text{if } n = 0 \\ 0, & \text{if } n > 0, \end{cases}$$

which are equivalent to (3.5.67). (3.5.68) follows from (3.5.67), if we replace v by $-pv$. □

Let us complete the proof of Proposition 3.5.19. As we have seen, it suffices to verify the relations (3.5.63). Using the formulae (3.5.57) for the numbers α_{ij} and the second identity of Lemma 3.5.21, we obtain the congruence

$$\begin{aligned}\sum_{j=0}^{i} \alpha_{ij} \frac{\varphi_{n-i}(p)}{\varphi_{n-i+j}(p)} v^j &= \sum_{j=0}^{i}\left(\sum_{t=0}^{j} \frac{(-p)^t}{\varphi_t(p)\varphi_{j-t}(p)}\right) v^j \\ &\equiv \left(\sum_{a=0}^{\infty} \frac{(-p)^a}{\varphi_a(p)} v^a\right)\left(\sum_{b=0}^{\infty} \frac{1}{\varphi_b(p)} v^b\right) \\ &= \left(\sum_{r=0}^{\infty} \frac{\mathrm{sm}_p(r,r)}{\varphi_r(p)} v^r\right)^{-1} \pmod{v^{i+1}}\end{aligned}$$

which implies the relations

$$\sum_{j,r\geq 0, j+r=i-k} \alpha_{ij} \frac{\varphi_{n-i}(p)\,\mathrm{sm}_p(r,r)}{\varphi_{n-i+j}(p)\varphi_r(p)} = \begin{cases} 1, & \text{if } i-k=0 \\ 0, & \text{if } 0 < i-k \leq i\text{'} \end{cases}$$

where $0 \leq k \leq i$. Since the right side of the relations is not zero only if $i = k$, the factor $\varphi_{n-i}(p)$ can be replaced by $\varphi_{n-k}(p)$. Then, setting $r = i - j - k$ and using (3.5.66), we obtain

$$\begin{aligned}\frac{\varphi_{n-k}(p)\,\mathrm{sm}_k(r,r)}{\varphi_{n-i+j}(p)\varphi_r(p)} &= \mathrm{sm}_p(i-j-k, i-j-k)\frac{\varphi_{n-k}(p)}{\varphi_{i-j-k}(p)\varphi_{n-k-(i-j-k)}(p)} \\ &= \mathrm{sm}_p(i-j-k, n-k).\end{aligned}$$ □

Exercise 3.5.22. *Show that the "negative squares" of the Frobenius elements* $\Pi_- = \Pi_-^n(p)$ *and* $\Pi_+ = \Pi_+^n(p) \in \mathcal{L}_p^n$ *are given by*

$$\Pi_-^{-2} = p^{-n(n+1)}[\mathbf{p}]^{-2} \sum_{j=0}^{i} \alpha_{ij}^n(p) \sum_{a=j}^{n} \Pi_{a,n-a}^{(a-j)},$$

$$\Pi_+^{-2} = p^{-n(n+1)}[\mathbf{p}]^{-2} \sum_{j=0}^{i} \alpha_{ij}^n(p) \sum_{a=j}^{n} \Pi_{a,0}^{(a-j)},$$

where d_{ij} are the numbers (3.5.57). *Use the formulae to prove that* $\Pi_-^{-d} \neq (\Pi_-^{-1})^d$ *and* $\Pi_+^{-d} \neq (\Pi_+^{-1})^d$ *if* $d > 1$.

3.5.5. A Symmetric Factorization of Polynomials $\mathbf{R}_p^n(v)$

Here we shall complete the factorization theory for the polynomials $\mathbf{R}_p^n(v)$. Using the factorizations (3.5.17), (3.5.18), the formulae (3.5.55), (3.5.56) and some additional consideration, we produce a factorization invariant under the anti-automorphism $*$. The polynomial $\mathbf{R}_p^n(v)$ of degree $2n$ will be decomposed into the product of three polynomials of degree n each, which is not so surprising, since the ring $\mathcal{L}_p^n$ contains zero divisors. It is surprising that such a factorization plays a key role in the applications to the theory of Hecke operators.

Theorem 3.5.23. *For every* $n \in \mathbb{N}$ *and* $p \in \mathbb{P}$, *the polynomial* $\mathbf{R}_p^n(v) \in \mathbf{L}_p^n[v]$ *defined by* (3.5.15) *has the following factorization over the ring* $\mathcal{L}_p^n$:

$$\mathbf{R}_p^n(v) = \left(\sum_{i=0}^{n} (-1)^i x_{-i} v^i\right)\left(\sum_{i=0}^{n} (-1)^i b_i v^i\right)\left(\sum_{i=0}^{n} (-1)^i x_i v^i\right), \tag{3.5.69}$$

where, in the notation of Lemma 3.3.32, *for* $i = 0, 1, \ldots, n$,

$$x_{-i} = x_{-i}^n(p) = (p^{\langle n\rangle}[\mathbf{p}])^{-1}\Pi_-\Pi_{n-i}, \tag{3.5.70}$$

$$x_i = x_i^n(p) = (p^{\langle n\rangle}[\mathbf{p}])^{-1}\Pi_i\Pi_+, \tag{3.5.71}$$

and

$$b_i = b_i^n(p) = p^{-\langle i\rangle - i(n-i)}[\mathbf{p}]^{-1}\sum_{j=0}^{i} \alpha_{ij}\Pi_{n,0}^{(i-j)} \tag{3.5.72}$$

with the numbers $\alpha_{ij} = \alpha_{ij}^n(p)$ *defined by* (3.5.57).

We first prove a lemma.

Lemma 3.5.24. *The following identities hold in the ring of formal power series over* $\mathcal{L}_p^n$:

$$\left(\sum_{i=0}^{n} (-1)^i x_{-i} v^i\right)^{-1} = \sum_{d=0}^{\infty} p^{-dn} t^-(p^d) v^d, \tag{3.5.73}$$

$$\left(\sum_{i=0}^{n} (-1)^i x_i v^i\right)^{-1} = \sum_{d=0}^{\infty} p^{-dn} t^+(p^d) v^d, \tag{3.5.74}$$

where x_{-i} *and* x_i *are the elements* (3.5.70) *and* (3.5.71),

$$t^-(p^d) = t_n^-(p^d) = \sum_{\substack{D \in \Lambda^n\backslash \mathrm{M}_n/\Lambda^n,\\ \det D = \pm p^d}} (U(D^*))_{\Gamma_0^n}, \tag{3.5.75}$$

and

$$t^+(p^d) = t_n^+(p^d) = \sum_{\substack{D \in \Lambda^n \backslash \mathsf{M}_n/\Lambda^n, \\ \det D = \pm p^d}} (U(D))_{\Gamma_0^n}. \tag{3.5.76}$$

Proof of the lemma. The proof of (3.5.73) is completely analogous to that of (3.5.74). Besides, the anti-automorphism $*$ carries (coefficient-wise) one of the identities to another (see (3.5.19)). Thus, it suffices to prove, for example, (3.5.74). It follows from (3.5.36) and Proposition 3.4.4 that all of the coefficients of the two sides in (3.5.74) are contained in the ring $\mathscr{C}_+ = \mathscr{C}_p^n$. Hence, by Theorem 3.5.5(2), to prove (3.5.74), we only have to verify the identity

$$\left(\sum_{i=0}^{\infty} (-1)^n \Phi(x_i) v^i\right)^{-1} = \sum_{d=0}^{\infty} p^{-dn} \Phi(t^+(p^d)) v^d, \tag{3.5.77}$$

where $\Phi = \Phi_p^n$. By Lemma 3.3.34 we have

$$\Phi(x_i) = \Phi(p^{\langle n \rangle}[\mathbf{p}])^{-1} \Phi(\Pi_i) \Phi(\Pi_+) = p^{\langle i \rangle} \pi_i^n(p).$$

To calculate $\Phi(t^+(p^d))$, we use the decompositions (3.4.22) of the corresponding double cosets, which gives us the formula

$$\Phi(t^+(p^d)) = \sum_{\substack{D \in \Lambda^n \backslash \mathsf{M}_n, \\ \det D = \pm p^d}} |\mathsf{S}_n(\mathbb{Z}) / {}^t D \mathsf{S}_n(\mathbb{Z}) D| (\Lambda^n D).$$

We now note that

$$|\mathsf{S}_n(\mathbb{Z}) / {}^t D \mathsf{S}_n(\mathbb{Z}) D| = m^{n+1}, \qquad \text{if } D \in \mathsf{M}_n \text{ and } |\det D| = m. \tag{3.5.78}$$

In fact, the index on the right side depends only on the double coset $\Lambda^n D \Lambda^n$ of D and so, by Lemma 3.2.2, one can replace D by its matrix of elementary divisors $\operatorname{ed}(D) = \operatorname{diag}(d_1, \ldots, d_n)$ for which the index is clearly equal to

$$\prod_{1 \le i \le j \le n} d_i d_j = (d_1 \cdots d_n)^{n+1} = m^{n+1}.$$

Thus, we have

$$\Phi(t^+(p^d)) = p^{d(n+1)} \sum_{\substack{D \in \Lambda \backslash \mathsf{M}_n/\Lambda, \\ \det D = \pm p^d}} (D)_\Lambda = p^{d(n+1)} t(p^d)$$

where $t(p^d)$ are the elements (3.2.10) of H^n. These formulae show that the identity (3.5.77) is, in fact, the identity of Proposition 3.2.22, where v is replaced by pv. □

Proof of Theorem 3.5.23. Let us set

$$X_-(v) = \sum_{i=0}^{n} (-1)^i x_{-i} v^i, \qquad X_+(v) = \sum_{i=0}^{n} (-1)^i x_i v^i \tag{3.5.79}$$

and define the polynomials

$$Y_-(v) = \sum_{i=0}^{n} (-1)^i \Pi_+^{-2} \Pi_{n-i} \Pi_+ v^i, \qquad Y_+(v) = \sum_{i=0}^{n} (-1)^i \Pi_- \Pi_i \Pi_-^{-2} v^i.$$

Then, by Proposition 3.5.10, we have

$$\mathbf{R}(v) = X_-(v)Y_+(v) = Y_-(v)X_+(v), \tag{3.5.80}$$

whence

$$X_-(v)^{-1}Y_-(v) = Y_+(v)X_+(v)^{-1}.$$

Denoting this formal power series by $B(v)$ and its coefficients by $(-1)^i b_i$, we obtain the identities

$$B(v) = \sum_{i=0}^{\infty} (-1)^i b_i v^i = X_-(v)^{-1}Y_-(v) = Y_+(v)X_+(v)^{-1}. \tag{3.5.81}$$

The identities (3.5.80) and (3.5.81) imply the factorization

$$\mathbf{R}(v) = X_-(v)B(v)X_+(v),$$

and so to prove the theorem, we only have to check that $B(v)$ is, in fact, a polynomial of degree n, whose coefficients are given by (3.5.72). For this we shall need some preparatory considerations.

The mapping

$$\mathscr{S}^n \supset M = \begin{pmatrix} A & B \\ 0 & D \end{pmatrix} \to h(M) = \det D \cdot (\det A)^{-1} \tag{3.5.82}$$

is clearly a homomorphism of the triangular subgroup (3.3.43) of S^n into the multiplicative group of positive rational numbers. If $M \in \mathscr{S}^n$ and p is a prime, then the number

$$\begin{aligned} \sigma_p(M) = \sigma_p((M)_{\Gamma_0}) &= v_p(h(M)) \\ &= v_p(\det D) - v_p(\det A), \end{aligned} \tag{3.5.83}$$

where $v_p(r)$ is the p-exponent of a rational r; will be called *the p-signature of the double coset* $(M)_{\Gamma_0}$. A non-zero linear combination $X \in \mathscr{L}^n$ of the double cosets will be called *p-homogeneous* if all of the double cosets in X have the same p-signature. The common p-signature of double cosets in a p-homogeneous element $X \in \mathscr{L}^n$ will be called *the p-signature of X* and denoted by $\sigma_p(X)$. Clearly, if X and Y are p-homogeneous and $XY \neq 0$, then XY is p-homogeneous and

$$\sigma_p(XY) = \sigma_p(X) + \sigma_p(Y). \tag{3.5.84}$$

Returning to the proof of Theorem 3.5.23, we consider the subspaces $\mathscr{L}^-$, $\mathscr{L}^+$, $\mathscr{L}^0$, $\mathscr{T}^-$, and $\mathscr{T}^+$ of $\mathscr{L} = \mathscr{L}_p^n$ consisting of all (finite) linear combinations of p-homogeneous elements whose p-signature, is non-positive, non-negative, zero, negative, and positive, respectively. From (3.5.84) it follows that $\mathscr{L}^-$, $\mathscr{L}^+$ and $\mathscr{L}^0$ are subrings of $\mathscr{L}$, and $\mathscr{T}^-$ (resp. $\mathscr{T}^+$) is a bilateral ideal in $\mathscr{L}^-$ (resp. $\mathscr{L}^+$). Let us consider the coefficients of the series in (3.5.81). It follows from the definitions of the corresponding elements that they are p-homogeneous with the p-signature given by

$$\left.\begin{aligned}&\sigma_p(\Pi_a) = 2a - n,\ \sigma_p(\Pi_{a,b}^{(r)}) = 2(a + 2b - n),\ \sigma_p([\mathbf{p}]) = 0\\&\sigma_p(t^-(p^d)) = -2d,\ \sigma_p(t^+(p^d)) = 2d\end{aligned}\right\} \tag{3.5.85}$$

In particular, all of the coefficients of $X_-(v)^{-1}$ (resp. $X_+(v)^{-1}$) are contained in $\mathscr{L}^-$ (resp. $\mathscr{L}^+$). On the other hand, these formulae together with the formulae (3.5.55) and (3.5.56) show that all of the coefficients of $Y_-(v)$ (resp. $Y_+(v)$) are contained in $\mathscr{L}^-$ (resp. $\mathscr{L}^+$) It follows from these remarks and the identities (3.5.81) that all of the coefficients of $B(v)$ are contained both in $\mathscr{L}^-$ and $\mathscr{L}^+$, and so

$$b_i \in \mathscr{L}^- \cap \mathscr{L}^+ = \mathscr{L}^0 \qquad (i = 0, 1, \dots). \tag{3.5.86}$$

Let us consider the coefficients b_i modulo ideal $\mathscr{T}^-$. Since, by (3.5.75) and (3.5.85), all of the coefficients of $X_-(v)^{-1}$, except the constant term, lie in $\mathscr{T}^-$, and the constant term equals $\mathbf{1}$, it follows that the series is coefficients-wise congruent modulo $\mathscr{T}^-$ to $\mathbf{1}$. Thus,

$$B(v) = X_-(v)^{-1} Y_-(v) \equiv Y_-(v) \pmod{\mathscr{T}^-},$$

whence

$$b_i \equiv \Pi_+^{-2} \Pi_{n-i} \Pi_+ \pmod{\mathscr{T}^-} \qquad \text{for } i = 0, 1, \dots, n, \tag{3.5.87}$$

and

$$b_i \equiv 0 \pmod{\mathscr{T}^-} \qquad \text{for } i > n. \tag{3.5.88}$$

Since $\mathscr{L}^0 \cap \mathscr{T}^- = \{0\}$, it follows from (3.5.86) and (3.5.88) that $b_i = 0$ for $i > n$, and so $B(v)$ is a polynomial whose degree is not greater than n. Further, the formulae (3.5.85) imply the inclusions $[p]^{-1}\Pi_{a,0}^{(r)} \in \mathscr{T}^-$ for $a < n$. Thus, from the congruences (3.5.87) and the formulae (3.5.56) we get the congruences

$$b_i \equiv p^{-\langle i \rangle - i(n-i)} [\mathbf{p}]^{-1} \sum_{j=0}^{i} \alpha_{ij} \Pi_{n,0}^{(i-j)} \pmod{\mathscr{T}^-},$$

for $i = 0, 1, \dots, n$. Since the two sides of each of the congruences are contained in $\mathscr{L}^0$ and $\mathscr{L}^0 \cap \mathscr{T}^- = \{0\}$, it follows that the two sides, in fact, equal. □

Chapter 4. Hecke Operators

Modular forms were introduced in order to study consequences of certain analytic and automorphic properties of theta-series, the generating functions for the numbers of integral representations of quadratic forms by quadratic forms. Therefore, from the point of view of the arithmetic of quadratic form, a study of modular forms primarily means the study of their Fourier coefficients. According to the general approach described in §§ 3.1.1 and 3.1.5, Hecke rings of the symplectic group operate as rings of linear operators, the Hecke operators, on spaces of modular forms. Acting on modular forms, the Hecke operators act on their Fourier coefficients. This allows us to reveal important multiplicative properties of the Fourier coefficients, which reflect multiplicative properties of the Hecke operators. The multiplicativity of the Fourier coefficients appears in the form of certain identities relating Dirichlet series formed by Fourier coefficients of an eigenfunction of Hecke operators on one side with Euler products formed by the corresponding eigenvalues, the zeta-functions of the eigenfunction, on the other.

§ 4.1. Hecke Operators for Congruence Subgroups of the Modular Group

In this section we shall define Hecke operators for a wide class of congruence subgroups of the modular group and consider some general properties of the operators.

4.1.1. Hecke Operators

In the same way as in § 3.3.1, speaking on Hecke rings of a congruence subgroup K of Γ^n, it is convenient to assume that K satisfies certain conditions. Besides, to define representations of the rings on modular forms, one has to restrict the set of congruence characters under consideration. Thus, let K be a congruence subgroup of Γ^n, and let χ be a character of K which is trivial on the principal congruence subgroup $\Gamma^n(q)$ contained in K. The pair K, χ will be called *q-regular of genus n* if K is q-symmetric (i.e. satisfies (3.3.4)), and the character χ can be extended to a homomorphism X of the group $S(K)$ into $\mathsf{GL}_1(\mathbb{C})$:

$$X\colon \mathsf{S}(K) \to \mathsf{GL}_1(\mathbb{C}), \qquad X|_K = \chi \tag{4.1.1}$$

For a q-regular pair K, χ of genus n and an integer k, we consider the space $\mathfrak{M} = \mathfrak{M}_k(K, \chi)$ of modular forms of weight k and character χ for K and define a linear representation of the Hecke ring $\mathsf{L}(K) = \mathsf{D}_{\mathbb{Q}}(K, \mathsf{S}(K))$ on the space. According to § 3.1.5, for this we need first a suitable factor of automorphy of $\mathsf{S}(K)$ on $\mathbb{H}_n$. For $M = \begin{pmatrix} A & B \\ C & D \end{pmatrix} \in \mathsf{S}(K)$ and $Z \in \mathbb{H}_n$, we set

$$\varphi_{k,X}(M, Z) = X(M)\det(CZ + D)^k, \tag{4.1.2}$$

where X is the mapping (4.1.1). According to Lemmas 1.4.2 and 1.4.1(2), the function $\varphi_{k,X}\colon \mathsf{S}(K) \times \mathbb{H}_n \to \mathsf{GL}_1(\mathbb{C})$ is a factor of automorphy of $\mathsf{S}(K)$ on $\mathbb{H}_n$ with values in $\mathsf{GL}_1(\mathbb{C})$. Then, by Lemma 1.4.1(3), we can define an action of $\mathsf{S}(K)$ on functions $F\colon \mathbb{H}_n \to \mathbb{C}$ by

$$\begin{aligned} \mathsf{S}(K) \supset M\colon F \to F|_{\varphi_{k,X}} M &= F|_{k,X} M \\ &= \varphi_{k,X}(M, Z)^{-1} F(M\langle Z \rangle) = X(M)^{-1} F|_k M, \end{aligned} \tag{4.1.3}$$

where $|_k M$ is the operator (2.3.14), which satisfies the rules

$$F|_{k,X} M|_{k,X} M_1 = F|_{k,X} M M_1 \qquad (M, M_1 \in \mathsf{S}(K)). \tag{4.1.4}$$

Since $X(M) = \chi(M)$ for $M \in K$, it follows that the condition (2.2.4) of the definition of modular forms of weight k and character χ for K can be rewritten as

$$F|_{k,X} M = F \qquad \text{for every } M \text{ of } K. \tag{4.1.5}$$

Thus, we can define an action of $\mathsf{L}(K)$ on $\mathfrak{M}$ in the same way as it was done in §§ 3.1.1, 3.1.5 for general Hecke rings. The only difference is that in the general situation automorphic forms were defined only by functional equations of the form (4.1.5), while the functions of $\mathfrak{M}$ must also satisfy certain analytic conditions. So, for $F \in \mathfrak{M} = \mathfrak{M}_k(K, \chi)$ and $T = \sum_i \alpha_i (KM_i) \in \mathsf{L}(K)$, we set

$$F|T = F|_{k,X} T = \sum_i \alpha_i F|_{k,X} M_i. \tag{4.1.6}$$

It is an immediate consequence of (4.1.4) and (4.1.5) that the function $F|T$ is independent of the choice of representatives M_i in the left cosets KM_i. Hence, by the definition of Hecke rings and (4.1.4), we have

$$\begin{aligned} (F|T)|_{k,X} M &= \sum_i \alpha_i F|_{k,X} M_i|_{k,X} M \\ &= \sum_i \alpha_i F|_{k,X} M_i M = F|T, \qquad \text{if } M \in K, \end{aligned}$$

i.e. the function $F|T$ along with F satisfies each of the functional equations (4.1.5) for modular forms of $\mathfrak{M}_k(K, \chi)$. It follows from (4.1.3), (4.1.6) and Proposition 2.3.6

that the operator $|T$ also preserves the analytic properties of modular forms. From these considerations, the definition of modular forms, and Proposition 3.1.10 we obtain

Proposition 4.1.1. *Let K, χ be a q-regular pair of genus $n \geq 1$, and let k be an integer. Then each of the operators $|_{k,X}T$ with $T \in \mathsf{L}(K)$ carries the space $\mathfrak{M}_k(K,\chi)$ into itself; the mapping $T \to |_{k,X}T$ is a linear homomorphism of $\mathsf{L}(K)$ into the ring of endomorphisms of $\mathfrak{M}_k(K,\chi)$; in particular, the following relations hold:*

$$F|T|T' = F|TT' \qquad (F \in \mathfrak{M}_k(K,\chi);\ T, T' \in \mathsf{L}(K)). \tag{4.1.7}$$

The operators $|_{k,X}T$ with $T \in \mathsf{L}(K)$ on the space $\mathfrak{M}_k(K,\chi)$ are called *Hecke operators.*

Our definition of Hecke operators $|_{k,X}T$ depends on the choice of the extension X of χ to $\mathsf{S}(K)$, which is generally not unique. Although a change of X does not substantially affect the operators (for example, this replaces the operator corresponding to a double coset $(M)_K$ with $M \in \mathsf{S}(K)$ by a constant multiple), it will be convenient to eliminate this ambiguity. First we shall describe all of the extensions of a character χ.

Lemma 4.1.2. *Let K, χ be a q-regular pair of genus n, and let ρ be a homomorphism of $\mathsf{S}(\Gamma^n(q))$ into $\mathsf{GL}_1(\mathbb{C})$ which is trivial on $\Gamma^n(q)$. Then there is a unique homomorphism $X = X_{\rho,\chi}$ of $\mathsf{S}(K)$ into $\mathsf{GL}_1(\mathbb{C})$ such that the restrictions of X to K and $\mathsf{S}(\Gamma^n(q))$ are equal to X and ρ, respectively.*

Proof. By the assumption on χ, there is an extension X_0 of the homomorphism $\chi\colon K \to \mathsf{GL}_1(\mathbb{C})$ to $\mathsf{S}(K)$. This implies that χ satisfies the following condition:

If $M\gamma = \gamma' M'$ with $M, M' \in \mathsf{S}(\Gamma^n(q))$ and $\gamma, \gamma' \in K$, then $\chi(\gamma) = \chi(\gamma')$. (4.1.8)

Really, if $M\gamma = \gamma' M'$, then $X_0(M)\chi(\gamma) = \chi(\gamma')X_0(M')$; since X_0 along with χ is trivial on $\Gamma^n(q)$, it follows that $X_0(M) = X_0(M')$, by Theorem 3.3.3(3) with $K_1 = \Gamma^n(q)$; hence, $\chi(\gamma) = \chi(\gamma')$.

According to (3.3.5), an arbitrary matrix $M \in \mathsf{S}(K)$ can be written in the form

$$M = \gamma N \qquad \text{with } \gamma \in K \text{ and } N \in \mathsf{S}(\Gamma^n(q)). \tag{4.1.9}$$

Then we set

$$X(M) = X_{\rho,\chi}(M) = \chi(\gamma)\rho(N). \tag{4.1.10}$$

If $M = \gamma N = \gamma_1 N_1$ are two decompositions of the form (4.1.9), then we have $\gamma_1^{-1}\gamma = N_1 N^{-1} \in K \cap \mathsf{S}(\Gamma^n(q)) = \Gamma^n(q)$, hence $\chi(\gamma_1) = \chi(\gamma)$ and $\rho(N_1) = \rho(N)$, and so $X(M)$ is independent of the representation of M in the form (4.1.9). It is clear that the restrictions of $X_{\rho,\chi}$ to K and $\mathsf{S}(\Gamma^n(q))$ are equal to χ and ρ, respectively. Let us check that X is a homomorphism. If $M = \gamma N$ and $M_1 = \gamma_1 N_1$ are two matrices of $\mathsf{S}(K)$ which are written in the form (4.1.9), then, by (3.3.4), we can write $N\gamma_1 = \gamma_1' N'$ with $\gamma_1' \in K$ and $N' \in \mathsf{S}(\Gamma^n(q))$. By (4.1.8), we have $\chi(\gamma_1') = \chi(\gamma_1)$.

By Theorem 3.3.3(3) with $K_1 = \Gamma^n(q)$, we conclude that $\rho(N') = \rho(N)$. Using these relations and (4.1.10), we obtain

$$\begin{aligned} X(MM_1) &= X(\gamma N\gamma_1 N_1) = X(\gamma\gamma_1' N'N_1) \\ &= \chi(\gamma\gamma_1')\rho(N'N_1) = \chi(\gamma)\chi(\gamma_1)\rho(N)\rho(N_1) = X(M)X(M_1). \end{aligned}$$

We have proved that $X_{\rho,\chi}$ exists. The uniqueness is clear. □

In view of the lemma, we may fix an extension X of χ to $\mathsf{S}(K)$ by choosing an arbitrary restriction ρ of X to $\mathsf{S}(\Gamma^n(q))$. We set

$$\rho_k(M) = \mu(M)^{-nk+\langle n\rangle} \ (M \in \mathsf{S}(\Gamma^n(q))), \tag{4.1.11}$$

and define *the normalized extension*

$$X_{k,\chi} = X_{\rho_k,\chi} \tag{4.1.12}$$

of a character χ of K to $\mathsf{S}(K)$, where K, χ is a q-regular pair, by the formula (4.1.10) with $\rho = \rho_k$. The corresponding factor of automorphy (4.1.2) we denote by $\varphi_{k,\chi}$:

$$\varphi_{k,\chi}(M, Z) = X_{k,\chi}(M)\det(CZ + D)^k, \tag{4.1.13}$$

where $M = \begin{pmatrix} A & B \\ C & D \end{pmatrix} \in \mathsf{S}(K)$ and $Z \in \mathbb{H}_n$, and, for a function F on $\mathbb{H}_n$, we set

$$F|_{k,\chi}M = F|_{\varphi_{k,\chi}}M = \varphi_{k,\chi}(M, Z)^{-1}F(M\langle Z\rangle). \tag{4.1.14}$$

Finally, Hecke operators (4.1.6) on $\mathfrak{M}_k(K, \chi)$ with $X = X_{k,\chi}$ will be called *the normalized Hecke operators* and denoted by $|_{k,\chi}T$, or just $|T$, if k and χ are clear:

$$|T = |_{k,\chi}T = |_{k,X_{k,\chi}}T \quad (T \in \mathsf{L}(K)). \tag{4.1.15}$$

This normalization turns out to be convenient in calculations.

We now show that the normalized Hecke operators are compatible with natural imbeddings of spaces of modular forms and isomorphisms of Hecke rings of Theorem 3.3.3(5). □

Proposition 4.1.3. *Let K, χ and K_1, χ_1 be two q-regular pairs of genus n, where $K_1 \subset K$ and the restriction of χ to K_1 equals χ_1. Then, for any modular form $F \in \mathfrak{M}_k(K, \chi) \subset \mathfrak{M}_k(K_1, \chi_1)$, where k is an integer, and for any $T \in \mathsf{L}(K)$, the following relation holds*:

$$F|_{k,\chi}T = F|_{k,\chi_1}\mathsf{i}(T),$$

where $\mathsf{i}\colon \mathsf{L}(K) \to \mathsf{L}(K_1)$ *is the isomorphism of Theorem* 3.3.3(5). *In particular, the subspace $\mathfrak{M}_k(K, \chi)$ of $\mathfrak{M}_k(K_1, \chi_1)$ is invariant under all Hecke operators of* $\mathsf{L}(K_1)$.

Proof. Let $T = \sum_i a_i(KM_i)$. By part (1) of Theorem 3.3.3, we may assume that each of the M_i is contained in $\mathsf{S}(K_1)$. Then the parts (4) and (5) of Theorem 3.3.3 imply that $\mathsf{i}(T) = \sum_i a_i(K_1M_i)$ (note that the mapping χ is of the form (3.1.14)).

Further, the restriction of $X_{k,\chi}$ to $\mathsf{S}(K_1)$ is equal to X_{k,χ_1}, since the two homomorphisms are equal on the subgroups K_1 and $\Gamma^n(q)$ generating $\mathsf{S}(K_1)$. This implies that $|_{k,\chi}M = |_{k,\chi_1}M$ for $M \in \mathsf{S}(K_1)$. Hence, for $F \in \mathfrak{M}_k(K, \chi)$, we have

$$F|_{k,\chi}T = \sum_i a_i F|_{k,\chi}M_i = \sum_i a_i F|_{k,\chi_1}M_i = F|_{k,\chi_1} \cdot \mathsf{i}(T). \qquad \square$$

4.1.2. Invariant Subspaces and Eigenfunctions

In this subsection we introduce some standard subspaces of spaces of modular forms which are invariant under Hecke operators; in particular, it will be shown that spaces of cusp forms are spanned by common eigenfunctions of all Hecke operators. We start with cusp forms.

Proposition 4.1.4. *Let K, χ be a q-regular pair of genus n with $n, q \in \mathbb{N}$, and let k be an integer. Then the subspace $\mathfrak{N}_k(K, \chi)$ of cusp forms in $\mathfrak{M}_k(K, \chi)$ is invariant under all of the Hecke operators $|_{k,\chi}T$ with $T \in \mathsf{L}(K)$.*

Proof. The assertion follows from Proposition 4.1.1, formulae (4.1.3), and Theorem 2.3.12(3). $\square$

In order to describe other invariant subspaces, we define a multiplicative family of operators on $\mathfrak{M}_k(K, \chi)$ commuting with Hecke operators. For $d \in \mathbb{N}_{(q)}$, we denote by $\tau(d) = \tau_n(d)$ one of the matrices satisfying

$$\tau(d) \in \Gamma^n, \qquad \tau(d) \equiv U(dE_n) \pmod{q}, \tag{4.1.16}$$

which exists by Lemma 3.3.2(1). If d is fixed modulo q, then it is clear that all such $\tau(d)$ belong to a fixed left (=right, =double) coset of Γ^n modulo $\Gamma^n(q)$. This implies that the operator

$$|_k\tau(d)\colon F \to F|_k\tau(d)$$

on $\mathfrak{M}_k(\Gamma^n(q), 1)$ is independent of the choice of the matrix $\tau(d)$ satisfying (4.1.16), depends only on d modulo q, and satisfies the rule

$$F|_k\tau(d)|_k\tau(d') = F|_k\tau(d)\tau(d') = F|_k\tau(dd'), \tag{4.1.17}$$

since $\tau(d)\tau(d') \equiv \tau(dd') \pmod{q}$.

Lemma 4.1.5. *Let K, χ be a q-regular pair of genus n, and let k be an integer. Then for each $d \in \mathbb{Z}$ with* g.c.d$(d, q) = 1$ *the following assertions hold:*

(1) *The mapping $\gamma_1 \to \gamma = \tau(d)^{-1}\gamma_1\tau(d)$ $(\gamma_1 \in K)$ is an automorphism of K which does not change the character χ, i.e. $\chi(\gamma) = \chi(\gamma_1)$.*

(2) *The subspaces $\mathfrak{M}_k(K, \chi)$ and $\mathfrak{N}_k(K, \chi)$ of $\mathfrak{M}_k(\Gamma^n(q), 1)$ are invariant under $|_k\tau(d)$;*

(3) *The operator $|_k\tau(d)$ on $\mathfrak{M}_k(K, \chi)$ commutes with each of the Hecke operators $|_{k,\chi}T$ with $T \in L(K)$.*

Proof. Since clearly $M = d\tau(d) \in \mathsf{S}(\Gamma^n(q))$ and $\Gamma^n(q)M\Gamma^n(q) = M\Gamma^n(q) = \Gamma^n(q)M$, it follows by Theorem 3.3.3(4) that $KMK = MK = KM$. Hence, given a $\gamma_1 \in K$, there is $\gamma \in K$ such that $\gamma_1 M = M\gamma$ and vice versa. Then, by (4.1.8), we have $\chi(\gamma) = \chi(\gamma_1)$. Since $\gamma = \tau(d)^{-1}\gamma_1\tau(d) = M^{-1}\gamma_1 M$, the first assertion follows.

Since the operator $|_k\tau(d)$ along with each of the operators $|_k\gamma$ for $\gamma \in \Gamma^n$ carries cusp forms to cusp forms, it suffices to prove the second assertion only for the space $\mathfrak{M}_k(K, \chi)$. In this case the assertion means that, for any function F in $\mathfrak{M}_k(\Gamma^n(q), 1)$ satisfying rules $F|_k\gamma = \chi(\gamma)F$ for each $\gamma \in K$, the function $F|_k\tau(d)$ also satisfies the rules. But this follows from the assertion (1) and (2.3.15): if $\gamma \in K$, then $\gamma = \tau(d)^{-1}\gamma_1\tau(d)$ with $\gamma_1 \in K$ and we have

$$\begin{aligned} F|_k\tau(d)|_k\gamma &= F|_k\tau(d)|_k\tau(d)^{-1}\gamma_1\tau(d) = F|_k\gamma_1|_k\tau(d) \\ &= \chi(\gamma_1)F|_k\tau(d) = \chi(\gamma)F|_k\tau(d). \end{aligned}$$

It easily follows from the assertion (1) and Lemma 4.1.2 that the mapping $\gamma_1 \to \gamma = \tau(d)^{-1}\gamma_1\tau(d)$ is an automorphism of $\mathsf{S}(K)$ which does not change the homomorphism $X = X_{k,\chi}$, i.e. $X(\gamma) = X(\gamma_1)$. Theorem 3.3.3(3) implies that matrices M and $\tau(d)^{-1}M\tau(d)$ with $M \in \mathsf{S}(\Gamma^n(q))$ belong to the same double coset modulo $\Gamma^n(q)$ and therefore they belong to the same double coset modulo K. Hence, by (3.3.4), we conclude that the automorphism carries each of the double cosets KMK with $M \in \mathsf{S}(K)$ into itself. Thus, if $M_1, \ldots, M_\nu$ is a set of representatives for the left cosets of KMK modulo K, then $\tau(d)^{-1}M_1\tau(d), \ldots, \tau(d)^{-1}M_\nu\tau(d)$ is again a set of representatives, and so, by (4.1.6), (4.1.3), (4.1.15), and (2.3.15), we have

$$\begin{aligned} F|_k\tau(d)|_{k,\chi}(M) &= \sum_{i=1}^{\nu} X(\tau(d)^{-1}M_i\tau(d))^{-1}F|_k\tau(d)|_k\tau(d)^{-1}M_i\tau(d) \\ &= \left(\sum_{i=1}^{\nu} X(M_i)^{-1}F|_k M_i\right)|_k\tau(d) = F|_{k,\chi}(M)|_k\tau(d), \end{aligned}$$

where $F \in \mathfrak{M}_k(K, \chi)$, which proves the assertion (3) with $T = (M)_K$. The general case follows by Lemma 3.1.5. □

Let V be the space $\mathfrak{M}_k(K, \chi)$ or $\mathfrak{N}_k(K, \chi)$. The mapping $d \to |_k\tau(d)$ defines a representation of the abelian group $\mathsf{GL}_1(\mathbb{Z}/q\mathbb{Z})$ on V. Thus, V is a direct sum of one-dimensional subspaces which are invariant under the representation. If $F|_k\tau(d) = \psi(d)F$ with $d \in \mathbb{N}_{(q)}$, then ψ is a character of $\mathsf{GL}_1(\mathbb{Z}/q\mathbb{Z})$. Hence, by Lemma 4.1.5, we obtain

Proposition 4.1.6. *Let K, χ be a q-regular pair of genus n, and let k be an integer. Then the following direct sum decompositions hold:*

$$\begin{aligned} \mathfrak{M}_k(K, \chi) &= \bigoplus_{\psi} \mathfrak{M}_k(K, \chi, \psi), \\ \mathfrak{N}_k(K, \chi) &= \bigoplus_{\psi} \mathfrak{N}_k(K, \chi, \psi), \end{aligned} \tag{4.1.18}$$

where ψ runs over the set of all characters of $\mathsf{GL}_1(\mathbb{Z}/q\mathbb{Z})$, and where

$$\begin{aligned}\mathfrak{M}_k(K,\chi,\psi) &= \{F\in\mathfrak{M}_k(K,\chi); F|_k\tau(d) = \psi(d)F, d\in\mathbb{N}_{(q)}\},\\ \mathfrak{N}_k(K,\chi,\psi) &= \mathfrak{N}_k(K,\chi)\cap\mathfrak{M}_k(K,\chi,\psi).\end{aligned} \tag{4.1.19}$$

The subspaces $\mathfrak{M}_k(K,\chi,\psi)$ and $\mathfrak{N}_k(K,\chi,\psi)$ are invariant under each of the Hecke operators $|_{k,\chi}T$ with $T\in\mathsf{L}(K)$.

In order to consider the action of Hecke operators on subspaces of the form (4.1.19), we use the scalar product of modular forms. Note first that Hecke operators respect the scalar product in the following sense.

Proposition 4.1.7. *Let K, χ be a q-regular pair, let ψ be a character modulo q, and let k be an integer. Suppose that at least one of forms F, $G\in\mathfrak{M}_k(K,\chi,\psi)$ is a cusp form. Then for each M of $\mathsf{S}(K)$ the following formula holds*

$$(F|_{k,\chi}(M)_K, G) = \psi(\mu(M))(F, G|_{k,\chi}(M)_K), \tag{4.1.20}$$

where $(\ ,\ )$ is the scalar product (2.5.7).

Proof. By Lemma 3.1.2, the number of the left cosets of KMK modulo K is equal to the index $[K: K_{(M)}]$. The number of the right cosets of KMK modulo K is clearly equal to the number of the left cosets in $KM^{-1}K$ and so equals $[K: K_{(M^{-1})}]$. Then, by Lemma 2.5.4, the number of the left cosets of KMK modulo K is equal to the number of the right cosets. Then, every right coset clearly meets every left coset. It follows that there is a set $M_1,\dots,M_\nu$ of common representatives for the left and right cosets of KMK modulo K. Using this set of representatives and properties of the scalar product listed in Theorem 2.5.3, we obtain

$$\begin{aligned}(F|_{k,\chi}(M)_K, G) &= \left(\sum_{i=1}^{\nu} F|_{k,\chi}M_i, G\right)\\ &= \sum_{i=1}^{\nu} X(M_i)^{-1}(F|_kM_i, G|_kM_i^{-1}|_kM_i)\\ &= \sum_{i=1}^{\nu} \mu^{-nk}X(M_i)^{-1}(F, G|_kM_i^{-1})\\ &= \left(F, \sum_{i=1}^{\nu} \mu^{-nk}\bar{X}(M_i)^{-1}G|_kM_i^{-1}\right),\end{aligned} \tag{4.1.21}$$

where $X = X_{k,\chi}$ and $\mu = \mu(M) = \mu(M_i)$. Since clearly $G|_k\lambda N = \lambda^{-nk}G|_kN$, we can write

$$\begin{aligned}\sum_{i=1}^{\nu} \mu^{-nk}\bar{X}(M_i)^{-1}G|_kM_i^{-1} &= \sum_{i=1}^{\nu} \bar{X}(M_i)^{-1}G|_k\tau(\mu)^{-1}\tau(\mu)\mu M_i^{-1}\\ &= \bar{\psi}(\mu)\sum_{i=1}^{\nu} \bar{X}(M_i)^{-1}G|_k\tau(\mu)\mu M_i^{-1}.\end{aligned} \tag{4.1.22}$$

Since $M_1, \dots, M_\nu$ is a set of representatives for the right cosets of KMK modulo K, it follows that $M_1^{-1}, \dots, M_\nu^{-1}$ is a set of representatives for the left cosets of $KM^{-1}K$ modulo K. Hence, by Lemma 4.1.5(1), we conclude that $\tau(\mu)\mu M_1^{-1}, \dots, \tau(\mu)\mu M_\nu^{-1}$ is a set of representatives for the left cosets of $K\tau(\mu)\mu M^{-1}K$ modulo K. We now show that

$$\bar{X}(M_i) = X(\tau(\mu)\mu M_i^{-1})$$

for each $i = 1, \dots, \nu$. If M_i is written in the form $M_i = \gamma_i N_i$ with $\gamma_i \in K$ and $N_i \in \mathsf{S}(\Gamma^n(q))$, then

$$\bar{X}(M_i) = \bar{X}_{k,\chi}(\gamma_i N_i) = \mu^{nk-\langle n\rangle}\bar{X}(\gamma_i),$$

by the definition of $X = X_{k,\chi}$. On the other hand, we can write M_i in the form $M_i = N_i'\gamma_i'$ with $\gamma_i' \in K$ and $N_i' \in \mathsf{S}(\Gamma^n(q))$, which gives us

$$\begin{aligned} X(\tau(\mu)\mu M_i^{-1} &= X_{k,\chi}(\tau(\mu)\mu)X_{k,\chi}(M_i^{-1}) \\ &= X_{k,\chi}(\tau(\mu)\mu)X_{k,\chi}((\gamma_i')^{-1}(N_i')^{-1}) \\ &= \mu^{2(nk-\langle n\rangle)}\chi((\gamma_i')^{-1})\mu^{-(nk-\langle n\rangle)} \\ &= \mu^{nk-\langle n\rangle}\bar{\chi}((\gamma_i') = \mu^{nk-\langle n\rangle}\bar{\chi}(\gamma_i), \end{aligned}$$

for $\mu(M_i) = \mu(N_i') = \mu$, and $\chi(\gamma_i') = \chi(\gamma_i)$, by (4.1.8). These considerations show that the sum (4.1.22) can be written in the form

$$= \bar{\psi}(\mu) \sum_{M_i' \in K\backslash K\tau(\mu)\mu M^{-1}K} X(M_i')^{-1}G|_k M_i' = \bar{\psi}(\mu)G|_{k,\chi}(\tau(\mu)\mu M^{-1})_K.$$

The matrices M and $\tau(\mu)\mu M^{-1}$ are both contained in $\mathsf{S}(\Gamma^n(q))$. It is clear that the matrices of symplectic divisors of the two matrices are equal. Hence, by Lemma 3.3.6, we have $\Gamma^n M\Gamma^n = \Gamma^n\tau(\mu)\mu M^{-1}\Gamma^n$. Then Theorem 3.3.3(3) implies that the matrices belong to the same double coset modulo $\Gamma^n(q)$; in particular,

$$(\tau(\mu)\mu M^{-1})_K = (M)_K \quad (M \in \mathsf{S}(K), \mu = \mu(M)). \tag{4.1.23}$$

Thus, we have

$$\bar{\psi}(\mu)G|_{k,\chi}(\tau(\mu)\mu M^{-1})_K = \bar{\psi}(\mu)G|_{k,\chi}(M)_K.$$

The above considerations prove (4.1.20). □

We can now prove the following theorem.

Theorem 4.1.8. *Let K, χ be a q-regular pair of genus n, and let k be an integer. Then each of the subspaces $\mathfrak{N}_k(K,\chi,\psi)$ of $\mathfrak{N}_k(K,\chi)$ has an orthogonal basis consisting of eigenfunctions for all operators $|_{k,\chi}T$ with $T \in \mathsf{L}(K)$; in particular, the whole space $\mathfrak{N}_k(K,\chi)$ has such a basis.*

Proof. By Theorem 2.4.5, each of the subspaces $\mathsf{V} = \mathfrak{N}_k(K,\chi,\psi)$ along with $\mathfrak{M}_k(K,\chi)$ is finite-dimensional. Thus, the ring of Hecke operators on V can be considered as a subring of a ring of matrices of finite order and so is finite-dimensional over $\mathbb{C}$. Since every T in $\mathsf{L}(K)$ is a linear combination of elements of the form $(M_i)_K$ with $M_i \in \mathsf{S}(K)$, it follows that there is a finite set $M_1, \ldots, M_d$ of matrices of $\mathsf{S}(K)$ such that each of the operators $|_{k,\chi}T$ $(T \in \mathsf{L}(K))$ on V is a linear combination of the operators $|_{k,\chi}(M_1), \ldots, |_{k,\chi}(M_d)$. Thus, it suffices to prove the theorem for a finite set of operators of the form $|_{k,\chi}(M_i)$ with $M_i \in \mathsf{S}(K)$. According to Theorem 3.3.3 and Lemma 3.3.5, the ring $\mathsf{L}(K)$ is isomorphic to the ring $\mathsf{L}^n(q) = \mathsf{L}(\Gamma_0^n(q))$ and so is commutative, in view of Theorem 3.3.7. Hence, the operators $|_{k,\chi}(M_i)$ commute with each other. □

Lemma 4.1.9. *Let* V *be a non-zero linear space of finite dimension over an algebraically closed field, and let* $T_1, \ldots, T_d$ *be a finite set of linear operators on* V *commuting with each other. Then* V *contains a non-zero common eigenvector for the operators* $T_1, \ldots, T_d$.

Proof of the lemma. The case $d = 1$ is trivial. Suppose the lemma is already proved for sets of $d - 1$ operators with some $d > 1$. Let λ_1 be an eigenvalue of T_1 on V, and let $\mathsf{V}' = \{v \in V; v|T_1 = \lambda_1 v\}$ be the corresponding subspace of eigenvectors. Since operators $T_2, \ldots, T_d$ commute with T_1, it follows that V' is invariant under $T_2, \ldots, T_d$. By the inductive hypothesis, V' contains a non-zero common eigenvector for $T_2, \ldots, T_d$. □

Returning to the proof of Theorem 4.1.8, we obtain that the space V contains a non-zero eigenfunction F_1 for the operators $|_{k,\chi}(M_1), \ldots, |_{k,\chi}(M_d)$, provided, of course, that $\mathsf{V} \neq \{0\}$. We set $\mathsf{V}_1 = \{cF_1; c \in \mathbb{C}\}$ and $\mathsf{V}_2 = \{G \in V; (F_1, G) = 0\}$. Since the scalar product on V is Hermitian and non-degenerate, it follows that V is the direct sum of V_1 and V_2, $\mathsf{V} = \mathsf{V}_1 \oplus \mathsf{V}_2$. By the definition, the space V_1 is invariant under $|(M_1), \ldots, |(M_d)$. Then the relations (4.1.20) imply that the same is true for V_2. Thus, if $\mathsf{V}_2 \neq \{0\}$, there is a non-zero eigenfunction $F_2 \in \mathsf{V}_2$. Repeating the same arguments for V_2 and F_2 in place of V and F_1, and so on, after finitely many steps we obtain an orthogonal basis of V consisting of eigenfunctions for all $|(M_1), \ldots, |(M_d)$. □

Exercise 4.1.10. *Let* K, χ *be a q-regular pair of genus n. Show that the subspace* $\mathfrak{E}_k(K, \chi) \subset \mathfrak{M}_k(K, \chi)$ *defined by* (2.5.10) *is invariant under all of the operators* $|_{k,\chi}T$ *with* $T \in \mathsf{L}(K)$.

Exercise 4.1.11. *Show that Eisenstein series* $G_k(z)$ *of Exercise* 2.5.6. *are eigenfunctions for all Hecke operators* $|_{k,1}T$ *with* $T \in \mathsf{L}^1(1)$. *Use this to prove that the space* $\mathfrak{M}_k$ *has a basis of eigenfunctions for all of the operators. Show that the functions* $G_{12}(z)$ *and* $\Delta'(z)$ *(see Exercise* 2.5.7) *form such a basis for* $\mathfrak{M}_{12}$.

Exercise 4.1.12. *Let us set*

$$G_k|_{k,1}T(m) = \lambda(m)G_k,$$

where $T(m)$ *are the elements* (3.3.19) *of* $\mathsf{L}^1(1)$. *Show that*

$$\sum_{m\in\mathbb{N}} \lambda(m)m^{-s} = \zeta(s)\zeta(s-k+1).$$

[Hint: see Exercise 3.3.19.]

§4.2. Action of Hecke Operators

In this section we consider the problem of calculation of action of Hecke operators on modular forms and their Fourier coefficients. The calculations are different for different congruence subgroups K of Γ^n. By the same reasons as in Chapter 3, we limit ourselves to the case $K = \Gamma_0^n(q)$.

4.2.1. Hecke Operators for $\Gamma_0^n(q)$

Lemma 4.2.1. *Let* χ *be a Dirichlet character modulo* q, *and let* $[\chi]$ *be the one-dimensional character* (2.3.49) *of* $\Gamma_0^n(q)$. *Then the pair* $\Gamma_0^n(q)$, $[\chi]$ *is* q*-regular, and, for any integral* k, *the normalized extension* $X_{k,[\chi]}$ *of* $[\chi]$ *to* $\mathsf{S}^n(q) = \mathsf{S}(\Gamma_0^n(q))$ *is given by*

$$X_{k,[\chi]}(M) = \mu(M)^{-nk+\langle n\rangle}\bar{\chi}(\det A), \qquad \left(M = \begin{pmatrix} A & B \\ C & D \end{pmatrix} \in \mathsf{S}^n(q)\right). \tag{4.2.1}$$

Proof. The group $\Gamma_0^n(q)$ is q-symmetric, by Lemma 3.3.5. It is clear that the restriction of the mapping (4.2.1) to $\Gamma_0^n(q)$ is $[\chi]$, and that to $\mathsf{S}(\Gamma^n(q))$ is ρ_k. Lemma 3.3.5 implies that the mapping is a homomorphism of $\mathsf{S}^n(q)$. □

In the notations of the lemma, for a function F on $\mathbb{H}_n$, we have

$$F|_{k,\chi}M = F|_{k,[\chi]}M = \mu(M)^{nk-\langle n\rangle}\chi(\det A)F|_k M \tag{4.2.2}$$

(see (4.1.14), (4.1.13), and (4.2.1)). For $T = \sum_i a_i(\Gamma_0^n(q)M_i) \in \mathsf{L}^n(q)$ we denote by $|T = |_{k,\chi}T$ the normalized Hecke operator

$$|\mathrm{T}: F \to F|T = \sum_i a_i F|_{k,\chi}M_i, \qquad (F \in \mathfrak{M}_k^n(q,\chi)) \tag{4.2.3}$$

on the space $\mathfrak{M}_k^n(q,\chi) = \mathfrak{M}_k(\Gamma_0^n(q),[\chi])$.

Acting on modular forms, Hecke operators act on their Fourier coefficients. According to (2.3.51), coefficients $f(A)$ of the Fourier expansion (2.3.50) of a modular form $F \in \mathfrak{M}_k^n(q,\chi)$ satisfy

$$f(A[V]) = \xi(\det V)f(A), \qquad (A \in \mathbb{A}_n, V \in \Lambda^n), \tag{4.2.4}$$

where ξ is the character of the group $\{\pm 1\}$ satisfying

$$\xi(-1) = (-1)^k\chi(-1). \tag{4.2.5}$$

Hence, we may consider the Fourier coefficients $f(A)$ as values of a function $f\colon \mathbb{A}_n \to \mathbb{C}$ satisfying (4.2.4). We let $\mathfrak{F}_k^n(q,\chi)$ be the space of all functions $f\colon \mathbb{A}_n \to \mathbb{C}$ corresponding to modular forms $F \in \mathfrak{M}_k^n(q,\chi)$ in this sense. If $f \in \mathfrak{F}_k^n(q,\chi)$ and $T \in \mathsf{L}^n(q)$, then we have

$$F = \sum_{A \in \mathbb{A}_n} f(A)e\{AZ\} \in \mathfrak{M}_k^n(q,\chi), \qquad \text{and } F|T \in \mathfrak{M}_k^n(q,\chi);$$

in particular, the function $F|T$ has a Fourier expansion, say,

$$F|T = \sum_{A \in \mathbb{A}_n} (f|T)(A)e\{AZ\}, \tag{4.2.6}$$

where $f|T$ can again be considered as a function of $\mathfrak{F}_k^n(q,\chi)$.

Lemma 4.2.2. *The mapping*

$$T \to |T\colon f \to f|T \qquad (T \in \mathsf{L}^n(q), f \in \mathfrak{F}_k^n(q,\chi))$$

is a linear representation of the ring $\mathsf{L}^n(q)$ *on the space* $\mathfrak{F}_k^n(q,\chi)$.

Proof. The lemma follows from Proposition 4.1.1. □

We call $\mathfrak{F}_k^n(q,\chi)$ *the space of Fourier coefficients of modular forms of* $\mathfrak{M}_k^n(q,\chi)$. The operators $|T$ with $T \in \mathsf{L}^n(q)$ on $\mathfrak{F}_k^n(q,\chi)$ will again be called *Hecke operators.*

4.2.2. Hecke Operators for Γ_0^n

In Chapter 3 a number of decompositions of elements of $\mathsf{L}^n(q)$ in terms of elements of Hecke rings of the group Γ_0^n was obtained. Generally speaking, Hecke operators for Γ_0^n do not map spaces $\mathfrak{M}_k^n(q,\chi)$ into themself. Therefore, in order to apply the decompositions to Hecke operators, we have to introduce certain extensions of the spaces $\mathfrak{M}_k^n(q,\chi)$.

For a character ξ of the group $\{\pm 1\}$, we define the character $[\xi]\colon \Gamma_0^n \to \{\pm 1\}$ by

$$[\xi]\left(\begin{pmatrix} A & B \\ 0 & D \end{pmatrix}\right) = \xi(\det D), \qquad \left(\begin{pmatrix} A & B \\ 0 & D \end{pmatrix} \in \Gamma_0^n\right) \tag{4.2.7}$$

The group Γ_0^n operates on the space $\mathfrak{M}(T_1^n)$ of modular forms for the group T_1^n (see (2.3.4)) by

$$M\colon F \to F|_0 M = F(M\langle Z\rangle) = F((AZ+B)D^{-1}), \qquad \left(M = \begin{pmatrix} A & B \\ 0 & D \end{pmatrix} \in \Gamma_0^n\right).$$

We set

$$\mathfrak{M}_\xi^n = \{F \in \mathfrak{M}(T_1^n); \quad F|_0 M = [\xi](M)F, M \in \Gamma_0^n\}. \tag{4.2.8}$$

It follows from definitions that

$$\mathfrak{M}_k^n(q,\chi) \subset \mathfrak{M}_\xi^n \qquad \text{if } \xi(-1) = (-1)^k \chi(-1). \tag{4.2.9}$$

Along the general scheme of § 3.1.5, we now define a representation of the global Hecke rings $\mathscr{L}^n(q) = D_{\mathbb{Q}}(\Gamma_0^n, \mathscr{S}^n(q))$ of Γ_0^n on $\mathfrak{M}_\xi^n$. In addition, we shall try to obtain operators compatible with the normalized Hecke operators on $\mathfrak{M}_k^n(q, \chi)$ relative to the imbeddings (3.4.4) and (4.2.9). For this we take the factor of automorphy of $\mathscr{S}^n(q)$ defined by

$$\varphi_{k,\chi}(M, Z) = \mu(M)^{\langle n \rangle - kn}\bar{\chi}(\det A)(\det D)^k, \quad \left(M = \begin{pmatrix} A & B \\ 0 & D \end{pmatrix} \in \mathscr{S}^n(q)\right) \tag{4.2.10}$$

i.e. the restriction of the factor (4.1.13) to $\mathscr{S}^n(q)$. Then the corresponding action of $\mathscr{S}^n(q)$ on functions $F\colon \mathbb{H}_n \to \mathbb{C}$ can be written in the form

$$\begin{aligned} \mathscr{S}^n(q) \ni M = \begin{pmatrix} A & B \\ 0 & D \end{pmatrix} &: F \to F|_{k,\chi} M = \varphi_{k,\chi}(M, Z)^{-1} F(M\langle Z \rangle) \\ &= \mu(M)^{nk - \langle n \rangle} \chi(\det A)(\det D)^{-k} F(M\langle Z \rangle). \end{aligned} \tag{4.2.11}$$

For $F \in \mathfrak{M}_\xi^n$ with $\xi(-1) = (-1)^k \chi(-1)$, and $T = \sum_i a_i(\Gamma_0^n M_i) \in \mathscr{L}^n(q)$ we set

$$F|_{k,\chi} T = \sum_i a_i F|_{k,\chi} M_i. \tag{4.2.12}$$

Since clearly the last function has the same analytic properties as F, and the conditions $F|_0 M = [\xi](M)F$ for $M \in \Gamma_0^n$ mean that $F|_{k,\chi} M = F$, it follows from Proposition 3.1.10 that $F|_{k,\chi} T \in \mathfrak{M}_\xi^n$. So we get the following

Lemma 4.2.3. *Let k be an integer, let χ be a Dirichlet character modulo $q \in \mathbb{N}$, and let ξ be the character of the group $\{\pm 1\}$ satisfying $\xi(-1) = (-1)^k \chi(-1)$. Then each of the operators $|_{k,\chi} T$ with $T \in \mathscr{L}^n(q)$ maps the space $\mathfrak{M}_\xi^n$ into itself. The mapping $T \to |_{k,\chi} T$ is a linear representation of $\mathscr{L}^n(q)$ on $\mathfrak{M}_\xi^n$.*

The operators $|_{k,\chi} T$ on $\mathfrak{M}_\xi^n$ will again be called *Hecke operators.*

Lemma 4.2.4. *With the above assumptions, the following relations hold*

$$F|_{k,\chi} T = F|_{k,\chi} \mathrm{i}(T), \quad (F = \mathfrak{M}_k^n(q, \chi), T \in \mathsf{L}^n(q)),$$

where i *is the imbedding* (3.4.4); *in particular, the subspace $\mathfrak{M}_k^n(q, \chi)$ of $\mathfrak{M}_\xi^n$ is invariant relative to all of the operators $|_{k,\chi} T$ with $T \in \mathbf{L}^n(q) = \mathrm{i}(\mathsf{L}^n(q))$.*

Proof. The lemma readily follows from definitions. □

By Theorem 2.3.1, every modular form F of $\mathfrak{M}_\xi^n$ has a Fourier expansion of the form

$$F(Z) = \sum_{A \in \mathbb{A}_n} f(A) e\{AZ\}. \tag{4.2.13}$$

Similarly to (4.2.4), the inclusion $F \in \mathfrak{M}_\xi^n$ implies the relations

$$f(A[V]) = \xi(\det V) f(A), \quad (A \in \mathbb{A}_n, V \in \Lambda^n). \tag{4.2.14}$$

Hence, we may consider the numbers $f(A)$ as values of a function $f\colon \mathbb{A}_n \to \mathbb{C}$ satisfying (4.2.14). We let $\mathfrak{F}_\xi^n$ be the space of all functions $f\colon \mathbb{A}_n \to \mathbb{C}$ corresponding to modular forms of $\mathfrak{M}_\xi^n$ in this sense. It is clear that

$$\mathfrak{F}_k^n(q,\chi) \subset \mathfrak{F}_\xi^n, \qquad \text{if } \xi(-1) = (-1)^k \chi(-1). \tag{4.2.15}$$

If $f \in \mathfrak{F}_\xi^n$, F is the function of $\mathfrak{M}_\xi^n$ with the Fourier coefficients $f(A)$ ($A \in \mathbb{A}_n$), and $T \in \mathscr{L}^n(q)$, then we denote by $(f|_{k,\chi}T)(A)$, where k, χ satisfy (4.2.5), the Fourier coefficients of the function $F|_{k,\chi}T$, so that

$$F|_{k,\chi}T = \sum_{A \in \mathbb{A}_n} (f|_{k,\chi}T)(A) e\{AZ\}.$$

It is not difficult to write $(f|_{k,\chi}T)(A)$ in an explicit way: If

$$T = \sum_i a_i \left(\Gamma_0^n \begin{pmatrix} \mu_i D_i^* & B_i \\ 0 & D_i \end{pmatrix} \right)$$

is an arbitrary element of $\mathscr{L}^n(q)$, then, using (4.2.11), (4.2.12), and (4.2.13), we have

$$\begin{aligned} F|_{k,\chi}T &= \sum_i a_i \mu_i^{nk-\langle n \rangle} \chi(\det(\mu_i D_i^*))(\det D_i)^{-k} F((\mu_i D_i^* Z + B_i) D_i^{-1}) \\ &= \sum_{i, A' \in \mathbb{A}_n} a_i \mu_i^{nk-\langle n \rangle} \chi(\det(\mu_i D_i^*))(\det D_i)^{-k} \\ &\quad \times f(A') e\{\mu_i A'[D_i^*] Z\} e\{A' B_i D_i^{-1}\}, \end{aligned}$$

whence, putting together all of the terms with $\mu_i A'[D_i^*] = A$ for an A of $\mathbb{A}_n$, we obtain the following formula

$$\begin{aligned} (f|_{k,\chi}T)(A) = &\sum_{i, \mu_i^{-1} A[{}^t D_i] \in \mathbb{A}_n} a_i \mu_i^{nk-\langle n \rangle} \chi(\det(\mu_i D_i^*))(\det D_i)^{-k} \\ &\times e\{\mu_i^{-1} A[{}^t D_i] B_i D_i^{-1}\} f(\mu_i^{-1} A[{}^t D_i]). \end{aligned} \tag{4.2.16}$$

By Lemmas 4.2.3 and 4.2.4, under the same assumptions, we have

Lemma 4.2.5. *The mapping*

$$T \to |_{k,\chi}T\colon f \to f|_{k,\chi}T \qquad (T \in \mathscr{L}^n(q), f \in \mathfrak{F}_\xi^n)$$

is a linear representation of $\mathscr{L}^n(q)$ on $\mathfrak{F}_\xi^n$. *If* $f \in \mathfrak{F}_k^n(q,\chi)$ *and* $T \in \mathsf{L}^n(q)$, *then*

$$f|_{k,\chi}T = f|_{k,\chi}\mathsf{i}(T), \tag{4.2.17}$$

where i *is the imbedding* (3.4.4); *in particular, the subspace* $\mathfrak{F}_k^n(q,\chi)$ *of* $\mathfrak{F}_\xi^n$ *is invariant relative to all of the operators* $|_{k,\chi}T$ *with* $T \in \mathbf{L}_1^n(q) = \mathsf{i}(\mathsf{L}^n(q))$.

We call $\mathfrak{F}_\xi^n$ the space of Fourier coefficients of functions of $\mathfrak{M}_\xi^n$. The operators $|_{k,\chi}T$ with $T \in \mathscr{L}^n(q)$ on $\mathfrak{F}_\xi^n$ are Hecke operators.

The formal background, we have considered, will allow us to use decompositions of Chapter 3 for calculation of action of Hecke operators on modular forms and their Fourier coefficients. The last three lemmas of this subsection give formulae for the action of certain elements of $\mathscr{L}^n$ appearing in the decompositions.

We again fix an integer k, a Dirichlet character χ modulo a $q \in \mathbb{N}$, and the character ξ of $\{\pm 1\}$ satisfying (4.2.5).

Lemma 4.2.6. *Let F be a function of $\mathfrak{M}_\xi^n$ with Fourier coefficients $f(A)$ ($A \in \mathbb{A}_n$). Then the images of F and f under the operator $|_{k,\chi}(M_0)_{\Gamma_0^n}$, where $M_0 = \begin{pmatrix} \mu D_0^* & B_0 \\ 0 & D_0 \end{pmatrix}$ is a matrix of $\mathcal{S}^n(q)$, are given by*

$$(F|_{k,\chi}(M_0)_{\Gamma_0})(Z) \tag{4.2.18}$$
$$= \mu^{nk-\langle n\rangle}\chi(\mu)^n \sum_{\substack{D \in \Lambda\backslash\Lambda D_0 \Lambda, \\ B \in \mathsf{B}(D, M_0)/\mathrm{mod}\, D}} \bar{\chi}(\det D)(\det D)^{-k} F(\mu Z[D^{-1}] + BD^{-1}),$$

$$(f|_{k,\chi}(M_0)_{\Gamma_0})(A) = \mu^{nk-\langle n\rangle}\chi(\mu)^n \sum_{D \in \Lambda\backslash\Lambda D_0\Lambda,\, A[{}^tD] \in \mu\mathbb{A}_n} \bar{\chi}(\det D)(\det D)^{-k}$$
$$\times\, s(\mu^{-1}A[{}^tD], D, M_0) f(\mu^{-1}A[{}^tD]), \tag{4.2.19}$$

where $\Lambda = \Lambda^n$, $\mathsf{B}(D, M_0)$ are the sets defined in Lemma 3.3.25, and where, for $Q \in \mathbb{A}_n$, we set

$$s(Q, D, M_0) = \sum_{B \in \mathsf{B}(D, M_0)/\mathrm{mod}\, D} e\{QBD^{-1}\}. \tag{4.2.20}$$

Proof. The formula (4.2.18) follows directly from formulae (4.2.12), (4.2.11) and the decomposition (3.3.44) of the double coset $\Gamma_0 M_0 \Gamma_0$ into left cosets. The formula (4.2.19) follows from (3.3.44) and (4.2.16). □

Lemma 4.2.7. *Let F be a function of $\mathfrak{M}_\xi^n$ with Fourier coefficients $f(A)$ ($A \in \mathbb{A}_n$), and let M_0 be a matrix in $\mathcal{L}^n(q)$ of the form $M_0 = U(\mu, D_0)$ satisfying $d_n(D_0)^2 | \mu$ (resp. $\mu | d_1(D_0)^2$), where $d_i(D_0)$ is the i-th elementary divisor of D_0. Then the following formulae hold:*

$$(F|_{k,\chi}(M_0)_{\Gamma_0})(Z)$$
$$= \mu^{nk-\langle n\rangle}\chi(\mu)^n \sum_{D \in \Lambda\backslash\Lambda D_0\Lambda} \bar{\chi}(\det D)(\det D)^{-k} F(\mu Z[D^{-1}]), \tag{4.2.21}$$

$$(f|_{k,\chi}(M_0)_{\Gamma_0})(A)$$
$$= \mu^{nk-\langle n\rangle}\chi(\mu)^n \sum_{\substack{D \in \Lambda\backslash\Lambda D_0\Lambda, \\ A[{}^tD] \in \mu\mathbb{A}_n}} \bar{\chi}(\det D)(\det D)^{-k} f(\mu^{-1}A[{}^tD]) \tag{4.2.22}$$

(resp.

$$(F|_{k,\chi}(M_0)_{\Gamma_0})(Z)$$
$$= \mu^{nk-n(n+1)}\chi(\mu)^n |\det D_0|^{n+1} \sum_{D \in \Lambda\backslash\Lambda D_0\Lambda} \bar{\chi}(\det D)(\det D)^{-k} \tag{4.2.23}$$
$$\times \sum_{A \in \mathbb{A}_n \cap \mu^{-1} D\mathbb{A}_n^t D} f(A) e\{\mu A[D^*] Z\},$$

$$(f|_{k,\chi}(M_0)_{\Gamma_0})(A) \tag{4.2.24}$$
$$= \mu^{nk-n(n+1)}\chi(\mu)^n|\det D_0|^{n+1} \sum_{D \in \Lambda\backslash\Lambda D_0 \Lambda} \bar{\chi}(\det D)(\det D)^{-k} f(\mu^{-1}A[{}^tD])).$$

Proof. Formulae (4.2.21) and (4.2.22) follow directly from (4.2.18) and (4.2.19), since in the case one can take $\mathsf{B}(D, M_0)/\mathrm{mod}\, D = \{0\}$ for each $D \in \Lambda D_0 \Lambda$, according to (3.4.21). But if $\mu | d_1(D_0)^2$, then it follows from (3.2.22) that one can take

$$\mathsf{B}(D, M_0)/\mathrm{mod}\, D = \{\mu D^* S; S \in \mathsf{S}_n(\mathbb{Z})/\mu^{-1} \cdot {}^tD\mathsf{S}_n(\mathbb{Z})D\}$$

for each $D \in \Lambda D_0 \Lambda$. Using this, we shall show that the sum (4.2.20) with $Q \in \mathbb{A}_n$ and $D \in \Lambda D_0 \Lambda$ is given by

$$s(Q, D, M_0) = \begin{cases} \mu^{-\langle n \rangle}|\det D_0|^{n+1} & \text{if } Q \in \mathbb{A}_n \cap \mu^{-1} D \mathbb{A}_n {}^tD \\ 0, & \text{if } Q \notin \mathbb{A}_n \cap \mu^{-1} D \mathbb{A}_n {}^tD. \end{cases} \tag{4.2.25}$$

Actually, by the definition, we have

$$s(Q, D, M_0) = \sum_{S \in \mathsf{S}_n(\mathbb{Z})/\mu^{-1}\,{}^tD\mathsf{S}_n(\mathbb{Z})D} e\{\mu Q[D^*]S\}.$$

Under our assumption the mapping $S \to e\{\mu Q[D^*]S\}$ is clearly a character of the finite additive group $\mathsf{S}_n(\mathbb{Z})/\mu^{-1}\,{}^tD\mathsf{S}_n(\mathbb{Z})D$. Hence, the sum is equal to the order of the group if the character is trivial, otherwise it equals zero. It is clear that the character is trivial if and only if $\mu Q[D^*] \in \mathbb{E}_n$, which, for $Q \in \mathbb{A}_n$, is equivalent to the condition $Q \in \mathbb{A}_n \cap \mu^{-1} D \mathbb{A}_n {}^tD$. Finally, to find the order of the group, we may obviously replace the matrix D by the matrix $\mathrm{ed}(D) = \mathrm{diag}(d_1, \ldots, d_n)$ of elementary divisors of D. Then the order equals

$$\prod_{1 \le i \le j \le n} (\mu^{-1} d_i d_j) = \mu^{-\langle n \rangle}(d_1 \cdots d_n)^{n+1} = \mu^{-\langle n \rangle}|\det D_0|^{n+1}.$$

We have proved (4.2.25). Formulae (4.2.23) and (4.2.24) follow from (4.2.18), (4.2.19), and (4.2.25). □

In Chapter 3 we often met elements of $\mathscr{L}^n$ of the form

$$\Pi_i^n(a) = (M_i^n(a))_{\Gamma_0} \qquad (0 \le i \le k, a \in \mathbb{N}) \tag{4.2.26}$$

with

$$M_i^n(a) = \begin{pmatrix} aD_i^n(a)^{-1} & 0 \\ 0 & D_i^n(a) \end{pmatrix}, \qquad \text{where } D_i^n(a) = \begin{pmatrix} E_{n-i} & 0 \\ 0 & aE_i \end{pmatrix},$$

and elements $[\mathbf{a}] = (aE_{2n})_{\Gamma_0} = (a\Gamma_0)$. The last lemma contains formulae for the action of the corresponding operators.

Lemma 4.2.8. *Let F be a function of $\mathfrak{M}_\xi^n$ with Fourier coefficients $f(A)$ $(A \in \mathbb{A}_n)$, and let $a \in \mathbb{N}_{(q)}$. Then the following formulae hold:*

$$(F|_{k,\chi}\Pi_i^n(a) \tag{4.2.27}$$
$$= a^{nk-\langle n\rangle+\langle i\rangle}\chi(a)^n \sum_{D\in\Lambda\backslash\Lambda D_i^n(a)\Lambda} \bar{\chi}(\det D)(\det D)^{-k}$$
$$\times \sum_{A\in\mathbb{A}_n\cap a^{-1}D\mathbb{A}_n^t D} f(A)e\{aA[D^*]Z\},$$

$$(f|_{k,\chi}\Pi_i^n(a))(A) \tag{4.2.28}$$
$$= a^{nk-\langle n\rangle+\langle i\rangle}\chi(a)^n \sum_{\substack{D\in\Lambda\backslash\Lambda D_i^n(a)\Lambda,\\ A[{}^tD]\in a\mathbb{A}_n}} \bar{\chi}(\det D)(\det D)^{-k}f(a^{-1}A[{}^tD]),$$

$$F|_{k,\chi}[\mathbf{a}] = a^{n(k-n-1)}\chi(a)^n F, \qquad f|_{k,\chi}[\mathbf{a}] = a^{n(k-n-1)}\chi(a)^n f. \tag{4.2.29}$$

Proof. We first show that

$$\mathsf{B}(D, M_0) = \underline{\mathsf{B}}(D), \qquad \text{if } M_0 = M_i^n(a) \text{ and } D\in\Lambda D_i^n(a)\Lambda,$$

where $\underline{\mathsf{B}}(D)$ is the set (3.3.60). Since the set $\Gamma_0 M_0 \Gamma_0$ consists of integral matrices, it follows that $\mathsf{B}(D, M_0) \subset \underline{\mathsf{B}}(D)$. Conversely, if $B\in\underline{\mathsf{B}}(D)$ and $D = \alpha D_0 \beta$, where $\alpha, \beta\in\Lambda$ and $D_0 = D_i^n(a)$, then

$$M = \begin{pmatrix} aD^* & B \\ 0 & D\end{pmatrix} = U(\alpha)\begin{pmatrix} aD_0^{-1} & B_0 \\ 0 & D_0\end{pmatrix}U(\beta)$$

with $B_0 = {}^t\alpha B\beta^{-1}$. By Lemma 3.3.33(1), the matrix B_0 is contained in $\underline{\mathsf{B}}(D_0)$, therefore it can be written in the form $B_0 = \begin{pmatrix} S_1 & aS_2 \\ {}^tS_2 & S_3\end{pmatrix}$ with $S_1\in\mathsf{S}_{n-i}(\mathbb{Z})$, $S_2\in\mathsf{M}_{n-i,i}$, and $S_3\in\mathsf{S}_i(\mathbb{Z})$. Hence, the matrix $\begin{pmatrix} aD_0^{-1} & B_0 \\ 0 & D_0\end{pmatrix} = \begin{pmatrix} E & T_1 \\ 0 & E\end{pmatrix}\begin{pmatrix} aD_0^{-1} & 0 \\ 0 & D_0\end{pmatrix}\begin{pmatrix} E & T_2 \\ 0 & E\end{pmatrix}$, where $T_1 = \begin{pmatrix} S_1 & S_2 \\ {}^tS_2 & 0\end{pmatrix}$ and $T_2 = \begin{pmatrix} 0 & 0 \\ 0 & S_3\end{pmatrix}$, is contained in $\Gamma_0 M_i^n(a)\Gamma_0$, and so $M\in\Gamma_0 M_i^n(a)\Gamma_0$. Thus, $B\in\mathsf{B}(D, M_0)$.

We now show that the sum (4.2.20) with $Q\in\mathbb{A}_n$, $M_0 = M_i^n(a)$, and $D\in\Lambda D_i^n(a)\Lambda$ is given by

$$s(Q, D, M_0) = \begin{cases} a^{\langle i\rangle}, & \text{if } Q\in\mathbb{A}_n\cap a^{-1}D\mathbb{A}_n{}^tD \\ 0, & \text{if } Q\notin\mathbb{A}_n\cap a^{-1}D\mathbb{A}_n{}^tD.\end{cases} \tag{4.2.30}$$

By the above and Lemma 3.3.33, writing D in the form $D = \alpha D_0\beta$ with $\alpha, \beta\in\Lambda$ and $D_0 = D_i^n(a)$, we have

$$s(Q, D, M_0) = \sum_{B\in\underline{\mathsf{B}}(D)/\mathrm{mod}\, D} e\{QBD^{-1}\} = \sum_{B'\in\underline{\mathsf{B}}(D_0)/\mathrm{mod}\, D_0} e\{Q\alpha^*B'\beta(\alpha D_0\beta)^{-1}\}$$
$$= \sum_{B'\in\underline{\mathsf{B}}(D_0)/\mathrm{mod}\, D_0} e\{Q[\alpha^*]B'D_0^{-1}\}.$$

It easily follows from Lemma 3.3.33(2) that the last sum is equal to $|\underline{\mathsf{B}}(D_0)/\mathrm{mod}\, D_0|$, if $Q[\alpha^*] = \begin{pmatrix} * & * \\ * & Q'\end{pmatrix}$ with $Q'\in a\mathbb{E}_i$, and otherwise it is equal to zero. Since $Q\in\mathbb{A}_n$, it follows that the condition $Q' \in a\mathbb{E}_i$ is equivalent to $aQ[\alpha^* D_0^{-1}] \in \mathbb{A}_n$, which is the same as $aQ[D^*] = aQ[\alpha^* D_0^{-1} \beta^*] \in \mathbb{A}_n$. The relations (4.2.30) follow. The formulae (4.2.27) and (4.2.28) follow from (4.2.18), (4.2.19), and (4.2.30); (4.2.29) follows directly from definitions. □

Exercise 4.2.9. *For $Q \in \mathbb{A}_n$ and $D \in \mathsf{M}_n$ with* $\det D \neq 0$, *let*

$$s(Q, D) = \sum_{B \in \underline{\mathrm{B}}(D)/\mathrm{mod}\, D} e\{QBD^{-1}\},$$

and let $\mathbb{A}_n(D) = \{Q \in \mathbb{A}_n; s(Q, D) \neq 0\}$. Show the following assertions:

(1) *If $\alpha, \beta \in \Lambda^n$, then*

$$s(Q, \alpha D \beta) = s(Q[\alpha^*], D) \qquad \text{and } \mathbb{A}_n(\alpha D \beta) = \alpha \mathbb{A}_n(D)^t_\alpha$$

(2) *if* $\operatorname{ed} D = \operatorname{diag}(d_1, \ldots, d_n)$, *then*

$$s(Q, D) = \begin{cases} d_1^n d_2^{n-1} \cdots dn, & \text{if } Q \in \mathbb{A}_n(D) \\ 0, & \text{if } Q \notin \mathbb{A}_n(D); \end{cases}$$

(3) *if $D = \operatorname{ed} D = \operatorname{diag}(d_1, \ldots, d_n)$ and $Q = ((1 + e_{ij}) q_{ij})$, where $(e_{ij}) = E_n$, then*

$$Q \in \mathbb{A}_n(D) \text{ if and only if } q_{ij} \equiv 0 \pmod{d_i} \text{ for } 1 \leq i \leq j \leq n;$$

(4) *If $n = 2$ and $D = d_1 D_0$,* where *D_0 is an integral matrix with coprime entries, then*

$$\mathbb{A}_2(D) = d_1 (d_2^{-1} D_0 \mathbb{A}_2 {}^t D_0 \cap \mathbb{A}_2),$$

where $d_2 = d_2(D) = |\det D_0|$.

Exercise 4.2.10. *Let k, χ, and ξ satisfy* (4.2.5). *Show that the images $\mathbf{T}(a) \in \mathscr{L}^n(q)$ of the elements* (3.3.22), *where $a \in \mathbb{N}_{(q)}$, operate on $\mathfrak{F}^n_\xi$ by*

$$(f|_{k,\chi} \mathbf{T}(a))(A) = a^{nk - \langle n \rangle} \chi(a)^n \sum_{\substack{d_1, \ldots, d_n \in \mathbb{N}, \\ d_1 | \cdots | d_n | a}} d_1^n d_2^{n-1} \cdots d_n \sum_{\substack{D \in \Lambda \backslash \Lambda \operatorname{diag}(d_1, \ldots, d_n) \Lambda, \\ A[{}^t D] \in a\mathbb{A}_n(D)}}$$
$$\times \bar{\chi}(\det D)(\det D)^{-k} f(a^{-1} A[{}^t D]);$$

in particular, if $n = 1$, then

$$(f|_{k,\chi} \mathbf{T}(a))(2a') = a^{k-1} \chi(a) \sum_{d | a, a/d | a'} d \bar{\chi}(d) d^{-k} f(2a^{-1} d^2 a')$$
$$= \sum_{\delta | a, a'} \delta^{k-1} \chi(\delta) f(2 \cdot a \cdot a'/\delta^2),$$

but if $n = 2$, then

$$(f|_{k,\chi} \mathbf{T}(a))(A) = \sum_{\substack{\delta_1 \delta_2 \delta_3 = a, \\ A \in \delta_3 \mathbb{A}_2}} \delta_2^{2k-2} \delta_3^{2k-3} \chi(\delta_2 \delta_3)^2 \sum_{\substack{D \in \Lambda \backslash \Lambda D_1^2(\delta_2) \Lambda, \\ A[{}^t D] \in \delta_2 \delta_3 \mathbb{A}_2}}$$
$$\times \bar{\chi}(\det D)(\det D)^{-k} f\left(\frac{\delta_1}{\delta_2 \delta_3} A[{}^t D]\right).$$

[Hint: For the first formula use Exercise 3.3.14, (4.2.19), and Exercise 4.2.9; for the last formula set $d_1 = \delta_1, d_2 = \delta_1\delta_2, a = \delta_1\delta_2\delta_3$ and use Exercise 4.2.9(4).]

4.2.3. Relations with Hecke Operators for GL_n

Formulae of the previous subsection often include the summation over representatives for the left cosets of a double coset $\Lambda^n D \Lambda^n$ with $D \in \mathsf{GL}_n(\mathbb{Q})$ modulo Λ^n. This suggests an idea to look at the formulae from the point of view of Hecke operators for GL_n. In this subsection we shall assume that $(-1)^k\chi(-1) = \xi(-1) = 1$. In the case $\xi(-1) = -1$ certain simple modifications are required, but we shall not use the case.

In order to define a suitable representation of $\mathsf{H}^n = \mathsf{D}_{\mathbb{Q}}(\Lambda^n, \mathsf{G}^n)$, we again use the scheme of §§ 3.3.1 and 3.3.5. The set $\mathsf{S}_n(\mathbb{Q})$ of rational symmetric matrices of order n will serve as the basic set H, on which G^n operates by $g\colon A \to g(A) = gA\,{}^tg$ $(g \in \mathsf{G}^n)$. We take $T = \mathbb{C}$ and the factor of automorphy $\varphi\colon \mathsf{G}^n \times S_n(\mathbb{Q}) \to \mathbb{C}$ which is identically equal to 1. Then the module $\mathfrak{M}$ of automorphic functions $f\colon \mathsf{S}_n(\mathbb{Q}) \to \mathbb{C}$ of weight φ for the subgroup Λ^n of G^n is the space

$$\mathsf{C}_n(\mathbb{Q}) = \{f\colon \mathsf{S}_n(\mathbb{Q}) \to \mathbb{C}; f(A[U]) = f(A)(A \in \mathsf{S}_n(\mathbb{Q}), U \in \Lambda^n)\} \tag{4.2.31}$$

of all functions on $\mathsf{S}_n(\mathbb{Q})$ which are constant on classes $\{A\} = \{A\}_{\mathbb{Z}}$ of matrices equivalent over $\mathbb{Z}$. Since $(f|_\varphi g)(A) = f(A[{}^tg])$, it follows that the Hecke operator $|t$ with a $t = \sum_i a_i(\Lambda g_i) \in \mathsf{H}^n$ on $\mathsf{C}_n(\mathbb{Q})$ takes the shape

$$|t\colon f \to (f|t)(A) = \sum_i a_i f(A[{}^tg_i]), \qquad (f \in \mathsf{C}_n(\mathbb{Q})); \tag{4.2.32}$$

in particular, if $t = (g) = (g)_\Lambda$ with $g \in \mathsf{G}^n$, then

$$(f|(g))(A) = \sum_{g_i \in \Lambda\backslash \Lambda g \Lambda} f(A[{}^tg_i]). \tag{4.2.33}$$

By Proposition 3.1.10, the mapping $t \to |t$ is a linear representation of H^n on $\mathsf{C}_n(\mathbb{Q})$. We now look at the "coefficients" of the representation. For $A, B \in \mathsf{S}_n(\mathbb{Q})$ and $g \in \mathsf{G}^n$ we set

$$r^*(A, B; \Lambda g) = \begin{cases} 1, & \text{if } A[{}^tg] \in \{B\} \\ 0, & \text{otherwise.} \end{cases} \tag{4.2.34}$$

It is clear that

$$r^*(A[{}^tU], B[{}^tV]; \Lambda g) = r^*(A, B; \Lambda g U), \qquad (U, V \in \Lambda); \tag{4.2.35}$$

in particular, $r^*(A, B; \Lambda g)$ depends only on the class $\{B\}$ of B. Further, for an element $t = \sum_i a_i(\Lambda g_i) \in \mathsf{L}_{\mathbb{Q}}(\Lambda^n, G^n)$ we define

$$r^*(A, B; t) = \sum_i a_i r^*(A, B; \Lambda g_i). \tag{4.2.36}$$

It follows from (4.2.35) that

$$r^*(A[{}^tU], B[{}^tV]; t) = r^*(A, B; tU), \quad (U, V \in \Lambda); \tag{4.2.37}$$

in particular, if $t \in \mathrm{H}^n$, then the function $r^*(A, B; t)$ depends only on the classes $\{A\}$ and $\{B\}$ of A and B. Putting together all of the terms in (4.2.32) with $A[{}^tg_i]$ belonging to a fixed class $\{B\} \subset \mathrm{S}_n(\mathbb{Q})$, we may write (4.2.32) in the form

$$(f|t)(A) = \sum_{\{B\}} r^*(A, B; t) f(B), \tag{4.2.38}$$

where the sum is taken over a set of representatives for all classes $\{B\} \subset \mathrm{S}_n(\mathbb{Q})$. Since t is a finite linear combination of left cosets, it follows that, for fixed t and $\{A\}$, the number $r^*(A, B; t)$ equals zero for all but a finite number of classes $\{B\}$. Hence, in fact, the sum (4.2.38) is finite.

The formula (4.2.38) shows that the numbers $r^*(A, B; t)$ can be considered as "coefficients" of the operator $|t$ on $\mathrm{C}_n(\mathbb{Q})$. We now consider some properties of the numbers.

Lemma 4.2.11. *Let A, $B \in \mathrm{S}_n(\mathbb{Q})$ and t, $t' \in \mathrm{H}^n$. Then*

$$r^*(A, B; tt') = r^*(A, B; t't) = \sum_{\{C\} \in \mathrm{S}_n(\mathbb{Q})} r^*(A, C; t) r^*(C, B; t').$$

Proof. For any $f \in \mathrm{C}_n(\mathbb{Q})$, by (4.2.38), we can write

$$(f|tt')(A) = \sum_{\{B\}} r^*(A, B; tt') f(B).$$

On the other hand, since $tt' = t't$ and $f|t't = f|t'|t$, it follows that

$$(f|tt')(A) = ((f|t')|t)(A) = \sum_{\{C\}} r^*(A, C; t)(f|t')(C)$$

$$= \sum_{\{C\},\{D\}} r^*(A, C; t) r^*(C, D; t') f(D).$$

Comparing the two expressions for the f equals 1 on the class $\{B\}$ and 0 outside $\{B\}$, we get the relation. □

If A is a positive definite matrix, then the numbers $r^*(A, B; t)$ can be interpreted in terms of solutions of the matrix equation $A[X] = B$. For A, $B \in \mathrm{S}_n(\mathbb{Q})$ and a subset X of $\mathrm{M}_n(\mathbb{Q})$, we set

$$\mathrm{R}(A, B; \mathrm{X}) = \{M \in \mathrm{X}; A[{}^tM] = B\}, \tag{4.2.39}$$

and denote by

$$r(A, B; \mathrm{X}) = |\mathrm{R}(A, B; \mathrm{X})| \tag{4.2.40}$$

the number of elements of $\mathrm{R}(A, B; X)$. If $A > 0$, and the set X satisfies $dX \subset \mathrm{M}_n$ with some $d \in \mathbb{N}$, then $r(A, B; \mathrm{X})$ is clearly finite. If $\det A \neq 0$, then the set $\mathrm{E}(A) = \mathrm{R}(A, A; \Lambda^n)$ is clearly a multiplicative group, *the group of units of* A. The order

$$e(A) = |\mathrm{E}(A)| = r(A, A; \Lambda^n) \tag{4.2.41}$$

of the group is called *the number of units of A*. It is clear that $e(A)$ depends only on the class $\{A\}$ of A. Assuming that $A > 0$, we now compare the numbers $r(A, B; \Lambda g)$ and $r^*(A, B; \Lambda g)$ for $g \in \mathsf{G}^n$. By the definition, $r(A, B; \Lambda g)$ equals the number of matrices $V \in \Lambda$ satisfying $VgA\,{}^tg\,{}^tV = B$; in particular, it equals zero, if $A[{}^tg] \notin \{B\}$. But if $A[{}^tg] \in \{B\}$, i.e. $A[{}^tg] = B[{}^tV_0]$ with a $V_0 \in \Lambda$, then the condition $VgA\,{}^tg\,{}^tV = B$ can be written in the form $VV_0B\,{}^tV_0{}^tV = B$, and so the number of such V is equal to $e(B)$. Thus, we have obtained the relation

$$r^*(A, B; \Lambda g) = e(B)^{-1} r(A, B; \Lambda g). \tag{4.2.42}$$

Similarly to (4.2.36), for a $t = \sum_i a_i(\Lambda g_i) \in \mathsf{L}_{\mathbb{Q}}(\Lambda^n, \mathsf{G}^n)$, we set

$$r(A, B; t) = \sum_i a_i r(A, B; \Lambda g_i). \tag{4.2.43}$$

Then the relations (4.2.42) and definitions imply the relations

$$r^*(A, B; t) = e(B)^{-1} r(A, B; t), \qquad (A, B \in \mathsf{S}_n(\mathbb{Q}), A > 0) \tag{4.2.44}$$

for every $t \in \mathsf{L}_{\mathbb{Q}}(\Lambda^n, \mathsf{G}^n)$.

In some cases the numbers (4.2.40) can be expressed in terms of numbers $r(A, B; t)$. For example, if X is a finite union of double cosets of G modulo Λ, say

$$\mathsf{X} = \bigcup_i \Lambda g_i \Lambda \qquad (g_i \in \mathsf{G}),$$

then it follows from definitions that

$$r(A, B; \mathsf{X}) = r\left(A, B; \sum_i (g_i)_\Lambda\right), \tag{4.2.45}$$

where we assume that the double cosets are pairwise distinct. We now consider the case A, $B \in \mathbb{A}_n^+$ and $\mathsf{X} = \mathsf{M}_n$. It is clear that an integral matrix M with $A[{}^tM] = B$ can only exist if $\det B/\det A = a^2$ with $a = \det M \in \pm\mathbb{N}$. Hence,

$$r(A, B) = r(A, B; \mathsf{M}_n) = \begin{cases} r(A, B; \mathsf{M}_n(\pm a)), & \text{if } \dfrac{\det B}{\det A} = a^2, a \in \pm\mathbb{N} \\ 0, & \text{otherwise,} \end{cases}$$

and so, by (4.2.45) and Lemma 3.2.6, we obtain the relations

$$r(A, B) = \begin{cases} r(A, B; t(a)), & \text{if } \dfrac{\det B}{\det A} = a^2, a \in \mathbb{N} \\ 0, & \text{otherwise,} \end{cases} \tag{4.2.46}$$

where $t(a) = t_n(a) \in \mathsf{H}^n$ are the elements (3.2.10). We call $r(A, B; t)$ (resp. $r^*(A, B; t)$) *the number* (*resp. reduced number*) *of representations of B by A over t*.

The relations (4.2.44) allow us to translate properties of reduced numbers of representations into the language of numbers of representations and vice versa. For example, by Lemma 4.2.11 and (4.2.44), we obtain the relations

$$r(A,B;tt') = \sum_{\{C\}\in \mathsf{S}_n(\mathbb{Q})} e(C)^{-1} r(A,C;t) r(C,B;t') \tag{4.2.47}$$

for A, $B \in \mathsf{S}_n(\mathbb{Q})$, $A > 0$ and t, $t' \in \mathsf{H}^n$. The following lemma goes the opposite direction.

Lemma 4.2.12. *Let A, $B \in \mathsf{S}_n(\mathbb{Q})$, and let X be a subset of G^n. Then*

$$r(A,B;\mathsf{X}) = r(B,A;\mathsf{X}^{-1}) \tag{4.2.48}$$

with $\mathsf{X}^{-1} = \{M^{-1}; M \in \mathsf{X}\}$; in particular,

$$r(A,B;(g)) = r(B,A;(g^{-1}))(g \in \mathsf{G}^n). \tag{4.2.49}$$

If A and B are positive definite, and $t \in \mathsf{H}^n$, then

$$r(A,B;t) = r(B,A;\mathsf{j}(t)), \tag{4.2.50}$$

and

$$e(A)^{-1} r^*(A,B;t) = e(B)^{-1} r^*(B,A;\mathsf{j}(t)), \tag{4.2.51}$$

where j: $\mathsf{H}^n \to \mathsf{H}^n$ is the anti-automorphism (3.1.17).

Proof. Since for invertible matrices M the relations $A[{}^tM] = B$ and $B[M^*] = A$ are equivalent, it follows that the mapping $M \to M^{-1}$ defines a one-one correspondence between the sets $\mathsf{R}(A,B;\mathsf{X})$ and $\mathsf{R}(B,A;\mathsf{X}^{-1})$, which implies (4.2.48). The relation (4.2.49) follows from (4.2.48) and (4.2.45). The relation (4.2.50) follows from (4.2.49) and the definition of j. Finally, (4.2.51) follows from (4.2.50) and (4.2.44). □

The above considerations allow us to write certain sums over left cosets of G^n modulo Λ^n in terms of numbers of representations.

Lemma 4.2.13. *Let $f \in \mathsf{C}_n(\mathbb{Q})$, and let $t = \sum_i a_i(\Lambda g_i) \in \mathsf{L}_{\mathbb{Q}}(\Lambda^n, \mathsf{G}^n)$. Then the following formulae hold:*

$$\sum_i a_i f(A[{}^tg_i]) = \sum_{\{B\}\in \mathsf{S}_n(\mathbb{Q})} r^*(A,B;t) f(B) \tag{4.2.52}$$

$$= \sum_{\{B\}\in \mathsf{S}_n(\mathbb{Q})} r(A,B;t) e(B)^{-1} f(B), \tag{4.2.53}$$

where $A \in \mathsf{S}_n(\mathbb{Q})$, and in (4.2.53) *it is assumed that $A > 0$. If f equals zero outside the subset $\mathbb{A}_n$ of $\mathsf{S}_n(\mathbb{Q})$, then, for every A of $\mathsf{S}_n(\mathbb{Q})$ and rational $\mu \neq 0$, the following formula holds:*

$$\sum_i a_i f(\mu^{-1} A[{}^tg_i]) = \sum_{\{B\}\in \mathbb{A}_n} r^*(A,\mu B;t) f(B), \tag{4.2.54}$$

which, if A is positive definite, can be written in the form

$$\sum_i a_i f(\mu^{-1} A[{}^tg_i]) = \sum_{\{B\}\in \mathbb{A}_n} r(A,\mu B;t) e(B)^{-1} f(B), \tag{4.2.55}$$

Proof. By (4.2.32), the left side in (4.2.52) can be written as $(f|t)(A)$, but then the formula is the same as (4.2.38). The formula (4.2.53) follows from (4.2.52) and (4.2.44). Writing the formulae (4.2.52) and (4.2.53) for the function $f'(S) = f(\mu^{-1}S)$ ($S \in \mathsf{S}_n(\mathbb{Q})$), we obtain the formulae (4.2.54) and (4.2.55), for $f'(S) = 0$, if $S \notin \mu\mathbb{A}_n$. □

Returning to formulae of the previous subsection, we may extend each function f of $\mathfrak{F}_\xi^n$ to a function on $\mathsf{S}_n(\mathbb{Q})$ by the convention that $f(S) = 0$, if $S \notin \mathbb{A}_n$. Then, by (4.2.14), we may assume that

$$\mathfrak{F}_\xi^n \subset \mathsf{C}_n(\mathbb{Q}), \quad \text{if } \xi(-1) = 1. \tag{4.2.56}$$

Thus, if $(-1)^k\chi(-1) = \xi(-1) = 1$, we can use formulae of Lemma 4.2.13 to write some formulae of action of Hecke operators on $\mathfrak{F}_\xi^n$ in terms of numbers of representations. For example, in this case we have

$$\bar{\chi}(\det D)(\det D)^{-k} = \bar{\chi}(\det D_0)(\det D_0)^{-k} \qquad \text{for } D \in \Lambda D_0 \Lambda,$$

and so, by (4.2.54), each of the sums on the right side in (4.2.22), (4.2.24), and (4.2.28) can be written in the form

$$\begin{aligned}\sum_{D \in \Lambda\backslash\Lambda D_0\Lambda} &\bar{\chi}(\det D)(\det D)^{-k} f(\mu^{-1}A[{}^tD]) \\ &= \bar{\chi}(\det D_0)(\det D_0)^{-k} \sum_{\{B\}} r^*(A, \mu B; (D_0)_\Lambda) f(B),\end{aligned} \tag{4.2.57}$$

where $A \in \mathsf{S}_n(\mathbb{Q})$, $f \in \mathfrak{F}_\xi^n$, and $f(S) = 0$ if $S \notin \mathbb{A}_n$.

Exercise 4.2.14. *Assuming that* $(-1)^k\chi(-1) = \xi(-1) = 1$, *show that the last formula of Exercise* 4.2.10 *can be written in the form*

$$(f|_{k,\chi}\mathbf{T}(a))(A) = \sum_{\substack{\delta_1\delta_2\delta_3 = a,\\ \delta_3|A}} \delta_2^{k-2}\delta_3^{2k-3}\chi(\delta_2\delta_3^2) \sum_{\{B\} \in \mathbb{A}_n} r^*(\delta_3^{-1}A, \delta_2 B; (D_1^2(\delta_2))_\Lambda) f(\delta_1 B),$$

where $D_1^2(\delta) = \operatorname{diag}(1, \delta)$, *and* $\delta|A$ *means that* $\delta^{-1}A \in \mathbb{E}_n$.

Exercise 4.2.15. *Let* $t = \sum_i a_i(\Lambda g_i)$, $t_1 \in \mathsf{H}^n$, *and let* $A, B \in \mathsf{S}_n(\mathbb{Q})$ *with* $A > 0$. *Prove the following formulae*:

$$\sum_i a_i r(A, B[{}^tg_i]; t_1) = r(A, B; \mathrm{j}(t)t_1),$$

where j *is the anti-automorphism* (3.1.17) *of* H^n;

$$\sum_i a_i r(A[{}^tg_i], B; t_1) = r(A, B; tt_1).$$

[Hint: Use formulae (4.2.53), (4.2.44), and (4.2.47).]

Exercise 4.2.16. *Let* $A, B \in \mathbb{A}_n^+$, *and let* $\det B = d^2 \det A$ *with* $d \in \mathbb{N}$. *Show that for each prime divisor* p *of* d *the function* $r(A, B)$ *with a fixed* A *satisfies the following recurrence relation*:

$$\sum_{i=0}^{n} (-1)^i p^{\langle i-1 \rangle} \sum_{D \in \Lambda \backslash \Lambda D_i(p) \Lambda} r(A, B[D^{-1}]) = 0,$$

where $D_i(p) = D_i^n(p)$ *are the matrices* (3.2.28).

[Hint: Since the inclusion $M \in \mathsf{R}(A, B[D^{-1}]; \mathsf{M}_n)$ with $D \in \Lambda D_i(p) \Lambda$ implies that $\det M = \pm d/p^i$, it follows that the sum under consideration equals

$$\sum_{\substack{i=0,1,\ldots,n,\\ p^i|d}} (-1)^i p^{\langle i-1 \rangle} \sum_{g \in \Lambda \backslash \Lambda D_i(p)^{-1} \Lambda} r(A, B[{}^t g]; t(d/p^i)),$$

which can be written in the form

$$= r\left(A, B; \sum_{0 \le i \le n,\, p^i|d} (-1)^i p^{\langle i-1 \rangle} \pi_i(p) t(d/p^i)\right),$$

by the first formula of Exercise 4.2.15, where the elements $t(a) = t_n(a)$ and $\pi_i(p) = \pi_i^n(p)$ are defined by (3.2.10) and (3.2.26). Setting $d = p^\nu d_1$ with $\nu > 0$ and $d_1 \in \mathbb{N}_{(p)}$, we have

$$\sum_{0 \le i \le \min(n,\nu)} (-1)^i p^{\langle i-1 \rangle} \pi_i(p) t(p^{\nu-i} d_1)$$

$$= \left(\sum_{0 \le i \le \min(n,\nu)} (-1)^i p^{\langle i-1 \rangle} \pi_i(p) t(p^{\nu-i})\right) t(d_1) = 0,$$

by (3.2.14) and Lemma 3.2.22.]

4.2.4. Hecke Operators and the Siegel Operator

In this subsection we shall consider relations between Hecke operators on spaces of the form $\mathfrak{M}_k^n(q,\chi)$ and $\mathfrak{M}_\xi^n$ and the Siegel operator Φ defined in § 2.3.4. The relations will allow us to reduce certain problem of the theory of Hecke operators for groups of genus n to similar problems for groups of a smaller genus.

The Siegel operator Φ was initially defined on spaces $\mathfrak{F}\mathfrak{S}_q^n$ of Fourier series of the form (2.3.5) which converge absolutely and uniformly on any subset $\mathbb{H}_n(\varepsilon) \subset \mathbb{H}_n$ with $\varepsilon > 0$. It follows from Theorem 2.3.1 that, for any ξ, the space $\mathfrak{M}_\xi^n$ is contained in $\mathfrak{F}\mathfrak{S}_1^n$. Hence, the operator Φ is defined on the spaces and therefore on their subspaces of the form $\mathfrak{M}_k^n(q,\chi)$. According to Theorem 2.3.1 and definitions, the subspace $\mathfrak{M}_\xi^n$ of $\mathfrak{F}\mathfrak{S}_1^n$ can be characterized as the subset of all functions in $\mathfrak{F}\mathfrak{S}_1^n$ whose Fourier coefficients satisfy (4.2.14). It easily follows from this remark and the expansion (2.3.29) that Fourier coefficients of the image $F|\Phi$ of any function F of $\mathfrak{M}_\xi^n$ satisfy (4.2.14) for genus $n-1$, and so

$$\Phi\colon \mathfrak{M}_\xi^n \to \mathfrak{M}_\xi^{n-1} \qquad (n \in \mathbb{N}), \tag{4.2.58}$$

where we set

$$\mathfrak{M}_\xi^0 = \begin{cases} \mathbb{C}, & \text{if } \xi(-1) = 1 \\ \{0\}, & \text{if } \xi(-1) = -1 \end{cases} \tag{4.2.59}$$

(if $\xi(-1) = -1$, then the constant term of the Fourier expansion of any function in $\mathfrak{M}_\xi^n$ is clearly equal to zero). Now let $q \in \mathbb{N}$ and let χ be a Dirichlet character modulo q. It is readily verified that the group $K^{(n-1)}$ defined by (2.3.34) for $K = \Gamma_0^n(q)$ is equal to $\Gamma_0^{n-1}(q)$, and the character $[\chi]^{(n-1)}$ of the group defined by (2.3.35) is the one-dimensional character associated to χ. Thus, by Proposition 2.3.9, we conclude that

$$\Phi\colon \mathfrak{M}_k^n(q,\chi) \to \mathfrak{M}_k^{n-1}(q,\chi), \qquad (n \in \mathbb{N}, k \in \mathbb{Z}), \tag{4.2.60}$$

where in accordance with earlier conventions we set

$$\mathfrak{M}_k^0(q,\chi) = \begin{cases} \mathbb{C}, & \text{if } (-1)^k\chi(-1) = 1 \\ \{0\}, & \text{if } (-1)^k\chi(-1) = -1. \end{cases} \tag{4.2.61}$$

We now turn to Hecke operators. We shall show that Hecke operators on spaces $\mathfrak{M}_\xi^n$ and $\mathfrak{M}_k^n(q,\chi)$ are compatible with the Siegel operator, in the sense that there is a homomorphism $X \to X'$ of the corresponding Hecke rings from genus n to genus $n-1$ such that $|_{k,\chi}X|\Phi = |\Phi|_{k,\chi}X'$. In this subsection we consider the case of local Hecke rings. The global situation will be examined in §5.1.3.

We fix arbitrary $n \in \mathbb{N}$ and $p \in \mathbb{P}$. If

$$X = \sum_i a_i\left(\Gamma_0\begin{pmatrix} p^{\delta_i}D_i^* & B_i \\ 0 & D_i\end{pmatrix}\right) \tag{4.2.62}$$

is an element of $\mathcal{L}_p^n$, then, by Lemma 3.2.7, representatives for the left cosets in (4.2.62) can be chosen in such a way that each of the matrices D_i is of the form

$$D_i = \begin{pmatrix} D_i' & * \\ 0 & p^{d_i}\end{pmatrix} \quad \text{with } D_i' \in \mathrm{G}_p^{n-1} \text{ and } d_i \in \mathbb{Z}. \tag{4.2.63}$$

With this choice of representatives we set

$$\Psi_p^n(X,u) = \sum_i a_i u^{-\delta_i}(up^{-n})^{d_i}\left(\Gamma_0^{n-1}\begin{pmatrix} p^{\delta_i}(D_i')^* & B_i' \\ 0 & D_i'\end{pmatrix}\right), \tag{4.2.64}$$

where u is an independent variable, and where B_i' is the matrix of order $n-1$ with $B_i = \begin{pmatrix} B_i' & * \\ * & * \end{pmatrix}$. If $n = 1$, then $D_i = p^{d_i}$, and we set

$$\Psi_p^1(X,u) = \sum_i a_i u^{-\delta_i}(up^{-1})^{d_i}. \tag{4.2.65}$$

It is easy to see that $\Psi(X,u)$ is independent of the choice of representatives. Thus, we obtain a linear mapping $X \to \Psi(X)$ of $\mathcal{L}_p^n$ into the module of left cosets of the pair $(\Gamma_0^{n-1}, \mathcal{S}_p^{n-1})$ over the ring $\mathbb{Q}[u^{\pm 1}]$ of polynomials in $u^{\pm 1}$ over $\mathbb{Q}$ (into $\mathbb{Q}[u^{\pm 1}]$, if $n = 1$).

Proposition 4.2.17. *The following assertions hold:*

(1) *The image $\Psi(X,u)$ of any $X \in \mathcal{L}_p^n$ is contained in the ring $\mathcal{L}_p^{n-1}[u^{\pm 1}] = \mathrm{D}_{\mathbb{Q}[u\pm 1]}(\Gamma_0^{n-1}, \mathcal{S}_p^{n-1})$, where we set $\mathcal{L}_p^0 = \mathbb{Q}$; the mapping*

$$\Psi = \Psi_p^n\colon \mathscr{L}_p^n \to \mathscr{L}_p^{n-1}[u^{\pm 1}] \tag{4.2.66}$$

is a ring homomorphism;

(2) *The mapping Ψ maps the subrings $\mathscr{C}_{-p}^n$, $\mathscr{C}_p^n$, and $\mathbf{L}_p^n$ of $\mathscr{L}_p^n$ into $\mathscr{C}_{-p}^n[u^{\pm 1}]$, $\mathscr{C}_p^n[u^{\pm 1}]$, and $\mathbf{L}_p^n[u^{\pm 1}]$, respectively, where we set $\mathscr{C}_{-p}^0 = \mathscr{C}_p^0 = \mathbf{L}_p^0 = \mathbb{Q}$;*

(3) *The following diagram*

$$\begin{array}{ccc} L_p^n & \xrightarrow{\Omega_p^n} & \mathbb{Q}[x_0^{\pm 1},\dots,x_n^{\pm 1}] \\ \Psi_p^n\big\downarrow & & \big\downarrow\Xi_n \\ L_p^{n-1}[u^{\pm 1}] & \xrightarrow{\bar{\Omega}_p^{n-1}} & \mathbb{Q}[x_0^{\pm 1},\dots,x_{n-1}^{\pm 1},u^{\pm 1}] \end{array} \tag{4.2.67}$$

where Ω_p^n is the spherical mapping (3.3.49), $\bar{\Omega}_p^{n-1}$ is the $\mathbb{Q}[u^{\pm 1}]$-linear extension of Ω_p^{n-1}, and where Ξ_n is a $\mathbb{Q}$-linear homomorphism satisfying

$$\Xi_n(x_0) = x_0 u^{-1}, \quad \Xi_n(x_n) = u, \quad \Xi_n(x_i) = x_i \quad (1 \le i \le n-1),$$

if $n > 1$, and $\Xi_1(x_0) = u^{-1}$, $\Xi_1(x_1) = u$, is commutative.

Proof. If $M' \in \Gamma_0^{n-1}$ and $M = \check{M}'$ (see near (2.3.32)), then $M \in \Gamma_0^n$, and the condition $XM = X$ for $X \in \mathscr{L}_p^n$ implies that $\Psi(X,u)M' = \Psi(X,u)$. Hence, $\Psi(X,u) \in \mathscr{L}_p^{n-1}[u^{\pm 1}]$. It easily follows from definitions that Ψ is a homomorphism.

From the definition of Ψ and decompositions (3.4.11) of the elements $\Pi_\pm^n(p)$, we have

$$\Psi(\Pi_-^n(p),u) = u^{-1}\Pi_-^{n-1}(p), \qquad \Psi(\Pi_+^n(p),u) = \Pi_+^{n-1}(p),$$

which implies that Ψ maps elements commuting with $\Pi_-^n(p)$ (resp. $\Pi_+^n(p)$) into elements commuting with $\Pi_-^{n-1}(p)$ (resp. $\Pi_+^{n-1}(p)$). Hence, $\Psi(\mathscr{C}_{-p}^n) \subset \mathscr{C}_{-p}^{n-1}[u^{\pm 1}]$ and $\Psi(\mathscr{C}_p^n) \subset \mathscr{C}_p^{n-1}[u^{\pm 1}]$, by (3.5.1). As to the image of $\mathbf{L}_p^n$, in the same way as (4.2.64), we may define a mapping $\Psi\colon \mathsf{L}_p^n \to \mathsf{L}_p^{n-1}[u^{\pm 1}]$ such that the diagram

$$\begin{array}{ccc} \mathsf{L}_p^n & \xrightarrow{\mathsf{i}} & \mathbf{L}_p^n \\ \Psi\big\downarrow & & \big\downarrow\Psi \\ \mathsf{L}_p^{n-1}[u^{\pm 1}] & \xrightarrow{\mathsf{i}} & \mathbf{L}_p^{n-1}[u^{\pm 1}] \end{array},$$

where i stand for imbeddings (3.1.14), is commutative. Hence, $\Psi(\mathbf{L}_p^n) = \Psi(\mathsf{i}(\mathsf{L}_p^n)) = \mathsf{i}(\Psi(\mathsf{L}_p^n))$ is contained in $\mathbf{L}_p^n[u^{\pm 1}]$.

To prove (3), we assume that each of the matrices D_i in (4.2.62) is of the form (3.2.36) with the diagonal entries $p^{d_{i1}}, \dots, p^{d_{in}}$. Then, by definitions, we have

$$\begin{aligned} \Xi_n(\Omega_p^n(X)) &= \Xi\left(\sum_i a_i x_0^{\delta_i} \prod_{j=1}^n (x_j p^{-j})^{d_{ij}}\right) \\ &= \sum_i a_i u^{-\delta_i}(up^{-n})^{d_{in}} x_0^{\delta_i} \prod_{j=1}^{n-1} (x_j p^{-j})^{d_{ij}} \end{aligned}$$

$$= \sum_i a_i u^{-\delta_i}(up^{-n})^{d_{in}} \omega_p^{n-1}\left(\Phi_p^{n-1}\left(\Gamma_0^{n-1}\begin{pmatrix} p^{\delta_i}(D_i')^* & B_i' \\ 0 & D_i' \end{pmatrix}\right)\right)$$

$$= \Omega_p^{n-1}(\Psi(X,u)). \qquad \square$$

We later show that the element X' satisfying $|_{k,\chi}X|\Phi = |\Phi|_{k,\chi}X'$ can be taken in the form $X' = \Psi(X, p^{n-k}\bar{\chi}(p))$. If $\chi(p) \notin \mathbb{Q}$, this element is not, in general, contained in $\mathcal{L}_p^{n-1}$. To avoid such an inconvenience, one can consider Hecke rings over an extension of $\mathbb{Q}$ containing the values of the character. Since the ground field is not important in our presentation of the theory, we may at once take the field $\mathbb{C}$ of complex numbers as a big extension of $\mathbb{Q}$.

For a Hecke ring D over $\mathbb{Q}$, we denote by

$$\tilde{\mathsf{D}} = \mathsf{D} \otimes_{\mathbb{Q}} \mathbb{C} \tag{4.2.68}$$

the complexification of D. We assume that all mappings of Hecke rings over $\mathbb{Q}$, including mappings into rings of linear operators, are extended to their complexifications by linearity.

Theorem 4.2.18. (*Žarkovskaya commutation relations*). *Let $q \in \mathbb{N}$, $p \in \mathbb{P}_{(q)}$, and $k \in \mathbb{Z}$. Let χ be a Dirichlet character modulo q, and let ξ be the character of $\{\pm 1\}$ satisfying* (4.2.5). *Then, for any F of $\mathfrak{M}_\xi^n$ and any X of $\mathcal{L}_p^n$ the following relation holds*

$$(F|_{k,\chi}X)|\Phi = (F|\Phi)|_{k,\chi}\Psi(X, p^{n-k}\bar{\chi}(p)), \tag{4.2.69}$$

where Φ is the Siegel operator, the element $\Psi(X, p^{n-k}\bar{\chi}(p)) \in \tilde{\mathcal{L}}_p^{n-1}$ is given by (4.2.64) *with $u = p^{n-k}\bar{\chi}(p)$, and where, for $n = 1$, the operator $|_{k,\chi}\Psi$ acts as multiplication by the number Ψ.*

Proof. Using the Fourier expansion (4.2.13) of F, and assuming that X is written in the form (4.2.62)–(4.2.63) with $a_i \in \mathbb{C}$, we obtain

$$F|_{k,\chi}X = \sum_i a_i\left(\sum_{A \in \mathbb{A}_n} f(A)e\{AZ\}\right)\Big|_{k,\chi}\begin{pmatrix} p^{\delta_i}D_i^* & B_i \\ 0 & D_i \end{pmatrix}$$

$$= \sum_{i,A} a_i p^{\delta_i(nk-\langle n\rangle)}\chi(\det p^{\delta_i}D_i^*)(\det D_i)^{-k}f(A)e\{p^{\delta_i}A[D_i^*]Z + AB_iD_i^{-1}\}.$$

If D_i is of the form (4.2.63) and $A = \begin{pmatrix} A' & * \\ * & a \end{pmatrix}$ is a matrix of $\mathbb{A}_n$, then the matrix $p^{\delta_i}A[D_i^*]$ has the form $\begin{pmatrix} * & * \\ * & p^{\delta_i - 2d_i}a \end{pmatrix}$. So, when we set $Z = \begin{pmatrix} Z' & 0 \\ 0 & i\lambda \end{pmatrix}$ with $Z' \in \mathbb{H}_{n-1}$ and tend λ to $+\infty$, then each of the summands of the last sum corresponding to a matrix A with $a > 0$ tends to zero. As to the other summands, since $A \geq 0$, it follows that $A = \begin{pmatrix} A' & 0 \\ 0 & 0 \end{pmatrix}$. Further, if $A = \begin{pmatrix} A' & 0 \\ 0 & 0 \end{pmatrix}$,

$Z = \begin{pmatrix} Z' & 0 \\ 0 & i\lambda \end{pmatrix}$, and D_i satisfies (4.2.63), then we obtain relations $p^{\delta_i}A[D_i^*]Z = p^{\delta_i}\begin{pmatrix} A'[(D_i')^*]Z' & 0 \\ 0 & 0 \end{pmatrix}$, and $AB_iD_i^{-1} = \begin{pmatrix} A'B_i'(D_i')^{-1} & 0 \\ 0 & 0 \end{pmatrix}$. Using the relations and the fact that the Fourier expansion converges uniformly on sets $\mathbb{H}_n(\varepsilon)$, we have

$$\begin{aligned}
&(F|_{k,\chi}X)|\Phi \\
&= \lim_{\lambda\to+\infty} (F|_{k,\chi}X)\begin{pmatrix} Z' & 0 \\ 0 & i\lambda \end{pmatrix} \\
&= \sum_{i, A' \in \mathbb{A}_{n-1}} a_i p^{\delta_i(nk-\langle n\rangle)}\chi(\det(p^{\delta_i}D_i^*))(\det D_i)^{-k} f\left(\begin{pmatrix} A' & 0 \\ 0 & 0 \end{pmatrix}\right) \\
&\quad \times e\{p^{\delta_i}A'[(D_i')^*]Z' + A'B_i'(D_i')^{-1}\} \\
&= \sum_i p^{\delta_i(k-n)}\chi(p^{\delta_i - d_i})p^{-d_ik}a_i\left(\sum_{A' \in \mathbb{A}_{n-1}} f\left(\begin{pmatrix} A' & 0 \\ 0 & 0 \end{pmatrix}\right)e\{A'Z'\}\right)\Bigg|_{k,\chi}\begin{pmatrix} p^{\delta_i}(D_i')^* & B_i' \\ 0 & D_i' \end{pmatrix} \\
&= (F|\Phi)|_{k,\chi}\Psi(X, p^{n-k}\bar{\chi}(p)),
\end{aligned}$$

for the sum in parenthesis is $(F|\Phi)(Z')$, by (2.3.29), and

$$p^{\delta_i(k-n)}\chi(p^{\delta_i-d_i})p^{-d_ik} = (p^{n-k}\bar{\chi}(p))^{-\delta_i}(p^{n-k}\bar{\chi}(p)p^{-n})^{d_i}. \qquad \square$$

We now consider the restriction of the mapping $X \to \Psi(X, p^{n-k}\bar{\chi}(p))$ to the complexification $\tilde{\mathbf{L}}_p^n$ of the ring (3.3.45). We shall see that the image of $\tilde{\mathbf{L}}_p^n$ is always contained in $\tilde{\mathbf{L}}_p^{n-1}$, but it is not necessarily equal to $\tilde{\mathbf{L}}_p^{n-1}$. In the last case the image equals the so-called even subring of $\tilde{\mathbf{L}}_p^{n-1}$. We start from the definition of even subring of Hecke rings. For a subgroup S of $\mathsf{S}^n = \mathsf{GSp}_n^+(\mathbb{Q})$, we define its *even subgroup* S^+ by

$$\mathsf{S}^+ = \{M \in \mathsf{S}; \mu(M) = t^2, t \in \mathbb{Q}\}. \tag{4.2.70}$$

If (Γ, S) is a Hecke pair, then the pair (Γ, S^+) is clearly again a Hecke pair; the Hecke ring

$$\mathsf{L}^+ = \mathsf{D}_\mathsf{K}(\Gamma, \mathsf{S}^+) \subset \mathsf{L} = \mathsf{D}_\mathsf{K}(\Gamma, \mathsf{S}) \tag{4.2.71}$$

of the pair (Γ, S^+) over a ring K will be called *the even subring of the ring* $\mathsf{L} = \mathsf{D}_\mathsf{K}(\Gamma, \mathsf{S})$. For a subring $\mathsf{L}_1 \subset \mathsf{L}$ the ring

$$\mathsf{L}_1^+ = \mathsf{L}_1 \cap \mathsf{L}^+ \tag{4.2.72}$$

will be called *the even subring of* L_1. The definitions imply, for example, the even subring of $\tilde{\mathbf{L}}_p^n$ is given by

$$(\tilde{\mathbf{L}}_p^n)^+ = \left\{\sum_i a_i(\Gamma_0 M_i) \in \tilde{\mathbf{L}}_p^n; \mu(M_i) = p^{\delta_i}, \delta_i \in 2\mathbb{Z}\right\} \tag{4.2.73}$$

Proposition 4.2.19. *In the notation of Theorem* 4.2.18, *the following assertions hold*:

(1) *If X is contained in $\tilde{\mathbf{L}}_p^n$ (resp. $(\tilde{\mathbf{L}}_p^n)^+$), then the element $\Psi(X, p^{n-k}\overline{\chi}(p))$ is contained in $\tilde{\mathbf{L}}_p^{n-1}$ (resp. $(\tilde{\mathbf{L}}_p^{n-1})^+$), where we set $\tilde{\mathbf{L}}_p^0 = (\tilde{\mathbf{L}}_p^0)^+ = \mathbb{C}$;*

(2) *The mapping*

$$\Psi(\dots, p^{n-k}\overline{\chi}(p)) \colon \tilde{\mathbf{L}}_p^n \to \tilde{\mathbf{L}}_p^{n-1} \tag{4.2.74}$$

is an epimorphism, except in the case

$$k = n > 1 \quad \text{and} \quad \chi(p) = -1. \tag{4.2.75}$$

In this case the image of $\tilde{\mathbf{L}}_p^n$ is equal to the even subring $(\tilde{\mathbf{L}}_p^{n-1})^+$ of $\tilde{\mathbf{L}}_p^{n-1}$;

(3) *The mapping*

$$\Psi(\dots, p^{n-k}\overline{\chi}(p)) \colon (\tilde{\mathbf{L}}_p^n)^+ \to (\tilde{\mathbf{L}}_p^{n-1})^+ \tag{4.2.76}$$

is always an epimorphism;

(4) *If the character χ is real, then the assertions* (1)–(3) *hold not only for the complexifications but also for the initial rings over $\mathbb{Q}$.*

Proof. If we replace each of the rings in (4.2.67) by its complexification and extend each of the mappings to the complexifications by linearity, then the new diagram is again commutative. Thus, instead of action of $\Psi(\dots, p^{n-k}\overline{\chi}(p))$ on $\tilde{\mathbf{L}}_p^n$, we may consider the action of $\Xi_n|_{u=p^{n-k}\overline{\chi}(p)}$ on $\Omega(\tilde{\mathbf{L}}_p^n)$. It follows from Theorem 3.3.30 that

$$\begin{aligned}\Omega(\tilde{\mathbf{L}}_p^n) &= \mathbb{C}[t, \rho_0^{\pm 1}, \rho_1, \dots, \rho_{n-1}] \\ &= \mathbb{C}[t, r_1, \dots, r_{n-1}, \rho_0^{\pm 1}]\end{aligned} \tag{4.2.77}$$

with the polynomials $r_a = r_a^n$, $\rho_a = \rho_a^n$, and $t = t^n$ defined by (3.3.52), (3.3.53), and (3.3.54), respectively. Since Ω is monomorphic on $\tilde{\mathbf{L}}_p^n$, it follows that Ω maps the ring (4.2.73) onto the ring of all polynomials in $\Omega(\tilde{\mathbf{L}}_p^n)$ of even degree with respect to x_0. Hence, since

$$\begin{aligned}(t)^2 &= x_0^2 x_1 \cdots x_n \prod_{i=1}^{n} (1 + x_i)(1 + x_i^{-1}) = \rho_0 \sum_{a=0}^{2n} r_a \\ &= \rho_0(2 + 2(r_1 + \dots + r_{n-1}) + r_n),\end{aligned} \tag{4.2.78}$$

it follows that

$$\begin{aligned}\Omega((\tilde{\mathbf{L}}_p^n)^+) &= \mathbb{C}[(t)^2, r_1, \dots, r_{n-1}, \rho_0^{\pm 1}] \\ &= \mathbb{C}[r_1, \dots, r_n, \rho_0^{\pm 1}].\end{aligned} \tag{4.2.79}$$

Let us consider the action of $\Xi = \Xi_n$ on generators of the rings (4.2.77) and (4.2.79). By definitions, we have

$$\Xi(\rho_0^n) = u^{-1} x_0^2 x_1 \cdots x_{n-1} = u^{-1} \rho_0^{n-1}; \qquad \Xi(t^n) = u^{-1}(1 + u) t^{n-1}. \tag{4.2.80}$$

Using (3.3.52), we obtain

$$\begin{aligned}\sum_{a=0}^{2n} (-1)^a \Xi(r_a^n) v^a &= (1 - uv)(1 - u^{-1}v) \prod_{i=1}^{n-1} (1 - x_i v)(1 - x_i^{-1} v) \\ &= (1 - (u + u^{-1})v + v^2) \sum_{a=0}^{2n-2} (-1)^a r_a^{n-1}(x_1, \dots, x_{n-1}) v^a,\end{aligned} \tag{4.2.81}$$

whence

$$\Xi(r_a^n) = r_a^{n-1} + (u + u^{-1})r_{a-1}^{n-1} + r_{a-2}^{n-1}. \tag{4.2.82}$$

These formulae imply that the mapping Ξ with any $u = c \in \mathbb{C}$ carries $\Omega(\tilde{\mathbf{L}}_p^n)$ into $\Omega(\tilde{\mathbf{L}}_p^{n-1})$ and $\Omega((\tilde{\mathbf{L}}_p^n)^+)$ into

$$\mathbb{C}[r_1^{n-1}, \ldots, r_{n-1}^{n-1}, (\rho_0^{n-1})^{\pm 1}] = \mathbb{C}[r_1^{n-1}, \ldots, r_{n-1}^{n-1}, (\rho_0^{n-1})^{\pm 1}] = \Omega((\tilde{\mathbf{L}}_p^{n-1})^+)$$

(clearly, $r_n^{n-1} = r_{n-2}^{n-1}$). This proves the assertion (1). (In the case of $\tilde{\mathbf{L}}_p^n$ it follows also from Proposition 4.2.17(2).)

To find out the images, we use relations (4.2.80) and (4.2.81). By (4.2.80), we have

$$\rho_0^{n-1} = u\Xi_n(\rho_0^n), \qquad t^{n-1} = \frac{u}{1+u}\Xi_n(t^n). \tag{4.2.83}$$

It follows from (4.2.81) that

$$r_a^{n-1} = \sum_{i=0}^{a} (-1)^i c_i(u) \Xi_n(r_{a-i}^n), \tag{4.2.84}$$

where $c_i(u)$ are given by

$$\sum_{i=0}^{\infty} c_i(u) v^i = \{(1 - uv)(1 - u^{-1}v)\}^{-1}.$$

Setting $u = p^{n-k}\bar{\chi}(p)$, we see that the image of $\Omega(\tilde{\mathbf{L}}_p^n)$ contains all of the generators of $\Omega(\tilde{\mathbf{L}}_p^{n-1})$, except in the case $p^{n-k}\bar{\chi}(p) = -1$ and $n > 1$, that is the case (4.2.75); in this case the image contains all of the elements r_a^{n-1} and $(\rho_0^{n-1})^{\pm 1}$, but does not contain t^{n-1}, and so is equal to $\Omega((\tilde{\mathbf{L}}_p^{n-1})^+)$, by (4.2.79) with $n - 1$ in place of n. The same formulae imply that the image of $\Omega((\tilde{\mathbf{L}}_p^n)^+)$ is always equal to $\Omega((\tilde{\mathbf{L}}_p^{n-1})^+)$. The last assertion is obvious. □

The Žarkovskaya relations allows us to reduce several problems of the theory of Hecke operators on modular forms of a given genus to similar problems for forms of a lesser genus provided the problems are already settled for cusp forms. The question of whether there is a basis of common eigenfunctions for all Hecke operators on a given invariant space is of this kind. Theorem 4.1.8 solves the problem for cusp forms. Here we shall consider invariant subspaces of the space

$$\mathfrak{M}_k^n = \mathfrak{M}_k(\Gamma^n, 1) = \mathfrak{M}_k^n(1, 1). \tag{4.2.85}$$

Theorem 4.2.20. *Any subspace* V *of* $\mathfrak{M}_k^n$ *which is invariant under all Hecke operators* $|_k T = |_{k,1} T$ *with* $T \in \mathsf{L}^n = \mathsf{L}^n(1)$ *has a basis of common eigenfunctions for all of the operators.*

Proof. If at least one of two forms $F, G \in \mathfrak{M}_k^n$ is a cusp form, and $T \in \mathsf{L}^n$, then, by (4.1.20), we have

$$(F|_k T, G) = (F, G|_k T). \tag{4.2.86}$$

Hence, the same arguments as those used in the proof of Theorem 4.1.8 show that the assertion is true if $\mathsf{V} \subset \mathfrak{N}_k^n = \mathfrak{N}_k(\Gamma^n, 1)$. In the general case we set

$$\mathsf{V}_2 = \mathsf{V} \cap \mathfrak{N}_k^n, \qquad \text{and } \mathsf{V}_1 = \{F \in \mathsf{V}; (F, G) = 0 \text{ for all } G \in \mathsf{V}_2\}.$$

It follows from the properties (3)–(5) of the scalar product listed in Theorem 2.5.3 and standard linear algebra that V is the direct sum of V_1 and V_2:

$$\mathsf{V} = \mathsf{V}_1 \oplus \mathsf{V}_2. \tag{4.2.87}$$

By the assumption and Proposition 4.1.4, the subspace V_2 is an intersection of invariant subspaces, and so V_2 is invariant under all Hecke operators. Then, by (4.2.86), the same is true for V_1. Since the subspace V_1 does not contain non-zero cusp forms, and cusp forms F of $\mathfrak{M}_k^n$ are characterized by $F|\Phi = 0$, it follows that the restriction of Φ to V_1 gives an isomorphism of V_1 onto the subspace $\mathsf{V}' = \Phi(\mathsf{V}_1)$ of $\mathfrak{M}_k^{n-1}$. Since the subspace V_1 is invariant, and the mapping (4.2.74) is epimorphic, it follows, by Theorem 4.2.18 and Lemma 4.2.4, that the subspace V' is invariant under all Hecke operators of L_p^{n-1} for each prime p. Hence, by Theorem 3.3.12, V' is invariant under all operators of L^{n-1}. Assuming that the theorem is already proved for invariant subspaces of $\mathfrak{M}_k^{n-1}$, we conclude that V' has a basis $F_1', \ldots, F_d'$ of common eigenfunctions for all operators of L^{n-1}. The inverse images $F_1 = \Phi^{-1}(F_1'), \ldots, F_d = \Phi^{-1}(F_d')$ of the functions $F_1', \ldots, F_d'$ in V_1 form clearly a basis for V_1. Furthermore, each of the forms $F_1, \ldots, F_d$ is an eigenfunction for all operators of L^n. To see this, we only have to consider the operators of the form $|_k T$ with $T \in \mathsf{L}_p^n$, where p is a prime, by Theorem 3.3.12; but for such a T, according to our assumptions and Lemma 4.2.4, we have

$$F_j'|_k \Psi(\mathrm{i}(T), p^{n-k}) = \lambda_j(T) F_j',$$

where $\lambda_j(T)$ is a scalar, whence, by (4.2.69),

$$(F_j|_k T - \lambda_j(T)F_j)|\Phi = F_j|_k \mathrm{i}(T)|\Phi - \lambda_j(T)F_j' = F_j'|_k \Psi(\mathrm{i}(T), p^{n-k}) - \lambda_j(T)F_j' = 0,$$

and so $F_j|_k T = \lambda_j(T)F_j$. The subspace V_2 has a basis of common eigenfunctions for all Hecke operators as an invariant subspace consisting of cusp forms. This basis together with the functions $F_1, \ldots, F_d$ form a basis of eigenfunctions for $\mathsf{V} = \mathsf{V}_1 \oplus \mathsf{V}_2$. To complete our induction, we only have to prove the theorem for $n = 1$. Again representing an invariant subspace V of $\mathfrak{M}_k^1$ in the form (4.2.87), we note that $\dim \mathsf{V}_1 = 0$ or 1, since $\Phi(\mathsf{V}_1) \subset \mathfrak{M}_k^0 = \mathbb{C}$. If $\dim \mathsf{V}_1 = 0$, then $\mathsf{V} \subset \mathfrak{N}_k^1$ and the assertion is already proved. If $\dim \mathsf{V}_1 = 1$, then $\mathsf{V}_1 = \mathbb{C} \cdot F$, and the form F is clearly an eigenfunction for all Hecke operators. The function F together with a basis of eigenfunctions for V_2 form such a basis for V. □

The reader will find another application of Žarkovskaya relations in the next subsection.

Exercise 4.2.21. *Let λ be a non-zero $\mathbb{Q}$-linear homomorphism of $\tilde{\mathbf{L}}_p^{n-1}$ into $\mathbb{C}$, where $n > 1$ and $p \in \mathbb{P}$, and let $\alpha_0, \alpha_1, \ldots, \alpha_{n-1}$ be the parameters of λ in the sense of*

Proposition 3.3.36. *Show that the numbers* $\alpha_0 u^{-1}, \alpha_1, \ldots, \alpha_{n-1}, u$ *can be taken as the parameters of the homomorphism*

$$\mathbf{T} \to \tilde{\lambda}(\Psi_p^n(\mathbf{T}, u)) \qquad (T \in \tilde{\mathbf{L}}_p^n)$$

of $\tilde{\mathbf{L}}_p^n$, *where* $\tilde{\lambda}$ *is the* $\mathbb{C}$-*linear extension of* λ *to* $\tilde{\mathbf{L}}_p^{n-1}$, *and where* u *is a non-zero complex number.*

Exercise 4.2.22. *Show that all eigenvalues of all Hecke operators on* $\mathfrak{M}_k^n$ *are real numbers.*

4.2.5. The Action of the Middle Factor of the Symmetric Factorization of Polynomials $\mathbf{R}_p^n(v)$

As it have been told above, we intend to study the action of Hecke operators of $\mathsf{L}^n(q)$ on modular forms for $\Gamma_0^n(q)$ and their Fourier coefficients. According to our approach, we replace the global ring $\mathsf{L}^n(q)$ by its local subrings $\mathsf{L}_p^n(q)$, which are usually realized as the subrings $\mathbf{L}_p^n$ of local Hecke rings of the group Γ_0^n. The imbeddings of the local rings into the Hecke rings of Γ_0^n will allow us to express their elements in terms of certain components with more simple action on modular forms. The most important expressions of this kind are contained in the factorization (3.5.69) of the polynomials $\mathbf{R}_p^n(v)$ (for example, among the coefficients of $\mathbf{R}_p^n(v)$ there are all generators of $\mathbf{L}_p^n$, exept $\mathbf{T}^n(p)$). The action of the coefficients of the first and last factors in (3.5.69) are given by Lemma 4.2.7. In this subsection we shall obtain formulae for the action of the middle factor considered as a polynomial with operational coefficients.

We again fix $k \in \mathbb{Z}$, n, $q \in \mathbb{N}$, $p \in \mathbb{P}_{(q)}$, a Dirichlet character χ modulo q, and denote by ξ the character of $\{\pm 1\}$ with $\xi(-1) = (-1)^k \chi(-1)$.

The following notation will be convenient. If

$$G(V) = \sum_{i \geq 0} g_i v^i \qquad \text{with } g_i \in \mathscr{L}_p^n$$

is a polynomial or a formal power series over $\mathscr{L}_p^n$, and $F \in \mathfrak{M}_\xi^n$ (resp. $f \in \mathfrak{F}_\xi^n$), then we set

$$F|_{k,\chi} G(v) = \sum_{i \geq 0} (F|_{k,\chi} g_i)(Z) v^i$$

(resp. (4.2.88)

$$f|_{k,\chi} G(v) = \sum_{i \geq 0} (f|_{k,\chi} g_i)(A) v^i),$$

where the right side is understood as a polynomial or formal power series in v. If the values of f are Fourier coefficients of F, then formally we have

$$F|_{k,\chi} G(v) = \sum_{A \in \mathbb{A}_n} (f|_{k,\chi} G(v))(A) e\{AZ\}. \tag{4.2.89}$$

An old Bourbaki fan may formalize the notation by extending the ground ring $\mathbb{Q}$ or $\mathbb{C}$ to a ring of polynomials or formal power series, extending mappings and operators by linearity, and so on. Instead of this, to save the time and space, we rely on common sense, for which it is clear that products of polynomials or series operate as products of corresponding operators, and so on.

We let

$$B(v) = B_p^n(v) = \sum_{i=0}^{n} (-1)^i b_i v^i \tag{4.2.90}$$

be the middle factor in (3.5.69). The formulae (3.5.72) express the coefficients $b_i = b_i^n(p) \in \mathscr{L}_p^n$ in terms of elements of the form $\Pi_{n,0}^{(r)}$ with $0 \le r \le n$. The following lemma describes the action of the elements on $\mathfrak{F}_\xi^n$ in terms of trigonometrical sums.

Lemma 4.2.23. *For each function f of $\mathfrak{F}_\xi^n$ the following formula holds:*

$$(f|_{k,\chi}\Pi_{n,0}^{(r)})(A) = p^{n(k-n-1)}\chi(p)^n s_p(r,n;A) f(A), \qquad (A \in \mathbb{A}_n),$$

where

$$s_p(r,n;A) = \sum_{S \in \mathsf{SM}_p(r,n)} e\{p^{-1}SA\}, \tag{4.2.91}$$

and $\mathsf{SM}_p(r,n)$ is the set defined in Lemma 3.5.20.

Proof. According to (3.3.62), 23 may write

$$\Pi_{n,0}^{(r)} = \sum_{\substack{B \in \mathsf{S}_n(\mathbb{Z}/p\mathbb{Z}),\\ r_p(B)=r}} \left(\Gamma_0 \begin{pmatrix} pE & B \\ 0 & pE \end{pmatrix}\right).$$

Let $F \in \mathfrak{M}_\xi^n$ be the function with Fourier coefficients $f(A)$. Then, by (4.2.11) and (4.2.12), we have

$$\begin{aligned} F|_{k,\chi}\Pi_{n,0}^{(r)} &= p^{2nk-n(n+1)-nk}\chi(p)^n \sum_{A,B} f(A) e\{A(Z + p^{-1}B)\} \\ &= p^{n(k-n-1)}\chi(p)^n \sum_A f(A)\cdot\left(\sum_B e\{p^{-1}AB\}\right) e\{AZ\}. \end{aligned}$$

The lemma follows, since the last sum over B is the sum (4.2.91). □

Proposition 4.2.24. *With the above notation and assumptions, for each $f \in \mathfrak{F}_\xi^n$, the following formula holds*

$$(f|_{k,\chi}B(v))(A) = B(v,A)f(A), \qquad (A \in \mathbb{A}_n) \tag{4.2.92}$$

where $B(v,A) = B_p^n(v,A) \in \mathbb{C}[v]$ is the polynomial given by

$$\mathrm{B}_p^n(v,A) = \sum_{i=0}^{n} (-1)^i p^{-\langle i\rangle - i(n-i)} \left(\sum_{j=0}^{i} \alpha_{ij} s_p(i-j,n;A)\right) v^i, \tag{4.2.93}$$

$\alpha_{ij} = \alpha_{ij}^n(p)$ being the numbers (3.5.57).

Proof. The formula follows directly from definitions, formula (3.5.72), and previous lemma. □

For the rest of the subsection we consider the problem of calculation of the polynomials $B_p^n(v, A)$. We start from the sums (4.2.91). For $0 < b \leq n$, we set

$$\mathsf{PM}_p(b, n) = \{M \in \mathsf{M}_{b,n}(\mathbb{Z}/p\mathbb{Z}); r_p(M) = b\}, \tag{4.2.94}$$

and

$$\mathsf{PM}_p(b, n; Q) = \{M \in \mathsf{PM}_p(b, n); Q[{}^tM] \equiv 0 \pmod{p}\} \tag{4.2.95}$$

with $Q \in \mathbb{E}_n$, where, for S, $S' \in \mathbb{E}_n$ and $d \in \mathbb{N}$, the congruence

$$S \equiv S' \pmod{d} \quad \text{means that } S - S' \in d\mathbb{E}_n. \tag{4.2.96}$$

We let

$$\mathrm{pm}_p(b, n) = |\mathsf{PM}_p(b, n)|, \mathrm{pm}_p(b, n; Q) = |\mathsf{PM}_p(b, n; Q)|$$

be the number of elements of the corresponding sets, and $\mathrm{pm}_p(0, n) = \mathrm{pm}_p(0, n; Q) = 1$.

Lemma 4.2.25. *Let $0 \leq r \leq n$, $A \in \mathbb{E}_n$, and let p be a prime number. Then the sum* (4.2.91) *can be written in the form*

$$s_p(r, n; A) = \varphi_{n-r}(p)^{-1} \sum_{a,b \geq 0, a+b=r} (-1)^a p^{\langle a-1 \rangle + b} \frac{\varphi_{n-b}(p)}{\varphi_a(p)\varphi_b(p)} \mathrm{pm}_p(b, n; A)$$

where $\varphi_i(v)$ are the polynomials (3.2.29).

Proof. If $r = 0$, then the formula is trivial. By Lemma 3.5.20, in the notation of the lemma, we have

$$\begin{aligned} s_p(r, n; A) &= \sum_{\mathbf{i} \in \mathsf{I}_{r,n}} \sum_{M \in \mathsf{SM}_p(r,n;\mathbf{i})} e\{p^{-1}AM\} \\ &= \sum_{\mathbf{i} \in \mathsf{I}_{r,n}} \sum_{\substack{S \in \mathsf{SM}_p(r,r), \\ V \in \mathsf{V}(\mathbf{i})}} \exp\left(\frac{\pi i}{p}\sigma(M(\mathbf{i})A\,{}^tM(\mathbf{i}) \cdot {}^t(E_r, V)S(E_r, V))\right) \\ &= \sum_{S \in \mathsf{SM}_p(r,r)} \sum_{\substack{\mathbf{i} \in \mathsf{I}_{r,n}, \\ X=(E_r, V)M(\mathbf{i}), V \in \mathsf{V}(\mathbf{i})}} \exp\left(\frac{\pi i}{p}\sigma(A[{}^tX]S)\right). \end{aligned}$$

To transform the last sum, we note that the matrix X runs over a set, which can be considered as a set of representatives for the orbits of the group $\mathsf{G} = \mathsf{GL}_r(\mathbb{Z}/p\mathbb{Z})$ acting on the set $\mathsf{PM}_p(r, n)$ by left multiplications:

$$\mathsf{G}\backslash\mathsf{PM}_p(r, n) = \bigcup_{\mathbf{i} \in \mathsf{I}_{r,n}} \{X = (E_r, V)M(\mathbf{i}); V \in \mathsf{V}(\mathbf{i})\}. \tag{4.2.97}$$

To see this, we let $\mathbf{i} = \mathbf{i}(T) = (i_1, \ldots, i_r) \in \mathsf{I}_{r,n}$ with $T \in \mathsf{PM}_p(r, n)$ be, in the sense of lexicographic order, the least system of indices of r linearly independent modulo

p columns of T. Since clearly

$$\mathbf{i}(gT) = \mathbf{i}(T), \quad \text{if } g \in \mathsf{G} \tag{4.2.98}$$

it follows that the matrix $T' = T_1^{-1}T$, where T_1 is the square matrix composed of the columns of T with indices $i_1, \ldots, i_r$, satisfies $\mathbf{i}(T') = \mathbf{i}(T)$, and the columns of T' with indices $i_1, \ldots, i_r$ form the unit matrix E_r. Since the system $\mathbf{i}$ is minimal, it follows that the entries t'_{α,j_β} of $T' = (t'_{ij})$, where $(j_1, \ldots, j_{n-r}) = \hat{\mathbf{i}} \in \mathsf{I}_{n-r,n}$ is the system complementary to $(i_1, \ldots, i_r)$ in $(1, 2, \ldots, n)$, satisfy $t'_{\alpha,j_\beta} \equiv 0 \pmod{p}$, if $i_\alpha > j_\beta$. Hence, assuming that all of the entries t'_{ij} of T' satisfy $0 \le t'_{ij} < p$ and using (3.5.41), we see that the matrix $T'M(\mathbf{i})^{-1} = T_1^{-1}TM(\mathbf{i})^{-1}$ is of the form (E_r, V) with $V \in \mathsf{V}(\mathbf{i})$ (see (3.5.42)). Thus, the right side in (4.2.97) contains representatives for all of the orbits on the left. If two matrices $X = (E_r, V)M(\mathbf{i})$ and $X' = (E_r, V')M(\mathbf{i}')$ with $V \in \mathsf{V}(\mathbf{i})$ and $V' \in \mathsf{V}(\mathbf{i}')$ belong to the same orbit, say $X' \equiv gX \pmod{p}$, then (4.2.98) and obvious relations $\mathbf{i}(X) = \mathbf{i}$, $\mathbf{i}(X') = \mathbf{i}'$ imply that $\mathbf{i} = \mathbf{i}'$, whence $g \equiv E_r \pmod{p}$, and $X = X'$. We have proved (4.2.97). Now, since the last expression for $s_p(r, n; A)$ contains a summation over the set $\mathsf{SM}_p(r, r)$, it follows that the matrix S may be replaced by any $S[g]$ with $g \in \mathsf{G}$. Hence, by (4.2.97) we get

$$\begin{aligned} s_p(r, n; A) &= |G|^{-1} \sum_{\substack{S \in \mathsf{SM}_p(r,r),\\ g \in G}} \; \sum_{X \in \mathsf{G} \backslash \mathsf{PM}_p(r,n)} \exp\left(\frac{\pi i}{p}\sigma(gXA\,{}^tX\,{}^tgS)\right) \\ &= \mathrm{pm}_p(r, r)^{-1} \sum_{S \in \mathsf{SM}_p(r,r)} G_p^*(S, A), \end{aligned} \tag{4.2.99}$$

where, for $S \in \mathsf{S}_r(\mathbb{Z})$ and $A \in \mathbb{E}_n$, we set

$$G_p^*(S, A) = \sum_{X \in \mathsf{PM}_p(r,n)} \exp\left(\frac{\pi i}{p}\sigma(AS[X])\right).$$

On the other hand, the numbers $\mathrm{pm}_p(b, n; A)$ can also be expressed in terms of the sums $G_p^*(S, A)$: the obvious relations

$$\sum_{S \in \mathsf{S}_b(\mathbb{Z}/p\mathbb{Z})} \exp\left(\frac{\pi i}{p}\sigma(SQ)\right) = \begin{cases} p^{\langle b \rangle}, & \text{if } Q \in p\mathbb{E}_b \\ 0, & \text{if } Q \notin p\mathbb{E}_b \end{cases}$$

with $Q \in \mathbb{E}_b$ imply the relations

$$\begin{aligned} p^{-\langle b \rangle} \sum_{S \in \mathsf{S}_b(\mathbb{Z}/p\mathbb{Z})} G_p^*(S, A) &= p^{-\langle b \rangle} \sum_{X \in \mathsf{PM}_p(b,n)} \; \sum_{S \in \mathsf{S}_b(\mathbb{Z}/p\mathbb{Z})} \exp\left(\frac{\pi i}{p}\sigma(XA\,{}^tXS)\right) \\ &= \mathrm{pm}_p(b, n; A) \end{aligned}$$

with $0 \le b \le n$ and $A \in \mathbb{E}_n$, whence we have

$$\mathrm{pm}_p(b, n; A) = p^{-\langle b \rangle} \sum_{t=0}^{b} \sum_{S \in \mathsf{SM}_p(t,b)} G^*(S, A).$$

Since each of the sums $G_p^*(S, A)$ is invariant under the replacement $S \to gS\,{}^tg$ with $g \in \mathsf{GL}_b(\mathbb{Z}/p\mathbb{Z})$, it follows, by (3.5.66), that

$$\sum_{S \in \mathsf{SM}_p(t,b)} G_p^*(S, A) = \frac{\varphi_b(p)}{\varphi_t(p)\varphi_{b-t}(p)} \sum_{S = \begin{pmatrix} S' & 0 \\ 0 & 0 \end{pmatrix} \in \mathsf{SM}_p(t,b)} G_p^*(S, A),$$

where $S^1 = S^{(t)}$ is the $(t \times t)$-block. Each matrix X of $\mathsf{PM}_p(b, n)$ can be written in the form $X = \begin{pmatrix} X_1 \\ X_2 \end{pmatrix}$ with $X_1 \in \mathsf{PM}_p(t, n)$. It is clear that the number of such X with a fixed X_1 does not depend upon X_1 and therefore is equal to $\mathrm{pm}_p(b, n)/\mathrm{pm}_p(t, n)$. Hence, each of the sums G_p^* in the last expression can be written in the form

$$\begin{aligned} G_p^*\left(\begin{pmatrix} S' & 0 \\ 0 & 0 \end{pmatrix}, A\right) &= \sum_{\binom{X_1}{X_2} \in \mathsf{PM}_p(b,n)} e\left\{p^{-1}\begin{pmatrix} X_1 \\ X_2 \end{pmatrix} A({}^tX_1, {}^tX_2) \cdot \begin{pmatrix} S' & 0 \\ 0 & 0 \end{pmatrix}\right\} \\ &= \frac{\mathrm{pm}_p(b, n)}{\mathrm{pm}_p(t, n)} \sum_{X_1 \in \mathsf{PM}_p(t,n)} e\{p^{-1} X_1 A\, {}^tX_1 S'\} \\ &= \frac{\mathrm{pm}_p(b, n)}{\mathrm{pm}_p(t, n)} G_p^*(S', A). \end{aligned}$$

From the above formulae and Lemma 3.5.18 we obtain the formula

$$\mathrm{pm}_p(b, n; A) = \sum_{t=0}^{b} p^{-\langle t-1\rangle - b} \frac{\varphi_b(p)\varphi_{n-t}(p)}{\varphi_t(p)\varphi_{b-t}(p)\varphi_{n-b}(p)} \sum_{S \in \mathsf{SM}_p(t,t)} G_p^*(S, A).$$

Hence, using the relations for the coefficients of series in (3.5.67), we see that the right side in the formula of the lemma is equal to

$$\begin{aligned} &\varphi_{n-r}^{-1} \sum_{a+b=r} (-1)^a p^{\langle a-1\rangle + b} \frac{\varphi_{n-b}}{\varphi_a \varphi_b} \sum_{t+t_1=b} p^{-\langle t-1\rangle - b} \frac{\varphi_b \varphi_{n-t}}{\varphi_t \varphi_{t_1} \varphi_{n-b}} \sum_{S \in \mathsf{SM}_p(t,t)} G_p^*(S, A) \\ &= \varphi_{n-r}^{-1} \sum_{t=0}^{r} \left(\sum_{a+t_1=r-t} (-1)^a p^{\langle a-1\rangle} \frac{1}{\varphi_a \varphi_{t_1}} \right) \cdot p^{-\langle t-1\rangle} \cdot \frac{\varphi_{n-t}}{\varphi_t} \sum_{S \in \mathsf{SM}_p(t,t)} G_p^*(S, A) \\ &= p^{-\langle r-1\rangle} \varphi_r^{-1} \sum_{S \in \mathsf{SM}_p(r,r)} G_p^*(S, A) \end{aligned}$$

with $\varphi_i = \varphi_i(p)$. By Lemma 3.5.18, $\mathrm{pm}_p(r, r) = p^{\langle r-1\rangle} \varphi_r(p)$. Hence, the last expression is equal to the expression (4.2.99) for the sum $s_p(r, n; A)$. □

We now use some results on quadratic spaces modulo p of Appendix 2.

Lemma 4.2.26. *Let $Q \in \mathbb{E}_n$, and let $\mathsf{q} = \mathsf{q}(x_1, \ldots, x_n)$ be the quadratic form* (1.1.4) *with the matrix Q. Then the number $\mathrm{pm}_p(b, n; Q)$, where $p \in \mathbb{P}$ and $0 < b \leq n$, is equal to the number $i(\mathsf{V}_p, f_p; b)$ of isotropic system consisting of b vectors in a quadratic space (V_p, f_p) over $\mathsf{K} = \mathbb{Z}/p\mathbb{Z}$ of type $\{\mathsf{q}\}_{\mathsf{K}}$ (see A.2.4).*

Proof. We fix a basis $\mathbf{e}_1, \ldots, \mathbf{e}_n$ for V_p such that

$$f_p(x_1\mathbf{e}_1 + \cdots + x_n\mathbf{e}_n) \equiv \mathsf{q}(x_1, \ldots, x_n) \pmod{p}.$$

Then, according to definitions, a system $\mathbf{m}_1, \ldots, \mathbf{m}_b$ with $\mathbf{m}_i = \sum_{j=1}^{n} m_{ij}\mathbf{e}_j$ is isotropic if and only if the matrix $M = (m_{ij})$ belongs to $\mathsf{PM}_p(b, n; Q)$. □

The formulae for the numbers $i(\mathsf{V}_p, f_p; Q)$ of A.2.4. will allow us to complete the calculation of the polynomials $B_p^n(v, A)$. To formulate the result, we shall need some definitions. We say that two matrices Q, Q_1 of $\mathbb{E}_n$ are *equivalent modulo* $d \in \mathbb{N}$, we write $Q \sim Q_1 \pmod{d}$, if there is a matrix M of M_n with g.c.d.$(\det M, d) = 1$ such that $Q_1 \equiv Q[M] \pmod{d}$, the congruence being understood in the sense of (4.2.96). We say that Q is *degenerate modulo d*, if it is equivalent modulo d to a matrix of the form $\begin{pmatrix} Q' & 0 \\ 0 & 0 \end{pmatrix}$ with $Q' \in \mathbb{E}_{n-1}$, otherwise *Q is non-degenerate modulo d.* If $d = p$ is a prime number, then it is clear that the condition $Q \sim Q_1 \pmod{p}$ means that the quadratic forms q and q_1 with matrices Q and Q_1 are equivalent over the field $\mathsf{F} = \mathbb{F}_p$ of p elements, and Q is degenerate (non-degenerate) modulo p if and only if the quadratic space of the type $\{\mathsf{q}\}_\mathsf{F}$ is degenerate (non-degenerate) in the sense of A.2.2.

Theorem 4.2.27. *The polynomials $B_p^n(v, A)$ given by* (4.2.93) *with $n \in \mathbb{N}$, $p \in \mathbb{P}$, and $A \in \mathbb{E}_n$ satisfy the following rules:*

(1) *$B_p^n(v, A) = B_p^n(v, A_1)$, if $A_1 \sim A \pmod{p}$;*

(2) *If A is degenerate modulo p, say*

$$A \sim \begin{pmatrix} A' & 0 \\ 0 & 0 \end{pmatrix} \pmod{p} \qquad \text{with } A' \in \mathbb{E}_{n-1},$$

then $B_p^n(v, A) = B_p^{n-1}(v, A')$;

(3) *If A is non-degenerate modulo p, then*

$$B_p^n(v, A) = \begin{cases} (1 + v)(1 - \chi_A(p)p^{-k}v) \prod_{i=1}^{k-1} (1 - p^{-2i}v^2), & \text{if } n = 2k \\ (1 + v) \prod_{i=1}^{k} (1 - p^{-2i}v^2), & \text{if } n = 2k + 1, \end{cases}$$

where, for even n, $\chi_A(p)$ denotes the sign (A.2.21) *of a quadratic space over $\mathbb{Z}/p\mathbb{Z}$ with quadratic form equivalent to the form with the matrix A modulo p.*

Proof. The first assertion follows from (4.2.93), since the sums $s_p(r, n; A)$ depend only on the equivalence class of A modulo p.

To prove the second assertion, we use the Siegel operator Φ and Žarkovskaya relations. Below we use the notation (4.2.88), and applying mappings of Hecke rings to polynomials over the rings, we assume that the mappings operate only on the coefficients of the polynomials. We first show that, in this sense, the following relation holds

$$\Psi(B_p^n(v), u) = B_p^{n-1}(v), \tag{4.2.100}$$

where $B_p^n(v)$ is the polynomial (4.2.90) over $\mathcal{L}_p^n$, and $\Psi\colon \mathcal{L}_p^n \to \mathcal{L}_p^{n-1}[u^{\pm 1}]$ is the mapping defined by (4.2.64)–(4.2.65). By the definition of the polynomials $\mathbf{R}_p^n(v)$ (see (3.5.16)), since the diagram (4.2.67) is commutative, it follows that

$$\begin{aligned}\Omega^{n-1}(\Psi(\mathbf{R}_p^n(v), u)) &= \Xi(\Omega^n \mathbf{R}_p^n)(v) \\ &= (1 - u^{-1}v)(1 - uv)(\Omega^{n-1}\mathbf{R}_p^{n-1})(v).\end{aligned}$$

By Proposition 4.2.17(2), all of the coefficients of $\Psi(\mathbf{R}_p^n(v), u)$ along with those of $\mathbf{R}_p^{n-1}(v)$ are contained in $\mathbf{L}_p^{n-1}$. By Theorem 3.3.30, the mapping Ω is monomorphic on $\mathbf{L}_p^{n-1}$. Hence, the last relation implies that

$$\Psi(\mathbf{R}_p^n(v), u) = (1 - u^{-1}v)(1 - uv)\mathbf{R}_p^{n-1}(v).$$

Now let $X_-^n(v)$ and $X_+^n(v)$ be the first and the last factors of the factorization (3.5.69). Using the commutative diagram (4.2.67) and Lemma 3.3.34, we have

$$\Omega^{n-1}(\Psi(X_-^n(v), u)) = \Xi(\Omega^n X_-^n)(v)$$

$$\Xi\left(\sum_{i=0}^{n}(-1)^i s_i(x_1^{-1}, \ldots, x_n^{-1})v^i\right) = (1 - u^{-1}v)(\Omega^{n-1}X_-^{n-1})(v).$$

Similarly,

$$\Omega^{n-1}(\Psi(X_+^n(v), u)) = (1 - uv)(\Omega^{n-1}X_+^{n-1})(v).$$

By (3.5.20), the coefficients of X_-^n are contained in the subring $\mathcal{C}_{-p}^n$ of $\mathcal{L}_p^n$. Then the duality relations (3.5.2) and (3.5.19) imply that the coefficients of X_+^n are contained in $\mathcal{C}_p^n$. Using Proposition 4.2.17(2) and the fact that the mapping Ω^{n-1} is monomorphic on $\mathcal{C}_{-p}^{n-1}$ and $\mathcal{C}_p^{n-1}$, we obtain the relations

$$\Psi(X_-^n(v), u) = (1 - u^{-1}v)X_-^{n-1}(v), \qquad \Psi(X_+^n(v), u) = (1 - uv)X_+^{n-1}(v).$$

Applying Ψ to the factorization (3.5.69), we have

$$\begin{aligned}(1 - u^{-1}v)(1 - uv)\mathbf{R}_p^{n-1}(v) &= \Psi(\mathbf{R}_p^n(v), u) = \Psi(X_-^n(v), u)\Psi(B_p^n(v), u)\Psi(X_-^n(v), u) \\ &= (1 - u^{-1}v)(1 - uv)X_-^{n-1}(v)\Psi(B^n(v), u)X_+^{n-1}(v),\end{aligned}$$

hence

$$\mathbf{R}_p^{n-1}(v) = X_-^{n-1}(v)\Psi(B_p^n(v), u)X_+^{n-1}(v).$$

On the other hand, we may write

$$\mathbf{R}_p^{n-1}(v) = X_-^{n-1}(v)B_p^{n-1}(v)X_+^{n-1}(v).$$

The relation (4.2.100) follows from these relations, since the polynomials X_-^{n-1}, X_+^{n-1} are invertible in the ring of formal power series over $\mathcal{L}_p^n$.

We now prove the assertion (2). By (1), we may assume that $A = \begin{pmatrix} A' & 0 \\ 0 & 0 \end{pmatrix}$, and the matrix A' can be arbitrarily chosen in its residue class modulo $p\mathbb{E}_{n-1}$.

Thus, we may assume without loss of generality that A' is positive definite (by Theorem A.1.5, $A' > 0$, if the diagonal entries of A' are big enough with respect to other entries). Then we set $Q = \begin{pmatrix} A' & 0 \\ 0 & 2 \end{pmatrix} \in \mathbb{A}_n^+$ and consider the theta-series $\theta^n(Z, Q)$ of genus n of Q. It follows from Proposition 1.1.3 that $\theta^n(Z, Q) \in \mathfrak{M}_\xi^n$, $\xi(-1) = 1$, since the theta-series is clearly invariant under the transformations $Z \to M\langle Z\rangle$ with $M \in \Gamma_0^n$. Let us take $k = 0$, $\chi = 1$, and calculate, in two ways, the Fourier coefficient $f(A') = f(A', v)$ of $(F|_{0,1} B_p^n(v))|\Phi$, where $F = \theta^n(Z, Q)$ and Φ is the Siegel operator. On one hand, by Proposition 4.2.24 and (2.3.29), we see that the coefficient is equal to

$$B_p^n\left(v, \begin{pmatrix} A' & 0 \\ 0 & 0 \end{pmatrix}\right) r\left(Q, \begin{pmatrix} A' & 0 \\ 0 & 0 \end{pmatrix}\right) = B_p^n(v, A) r(Q, A').$$

On the other hand, by (4.2.69), (2.3.29), (4.2.100), and Proposition 4.2.24 for the function $F|\Phi$, we conclude that the coefficient equals

$$B_p^{n-1}(v, A') r\left(Q, \begin{pmatrix} A' & 0 \\ 0 & 0 \end{pmatrix}\right) = B_p^{n-1}(v, A') r(Q, A').$$

Since clearly $r(Q, A') \neq 0$, the assertion (2) follows.

Suppose, now, that A is non-degenerate modulo p. Let $\mathsf{a}(x_1, \ldots, x_n)$ be the quadratic form with matrix A, and let (V_p, f_p) be a quadratic space over $\mathsf{F} = \mathbb{F}_p$ of the type $\{\mathsf{a}\}_\mathsf{F}$. Since A is non-degenerate modulo p, it follows that (V_p, f_p) is non-degenerate. By Lemma 4.2.26, we can write the formula of Lemma 4.2.25 in the form

$$s_p(r, n; A) = \varphi_{n-r}(p)^{-1} \sum_{a+b=r} (-1)^a p^{b+\langle a-1\rangle} \frac{\varphi_{n-b}(p)}{\varphi_a(p)\varphi_b(p)} i(b),$$

where $i(b) = i(\mathsf{V}_p, f_p, b)$ is the number of isotropic system of b vectors in V_p. Substituting this in (4.2.93), we have

$$B_p^n(v, A) = \sum_{i=0}^{n} (-1)^i p^{-\langle i\rangle - i(n-i)} \frac{1}{\varphi_{n-i}} B_i v^i,$$

where $\varphi_\alpha = \varphi_\alpha(p)$, and

$$\begin{aligned} B_i &= \sum_{j+r=i} \varphi_{n-r} \sum_{c+d=j} \frac{(-p)^c}{\varphi_c \varphi_d} \cdot \varphi_{n-r}^{-1} \sum_{a+b=r} (-1)^a p^{b+\langle a-1\rangle} \frac{\varphi_{n-b}}{\varphi_a \varphi_b} i(b) \\ &= \sum_{c+d+a+b=i} \frac{(-p)^c}{\varphi_c} \cdot \frac{(-1)^a p^{\langle a-1\rangle}}{\varphi_d \varphi_a} \cdot \frac{p^b \varphi_{n-b}}{\varphi_b} \cdot i(b). \end{aligned}$$

In the last formula we sum up over all non-negative integers d and a with $d + a = i - c - b$, where c and b are fixed. Then, by (3.5.67), we get

$$B_i = p^i \sum_{c+b=i} \frac{(-1)^c \varphi_{n-b}}{\varphi_c \varphi_b} i(b).$$

Let λ be the dimension of the maximal isotropic subspace of (V_p, f_p). According to Corollary A.2.15, $\lambda = k$, if $n = 2k$ and $\chi_A(p) = 1$; $\lambda = k - 1$, if $n = 2k$ and $\chi_A(p) = -1$; and $\lambda = k$, if $n = 2k + 1$. This implies that the right side of the formula for $B_p^n(v, A)$, which we want to prove, can always be written in the form

$$\left\{\prod_{i=0}^{n-\lambda-1} (1 + p^{-i}v)\right\}\left\{\prod_{i=0}^{\lambda} (1 - p^{-i}v)\right\}. \tag{4.2.101}$$

Since $i(b) = 0$ for $b > \lambda$, it follows that

$$\begin{aligned} B_p^n(v, A) &= \sum_{i=0}^{n} (-1)^i p^{\langle i \rangle - in} \varphi_{n-i}^{-1} \left(\sum_{c+b=i} \frac{(-1)^c \varphi_{n-b}}{\varphi_c \varphi_b} i(b) \right) v^i \\ &\overset{n-i=a}{=} \sum_{a+b+c=n} (-1)^{b+2c} p^{\langle b \rangle + \langle c \rangle + bc - (b+c)n} i(b) \frac{\varphi_{a+c}}{\varphi_a \varphi_b \varphi_c} v^{b+c} \\ &= \sum_{b=0}^{\lambda} (-1)^b p^{\langle b \rangle - bn} i(b) \varphi_b^{-1} v^b \sum_{a+c=n-b} \frac{p^{\langle c \rangle} \varphi_{a+c}}{\varphi_a \varphi_c} (p^{b-n} v)^c. \end{aligned}$$

By (3.2.34), we have

$$\begin{aligned} \sum_{a+c=n-b} \frac{p^{\langle c \rangle} \varphi_{a+c}}{\varphi_a \varphi_c} (p^{b-n} v)^c &= \prod_{i=1}^{n-b} (1 + p^{b-n+i} v) \\ &= \left\{\prod_{j=0}^{n-\lambda-1} (1 + p^{-j}v)\right\}\left\{\prod_{i=1}^{\lambda-b} (1 + p^{i+b-n} v)\right\}, \end{aligned}$$

where the second factor equals 1, when $b = \lambda$. Hence,

$$B_p^n(v, A) = \left\{\prod_{j=0}^{n-\lambda-1} (1 + p^{-j}v)\right\} \cdot G(v),$$

where

$$G(v) = \sum_{b=0}^{\lambda} (-1)^b p^{\langle b \rangle - bn} i(b) \varphi_b^{-1} \cdot v^b \cdot \prod_{i=1}^{\lambda-b} (1 + p^{i+b-n} v).$$

Thus, to complete the proof, we only have to show that the polynomial $G(v)$ is equal to the second factor of the product (4.2.101). Since $G(0) = 1$, and degree of G does not exceed λ, it suffices to show that $G(p) = G(p^2) = \cdots = G(p^\lambda) = 0$. For integers j with $j \geq -1$, we define numbers $\kappa_j = \kappa_j(p)$ by setting $\kappa_{-1}(p) = 1/2$, and $\kappa_j(p) = \varphi_j(p^2)\varphi_j(p)^{-1}$, if $j \geq 0$. It is easy to see that

$$\frac{\kappa_{g-1}(p)}{\kappa_{h-1}(p)} = \begin{cases} 1, & \text{if } g = h \geq 0 \\ \prod_{j=h}^{g-1} (p^j + 1), & \text{if } g > h \geq 0. \end{cases} \tag{4.2.102}$$

Substituting $v = p^\mu$ with $1 \le \mu \le \lambda$ to one of the products occuring in $G(v)$, we obtain

$$\prod_{i=1}^{\lambda-b}(1 + p^{i+b+\mu-n}) = \prod_{i=1}^{\lambda-b} p^{i+b+\mu-n}(p^{n-(i+b+\mu)} + 1)$$

$$= p^{\langle\lambda-b\rangle+(\lambda-b)(b+\mu-n)} \prod_{j=n-\lambda-\mu}^{n-b-\mu-1}(p^j + 1)$$

$$= p^{\langle\lambda\rangle+\lambda\mu-\lambda n-b\mu-\langle b\rangle+bn}\kappa_{n-b-\mu-1}\kappa_{n-\lambda-\mu-1}^{-1}$$

(recall that the product equals 1, if $b = \lambda$). Hence, we have

$$G(p^\mu) = p^{\langle\lambda\rangle+\lambda\mu-\lambda n}\kappa_{n-\lambda-\mu-1}^{-1}\sum_{b=0}^{\lambda}(-1)^b i(b)\varphi_b^{-1}\kappa_{n-b-\mu-1}.$$

We now use the formulae for $i(b)$ obtained in A.2.4, but first we write the formulae in a more convinient form. By (A.2.24) and (A.2.25), we have

$$i(b) = p^{\langle b-1\rangle}(p^\lambda - 1)(p^{\lambda-b} - 1)\varphi_{\lambda-1}(p^2)\varphi_{\lambda-b}(p^2)^{-1}$$

$$= \frac{(p^\lambda - 1)\varphi_{\lambda-1}(p^2)}{\varphi_\lambda(p)}\cdot\frac{1}{\kappa_{\lambda-b-1}(p)}\cdot\frac{p^{\langle b-1\rangle}\varphi_\lambda(p)}{\varphi_{\lambda-b}(p)},$$

if $n = 2k$ and $\chi_A(p) = 1$;

$$i(b) = p^{\langle b-1\rangle}(p^{\lambda+1} + 1)(p^{\lambda+1-b} - 1)\varphi_\lambda(p^2)\varphi_{\lambda+1-b}(p^2)^{-1}$$

$$= \frac{(p^{\lambda+1} + 1)\varphi_\lambda(p^2)}{\varphi_\lambda(p)}\cdot\frac{1}{\kappa_{\lambda-b+1}(p)}\cdot\frac{p^{\langle b-1\rangle}\varphi_\lambda(p)}{\varphi_{\lambda-b}(p)},$$

if $n = 2k$ and $\chi_A(p) = -1$; and

$$i(b) = \frac{\varphi_\lambda(p^2)}{\varphi_\lambda(p)}\cdot\frac{1}{\kappa_{\lambda-b}(p)}\cdot\frac{p^{\langle b-1\rangle}\varphi_\lambda(p)}{\varphi_{\lambda-b}(p)},$$

if $n = 2k + 1$. According to these formulae, in each of the cases we can write $i(b)$ in the form

$$i(b) = c\kappa_{n-b-\lambda-1}^{-1}p^{\langle b-1\rangle}\varphi_\lambda\varphi_{\lambda-b}^{-1}$$

with a constant c independent of b. Hence,

$$G(p^\mu) = c'\sum_{b=0}^{\lambda}\frac{\kappa_{n-b-\mu-1}}{\kappa_{n-b-\lambda-1}}\cdot\frac{(-1)^b p^{\langle b-1\rangle}\varphi_\lambda}{\varphi_b\varphi_{\lambda-b}}.$$

It is easy to see that one can write

$$\kappa_{n-b-\mu-1}\cdot\kappa_{n-b-\lambda-1}^{-1} = \sum_{i=0}^{\lambda-\mu} c_i p^{-b_i},$$

where the coefficients c_i are independent of b. In fact, this is clear if $\mu = \lambda$, but if $\mu < \lambda$, then, by (4.2.102), we have

$$\kappa_{n-b-\mu-1}\kappa_{n-b-\lambda-1}^{-1} = \prod_{j=n-b-\lambda}^{n-b-\mu-1} (p^j + 1) = \prod_{i=1}^{\lambda-\mu} (p^{i+n-\lambda-1} \cdot p^{-b} + 1)$$

$$= \sum_{i=0}^{\lambda-\mu} s_i(p^{1+n-\lambda-1}, \ldots, p^{\lambda-\mu+n-\lambda-1}) p^{-b_i},$$

s_i being the elementary symmetric polynomials. Substituting these expressions in $G(p^\mu)$ and inverting the order of summation, we obtain

$$G(p^\mu) = c' \sum_{i=0}^{\lambda-\mu} c_i \sum_{b=0}^{\lambda} (-p^{-i-1})^b \frac{p^{\langle b \rangle} \varphi_\lambda(b)}{\varphi_b(p)\varphi_{\lambda-b}(p)}$$

$$= c' \sum_{i=0}^{\lambda-\mu} c_i \prod_{t=1}^{\lambda} (1 - p^{t-i-1}),$$

where the inner sum is transformed into a product by (3.2.34). Since $\mu \geq 1$, it follows that each of the numbers $i = 0, 1, \ldots, \lambda - \mu$ satisfies $1 \leq i + 1 \leq \lambda$. Hence, each of the products in the last sum equals zero, and so $G(p^\mu) = 0$. □

§ 4.3. Multiplicative Properties of Fourier Coefficients

The time has come to gather flowers in our garden. In this section we shall apply the machinery of Hecke operators to investigation of multiplicative properties of Fourier coefficients of modular forms. The scheme of the investigation looks as follows: Matrices of Hecke operators in a fixed basis for an invariant space of modular forms satisfy the same relations as the corresponding elements of Hecke rings; on the other hand, Hecke operators act on Fourier coefficients of modular forms, and their matrices appear in formulae describing the action; in some cases the formulae allows us to express certain Fourier coefficients or their combinations in terms of the matrices and thus observe how the relations in Hecke rings affect the Fourier coefficients. In the case of modular forms of several variables one can not usually express individual Fourier coefficients in terms of the matrices of Hecke operators, and relations between them take the shape of identities between the Dirichlet series constructed, on one hand, from the Fourier coefficients, and, on the other hand, from the matrices of operators. The latter Dirichlet series, the so-called zeta-functions, often have the Euler product expansion, due to multiplicative properties of Hecke operators. Thus, multiplicative properties of Fourier coefficients appear in the shape of Euler product expansions of certain Dirichlet series constructed from the coefficients. In the analytic theory of the zeta-functions, which we do not touch in this book, the same identities work in other direction: they enable us to express the zeta-functions in terms of modular forms (usually, by means of an integral transformation) and hence study such their properties as analytic continuation and functional equations (see Exercises 4.3.9–4.3.12 and Notes on §§ 4.3.2. and 4.3.3.).

In this section we consider the case of one-dimensional invatiant subspaces, that is the case of eigenfunctions of Hecke operators. Actually, this does not diminish the generality, since invariant subspaces are, as a rule, spanned by eigenfunctions. Besides, the general case can be treated almost in the same ways if one uses the language of vector-valued modular forms. We shall use the vector language in the next chapter.

4.3.1. Modular Forms of One Variable

Let us consider a modular form

$$F(z) = \sum_{A \in \mathbb{A}_1} f(A)e\{Az\}$$

$$= \sum_{a=0}^{\infty} f(2a)e^{2\pi i a z} \in \mathfrak{M}_k(q, \chi) = \mathfrak{M}_k^1(q, \chi)$$

of weight k and character χ for the group $\Gamma_0(q) = \Gamma_0^1(q)$, which is an eigenfunction for all of the Hecke operators $|T = |_{k,\chi} T$ with $T \in \mathsf{L}(q) = \mathsf{L}^1(q)$, say

$$F|T = \lambda(T)F \qquad (T \in \mathsf{L}(q)),$$

where $\lambda(T) = \lambda(T, F)$ are the corresponding eigenvalues. In particular, for elements T of the form (3.3.19), we have

$$F|T(m) = \lambda(m)F \qquad (m \in \mathbb{N}_{(q)}) \tag{4.3.1}$$

Calculating the Fourier coefficients $(f|T(m))(2a)$ of $F|T(m)$, by (3.4.31) and Lemma 4.2.7 or 4.2.8, we obtain

$$(f|T(m))(2a) = \sum_{d_1 d_2 = m, (d_i \in \mathbb{N})} ((f|\Pi_+^1(d_1))|\Pi_-^1(d_2))(2a)$$

$$= \sum_{d|m, a} d^{k-1}\chi(d)(f|\Pi_+^1(m/d))(2 \cdot a/d)$$

$$= \sum_{d|m, a} d^{k-1}\chi(d)f(2 \cdot ma/d^2).$$

On the other hand, by (4.3.1), the same coefficient is equal to $\lambda(m)f(2a)$, and so we have the relations

$$\sum_{d|m, a} d^{k-1}\chi(d)f(2ma/d^2) = \lambda(m)f(2a). \tag{4.3.2}$$

On replacing of m by m/δ, and a by a/δ with a common positive divisor δ of m and a, we get

$$\sum_{d|m/\delta, a/\delta} d^{k-1}\chi(d)f(2ma/(\delta d)^2) = \lambda(m/\delta)f(2a/\delta).$$

Multiplying both sides by $\delta^{k-1}\chi(\delta)\mu(\delta)$, where μ is the Möbius function, and summing over all common positive divisors δ of m and a, we have

$$\sum_{\delta, d \in \mathbb{N}, \delta d | m, a} (\delta d)^{k-1} \chi(\delta d) \mu(\delta) f(2ma/(\delta d)^2)$$
$$= \sum_{\delta | m, a} \delta^{k-1} \chi(\delta) \mu(\delta) \lambda(m/\delta) f(2a/\delta).$$

Since, for $b \in \mathbb{N}$,

$$\sum_{\delta | b} \mu(\delta) = \begin{cases} 1, & \text{if } b = 1 \\ 0, & \text{if } b > 1, \end{cases} \tag{4.3.3}$$

it follows that the left side of the last formula equals

$$\sum_{b | m, a} b^{k-1} \chi(b) f(2ma/b^2) \sum_{\delta | b} \mu(\delta) = f(2ma).$$

Thus, we obtain the following multiplicative relations for the function f:

$$f(2ma) = \sum_{d \in \mathbb{N}, d | m, a} d^{k-1} \chi(d) \mu(d) \lambda(m/d) f(2a/d) \tag{4.3.4}$$

which, in fact, are equivalent with (4.3.2). If $m = m(a)$ is the greatest divisor of a contained in $\mathbb{N}_{(q)}$, then m is prime to a/m, and so, by (4.3.2) or (4.3.4), we have the relation

$$f(2a) = \lambda(m) f(2a/m), \tag{4.3.5}$$

which to some extent reduces the study of Fourier coefficients of an eigenfunction to that of corresponding eigenvalues. The following theorem, which is based on the same relations, gives, in particular, a complete description of multiplicative properties of the eigenvalues $\lambda(m)$.

Theorem 4.3.1. *The eigenvalues $\lambda(m) = \lambda(m, F)$ of the Hecke operators $T(m)$ corresponding to a non-zero eigenfunction F of the space $\mathfrak{M}_k(q, \chi)$ satisfy the following multiplicative relations*:

$$\lambda(m)\lambda(m_1) = \sum_{d | m, m_1} d^{k-1} \chi(d) \lambda(mm_1/d^2), \tag{4.3.6}$$

$$\lambda(mm_1) = \sum_{d | m, m_1} d^{k-1} \chi(d) \mu(d) \lambda(m/d) \lambda(m_1/d) \tag{4.3.7}$$

for every $m, m_1 \in \mathbb{N}_{(q)}$.

For $k > 0$, the eigenvalues satisfy

$$|\lambda(m)| \leq c_F m^k \quad (m \in \mathbb{N}_{(q)}) \tag{4.3.8}$$

(respectively,

$$|\lambda(m)| \leq c_F m^{k/2} \quad (m \in \mathbb{N}_{(q)}), \tag{4.3.9}$$

if F is a cusp form), where c_F depends only on F.

Proof. Let $f(2a)$ with $a = 0, 1, \ldots$ be the Fourier coefficients of F. Let us consider first the case $f(0) \neq 0$. In this case, if we divide both sides of the relation (4.3.2) with $a = 0$ by $f(0)$, we get

$$\lambda(m) = \sum_{d|m} d^{k-1}\chi(d) \qquad (m \in \mathfrak{N}_{(q)}), \tag{4.3.10}$$

which implies (4.3.6) and (4.3.7), by standard arguments of the elementary number theory, left to the reader. We now assume that $f(2a) \neq 0$ with a positive integer a, and denote by $\kappa = \kappa(F)$ the smallest of such integers. Let d be the greatest divisor of κ contained in $\mathbb{N}_{(q)}$. Since d, κ/d are coprime, it follows from (4.3.2) that $f(2\kappa) = \lambda(d)f(2\kappa/d)$. Hence, $f(2\kappa/d) \neq 0$, and so $d = 1$, i.e.

$$\kappa(F)|q^\infty; \qquad \text{in particular, } \kappa(F) = 1, \qquad \text{if } q = 1. \tag{4.3.11}$$

Hence, by (4.3.2) with $a = \kappa$, we obtain the relations

$$\lambda(m) = f(2m\kappa)/f(2\kappa) \qquad (m \in \mathbb{N}_{(q)}). \tag{4.3.12}$$

By (4.3.12) and (4.3.2), we have

$$\begin{aligned}\lambda(m)\lambda(m_1) &= \lambda(m)f(2m_1\kappa)/f(2\kappa)\\ &= \sum_{d|m,\, m_1\kappa} d^{k-1}\chi(d)f\left(\frac{2mm_1 \cdot \kappa}{d^2}\right)\Big/ f(2\kappa)\\ &= \sum_{d|m,\, m_1} d^{k-1}\chi(d)\lambda(mm_1/d^2),\end{aligned}$$

for common divisors of m and $m_1\kappa$ must divide m_1, in view of (4.3.11). Similarly, by (4.3.12), (4.3.4), and (4.3.11), we get

$$\begin{aligned}\lambda(mm_1) &= f(2mm_1\kappa)/f(2\kappa)\\ &= \sum_{d|m,\, m_1\kappa} d^{k-1}\chi(d)\mu(d)\lambda(m/d)f(2m_1\kappa/d)/f(2\kappa)\\ &= \sum_{d|m,\, m_1} d^{k-1}\chi(d)\mu(d)\lambda(m/d)\lambda(m_1/d).\end{aligned}$$

If $k > 0$, then F is not a constant. Hence, the number $\kappa = \kappa(F)$ exists, and we may represent the eigenvalues $\lambda(m)$ in the form (4.3.12). Applying (2.3.20) (resp. (2.3.43) to $f(2m\kappa)$, we obtain (4.3.8) (resp. (4.3.9)). □

The relations (4.3.7) imply that the eigenvalues $\lambda(m)$ with $m \in \mathbb{N}_{(q)}$ can be expressed as a polynomial in $\lambda(p)$, where p runs over prime divisors of m.

The following theorem gives an elegant reformulation of the relations (4.3.4) and relations of Theorem 4.3.1 in terms of Dirichlet series.

Theorem 4.3.2. *Let*

$$F = \sum_{a=0}^{\infty} f(2a)e^{2\pi i n z} \in \mathfrak{M}_k(q, \chi)$$

with a $k \in \mathbb{N}$ and a Dirichlet character χ modulo q be a non-zero eigenfunction for all Hecke operators $|_{k,\chi}T(m)$ with $m \in \mathbb{N}_{(q)}$, and let $\lambda(m) = \lambda(m, F)$ be the *corresponding eigenvalues. Then, for each* $a \in \mathbb{N}$, *the Dirichlet series*

$$D(s, a; F) = \sum_{m \in \mathbb{N}(q)} \frac{f(2am)}{m^s} \tag{4.3.13}$$

converges absolutely and uniformly in any right half-plane $\operatorname{Re} s \geq k + 1 + \varepsilon$ *(resp.* $\operatorname{Re} s \geq (k/2) + 1 + \varepsilon$, *if F is a cusp form) with* $\varepsilon > 0$ *of the complex variable s and has a factorization*

$$D(s, a; F) = \left(\sum_{d \in \mathbb{N}_{(q)}, d|a} \chi(d)\mu(d)f(2a/d)d^{k-1-s} \right) \xi(s, F) \tag{4.3.14}$$

where

$$\zeta(s, F) = \zeta(s, F, q) = \sum_{m \in \mathbb{N}_{(q)}} \frac{\lambda(m)}{m^s} \tag{4.3.15}$$

is the zeta-function of F.

In each of the half-planes indicated above, the zeta-function $\zeta(s, F)$ *converges absolutely and uniformly and has the Euler product expansion of the form*

$$\zeta(s, F) = \prod_{p \in \mathbb{P}_{(q)}} (1 - \lambda(p)p^{-s} + \chi(p)p^{k-1-2s})^{-1} \tag{4.3.16}$$

Proof. The assertions concerning convergence follow from estimates (2.3.20), (2.3.43), and those of Theorem 4.3.1.

By (4.3.4), we have

$$\begin{aligned} D(s, a; F) &= \sum_{m \in \mathbb{N}_{(q)}} \frac{1}{m^s} \sum_{d|m, a} d^{k-1}\chi(d)\mu(d)\lambda(m/d)f(2a/d) \\ &\overset{m \to md}{=} \sum_{d, m \in \mathbb{N}_{(q)}, d|a} \frac{1}{(dm)^s} d^{k-1}\chi(d)\mu(d)f(2a/d)\lambda(m) \\ &= \left(\sum_{d \in \mathbb{N}_{(q)}, d|a} \chi(d)\mu(d)f(2a/d)d^{k-1-s} \right) \sum_{m \in \mathbb{N}_{(q)}} \frac{\lambda(m)}{m^s}, \end{aligned}$$

which proves (4.3.14).

In order to prove the Euler expansion (4.3.16), we first note that, by (4.3.6),

$$\lambda(m)\lambda(m_1) = \lambda(mm_1), \qquad \text{if g.c.d.}(m, m_1) = 1. \tag{4.3.17}$$

Hence we have

$$\sum_{m \in \mathbb{N}_{(q)}} \frac{\lambda(m)}{m^s} = \prod_{p \in \mathbb{P}_{(q)}} \sum_{\delta=0}^{\infty} \frac{\lambda(p^\delta)}{p^{\delta s}} = \prod_{p \in \mathbb{P}_{(q)}} \zeta_p(s), \tag{4.3.18}$$

where, for a prime p,

$$\zeta_p(s) = \zeta_p(s, F) = \sum_{\delta=0}^{\infty} \lambda(p^\delta)v^\delta \qquad (v = p^{-s}) \tag{4.3.19}$$

is *a local zeta-function of F*. To sum up the power series (4.3.19), we use the relation (4.3.6) with $m = p$ and $m_1 = p^\delta$, that is

$$\lambda(p)\lambda(p^\delta) = \lambda(p^{\delta+1}) + p^{k-1}\chi(p)\lambda(p^{\delta-1}) \qquad (\delta \geq 1).$$

Using the relations, we have

$$\begin{aligned}\lambda(p)\zeta_p(s) &= \lambda(p) + \sum_{\delta=1}^{\infty} \lambda(p)\lambda(p^\delta)v^\delta \\ &= \lambda(p) + \sum_{\delta=1}^{\infty} \lambda(p^{\delta+1})v^\delta + p^{k-1}\chi(p)\sum_{\sigma=1}^{\infty}\lambda(p^{\delta-1})v^\delta \\ &= (\zeta_p(s) - 1)v^{-1} + p^{k-1}\chi(p)v\zeta_p(s),\end{aligned}$$

whence

$$\zeta_p(s) = (1 - \lambda(p)v + p^{k-1}\chi(p)v^2)^{-1}, \qquad (v = p^{-s}), \tag{4.3.20}$$

which along with (4.3.18) proves (4.3.16). □

Remark. The Euler expansion (4.3.16) follows also from the properties of the elements $\mathbf{T}(m) \in \mathscr{L}^n$ considered in § 3.3. In fact, the relations (4.3.17), which imply (4.3.18), follow directly from (3.3.20); the first identity of Proposition 3.3.35 implies the identity (4.3.20).

The expansions of Theorem 4.3.2 do not contain any new information on multiplicative properties of Fourier coefficients of modular forms in comparison with relations (4.3.2) or (4.3.4). But the relations allow us to express the zeta-function $\zeta(s, F)$ in terms of the form F, which makes it possible to study analytic properties of the zeta-function (see Exercise 4.3.11). For modular forms of genus $n > 1$, perhaps, there are no universal relations expressing individual Fourier coefficients of an eigenfunction of Hecke operators in terms of the corresponding eigenvalues or vice versa. Note that, for $n > 1$, the Fourier coefficients and the eigenvalues are even indexed by the sets, which have no visible connection, that is the set of classes under the equivalence (or proper equivalence) over $\mathbb{Z}$ of matrices in $\mathbb{A}_n$, on one hand, and the set of diagonal matrices of symplectic elementary divisors, on the other. In the general situation, however, it proves sometimes to be possible to express some Dirichlet series constructed from Fourier coefficients of eigenfunctions in terms of zeta-functions constructed from the corresponding eigenvalues. This reveals multiplicative properties of the Fourier coefficients and gives a way to attack analytic properties of the zeta-functions. Unfortunately, at present we can only generalize the relations of Theorem 4.3.2 to the case of modular forms of genus $n = 2$. The generalization will be considered in the next subsection. The following theorem contains more complicated relations, which, however, can be generalized to the case of arbitrary n. The generalization will be treated in the third subsection.

We first define a new Euler product associated to an eigenfunction $F \in \mathfrak{M}_k(q, \chi)$. Let $\lambda(T) = \lambda(T, F)$ with $T \in \mathsf{L}(q)$ be the corresponding eigenvalues, and let $p \in \mathbb{P}_{(q)}$.

Then the mapping

$$\mathbf{T} \to \lambda(\mathbf{T}) = \lambda(\mathrm{i}_q^{-1}(\mathbf{T})) \quad (T \in \mathbf{L}_p^1 = \mathrm{i}_q(\mathrm{L}_p^1(q)))$$

(see Lemma 3.3.26 and (3.3.45)) is a non-zero $\mathbb{Q}$-linear homomorphism of $\mathbf{L}_p^1$ into $\mathbb{C}$. We let $\alpha_0(p) = \alpha_0(p, F)$, $\alpha_1(p) = \alpha_1(p, F)$ be the parameters of the homomorphism in the sense of Proposition 3.3.36, which will be called *the local parameters* or *p-parameters of the eigenfunction F*. The formulae

$$\Omega(\mathbf{T}^1(p)) = x_0 + x_0 x_1, \quad \Omega([\mathbf{p}]_1) = p^{-1} x_0^2 x_1,$$

which follow from (3.3.58) and Lemma 3.3.34, and the definition of local parameters imply the relations

$$\alpha_0(p) + \alpha_0(p)\alpha_1(p) = \lambda(\mathbf{T}^1(p)) = \lambda(T(p)) = \lambda(p),$$

and (4.3.21)

$$\alpha_0(p)^2 \alpha_1(p) = \lambda([\mathbf{p}]_1) = \lambda([p]) = p^{k-1}\chi(p)$$

(see also (4.2.29)). In view of the relations, the local zeta-function (4.3.20) has the following expression in terms of the local parameters:

$$\zeta_p(s, F) = [(1 - \alpha_0(p)v)(1 - \alpha_0(p)\alpha_1(p)v)]^{-1}. \tag{4.3.22}$$

According to (4.3.19) and (4.3.8) (resp. (4.3.9), if F is a cusp form), the function $\zeta_p(s, F)$ is holomorphic in the disk $|v| < p^{-k}$ (resp. $|v| < p^{-k/2}$). Then the formula (4.3.22) implies that the local parameters satisfy

$$\max(|\alpha_0(p)|, |\alpha_0(p)\alpha_1(p)|) \le p^k \tag{4.3.23}$$

(resp.

$$\max(|\alpha_0(p)|, |\alpha_0(p)\alpha_1(p)|) \le p^{k/2}, \tag{4.3.24}$$

if F is a cusp form). We define *the even zeta-function* $\zeta^+(s, F)$ *of* F *by*

$$\zeta^+(s, F) = \prod_{p \in \mathbb{P}(q)} \zeta_p^+(s, F), \tag{4.3.25}$$

where *the local zeta-functions* $\zeta_p^+(s, F)$ *of* F are defined by

$$\zeta_p^+(s, F) = [(1 - \alpha_1(p)^{-1}\chi(p)v)(1 - \alpha_1(p)\chi(p)v)]^{-1}, \quad (v = p^{-s}). \tag{4.3.26}$$

According to (4.3.21), the denominator of the local zeta-function has the following expressions

$$\begin{aligned}
&(1 - \alpha_1(p)^{-1}\chi(p)v)(1 - \alpha_1(p)\chi(p)v) \\
&= (1 - \alpha_0(p)^2 p^{1-k} v)(1 - (\alpha_0(p)\alpha_1(p))^2 p^{1-k} v) \\
&= 1 - (\lambda(p)^2 - 2p^{k-1}\chi(p))p^{1-k} v + \chi(p^2)v^2 \\
&= 1 - (\lambda(p^2) - p^{k-1}\chi(p))p^{1-k} v + \chi(p^2)v^2
\end{aligned} \tag{4.3.27}$$

(see (4.3.6)), which, in particular, show that the definition of even zeta-functions is independent of the choice of local parameters.

Theorem 4.3.3. *In the notation of Theorem 4.3.2, each of the Dirichlet series*

$$D^+(s,a,F) = \sum_{m\in\mathbb{N}_{(q)}} \frac{f(2am^2)}{m^{s+k-1}} \tag{4.3.28}$$

with $a\in\mathbb{N}$, and the Euler product $\zeta^+(s,F)$ converge absolutely and uniformly in every right half-plane of the form $\operatorname{Re} s \geq k+2+\varepsilon$ *(resp.* $\operatorname{Re} s \geq 2+\varepsilon$*, if F is a cusp form) with $\varepsilon>0$ and satisfy the relation*

$$D^+(s,a,F) = \left(\sum_{d\in\mathbb{N}_{(q)},\, d^2|a} \chi(d^2)\mu(d)f(2a/d^2)d^{k-s-1}\right)\left(\prod_{p\in\mathbb{P}_{(qa)}}(1+\chi(p)p^{-s})\right)\zeta^+(s,F). \tag{4.3.29}$$

Proof. The estimates (2.3.20) and (2.3.43) (resp. (4.3.23) and (4.3.24)) imply that the series $D^+(s,a,F)$ (resp. the product $\zeta^+(s,F)$) converges absolutely and uniformly in the half-planes (for the product, see the second formula in (4.3.27) for denominators of p-factors).

Let $m\in\mathbb{N}_{(q)}$, and let δ^2 be a common square divisor of m^2 and a. Then, by (4.3.2), we have

$$\sum_{d|m^2/\delta^2,\, a/\delta^2} d^{k-1}\chi(d) f\left(\frac{2m^2a}{\delta^4 d^2}\right) = \lambda(m^2/\delta^2)f(2a/\delta^2).$$

Multiplying both sides by $\delta^{2(k-1)}\chi(\delta^2)\mu(\delta)$, μ being the Möbius function, and summing up over all $\delta\in\mathbb{N}$ with $\delta^2|m^2$, a, we obtain

$$\begin{aligned}&\sum_{d\delta^2|m^2,\, a} (d\delta^2)^{k-1}\chi(d\delta^2)f(2m^2a/(d\delta^2)^2)\mu(\delta)\\ &\quad = \sum_{\delta^2|m^2,\, a} \delta^{2(k-1)}\chi(\delta^2)\mu(\delta)\lambda(m^2/\delta^2)f(2a/\delta^2).\end{aligned}$$

If $\Delta = \Delta_0 t$ with $\Delta_0, t\in\mathbb{N}$, where Δ_0 is square-free, then, by (4.3.3), one may write the relations

$$\sum_{d\delta^2=\Delta} \mu(\delta) = \sum_{\delta|t}\mu(\delta) = \begin{cases}1, & \text{if } \Delta=\Delta_0\\ 0, & \text{if } \Delta\neq\Delta_0.\end{cases} \tag{4.3.30}$$

Hence, the last relation can be written in the form

$$\sum_{\substack{\Delta|m^2,\, a\\ \Delta \text{ is square-free}}} \Delta^{k-1}\chi(\Delta)f(2m^2a/\Delta^2) = \sum_{\delta^2|m^2,\, a} \delta^{2(k-1)}\chi(\delta^2)\mu(\delta)\lambda(m^2/\delta^2)f(2a/\delta^2).$$

Dividing both sides by m^{s+k-1} and summing up over all $m\in\mathbb{N}_{(q)}$, we get a formal identity

$$\begin{aligned}&\sum_{\substack{m\in\mathbb{N}_{(q)};\, \Delta|m^2,\, a\\ \Delta \text{ is square-free}}} m^{-(s+k-1)}\Delta^{k-1}\chi(\Delta)f(2m^2a/\Delta^2)\\ &\quad = \sum_{m\in\mathbb{N}_{(q)};\, \delta^2|m^2,\, a} m^{-s-k+1}\delta^{2(k-1)}\chi(\delta^2)\mu(\delta)\lambda(m^2/\delta^2)f(2a/\delta^2).\end{aligned} \tag{4.3.31}$$

Since the condition $\Delta|m^2$ with a square-free Δ is equivalent to $\Delta|m$, it follows that the m on the left side in (4.3.31) can be replaced by $m\Delta$, which gives us the series

$$\sum_{\substack{m,\Delta\in\mathbb{N}_{(q)},\Delta|a,\\ \Delta \text{ is square-free}}} (m\Delta)^{-s-k+1}\Delta^{k-1}\chi(\Delta)f(2m^2a)$$

$$= \left(\sum_{\substack{\Delta\in\mathbb{N}_{(q)},\Delta|a,\\ \Delta \text{ is square-free}}} \Delta^{-s}\chi(\Delta)\right)\sum_{m\in\mathbb{N}_{(q)}} f(2m^2a)/m^{s+k-1}$$

$$= \left\{\prod_{p\in\mathbb{P}_{(q)},\,p|a}(1+\chi(p)p^{-s})\right\}D^+(s,a,F).$$

Since the condition $\delta^2|m^2$ is equivalent to $\delta|m$, it follows that the m on the right side in (4.3.31) can be replaced by $m\delta$, which gives us the series

$$\sum_{m,\delta\in\mathbb{N}_{(q)},\delta^2|a} (m\delta)^{-s-k+1}\delta^{2(k-1)}\chi(\delta^2)\mu(\delta)\lambda(m^2)f(2a/\delta^2)$$

$$= \sum_{\delta\in\mathbb{N}_{(q)},\delta^2|a} \delta^{-s+k-1}\chi(\delta^2)\mu(\delta)f(2a/\delta^2)\sum_{m\in\mathbb{N}_{(q)}}\frac{\lambda(m^2)}{m^{s+k-1}}.$$

Further, by (4.3.17), we obtain the Euler expansion

$$\sum_{m\in\mathbb{N}_{(q)}}\frac{\lambda(m^2)}{m^{s+k-1}} = \prod_{p\in\mathbb{P}_{(q)}}\sum_{\beta=0}^{\infty}\lambda(p^{2\beta})/p^{\beta(s+k-1)}. \tag{4.3.32}$$

To sum up the p-factors of this product, we note that

$$\sum_{\beta=0}^{\infty}\lambda(p^\beta)v^\beta = [(1-\alpha_0(p)v)(1-\alpha_0(p)\alpha_1(p)v)]^{-1},$$

by (4.3.22), whence

$$\sum_{\beta=0}^{\infty}\lambda(p^{2\beta})v^{2\beta} = \tfrac{1}{2}\{[(1-\alpha_0(p)v)(1-\alpha_0(p)\alpha_1(p)v)]^{-1}$$
$$+ [(1+\alpha_0(p)v)(1+\alpha_0(p)\alpha_1(p)v)]^{-1}\}$$
$$= (1+\chi(p)p^{k-1}v^2)[(1-\alpha_0(p)^2v^2)(1-(\alpha_0(p)\alpha_1(p))^2v^2)]^{-1}.$$

Thus, the relation (4.3.32) can be written in the form

$$\sum_{m\in\mathbb{N}_{(q)}}\frac{\lambda(m^2)}{m^{s+k-1}}$$
$$= \prod_{p\in\mathbb{P}_{(q)}}\left(1+\frac{\chi(p)}{p^s}\right)\left[\left(1-\frac{\alpha_0(p)^2p^{1-k}}{p^s}\right)\left(1-\frac{(\alpha_0(p)\alpha_1(p))^2p^{1-k}}{p^s}\right)\right]^{-1}$$
$$= \left\{\prod_{p\in\mathbb{P}_{(q)}}\left(1+\frac{\chi(p)}{p^s}\right)\right\}\zeta^+(s,F).$$

The formulae obtained above prove (4.3.29). □

The reader will find another proof of (4.3.29), which is based on factorization of suitable symplectic polynomials, in the proof of Theorem 4.3.19.

Exercise 4.3.4. *Show that*

(1) *A form* $F \in \mathfrak{M}_k(q, \chi)$ *is an eigenfunction for all operators* $|T$ *with* $T \in \mathsf{L}(q)$ *if it is true for all* T *of the form* $T = T(p)$ *with* $p \in \mathbb{P}_{(q)}$;

(2) *Theorem* 4.3.3 *remains correct if one only assumes that* F *is an eigenfunction for all operators* $|T(p^2)$ *with* $p \in \mathbb{P}_{(q)}$.

Exercise 4.3.5. *Show that the eigenvalues* $\lambda(m)$ *of operators* $|T(m)$ *corresponding to the eigenfunction* $G_k \in \mathfrak{M}_k$ *(see Exercises* 2.5.6 *and* 4.1.11) *are given by*

$$\lambda(m) = \sum_{d|m} d^{k-1}, \qquad (m \in \mathbb{N});$$

Use this to prove that the Fourier coefficients $f(2a)$ *of* G_k *can be written in the form*

$$f(2a) = f(2) \sum_{d|a} d^{k-1}, \qquad (a \in \mathbb{N}).$$

[Hint: Show that $f(0) = \lim_{z \to i\infty} f(z) \neq 0$, and then use the formulae (4.3.10)–(4.3.12).]

Exercise 4.3.6. *Let* $F_1, \ldots, F_h$ *be a basis for a subspace of* $\mathfrak{M}_k(q, \chi)$, *which is invariant under Hecke operators* $|T$ *with* $T \in \mathsf{L}(q)$, *so that*

$$F_i|T = \sum_{j=1}^{h} \lambda_{ij}(T) F_j \qquad (i = 1, \ldots, h).$$

We let $f(2a)$ *with* $a = 0, 1, \ldots$ *be the h-column, whose i-th entry is the Fourier coefficient* $f_i(2a)$ *of* F_i $(i = 1, \ldots, h)$, *and let* $\Lambda(m)$ *with* $m \in \mathbb{N}_{(q)}$ *be the* $(h \times h)$*-matrix with entries* $\lambda_{ij}(T(m))$. *Prove the following relations*:

$$\sum_{d|m,a} d^{k-1} \chi(d) f(2ma/d^2) = \Lambda(m) f(2a), \qquad (m \in \mathbb{N}_{(q)}, a = 0, 1, \ldots);$$

$$f(2ma) = \sum_{d|m,a} d^{k-1} \chi(d) \mu(d) \Lambda(m/d) f(2a/d), \qquad (m \in \mathbb{N}_{(q)}, a = 0, 1, \ldots);$$

$$\Lambda(m)\Lambda(m_1) = \sum_{d|m,m_1} d^{k-1} \chi(d) \Lambda(mm_1/d^2), \qquad (m, m_1 \in \mathbb{N}_{(q)});$$

$$\Lambda(mm_1) = \sum_{d|m,m_1} d^{k-1} \chi(d) \mu(d) \Lambda(m/d) \Lambda(m_1/d), \qquad (m, m_1 \in \mathbb{N}_{(q)});$$

$$\sum_{m \in \mathbb{N}_{(q)}} m^{-s} f(2am) = \left(\sum_{m \in \mathbb{N}_{(q)}} m^{-s} \Lambda(m) \right) \left(\sum_{d \in \mathbb{N}_{(q)}, d|a} \chi(d) \mu(d) d^{k-1-s} f(2a/d) \right);$$

$$\sum_{m \in \mathbb{N}_{(q)}} m^{-s} \Lambda(m) = \prod_{p \in \mathbb{P}_{(q)}} (E_h - p^{-s} \Lambda(p) + \chi(p) p^{k-1-2s} E_h)^{-1};$$

where the last two relations are considered as formal identities. Formulate and prove a matrix form of (4.3.29).

Exercise 4.3.7. *In the notation of Exercise* 4.3.6, *show that, for $k > 0$, the estimates hold*:

$$\max_{i,j} |\lambda_{ij}(T(m))| \leq cm^k, \qquad (m \in \mathbb{N}_{(q)}),$$

and

$$\max_{i,j} |\lambda_{ij}(T(m))| \leq cm^{k/2}, \qquad (m \in \mathbb{N}_{(q)}),$$

if $F_1, \ldots, F_h$ are cusp forms, where c is independent of m. Use the estimates to investigate convergence of the matrix series and products of Exercise 4.3.6.

[Hint: If $a_1, \ldots, a_h \in \mathbb{N}$ are chosen in such a way that the matrix $A = (f_i(2a_j))$ is invertible, then $\Lambda(m) = B(m)A^{-1}$, where $B(m)$ is the matrix with columns $\sum_{d|m, a_j} d^{k-1}\chi(d)f(2ma_j/d^2)$.]

Exercise 4.3.8. *Let $\mathsf{q}(x) = \mathsf{q}(x_1, \ldots, x_m)$ be an integral positive definite quadratic form with the matrix* (1.1.3) *of determinant* 1. *Prove that there are functions $\lambda_1, \ldots, \lambda_h$: $\mathbb{N} \to \mathbb{C}$, where $h = h(q) < \infty$, satisfying*

$$\lambda_i(ab) = \sum_{d|a,b} d^{m/2-1}\lambda_i(ab/d^2), \qquad (a, b \in \mathbb{N}),$$

and such that the number $r(q, a)$ of integral solutions of the equation $\mathsf{q}(X) = a$ can be written in the form

$$r(\mathsf{q}, a) = \sum_{i=1}^{h} \alpha_i \lambda_i(a), \qquad \textit{for all } a \in \mathbb{N},$$

with constant coefficients α_i. Use this to show that any power series of the form

$$\sum_{\delta=0}^{\infty} r(q, ap^\delta)v^\delta$$

with $a \in \mathbb{N}$ and $p \in \mathbb{P}$ is equal to a rational fraction with the polynomial

$$\prod_{i=1}^{h} (1 - \lambda_i(p)v + p^{(m/2)-1}v^2)$$

as a denominator.

Exercise 4.3.9. *Let*

$$F(z) = \sum_{a=0}^{\infty} f(2a)e^{2\pi i a z} \in \mathfrak{M}_k(q, \chi).$$

Show that, for $\operatorname{Re} s > k$, *the following integral representation holds*:

$$\psi(s, F) = (2\pi)^{-s}\Gamma(s) \sum_{a \in \mathbb{N}} \frac{f(2a)}{a^s} = \int_0^\infty t^{s-1}(F(it) - f(0))\,dt$$

[Hint: Use the integral representation of the gamma-function

$$\Gamma(s) = \int_0^\infty t^{s-1}e^{-t}\,dt \qquad (\operatorname{Re} s > 0).]$$

Exercise 4.3.10. *In the notation of Exercise 4.3.9, let us assume that* $q = 1$ *and* $\chi = 1$. *Prove then the following formula*

$$\psi(s, F) = \int_1^\infty t^{s-1}(F(it) - f(0))\,dt$$

$$+ i^k \int_1^\infty t^{k-s-1}(F(it) - f(0))\,dt - f(0)\left(\frac{i^k}{k-s} + \frac{1}{s}\right), \qquad (\mathrm{Re}\, s > k).$$

Use the formula to prove the following assertions: the function $\Psi(s, F)$ *can be meromorphically continued to all of the s-plane; the function*

$$\psi(s, F) + f(0)\left(\frac{i^k}{k-s} + \frac{1}{s}\right)$$

is an entire function; $\psi(s, F)$ *satisfies the functional equation*

$$\psi(k - s, F) = i^k \psi(s, F).$$

[Hint: Write the integral of Exercise 4.3.9. as the sum of integrals over $[1, \infty]$ and $[0, 1]$. Then, in the second integral replace t by $1/t$ and use the relation $F(i/t) = F(-1/it) = (it)^k F(it)$.]

Exercise 4.3.11. *Prove that the zeta-function* $\zeta(s, F)$ *of a nonzero eigenfunction* F *of* $\mathfrak{M}_k$ *has the following properties*:

(1) *The function* $\zeta(s, F)$ *can be meromorphically continued to all of the s-plane*;

(2) *The function*

$$\varphi(s, F) + \frac{f(0)}{f(2)}\left(\frac{i^k}{k-s} + \frac{1}{s}\right) \qquad \text{with } \varphi(s, F) = (2\pi)^{-s}\Gamma(s)\zeta(s, F),$$

where $f(0), f(2), \ldots$ *are the Fourier coefficients of* F, *is an entire function*;

(3) $\zeta(s, F)$ *satisfies the functional equation*

$$\varphi(k - s; F) = i^k \varphi(s, F).$$

Exercise 4.3.12. *Prove the following assertions*:

(1) $H_q^{-1}\Gamma_0^1(q)H_q = \Gamma_0^1(q)$, *where* $H_q = \begin{pmatrix} 0 & -1 \\ q & 0 \end{pmatrix}$;

(2) *The mapping* $F \to F|_k H_q = (qz)^{-k}F(-1/qz)$ *defines an isomorphism of* $\mathfrak{M}_k(q, \chi)$ *onto* $\mathfrak{M}_k(q, \bar{\chi})$.

(3) *If the character* χ *is real, then*

$$\mathfrak{M}_k(q, \chi) = \mathfrak{M}_k^+(q, \chi) \oplus \mathfrak{M}_k^-(q, \chi),$$

where $\mathfrak{M}_k^\pm(q, \chi) = \{F \in \mathfrak{M}_k(q, \chi); F|_k H_q = \pm q^{-k/2} i^k F\}$;

(4) *If* $F \in \mathfrak{M}_k(q, \chi)$ *and* $m \in \mathbb{N}_{(q)}$, *then*

$$(F|_k H_q)|_{k,\bar{\chi}} T(m) = \bar{\chi}(m)(F|_{k,\chi} T(m))|_k H_q, \qquad (m \in \mathbb{N}_{(q)}),$$

in particular, the subspaces $\mathfrak{M}_k^{\pm}(q, \chi)$ *with a real character* χ *are invariant under all of the operators* $|T(m)$;

(5) *Let* $F \in \mathfrak{M}_k^{\pm}(q, \chi)$ *with a real character* χ. *In the notation of Exercise* 4.3.9, *show that, for* $\operatorname{Re} s > k$,

$$\Psi(s, F) = \int_{q^{-1/2}}^{\infty} t^{s-1}(F(it) - f(0))\, dt$$

$$\pm q^{(k/2)-s} \int_{q^{-1/2}}^{\infty} t^{k-s-1}(F(it) - f(0))\, dt - f(0)q^{-s/2}\left(\frac{1}{s} \pm \frac{1}{k-s}\right).$$

Use this to prove that the function $\psi(s, F)$ *can be meromorphically continued to all of the s-plane, the function*

$$\psi(s, F) + f(0)q^{-s/2}\left(\frac{1}{s} \pm \frac{1}{k-s}\right)$$

is an entire function, and $\psi(s, F)$ *satisfies the functonal equation*

$$\psi(k - s, F) = \pm q^{s-k/2}\psi(s, F).$$

(6) *Let* $F \in \mathfrak{M}_k(q, \chi)$, *and let* $f(0), f(2), \ldots$ *be Fourier coefficients of* F. *Let us assume that* F *satisfies* $F|T(m) = \lambda(m)F$ *for all* $m \in \mathbb{N}_{(q)}$. *Then the following identity holds*

$$\sum_{a \in \mathbb{N}} f(2a)a^{-s} = \left(\sum_{a \in \mathbb{N}, a|q^\infty} f(2a)a^{-s}\right)\left(\sum_{m \in \mathbb{N}_{(q)}} \lambda(m)m^{-s}\right), \qquad (\operatorname{Re} s > k).$$

4.3.2. Modular Forms of Genus 2, Gaussian Composition, and Zeta-Functions

In this subsection we shall obtain relations similar to (4.3.14) in the case of modular forms of genus 2. The specific feature of forms of genus 2 is that their Fourier coefficients are indexed by matrices of binary quadratic forms which can sometimes be multiplied in the sense of Gaussian composition of quadratic forms. This proves to be of fundamental importance to an understanding of the multiplicative nature of the Fourier coefficients. At present no satisfactory generalization of the relations (4.3.14) to modular forms of genus greater than 2 is known.

Let

$$F(Z) = \sum_{A \in \mathbb{A}_2} f(A)e\{AZ\} \in \mathfrak{M}_k^2(q, \chi)$$

be a modular form of integral weight k and one-dimensional character $[\chi]$ for the group $\Gamma_0^2(q)$. We assume that F is an eigenfunction for all of the operators

$|T = |_{k,\chi}T$ with $T \in \mathsf{L}^2(q)$, say

$$F|T = \lambda(T)F,$$

where $\lambda(T) = \lambda(T, F)$ are the corresponding eigenvalues.

In the following theory the part of the Dirichlet series constructed from Fourier coefficients of the eigenfunction F will be played by linear combinations of the series of the form

$$D(s, A, \gamma, F) = \sum_{m \in \mathbb{N}_{(q)}} \frac{f(mA)\gamma(m)}{m^s} \tag{4.3.33}$$

with $A \in \mathbb{A}_2^+$, where $\gamma\colon \mathbb{N}_{(q)} \to \mathbb{C}$ is a *completely multiplicative function*, i.e. a function satisfying

$$\gamma(mm_1) = \gamma(m)\gamma(m_1) \qquad \text{for all } m, m_1 \in \mathbb{N}_{(q)}. \tag{4.3.34}$$

The part of the Dirichlet series constructed from the eigenvalues will be played by the Euler products of the form

$$\zeta(s, \gamma, F) = \prod_{p \in \mathbb{P}_{(q)}} Q_p(\gamma(p)p^{-s}; F)^{-1} \tag{4.3.35}$$

with the same γ as above, where, for $p \in \mathbb{P}_{(q)}$, we set

$$Q_p(v, F) = \sum_{j=0}^{4} (-1)^j \lambda(q_j(p))v^j, \tag{4.3.36}$$

$q_j(p) = q_j^2(p)$ being the inverse images in $\mathsf{L}_p^2(q)$ of the elements (3.3.77) under the isomorphism $\mathsf{i}_q\colon \mathsf{L}_p^2(q) \to \mathbf{L}_p^2$ (see Lemma 3.3.26). We call $\zeta(s, \gamma, F)$ *the zeta-function of F with the "character"* γ; the p-factors

$$\zeta_p(s, \gamma, F) = Q_p(\gamma(p)p^{-s}, F)^{-1} \tag{4.3.37}$$

are called *the local zeta-functions of F with the "character"* γ.

Our journey in search of "global" relations between the series (4.3.33) and the products (4.3.35) we start from corresponding "local" relations, i.e. relations between the power series of the form

$$D_p(s, A, \gamma, F) = \sum_{\delta=0}^{\infty} f(p^\delta A)v^\delta \qquad \text{with } v = \gamma(p)p^{-s} \tag{4.3.38}$$

and the local zeta-functions $\zeta_p(s, \gamma, F)$. We shall make use of the techniques of Chapter 3 based on calculations in certain extensions of Hecke rings of the group $\Gamma_0^2(q)$. Therefore, according to the ideology of §4.2.2, we consider the space $\mathfrak{M}_k^2(q, \chi)$ as an subspace of $\mathfrak{M}_\xi^2$ with $\xi(-1) = (-1)^k\chi(-1)$, which is invariant under all Hecke operators of $\mathbf{L}^2(q) = \mathsf{i}_q(\mathsf{L}^2(q))$, and we consider the space of Fourier coefficients $\mathfrak{F}_k^2(q, \chi)$ as an invariant subspace of $\mathfrak{F}_\xi^n$. Besides, we systematically make use of the notation (4.2.88) and the relations (4.2.89). We fix q, k, χ, and ξ with $\xi(-1) = (-1)^k\chi(-1)$.

Lemma 4.3.13. *Let*

$$F = \sum_{A \in \mathbb{A}_2} f(A) e\{AZ\} \in \mathfrak{M}_\xi^2$$

satisfies $F|_{k,\chi} T = \lambda(T) F$ *for all* $T \in \mathbf{L}_p^2$ *with a prime p, then, for each* $B \in \mathbb{A}_2$, *the power series*

$$d_p(v, B) = \sum_{\delta=0}^{\infty} f(p^\delta B) v^\delta$$

is formally equal to the rational fraction

$$Q_p(v, F)^{-1}(f|_{k,\chi}(1 - \Pi_- v)(1 - \Pi v + p(\Pi_{2,0}^{(1)} + \Pi_{2,0}^{(0)})v^2))(B),$$

where, similarly to (4.3.36), *we set*

$$Q_p(v, F) = \sum_{j=0}^{4} (-1)^j \lambda(\mathbf{q}_j^2(p)) v^j, \tag{4.3.39}$$

and where $\Pi_- = \Pi_-^2(p) = \Pi_0^2(p)$, $\Pi = \Pi_1^2(p)$, *and* $\Pi_{2,0}^{(r)} = \Pi_{2,0}^{(r)}(p)$ *are the elements defined in* (3.3.59) *and* (3.3.62) *with* $n = 2$.

Proof. The function $f \in \mathfrak{F}_\xi^2$ is clearly an eigenfunction for all Hecke operators of $\mathbf{L}_p^2$ with the same eigenvalues as F. Hence, for every B' of $\mathbb{A}_2$, we may write

$$\begin{aligned} Q_p(v, F) f(B') &= \sum_{j=0}^{4} (-1)^j (f|\mathbf{q}_j(p))(B') v^j \\ &= (f|\mathbf{Q}_p(v))(B'), \end{aligned}$$

where $\mathbf{Q}_p = \mathbf{Q}_p^2$ is the polynomial (3.3.78). Further, by (4.2.28), for any $g \in \mathfrak{F}_\xi^2$ and $B \in \mathbb{A}_2$, one can write

$$g(p^\delta B) = (g|\Pi_2^2(p^\delta))(B) = (g|\Pi_+^\delta)(B), \qquad (\delta \geq 0)$$

with $\Pi_+ = \Pi_+^2(p) = \Pi_2^2(p)$. Using these formulae and the factorization of the polynomial $\mathbf{Q}_p^2(v)$ of Proposition 3.5.14, we have

$$\begin{aligned} Q_p(v, F) d_p(v, B) &= \sum_{j,\delta} (-1)^j (f|\mathbf{q}_j^2(p))(p^\delta B) v^{j+\delta} \\ &= \sum_{j,\delta} (-1)^j (f|\mathbf{q}_j^2(p)|\Pi_+^\delta)(B) v^{j+\delta} \\ &= \left(f|\mathbf{Q}_p^2(v)|\sum_{\delta=0}^{\infty} \Pi_+^\delta v^\delta\right)(B) \\ &= (f|(1 - \Pi_- v)(1 - \Pi v + p(\Pi_{2,0}^{(1)} + \Pi_{2,0}^{(0)})v^2))(B). \qquad \square \end{aligned}$$

As we have seen, each of the series $d_p(v, B)$ is a rational fraction with the common denominator $Q_p(v, F)$. In order to calculate their numerators, we need some formulae of § 4.2, which we now list. First, for $f \in \mathfrak{F}_\xi^2$, by (4.2.28) with $n = 2$ and $a = 0, 1$ we have

$$(f|\Pi_-)(B) = \begin{cases} p^{2k-3}\chi(p)^2 f(p^{-1}B), & \text{if } B \in p\mathbb{A}_2 \\ 0, & \text{if } B \notin p\mathbb{A}_2 \end{cases} \tag{4.3.40}$$

and

$$\begin{aligned}(f|\Pi)(B) &= p^{2k-2}\chi(p)^2 \sum_{\substack{D \in \Lambda\backslash\Lambda D_1^2(p)\Lambda,\\ B[{}^tD] \in p\mathbb{A}_2}} \bar{\chi}(\det D)(\det D)^{-k} f(p^{-1}B[{}^tD]) \\ &= p^{k-2}\chi(p) \sum_{\substack{D \in \Lambda_+\backslash\Lambda_+ D_1^2(p)\Lambda_+,\\ B[{}^tD] \in p\mathbb{A}_2}} f(p^{-1}B[{}^tD])\end{aligned} \tag{4.3.41}$$

where $\Lambda_+ = \Lambda_+^2 = \mathsf{SL}_2(\mathbb{Z})$. Further, by Lemma 4.2.23, we have

$$(f|\Pi_{2,0}^{(1)})(B) = p^{2(k-3)}\chi(p)^2 s_p(1,2;B) f(B) \tag{4.3.42}$$

with

$$s_p(1,2;B) = \sum_{S \in \mathsf{SM}_p(1,2)} e\{p^{-1}BS\},$$

and

$$(f|\Pi_{2,0}^{(0)})(B) = (f|[\mathbf{p}])(B) = \mathrm{p}^{2(k-3)}\chi(p)^2 f(B). \tag{4.3.43}$$

To calculate the sum $s_p(1,2;B)$, we make use of the geometric language of Appendix 2. By Lemma 4.2.25, one can write

$$\begin{aligned}s_p(1,2;B) &= (p-1)^{-1}(p\cdot \mathrm{pm}_p(1,2;B) - (p^2-1)\,\mathrm{pm}_p(0,2;B)) \\ &= (p-1)^{-1}p\cdot \mathrm{pm}_p(1,2;B) - (p+1),\end{aligned}$$

where, for $B = \begin{pmatrix} 2b_1 & b_2 \\ b_2 & 2b_3 \end{pmatrix}$, $\mathrm{pm}_p(1,2;B)$ is the number of non-trivial solutions of the congruence

$$\mathsf{b}(x,y) = b_1x^2 + b_2xy + b_3y^2 \equiv 0 \pmod p. \tag{4.3.44}$$

By Lemma 4.2.26, the last number equals the number $i(\mathsf{V}_p, f_p, 1)$ of isotropic vectors in the two-dimensional quadratic space (V_p, f_p) over $\mathsf{F} = \mathbb{F}_p$ with the quadratic form $\mathsf{b}_p \equiv \mathsf{b}(x,y)$ modulo p. If the space (V_p, f_p) is non-degenerate, then, by (A.2.24), $i(\mathsf{V}_p, f_p, 1) = (p-\varepsilon)(1+\varepsilon)$, where $\varepsilon = \varepsilon(\mathsf{V}_p, f_p)$ is the sign (A.2.21) of (V_p, f_p). Since $\varepsilon = \pm 1$, it follows that

$$i(\mathsf{V}_p, f_p, 1) = (\varepsilon + 1)(p-1). \tag{4.3.45}$$

This formula remains true for two-dimensional degenerate spaces over F if we define their "*sign*" by

$$\varepsilon = \varepsilon(\mathsf{V}_p, f_p) = \begin{cases} 0, & \text{if } (\mathsf{V}_p, f_p) \text{ is degenerate, but } f_p \neq 0 \\ p, & \text{if } f_p = 0. \end{cases} \tag{4.3.46}$$

In fact, in the first case the space (V_p, f_p) is a sum of two one-dimensional subspaces, one of which has the zero quadratic form and other is non-degenerate, and so each of the isotropic vectors must belong to the first subspace and their number is $p-1$; in the second case each of the non-zero vectors is isotropic, and so their number is p^2-1. In the above notation, we define *the p-sign* $\varepsilon_p(B)$ *of a matrix* $B \in \mathbb{A}_2$ by

$$\varepsilon_p(B) = \varepsilon(\mathsf{V}_p, f_p). \tag{4.3.47}$$

Then we can write the formula

$$s_p(1,2;B) = p(\varepsilon_p(B)+1) - (p+1) = p\varepsilon_p(B) - 1.$$

Hence, by (4.3.42) and (4.3.43), we have finally that

$$(f|(\Pi_{2,0}^{(1)} + \Pi_{2,0}^{(0)}))(B) = p^{2k-5}\chi(p)^2\varepsilon_p(B)f(B). \tag{4.3.48}$$

With the above formulae, in order to make use of the identities of Lemma 4.3.13, we only have to understand the meaning of the right side in (4.3.41). If $d = \det B \neq 0$, then each of the matrices $p^{-1}B[{}^tD]$ appearing in (4.3.41) is the matrix of an integral positive definite binary quadratic form of discriminant $\Delta = -d$, which are naturally related with modules in the imaginary quadratic field $\mathbb{Q}(\Delta^{1/2})$ (see (A.3.3)). This phenomenon will allow us to interpret the right side in (4.3.41) in terms of Gaussian composition of quadratic forms.

We first examine the set of summation in (4.3.41). By Lemma 3.1.2, one can take

$$\Lambda_+ \backslash \Lambda_+ D_1^2(p)\Lambda_+ = \left\{D_1^2(p)\begin{pmatrix} u_1 & u_2 \\ v_1 & v_2\end{pmatrix} = \begin{pmatrix} u_1 & u_2 \\ pv_1 & pv_2\end{pmatrix}; \begin{pmatrix} u_1 & u_2 \\ v_1 & v_2\end{pmatrix} \in \Gamma\backslash\Lambda_+\right\},$$

where

$$\Gamma = \Lambda_+ \cap D_1^2(p)^{-1}\Lambda_+ D_1^2(p) = \left\{\begin{pmatrix} \alpha & \beta \\ \gamma & \delta\end{pmatrix} \in \Lambda_+, \beta \equiv 0 \pmod p\right\} \tag{4.3.49}$$

It is clear that two matrices of Λ_+ with the first rows (u_1, u_2) and (u_1', u_2') belong to one left coset modulo Γ if and only if $u_1'u_2 \equiv u_2'u_1 \pmod p$, which means that the rows (u_1, u_2), (u_1', u_2') are proportional modulo p in the sense that

$$(u_1', u_2') \equiv \alpha(u_1, u_2) \pmod p \quad \text{with a } \alpha \not\equiv 0 \pmod p. \tag{4.3.50}$$

Thus, one can take

$$\Lambda_+ \backslash \Lambda_+ D_1^2(p)\Lambda_+ = \left\{\begin{pmatrix} 1 & 0 \\ 0 & p\end{pmatrix}U; U = \begin{pmatrix} u_1 & u_2 \\ v_1 & v_2\end{pmatrix} \in \Lambda_+, (u_1, u_2) \in \mathbb{P}^1(\mathbb{Z}/p\mathbb{Z})\right\}, \tag{4.3.51}$$

where $\mathbb{P}^1(\mathbb{Z}/p\mathbb{Z})$ stands for a system of representatives for classes of all pairs of coprime integers modulo the equivalence (4.3.50) (the projective line over $\mathbb{Z}/p\mathbb{Z}$).

Let, now, $B = \begin{pmatrix} 2b_1 & b_2 \\ b_2 & 2b_3\end{pmatrix} \in \mathbb{A}_2$, and let $\mathsf{b}(x,y) = b_1x^2 + b_2xy + b_3y^2$ be the corresponding quadratic form. If $D = \begin{pmatrix} 1 & 0 \\ 0 & p\end{pmatrix}U$ with $U = \begin{pmatrix} u_1 & u_2 \\ v_1 & v_2\end{pmatrix}$, then we easily obtain that

$$B[{}^tD] = \begin{pmatrix} 2\mathsf{b}(u_1, u_2) & pb_2(U) \\ pb_2(U) & 2p^2\mathsf{b}(v_1, v_2) \end{pmatrix} \tag{4.3.52}$$

with $b_2(U) = 2b_1u_1v_1 + b_2(u_1v_2 + u_2v_1) + 2b_3u_2v_2$.

In particular, the condition $B[{}^tD] \in p\mathbb{A}_2$ with an integral matrix $U = \begin{pmatrix} u_1 & u_2 \\ v_1 & v_2 \end{pmatrix}$ is equivalent to the congruence $\mathsf{b}(u_1, u_2) \equiv 0 \pmod p$. Thus, we can rewrite the formula (4.3.41) in the form

$$(f|\Pi)(B) = p^{k-2}\chi(p) \sum_{\substack{(u_1, u_2) \in \mathbb{P}^1(\mathbb{Z}/p\mathbb{Z}), \\ \mathsf{b}(u_1, u_2) \equiv 0 \pmod p}} f\left(\begin{pmatrix} 2\mathsf{b}(u_1, u_2)/p & b_2(U) \\ b_2(U) & 2p\mathsf{b}(v_1, v_2) \end{pmatrix}\right) \tag{4.3.53}$$

where, for a pair of coprime integers (u_1, u_2), (v_1, v_2) is a pair satisfying $U = \begin{pmatrix} u_1 & u_2 \\ v_1 & v_2 \end{pmatrix} \in \Lambda_+$.

The next proposition interprets the right side in (4.3.53) in terms of the composition (A.3.17) of matrices of quadratic forms and modules. This is the corner-stone of the following consideration.

For brevity, we call complete modules in an algebraic number field contained in their ring of multipliers the regular ideals of the ring.

Proposition 4.3.14. *Let* $A = \begin{pmatrix} 2a & b \\ b & 2c \end{pmatrix}$ *be a matrix of* $\mathbb{A}_2^+$ *with* g.c.d.$(a, b, c) = 1$, *and let* $\Delta = b^2 - 4ac$ *be the discriminant of* A. *We let* d *denote the discriminant of the imaginary quadratic field* $\mathsf{F} = \mathbb{Q}(\Delta^{1/2})$ *and let* O_l *be the subring of the index* $l = (\Delta/d)^{1/2}$ *of the ring of all integral numbers in* F. *Then, for every* $p \in \mathbb{P}$, $m \in \mathbb{N}_{(p)}$, *and* $f \in \mathfrak{F}_\xi^2$, *the following assertions hold*:

(1) *If the p-sign* $\varepsilon_p(A)$ *of* A *equals* 1, *then the ring* O_l *contains precisely two regular ideals of norm* p, *say* $\mathfrak{p}$ *and* $\mathfrak{p}'$, *the ideals are conjugate with each other*, $\mathfrak{p}' = \bar{\mathfrak{p}}$, *and the formula holds*

$$(f|_{k,\chi}\Pi_1^2(p))(mA) = p^{k-2}\chi(p)(f(m\cdot(A \times \mathfrak{p})) + f(m(A \times \bar{\mathfrak{p}})));$$

(2) *If* $\varepsilon_p(A) = -1$, *then the ring* O_l *contains no regular ideals of norm* p, *and the formula holds*

$$(f|_{k,\chi}\Pi_1^2(p))(mA) = 0;$$

(3) *If* $\varepsilon_p(A) = 0$ *and* $p \nmid l$, *then there is a single regular ideal* $\mathfrak{p}$ *of* O_l *of norm* p, *the ideal satisfies* $\bar{\mathfrak{p}} = \mathfrak{p}$, *and the formula holds*

$$(f|_{k,\chi}\Pi_1^2(p))(mA) = p^{k-2}\chi(p)f(m\cdot(A \times \mathfrak{p}));$$

(4) *If* $\varepsilon_p(A) = 0$ *and* $p|l$, *then the formula holds*

$$(f|_{k,\chi}\Pi_1^2(p))(mA) = p^{k-2}\chi(p)f(mp\cdot(A \times \mathsf{O}_{l/p})),$$

where $\mathsf{O}_{l/p}$ *is the subring of index* l/p *of the ring of all integral numbers of* F.

The composition $\times$ *of matrices and modules is understood in the sense of* (A.3.17).

Proof. Let $\mathsf{q}(x,y) = ax^2 + bxy + cy^2$ be the quadratic form with the matrix A, and let $\mathsf{V} = (\mathsf{V}_p, f_p)$ be a two-dimensional quadratic space over $\mathbb{F}_p = \mathbb{Z}/p\mathbb{Z}$ of the type $\{\mathsf{q}\}_{\mathbb{F}_p}$. By the definition,

$$\varepsilon_p(A) = \varepsilon_p(\mathsf{V}) = \left(\frac{-\det A}{p}\right) = \left(\frac{\Delta}{p}\right), \quad \text{if } p \neq 2 \text{ and } A \notin p\mathbb{E}_2. \tag{4.3.54}$$

Hence, the condition $\varepsilon_p(A) = 1$ means that $p \nmid \Delta$ and the congruence $x^2 \equiv \Delta = dl^2 \pmod p$ has a solution; since $d \equiv 0, 1 \pmod 4$, it follows that the congruence $x^2 \equiv d \pmod{4p}$ also has a solution. Similarly, the condition $\varepsilon_p(A) = -1$ means that $p \nmid \Delta$ and the congruences $x^2 \equiv \Delta \pmod p$ and $x^2 \equiv d \pmod{4p}$ have no solutions. Finally, the condition $\varepsilon_p(A) = 0$ means that $p|\Delta$, whence $p|l$, or $p \nmid l$ and $p|d$; in the last case the congruence $x^2 \equiv d \pmod{4p}$ is obviously solvable. Now let $p = 2$. Since the square of any odd integer is congruent to 1 modulo 8, it follows that

$$\varepsilon_2(A) = \begin{cases} 1, & \text{if } \det A \equiv -1 \pmod 8 \\ -1, & \text{if } \det A \equiv 3 \pmod 8 \\ 0, & \text{if } \det A \equiv 0 \pmod 8, \quad \text{but } A \notin 2\mathbb{E}_2 \end{cases} \tag{4.3.55}$$

Hence, in the first case $d \equiv -l^{-2}\det A \equiv 1 \pmod 8$, and the congruence $x^2 \equiv d \pmod 8$ has a solution; in the second case $d \equiv -l^{-2}\det A \equiv 5 \pmod 8$, and the congruence $x^2 \equiv d \pmod 8$ has no solutions; and in the third case, if l is odd, then $d \equiv 0$ or $4 \pmod 8$ and the congruence is solvable.

According to Proposition A.3.5, the remarks stated above prove all of the assertions of the proposition, exept for the formulae. In particular, in the cases (1)–(3) the number of regular ideals of O_l of norm p equals $\varepsilon_p(A) + 1$.

The number $\varepsilon_p(A) + 1$ appears also in the formula (4.3.45) for the number of isotropic vectors in V_p, which can obviously be interpreted as the number of different modulo p solutions of the congruence $\mathsf{q}(u_1, u_2) \equiv 0 \pmod p$ in coprime pairs of integers. Dividing the set of the solutions into classes modulo the equivalence (4.3.50), we see that each of the classes contains $p - 1$ elements, and hence the number of terms in the sum (4.3.53), for $B = mA$ with $m \not\equiv 0 \pmod p$ equals

$$\begin{aligned} &|(u_1, u_2) \in \mathbb{P}^1(\mathbb{Z}/p\mathbb{Z}); \mathsf{q}(u_1, u_2) \equiv 0 \pmod p| \\ &\quad = (p-1)^{-1} i(\mathsf{V}_p, f_p, 1) = \varepsilon_p(A) + 1. \end{aligned} \tag{4.3.56}$$

Thus, it would be natural to expect that in the cases (1)–(3) the terms in (4.3.53) are directly related to the regular ideals of O_l of norm p. Let $\varepsilon_p(A) = 1$. Then, by (4.3.56), there are precisely two solutions of $\mathsf{q}(x, y) \equiv 0 \pmod p$ in coprime pairs of integers, say (u_1, u_2) and (u_1', u_2'), which are not proportional modulo p. We set $\mathfrak{q}(u_1, u_2) = pa_1$, $\mathfrak{q}(u_1', u_2') = pa_2$. Let us choose integers v_1, v_2, v_1', and v_2' so that

$$U_1 = \begin{pmatrix} u_1 & u_2 \\ v_1 & v_2 \end{pmatrix}, \quad U_2 = \begin{pmatrix} u_1' & u_2' \\ v_1' & v_2' \end{pmatrix} \in \Lambda_+,$$

and set

$$A_i = A[{}^tU_i] = \begin{pmatrix} 2pa_i & b_i \\ b_i & 2c_i \end{pmatrix}, \quad A_i' = \begin{pmatrix} 2a_i & b_i \\ b_i & 2pc_i \end{pmatrix}$$

with $i = 1, 2$. By (4.3.53), we have

$$(f|\Pi)(mA) = p^{k-2}\chi(p)(f(mA_1') + f(mA_2')). \tag{4.3.57}$$

Let us consider complete modules in F given by

$$\mathsf{M}_1 = \left\{p, \frac{b_1 - \Delta^{1/2}}{2}\right\}, \quad \mathsf{M}_2 = \left\{p, \frac{b_2 - \Delta^{1/2}}{2}\right\}$$

and show that the modules are regular ideals of O_l of norm p. Speaking, for example, of M_1, we note that the number $\gamma_1 = \dfrac{b_1 - \Delta^{1/2}}{2p}$ is obviously a root of the polynomial $pv^2 - b_1 v + a_1 c_1$. Since $\Delta = b^2 - 4ac = b_1^2 - 4pa_1c_1$ is not divisible by p, it follows that b_1 is not divisible by p, and hence the coefficient of the polynomial are coprime. Then, by Lemma A.3.2, the ring of multipliers of the module $\mathsf{M}_1 = p\{1, \gamma_1\}$ is O_l, and $N(\mathsf{M}_1) = N(p)/p = p$. It is also clear that $\mathsf{M}_1 \subset \mathsf{O}_l$. The module M_2 is considered in the same way. We now show that $\mathsf{M}_1 \neq \mathsf{M}_2$. To prove this, it suffices to check that $\mathsf{M}_1 + \mathsf{M}_2 = \mathsf{O}_l$. The numbers $(b_2 - b_1)/2$ and p are obviously contained in $\mathsf{M}_1 + \mathsf{M}_2$. Hence, every integral linear combination of the numbers is contained in $\mathsf{M}_1 + \mathsf{M}_2$; in particular, $\sigma = \text{g.c.d.}\left(\dfrac{b_1 - b_2}{2}, p\right)$ is contained in $\mathsf{M}_1 + \mathsf{M}_2$. Thus, we only have to show that $\sigma = 1$. Let us set

$$U_2 U_1^{-1} = \begin{pmatrix} u_1' & u_2' \\ v_1' & v_2' \end{pmatrix}\begin{pmatrix} v_2 & -u_2 \\ -v_1 & u_1 \end{pmatrix} = \begin{pmatrix} t_1 & t_2 \\ t_3 & t_4 \end{pmatrix} = T \in \Lambda_+.$$

Since the pairs (u_1, u_2) and (u_1', u_2') are not proportional modulo p, it follows that $t_2 = u_1 u_2' - u_2 u_1' \not\equiv 0 \pmod p$. By comparing the corresponding entries in the obvious relation $A_1[{}^tT] = A_2$, we obtain the relations

$$pa_1 t_1^2 + b_1 t_1 t_2 + c_1 t_2^2 = pa_2 \quad \text{and} \quad 2pa_1 t_1 t_3 + b_1(t_1 t_4 + t_2 t_3) + 2c_1 t_2 t_4 = b_2.$$

Since $t_2 \not\equiv 0 \pmod p$, the first relation implies that $b_1 t_1 + c_1 t_2 \equiv 0 \pmod p$. Since $t_1 t_4 = 1 + t_2 t_3$, the second relation implies that $(b_2 - b_1)/2 \equiv t_2(b_1 t_3 + c_1 t_4) \pmod p$. Thus, if $(b_2 - b_1)/2$ is divisible by p, then we have $b_1 t_1 + c_1 t_2 \equiv b_1 t_3 + c_1 t_4 \equiv 0 \pmod p$, which implies the congruence $b_1 \equiv c_1 \equiv 0 \pmod p$, contrary to the assumption that $\Delta = b_1^2 - 4pa_1c_1$ is not divisible by p. We have proved that $\mathsf{M}_1 = \mathfrak{p}$ and $\mathsf{M}_2 = \bar{\mathfrak{p}}$ are the only regular ideals of O_l of norm p. We let q_1, q_2, q'_1, and q'_2 be the quadratic forms with the matrices A_1, A_2, A'_1, and A'_2, respectively, and let $\mathsf{M}(\mathsf{q}_1)$, $\mathsf{M}(\mathsf{q}_2)$, $\mathsf{M}(\mathsf{q}'_1)$, and $\mathsf{M}(\mathsf{q}'_2)$ be the modules (A.3.12) of the forms. We set $\delta_1 = (b_1 - \Delta^{1/2})/2$. Since clearly $\delta_1^2 = b_1\delta_1 - pa_1c_1$ and, by the above, b_1 is prime to p, it follows that

$$\begin{aligned}\mathsf{M}(\mathsf{q}'_1)\mathfrak{p} &= \{a_1, \delta_1\}\{p, \delta_1\} = \{a_1 p, a_1\delta_1, p\delta_1, b_1\delta_1 - pa_1c_1\} \\ &= \{pa_1, \delta_1\} = \mathsf{M}(\mathsf{q}_1).\end{aligned}$$

Hence, by Proposition A.3.3, we have

$$\mathsf{M}(\mathsf{q}'_1) = \frac{1}{p}\mathsf{M}(\mathsf{q}'_1)\mathfrak{p}\bar{\mathfrak{p}} = \frac{1}{p}\mathsf{M}(\mathsf{q}_1)\bar{\mathfrak{p}}.$$

Since the form q_1 is properly equivalent to q, it follows that the last module is equivalent to $\mathsf{M}(\mathsf{q})\bar{\mathfrak{p}}$. Thus, the matrix A'_1 of q'_1 is properly equivalent to the matrix $A \times \bar{\mathfrak{p}}$, and $f(mA'_1) = f(m\cdot(A \times \bar{\mathfrak{p}}))$. Similarly, A'_2 is properly equivalent to $A \times \mathfrak{p}$, and so $f(mA'_2) = f(m\cdot(A \times \mathfrak{p}))$. These relations along with (4.3.57) prove the formula of the assertion (1).

The formula of the assertion (2) directly follows from (4.3.53) and (4.3.56).

Now let $\varepsilon_p(A) = 0$. Then, by (4.3.56), each of the pairs of coprime integers (u, u') with $\mathsf{q}(u, u') \equiv 0 \pmod p$ is proportional modulo p to one of the pairs, say (u_1, u_2). We set $\mathsf{q}(u_1, u_2) = pa_1$. Let us choose integers v_1, v_2 such that $U = \begin{pmatrix} u_1 & u_2 \\ v_1 & v_2 \end{pmatrix} \in \Lambda_+$, and set

$$A_1 = A[{}^tU] = \begin{pmatrix} 2pa_1 & b_1 \\ b_1 & 2c_1 \end{pmatrix} \quad \text{and} \quad A'_1 = \begin{pmatrix} 2a_1 & b_1 \\ b_1 & 2pc_1 \end{pmatrix}.$$

Then, by (4.3.53), we have

$$(f|\Pi)(mA) = p^{k-2}\chi(p)f(mA'_1). \tag{4.3.58}$$

As we have seen, in this case the discriminant $\Delta = b^2 - 4ac = b_1^2 - 4pa_1c_1$ is divisible by p, which implies the congruences

$$b_1 \equiv 0 \pmod p \quad \text{and} \quad \Delta \equiv -4pa_1c_1 \pmod{p^2}. \tag{4.3.59}$$

First we assume that $p \nmid l$. Then $p | d$. Let us show that a_1c_1 is not divisible by p. If $p \neq 2$, this follows at once from the second congruence in (4.3.59), for Δ along with d is not divisible by p^2. If $p = 2$, then $d = 4d_0$, where $d_0 \equiv 2, 3 \pmod 4$. Let us set $b_1 = 2b_0$. Then $\Delta = 4d_o l^2 = 4(b_0^2 - 2a_1c_1)$, whence

$$2a_1c_1 = b_0^2 - d_0 l^2 \equiv b_0^2 - d_0 \pmod 4.$$

The number $b_0^2 - d_0$ is not divisible by 4, since $b_0^2 \equiv 0, 1 \pmod 4$ and $d_0 \equiv 2, 3 \pmod 4$. Hence, a_1c_1 is odd. Further, since the quadratic form q_1 with the matrix A_1 is primitive (along with q) and $p \nmid a_1$, it follows that the form q'_1

with the matrix A'_1 is also primitive. Hence, one may define the module $\mathsf{M}(\mathsf{q}'_1)$ associated to q'_1. Let us consider the modul $\mathsf{M} = \{p, (b_1 - \Delta^{1/2})/2\}$. Since $\gamma_1 = (b_1 - \Delta^{1/2})/2p$ is a root of the polynomial $pv^2 - b_1 v + a_1 c_1$ and $a_1 c_1$ is not divisible by p, it follows, by Lemma A.3.2, that the ring of multipliers of $\mathsf{M} = p\{1, \gamma_1\}$ is O_l, and $N(\mathsf{M}) = N(p)/p = p$. Since $\mathsf{M} \subset \mathsf{O}_l$, it follows that $\mathsf{M} = \mathfrak{p}$ is the only regular ideal of O_l of norm p. Let $\delta_1 = p\gamma_1 = (b_1 - \Delta^{1/2})/2$. Since $\delta_1^2 = b_1\delta_1 - pa_1c_1$ and $p \nmid a_1$, it follows that

$$\begin{aligned}\mathsf{M}(\mathsf{q}'_1)\cdot\mathfrak{p} &= \{a_1, \delta_1\}\{p, \delta_1\} = \{a_1 p, a_1\delta_1, p\delta_1, b_1\delta_1 - pa_1c_1\}\\ &= \{a_1 p, \delta_1\} = \mathsf{M}(\mathsf{q}_1),\end{aligned}$$

where q_1 is the form with matrix A_1. Hence, by $\bar{\mathfrak{p}} = \mathfrak{p}$ and Proposition A.3.3, we have

$$\mathsf{M}(\mathsf{q}'_1) = p^{-1}\mathsf{M}(\mathsf{q}'_1)\mathfrak{p}\bar{\mathfrak{p}} = p^{-1}\mathsf{M}(\mathsf{q}_1)\cdot\bar{\mathfrak{p}} = p^{-1}\mathsf{M}(\mathsf{q}_1)\mathfrak{p}.$$

Since the form q_1 is properly equivalent to q, it follows that the last module is equivalent to $\mathsf{M}(\mathsf{q})\mathfrak{p}$, and so the matrix A'_1 is properly equivalent to $A \times \mathfrak{p}$. Thus, $f(mA'_1) = f(m\cdot(A \times \mathfrak{p}))$, which along with (4.3.58) completes the proof of the assertion (3).

Finally, let $p|l$. Since $p|b_1$ and the form q_1 is primitive, it follows that c_1 is not divisible by p. Let us show that $p|a_1$. If $p \neq 2$, then this follows from the second congruence in (4.3.59). If $p = 2$, then, by $d \equiv 0, 1 \pmod 4$, we conclude that $\Delta = dl^2 = b_1^2 - 8a_1c_1 \equiv 0, 4 \pmod{16}$, whence $(b_1/2)^2 - 2a_1c_1 \equiv 0, 1 \pmod 4$, and so $2a_1c_1 \equiv 0 \pmod 4$, and $a_1 \equiv 0 \pmod 2$. Let us set $A'_1 = pA_2$. The above considerations show that $A_2 \in \mathbb{E}_2$. Since q_1 is primitive, it follows that the form q_2 with the matrix A_2 is primitive. The number $\gamma_1 = (b_1 - \Delta^{1/2})/2p$ is a root of the polynomial $v^2 - (b_1/p)v + a_1c_1/p$ with coefficients in $\mathbb{Z}$. Then, by Lemma A.3.2, the module $\{1, \gamma_1\}$ is an order in F with the discriminant $(b_1/p)^2 - 4pa_1c_1/p^2 = d(l/p)^2$. Hence $\{1, \gamma_1\} = \mathsf{O}_{l/p}$, and we have

$$\begin{aligned}\mathsf{M}(\mathsf{q}_1)\mathsf{O}_{l/p} &= \{pa_1, p\gamma_1\}\{1, \gamma_1\}\\ &= \{pa_1, p\gamma_1, pa_1\gamma_1, p(b_1\gamma_1/p - a_1c_1/p)\} = \{a_1c_1, pa_1, p\gamma_1\}\\ &= \{a_1, p\gamma_1\} = p\mathsf{M}(\mathsf{q}_2).\end{aligned}$$

As before, the module $\mathsf{M}(\mathsf{q}_1)$ is equivalent to $\mathsf{M}(\mathsf{q})$, hence $\mathsf{M}(\mathsf{q}_2)$ is equivalent to $\mathsf{M}(\mathsf{q})\cdot\mathsf{O}_{l/p}$, and the matrix A_2 is properly equivalent to $A \times \mathsf{O}_{l/p}$. Thus, $f(mA'_1) = f(\mathrm{pm}\, A_2) = f(\mathrm{pm}\cdot(A \times \mathsf{O}_{l/p}))$. □

We now return to relations of Lemma 4.3.13. Using the above formulae, we shall calculate numerators of the rational fractions in the right side of the relations. Taking $B = mA$ with $m \in \mathbb{N}_{(p)}$ and a matrix $A = \begin{pmatrix} 2a & b \\ b & 2c \end{pmatrix}$ of $\mathbb{A}_2$ satisfying g.c.d.$(a, b, c) = 1$, by (4.3.48) and (4.3.40), we obtain the relations

$$
\begin{aligned}
&Q_p(v,F)d_p(v,mA)\\
&\quad = (f|(1-\Pi v + p(\Pi_{2,0}^{(1)}+\Pi_{2,0}^{(0)})v^2 - \Pi v + \Pi_-\Pi v^2\\
&\qquad - p\Pi_-(\Pi_{2,0}^{(1)}+\Pi_{2,0}^{(0)})v^3))(mA)\\
&\quad = f(mA) - (f|\Pi)(mA)v + p^{2k-4}\chi(p)^2\varepsilon_p(A)f(mA)v^2 + (f|\Pi_-\Pi)(mA)v^2,
\end{aligned}
\tag{4.3.60}
$$

for, clearly, $mA \notin p\mathbb{A}_2$ and $\varepsilon_p(mA) = \varepsilon_p(A)$. The further calculation of the numerators of the series $d_p(v, mA)$ is based on formulae of Proposition 4.3.14, but it turns out to be more convenient to apply the formulae to certain linear combinations of the series, rather than to individual series.

Let us fix a negative integer Δ, and let $A_1, \ldots, A_{h(\Delta)}$ be a system of representatives for classes of matrices $A = \begin{pmatrix} 2a & b \\ b & 2c \end{pmatrix} \in \mathbb{A}_2$ with g.c.d.$(a, b, c) = 1$ and $b^2 - 4ac = \Delta$ modulo the proper equivalence (A.3.9). According to Theorem A.3.9, these classes form a finite abelian group $\mathsf{H}(\Delta)$ with respect to the composition (A.3.15). We fix a character η of the group $\mathsf{H}(\Delta)$, and define, for $p \in \mathbb{P}$, $m \in \mathbb{N}$, and $f \in \mathfrak{F}_\xi^2$, the formal power series

$$
\begin{aligned}
d_p(v, m, \eta, \Delta) &= \sum_{i=1}^{h(\Delta)} \eta(A_i) d_p(v, mA_i)\\
&= \sum_{\delta=0}^{\infty} f(p^\delta m, \eta) v^\delta,
\end{aligned}
\tag{4.3.61}
$$

where

$$
f(t, \eta) = \sum_{i=1}^{h(\Delta)} \eta(A_i) f(tA_i).
$$

Before calculating the series (4.3.61), we note that, by (4.3.54) and (4.3.55), the p-sign $\varepsilon_p(A_i)$ of A_i is independent of i, but depends only on Δ. Thus, we can set

$$
\varepsilon_p(\Delta) = \varepsilon_p(A_i) = \left(\frac{\Delta}{p}\right), \qquad \text{if } p \neq 2 \tag{4.3.62}
$$

and

$$
\varepsilon_2(\Delta) = \varepsilon_2(A_i) = \begin{cases} 1, & \text{if } \Delta \equiv 1 \pmod 8 \\ -1, & \text{if } \Delta \equiv 5 \pmod 8 \\ 0, & \text{if } \Delta \equiv 0 \pmod 2 \end{cases} \tag{4.3.63}
$$

Lemma 4.3.15. *With the above notation and assumptions including those of Lemma* 4.3.13 *and Proposition* 4.3.14, *the power series* $d_p(v) = d_p(v, m, \eta, \Delta)$ *is formally equal to a rational fraction with the denominator* $Q_p(v, F)$, *say*

$$
d_p(v) = Q_p(v, F)^{-1} P_p(v),
$$

where the numerator $P_p(v)$ *is a polynomial, which can be given by the following formulae*:

(1) *If* $\varepsilon_p(\Delta) = 1$, *then*

$$P_p(v) = (1 - (N\mathfrak{p})^{k-2}\chi(N\mathfrak{p})\eta(\mathfrak{p})v)(1 - (N\bar{\mathfrak{p}})^{k-2}\chi(N\bar{\mathfrak{p}})\eta(\bar{\mathfrak{p}})v)f(m, \eta),$$

where $\mathfrak{p}$ *and* $\bar{\mathfrak{p}}$ *are the only regular ideals of* O_l *of norm* p;

(2) *If* $\varepsilon_p(\Delta) = -1$, *then*

$$P_p(v) = (1 - (N\mathfrak{p})^{k-2}\chi(N\mathfrak{p})\eta(\mathfrak{p})v^2)f(m, \eta),$$

where $\mathfrak{p} = p\mathsf{O}_l$ *is the only regular ideal of* O_l *of norm* p^2;

(3) *If* $\varepsilon_p(\Delta) = 0$ *and* $p \nmid l$, *then*

$$P_p(v) = (1 - (N\mathfrak{p})^{k-2}\chi(N\mathfrak{p})\eta(\mathfrak{p})v)f(m, \eta),$$

where $\mathfrak{p} = \bar{\mathfrak{p}}$ *is the only regular ideal of* O_l *of norm* p;

(4) *If* $\varepsilon_p(\Delta) = 0$ *and* $p \mid l$, *then*

$$\begin{aligned} P_p(v) &= (f|(1 - \Pi_- v)(1 - \Pi v))(m, \eta) \\ &= \sum_{i=1}^{h(\Delta)} \eta(A_i)(f|(1 - \Pi_- v)(1 - \Pi v))(mA_i). \end{aligned}$$

Proof. We first assume that $p \nmid l$. Then it follows from Proposition 4.3.14 and (4.3.40) that $(f|\Pi_-\Pi)(mA_i) = 0$ for any $i = 1, \ldots, h(\Delta)$. Hence, by (4.3.60), (4.3.62), and (4.3.63), we get

$$\begin{aligned} P_p(v) &= Q_p(v, F)d_p(v) \\ &= \sum_{i=1}^{h(\Delta)} (\eta(A_i)f(mA_i) - \eta(A_i)(f|\Pi)(mA_i)v + p^{2k-4}\eta(A_i)\chi(p^2)\varepsilon_p(\Delta)f(mA_i)v^2) \\ &= f(m, \eta) - (f|\Pi)(m, \eta)v + p^{2k-4}\chi(p^2)\varepsilon_p(\Delta)f(m, \eta)v^2. \end{aligned}$$

If $\varepsilon_p(\Delta) = 1$, then, by Proposition 4.3.14(1), we have

$$\begin{aligned} (f|\Pi)(m, \eta) &= \sum_{i=1}^{h(\Delta)} \eta(A_i)(f|\Pi)(mA_i) \\ &= p^{k-2}\chi(p)\sum_i \eta(A_i)(f(m\cdot(A_i \times \mathfrak{p})) + f(m\cdot(A_i \times \bar{\mathfrak{p}}))) \\ &= p^{k-2}\chi(p)\Big[\eta(\bar{\mathfrak{p}})\sum_i \eta(A_i \times \mathfrak{p})f(m\cdot(A_i \times \mathfrak{p})) \\ &\qquad + \eta(\mathfrak{p})\sum_i \eta(A_i \times \bar{\mathfrak{p}})f(m\cdot(A_i \times \bar{\mathfrak{p}}))\Big] \\ &= p^{k-2}\chi(p)(\eta(\mathfrak{p}) + \eta(\bar{\mathfrak{p}}))f(m, \eta), \end{aligned}$$

because $\eta(\mathfrak{p})\eta(\bar{\mathfrak{p}}) = \eta(\mathfrak{p}\bar{\mathfrak{p}}) = \eta(p\mathsf{O}_l) = 1$, and the matrix $A_i \times \mathfrak{p}$ (resp. $A_i \times \mathfrak{p}$) runs over a set of representatives for the classe in $\mathsf{H}(\Delta)$, when A_i does. Hence, in this case we obtain

$$D_p(v) = (1 - p^{k-2}\chi(p)(\eta(\mathfrak{p}) + \eta(\bar{\mathfrak{p}}))v + p^{2k-4}\chi(p^2)v^2)f(m,\eta)$$
$$= (1 - p^{k-2}\chi(p)\eta(\mathfrak{p})v)(1 - p^{2k-2}\chi(p)\eta(\bar{\mathfrak{p}})v)f(m,\eta).$$

If $\varepsilon_p(\Delta) = -1$, then, by Proposition 4.3.14(2), we have

$$(f|\Pi)(m,\eta) = \sum_{i=1}^{h(\Delta)} \eta(A_i)(f|\Pi)(mA_i) = 0,$$

whence

$$P_p(v) = (1 - p^{2k-4}\chi(p^2)v^2)f(m,\eta).$$

If $\varepsilon_p(\Delta) = 0$ and $p \nmid l$, then, by Proposition 4.3.14(3), we get

$$(f|\Pi)(m,\eta) = p^{k-2}\chi(p) \sum_{i=1}^{h(\Delta)} \eta(A_i) f(m\cdot(A_i \times \mathfrak{p}))$$
$$= p^{k-2}\chi(p)\eta(\bar{\mathfrak{p}}) \sum_i \eta(A_i \times \mathfrak{p}) f(m\cdot(A_i \times \mathfrak{p}))$$
$$= p^{k-2}\chi(p)\eta(\mathfrak{p})f(m,\eta),$$

since $\bar{\mathfrak{p}} = \mathfrak{p}$, whence

$$P_p(v) = (1 - p^{k-2}\chi(p)\eta(\mathfrak{p})v)f(m,\eta).$$

Finally, let $\varepsilon_p(\Delta) = 0$ and $p|l$. Since $(f|\Pi_-)(mA_i) = 0$, the relation (4.3.60) with $A = A_i$ can be written in the form

$$Q_p(v,F)d_p(v,mA_i) = f(mA_i) - (f|\Pi)(mA_i)v + (f|\Pi_-\Pi)(mA_i)v^2$$
$$= (f|(1 - \Pi_- v)(1 - \Pi v))(mA_i).$$

Multiplying the two sides by $\eta(A_i)$ and taking the sum in i, we obtain the last formula. □

Returning to our consideration of multiplicative properties of Fourier coefficients of modular forms of genus 2, we only have to join local pieces into a global picture.

Theorem 4.3.16. *Let*

$$F(Z) = \sum_{A \in \mathbb{A}_2} f(A)e\{AZ\} \in \mathfrak{M}_k^2(q,\chi)$$

with $k \in \mathbb{Z}$ and a Dirichlet character χ modulo q be a non-zero eigenfunction for all of the Hecke operators $|_{k,\chi}T$ with $T \in \mathsf{L}^2(q)$. For a negative rational integer Δ, let d denotes the discriminant of the field $\mathsf{F} = \mathbb{Q}(\Delta^{1/2})$, $l = (\Delta/d)^{1/2} \in \mathbb{N}$, O_l the subring of index l in the ring O of all integers of F, and let η be a character of the class group $\mathsf{H}(\Delta)$. Then, for any $a \in \mathbb{N}$ with $a|q^\infty$ and any completely multiplicative function γ: $\mathbb{N}_{(q)} \to \mathbb{C}$ the following formal identity holds

$$\sum_{\{A_i\}_+\in \mathsf{H}(\Delta)} \eta(A_i) \sum_{m\in\mathbb{N}_{(q)}} \frac{f(maA_i)\gamma(m)}{m^s}$$
$$= \rho(s)\left[\prod_{\mathfrak{p},\, N\mathfrak{p}\nmid(ql)^2}\left(1-\frac{\chi(N\mathfrak{p})\gamma(N\mathfrak{p})\eta(\mathfrak{p})}{(N\mathfrak{p})^{s-k+2}}\right)\right]\zeta(s,\gamma;F) \tag{4.3.64}$$

where the group $\mathsf{H}(\Delta)$ *in the left side is realized as a group of proper classes* $\{A\}_+\subset\mathbb{A}_2^+$ *and* A_i *runs a set of representatives for the classes; where in the right side* $\rho(s)=\rho(s,\Delta,\eta,a,\gamma,F)$ *is a finite sum given by*

$$\rho(s)=\sum_{\{A_i\}_+\in\mathsf{H}(\Delta)}\eta(A_i)\left(f|_{k,\chi}\prod_{\substack{p\in\mathbb{P}_{(q)},\\ p|l}}\left(1-\frac{\gamma(p)}{p^s}\Pi_0^2(p)\right)\left(1-\frac{\gamma(p)}{p^s}\Pi_1^2(p)\right)(aA_i)\right.$$
$$=\sum_{\{A_i\}_+}\eta(A_i)\sum_{\substack{\varkappa,\varkappa_1\in\mathbb{N}_{(q)}\\ \varkappa|\varkappa_1|l}}\frac{\chi(\varkappa^2\varkappa_1)\gamma(\varkappa\varkappa_1)\mu(\varkappa)\mu(\varkappa_1)}{\varkappa^{s-2k+3}\varkappa_1^{s-k+2}}f\left(\frac{a\varkappa_1}{\varkappa}\left(A_i\times\mathsf{O}_{l/\varkappa_1}\right)\right) \tag{4.3.65}$$

with the Möbius function μ, *the subrings* $\mathsf{O}_t\subset\mathsf{O}$ *of index* t, *and the composition* $\times$ *defined by* (A.3.17), *where in the middle factor* $\mathfrak{p}$ *runs over all regular prime ideals of* O_l *whose norm is prime to* ql, *and where* $\zeta(s,\gamma;F)$ *is the Euler product* (4.3.35).

Furthermore, if $k>0$ *and* γ *satisfies* $|\gamma(m)|\le cm^\sigma$ *with some constants* c *and* σ, *then the Dirichlet series on the left in* (4.3.64) *and the infinite products on the right converge absolutely and uniformly in each of the half-planes of the form* $\operatorname{Re}s>2\tau k+\sigma+1+\varepsilon$ *with* $\varepsilon>0$, *where* $\tau=1$ *in the general case, and* $\tau=\frac{1}{2}$ *if* F *is a cusp form.*

Proof. Let $p_1,\ldots,p_b$ be all of the prime divisors of l, which do not divide q, and let $\mathbb{P}_{(ql)}=\{p_{b+1},p_{b+2},\ldots\}$, where we assume that $p_i<p_{i+1}$. We first prove, by induction on $c\in\mathbb{N}$, the following formal identity

$$\sum_{m\in\mathbb{N}_{(q)}}\frac{f(am,\eta)\gamma(m)}{m^s}$$
$$=\left[\sum_{m\in\mathbb{N}_{(q\cdot q(c))}}\frac{f(am,\eta)\gamma(m)}{m^s}\right]\left[\prod_{\mathfrak{p},\,N\mathfrak{p}|q(c)^2}\left(1-\frac{\chi(N\mathfrak{p})\gamma(N\mathfrak{p})\eta(\mathfrak{p})}{(N\mathfrak{p})^{s-k+2}}\right)\right]$$
$$\times\prod_{p\in\mathbb{P},\,p|q(c)}Q_p(\gamma(p)p^{-s},F)^{-1} \tag{4.3.66}$$

where

$$q(c)=p_{b+1}\cdots p_{b+c},\qquad f(am,\eta)=\sum_i\eta(A_i)f(amA_i),$$

and where $\mathfrak{p}$ runs over regular prime ideals of O_l with $N\mathfrak{p}|q(c)^2$. If $c=1$, we set $p_{b+1}=p$, $\gamma(p)p^{-s}=v$ and make use of Lemma 4.3.15. By the lemma, we obtain

$$Q_p(\gamma(p)p^{-s}, F) \sum_{m \in \mathbb{N}_{(q)}} \frac{f(am,\eta)\gamma(m)}{m^2}$$

$$= \sum_{m \in \mathbb{N}_{(qp)}} \frac{\gamma(m)}{m^s} Q_p(v, F) \sum_{\delta=0}^{\infty} f(amp^\delta, \eta) v^\delta \tag{4.3.67}$$

$$= \left[\sum_{m \in \mathbb{N}_{(q\cdot p)}} \frac{f(am,\eta)\gamma(m)}{m^s}\right] \prod_{\mathfrak{p}, N\mathfrak{p}|p^2} \left(1 - (N\mathfrak{p})^{k-2}\chi(N\mathfrak{p})\eta(\mathfrak{p})\frac{\gamma(N\mathfrak{p})}{(N\mathfrak{p})^s}\right),$$

where in the last product $\mathfrak{p}$ runs over all regular prime ideals of O_l of norm p or p^2, by Proposition A.3.5, which proves the identity (4.3.66) with $c = 1$. Suppose the identity is already proved for some $c \geq 1$. We set $p_{b+c+1} = \mathfrak{p}$ and use the identity (4.3.67) with $q \cdot q(c)$ in place of q. We get

$$Q_{\mathfrak{p}}(\gamma(\mathfrak{p})\mathfrak{p}^{-s}, F) \sum_{m \in \mathbb{N}_{(q)}} \frac{f(am,\eta)\gamma(m)}{m^s}$$

$$= Q_{\mathfrak{p}}(\gamma(\mathfrak{p})\mathfrak{p}^{-s}, F)\left[\sum_{m \in \mathbb{N}_{(q\cdot q(c))}} \frac{f(am,\eta)\gamma(m)}{m^s}\right]$$

$$\times \left[\prod_{\mathfrak{p}, N\mathfrak{p}|q(c)^2} \left(1 - \frac{\chi(N\mathfrak{p})\gamma(N\mathfrak{p})\eta(\mathfrak{p})}{(N\mathfrak{p})^{s-k+2}}\right)\right] \prod_{p|q(c)} Q_p(\gamma(p)p^{-s}, F)^{-1}$$

$$= \left[\sum_{m \in \mathbb{N}_{(q\cdot q(c+1))}} \frac{f(am,\eta)\gamma(m)}{m^s}\right]\left[\prod_{\mathfrak{p}, N\mathfrak{p}|q(c+1)^2} \left(1 - \frac{\chi(N\mathfrak{p})\gamma(N\mathfrak{p})\eta(\mathfrak{p})}{(N\mathfrak{p})^{s-k+2}}\right)\right]$$

$$\times \prod_{p|q(c)} Q_p(\gamma(p)p^{-s}, F)^{-1},$$

which proves (4.3.66) for $c + 1$.

Taking the coefficient-wise limit of the Dirichlet series on both sides in (4.3.66) as $c \to \infty$, we obtain the formal identity

$$\sum_{m \in \mathbb{N}_{(q)}} \frac{f(am,\eta)\gamma(m)}{m^s}$$

$$= \left[\sum_{m|(p_1\cdots p_b)^\infty} \frac{f(am,\eta)\gamma(m)}{m^s}\right]\left[\prod_{\mathfrak{p}, N\mathfrak{p}\nmid(ql)^2} \left(1 - \frac{\chi(N\mathfrak{p})\gamma(N\mathfrak{p})\eta(\mathfrak{p})}{(N\mathfrak{p})^{s-k+2}}\right)\right]$$

$$\times \prod_{p \in \mathbb{P}_{(ql)}} Q_p(\gamma(p)p^{-s}, F)^{-1}.$$

Thus, in order to prove (4.3.64) with the factor $\rho(s)$ given by the first formula in (4.3.65), we only have to check that

$$\left[\prod_{p|p_1\cdots p_r} Q_p(\gamma(p)p^{-s}, F)\right] \cdot \sum_{m|(p_1\cdots p_r)^\infty} \frac{f(am,\eta)\gamma(m)}{m^s} = \rho_r(s), \tag{4.3.68}$$

where $1 \leq r \leq b$, and

$$\rho_r(s) = \sum_i \eta(A_i)\left(f\,\Big|\prod_{p|p_1\cdots p_r}\left(1 - \frac{\gamma(p)}{p^s}\Pi_0^2(p)\right)\left(1 - \frac{\gamma(p)}{p^s}\Pi_1^2(p)\right)\right)(aA_i).$$

If $r = 1$, then (4.3.68) is true, by Lemma 4.3.15(4). Suppose the relation is already proved for some r with $1 \le r < b$. Set $p_{r+1} = \mathrm{p}$. By the inductive hypothesis, we obtain

$$\left[\prod_{p|p_1\cdots p_r} Q_p(\gamma(p)p^{-s}, F)\right]\cdot Q_{\mathrm{p}}(\gamma(\mathrm{p})\mathrm{p}^{-s}, F)\cdot \sum_{m|(p_1\cdots p_r\mathrm{p})^\infty} \frac{f(am,\eta)\gamma(m)}{m^s}$$

$$= Q_{\mathrm{p}}(\gamma(\mathrm{p})\mathrm{p}^{-s}, F)\sum_{\delta=0}^{\infty}\left\{\left[\prod_{p|p_1\cdots p_r} Q_p(\gamma(p)p^{-s}, F)\right]\sum_{m|(p_1\cdots p_r)^\infty}\frac{f(am\mathrm{p}^\delta,\eta)\gamma(m)}{m^s}\right\}$$

$$\times\left(\frac{\gamma(\mathrm{p})}{\mathrm{p}^s}\right)^\delta$$

$$= Q_{\mathrm{p}}(\gamma(\mathrm{p})\mathrm{p}^{-s}, F)\sum_{\delta=0}^{\infty} f'(a\mathrm{p}^\delta,\eta)(\gamma(\mathrm{p})\mathrm{p}^{-s})^\delta, \tag{4.3.69}$$

where

$$f'(a\mathrm{p}^\delta,\eta) = \sum_i \eta(A_i) f'(a\mathrm{p}^\delta A_i),$$

and

$$f'(A) = \left(f|_{k,\chi}\prod_{p|p_1\cdots p_r}\left(1 - \frac{\gamma(p)}{p^s}\Pi_0^2(p)\right)\left(1 - \frac{\gamma(p)}{p^s}\Pi_1^2(p)\right)\right)(A), \qquad (A \in \mathbb{A}_2).$$

It is clear that, for any fixed $s \in \mathbb{C}$, the function f' belongs to $\mathfrak{F}_\xi^2$. According to Proposition 3.4.11, each element of $\mathbf{L}_p^2$ commutes with each of the elements $\Pi_0^2(p_j) \in \mathscr{C}_{-p_j}^2$ and $\Pi_2^2(p_j) \in \mathscr{C}_{p_j}^2$ for $j = 1, 2, \ldots, r$. Since the global ring $\mathbf{L}^2$ is commutative, it follows that elements of $\mathbf{L}_{\mathrm{p}}^2$ commute with elements of $\mathbf{L}_{p_j}^2$; in particular, they commute with the elements $\mathbf{T}^2(p_j) = \Pi_0^2(p_j) + \Pi_1^2(p_j) + \Pi_2^2(p_j)$ (see (3.4.32)). Hence, each element of $\mathbf{L}_p^2$ commutes with each of the elements $\Pi_1^2(p_j) = \mathbf{T}^2(p_j) - \Pi_0^2(p_j) - \Pi_2^2(p_j)$ for $j = 1, 2, \ldots, r$. These considerations imply that the function f' along with f is an eigenfunction for all Hecke operators of $\mathbf{L}_{\mathrm{p}}^2$ with the same eigenvalues as f. Thus, in order to calculate the expression (4.3.69), we may apply Lemma 4.3.15(4), which shows that the expression is equal to

$$\left(f'\Big|\left(1 - \frac{\gamma(\mathrm{p})}{\mathrm{p}^s}\Pi_0^2(\mathrm{p})\right)\left(1 - \frac{\gamma(\mathrm{p})}{\mathrm{p}^s}\Pi_1^2(\mathrm{p})\right)\right)(a,\eta).$$

The relations (4.3.68) are proved.

In order to prove the second formula for $\rho(s) = \rho_b(s)$, we note that, by Proposition 3.4.11, elements $\Pi_0^2(p)$ commute with $\Pi_1^2(p') = \mathbf{T}^2(p') - \Pi_0^2(p') - \Pi_2^2(p')$, if $p \neq p'$. Hence,

$$\prod_{p|p_1\cdots p_b}\left(1-\frac{\gamma(p)}{p^s}\Pi_0^2(p)\right)\left(1-\frac{\gamma(p)}{p^s}\Pi_1^2(p)\right)$$
$$=\left[\prod_{p|p_1\cdots p_b}\left(1-\frac{\gamma(p)}{p^s}\Pi_0^2(p)\right)\right]\left[\prod_{p|p_1\cdots p_b}\left(1-\frac{\gamma(p)}{p^s}\Pi_1^2(p)\right)\right]$$
$$=\left[\sum_{\varkappa|p_1\cdots p_b}\frac{\mu(\varkappa)\gamma(\varkappa)}{\varkappa^s}\Pi_0(\varkappa)\right]\left[\sum_{\varkappa_1|p_1\cdots p_b}\frac{\mu(\varkappa_1)\gamma(\varkappa_1)}{\varkappa_1^s}\Pi(\varkappa_1)\right]$$
$$=\sum_{\varkappa,\varkappa_1|p_1\cdots p_b}\frac{\mu(\varkappa)\mu(\varkappa_1)\gamma(\varkappa\varkappa_1)}{(\varkappa\varkappa_1)^s}\Pi_-(\varkappa)\Pi(\varkappa_1),$$

where $\Pi_-(\varkappa)$ are the elements (3.4.10) (see (3.4.12)), and where, for a square-free $\varkappa_1$, we set $\Pi(\varkappa_1)=\prod_{p|\varkappa_1}\Pi_1^2(p)$ (the order of prime numbers p makes no difference, in view of Proposition 3.4.11 and formulae (3.4.32) with prime a; one can show that the notation is consistent with (3.4.34)). The formula of Proposition 4.3.14(4) implies, by induction on the number of prime divisors of a square-free $\varkappa_1$ with $\varkappa_1|l$, the formula

$$(f|_{k,\chi}\Pi(\varkappa_1))(aA_i)=\varkappa_1^{k-2}\chi(\varkappa_1)f(a\varkappa_1\cdot(A_i\times \mathsf{O}_{l/\varkappa_1}))$$

(see (A.3.6)), hence, by (4.3.40), we have

$$(f|_{k,\chi}\Pi_-(\varkappa)\Pi(\varkappa_1))(aA_i)=((f|\Pi_-(\varkappa))|\Pi(\varkappa_1))(aA_i)$$
$$=\begin{cases}\varkappa^{2k-3}\varkappa_1^{k-2}\chi(\varkappa^2\varkappa_1)f\left(\dfrac{a\varkappa_1}{\varkappa}\cdot(A_i\times \mathsf{O}_{l/\varkappa_1})\right), & \text{if } \varkappa|\varkappa_1\\ 0, & \text{if } \varkappa\nmid\varkappa_1.\end{cases}$$

These considerations show that the second formula for $\rho(s)$ follows from the first formula.

It remains to prove the assertions concerning convergence. By (2.3.20) and (2.3.43), we have

$$|f(amA_i)|\le c(-\Delta)^{k\tau}a^{2k\tau}m^{2k\tau},\qquad (m\in\mathbb{N}),$$

where c depends only on F. Hence, the Dirichlet series in the left side of (4.3.64) converges absolutely and uniformly in the half-planes indicated in the theorem, by the standard arguments. The product over $\mathfrak{p}$ in the right side converges absolutely and uniformly in $\operatorname{Re} s\ge k+\sigma-1+\varepsilon$ with $\varepsilon>0$, since

$$\sum_{\mathfrak{p},N\mathfrak{p}\nmid(ql)^2}\left|\frac{\chi(N\mathfrak{p})\gamma(N\mathfrak{p})\eta(\mathfrak{p})}{(N\mathfrak{p})^{s-k+2}}\right|\le\sum_{\mathfrak{p},N\mathfrak{p}=p}p^{\sigma-\operatorname{Re}s+k-2}+\sum_{\mathfrak{p},N\mathfrak{p}=p^2}p^{2\sigma-2\operatorname{Re}s+2k-4}$$
$$\le 2\sum_{p\in\mathbb{P}}p^{-\operatorname{Re}s+k+\sigma-2}+\sum_{p\in\mathbb{P}}p^{-2\operatorname{Re}s+2k+2\sigma-4}$$

in view of Proposition A.3.5 (note that the norm of any prime ideal of O_l is of the form p or p^2 with $p\in\mathbb{P}$). Moreover, in each of the half-planes the absolute value of the product is bounded from below by a positive constant. These arguments along with the formal identity

$$\rho(s)\zeta(s,\gamma;F) = \left[\prod_{\mathfrak{p}}\left(1 - \frac{\chi(N\mathfrak{p})\gamma(N\mathfrak{p})\eta(\mathfrak{p})}{(N\mathfrak{p})^{s-k+2}}\right)\right]^{-1} \sum_{m\in\mathbb{N}_{(q)}} \frac{f(am,\eta)\gamma(m)}{m^s} \tag{4.3.70}$$

imply that the Dirichlet series formally representing the product $\rho(s)\zeta(s,\gamma;F)$ converges absolutely and uniformly in $\operatorname{Re} s > 2\tau k + \delta + 1 + \varepsilon$, $(\varepsilon > 0)$. To show that the same is true for $\zeta(s,\gamma;F)$, it suffices to check that, for some Δ, η, and a, the factor $\rho(s) = \rho(s,\Delta,\eta,a,\gamma,F)$ equals to a non-zero constant (note that $\zeta(s,\gamma;F)$ is independent of Δ, η, and a). Since F is a non-zero form of positive weight, it follows that F is not a constant. By Theorem 2.3.16, the space $\mathfrak{M}_k^2(q,\chi)$ does not contain singular forms. Thus, there is a negative integer Δ, a character η of $\mathsf{H}(\Delta)$, $m\in\mathbb{N}_{(q)}$, and $a\in\mathbb{N}$ with $a|q^\infty$ such that

$$f(am,\eta) = \sum_{\{A_i\}_+\in \mathsf{H}(\Delta)} \eta(A_i)f(amA_i) \neq 0 \tag{4.3.71}$$

We let Δ_0, η_0, a_0 be a triple satisfying (4.3.71) with the greatest $\Delta = \Delta_0$. Then, by (A.3.6) and Theorem A.3.1, for each primitive A_i with $\det A_i = -\Delta_0$, we have $\det(A_i \times \mathsf{O}_{l/\varkappa_1}) = -\Delta_0/\varkappa_1^2$. Hence, by the maximality of Δ_0, we have

$$\rho_0 = \rho(s,\Delta_0,\eta_0,a_0,\gamma;F) = \sum_i \eta(A_i)f(a_0A_i)$$

If this number equals zero, then all of the coefficients of the series in the left side of (4.3.64) with $\Delta = \Delta_0$, $\eta = \eta_0$, and $a = a_0$ equal zero, contrary to (4.3.71). So $\rho_0 \neq 0$, and the Dirichlet series for $\zeta(s,\gamma;F)$ as well as the Euler product has the same analytic properties as the series in the right side of (4.3.70). □

Exercise 4.3.17. *Let*

$$F = \sum_{A\in\mathbb{A}_n} f(A)e\{AZ\} \in \mathfrak{M}_\xi^n$$

be a non-zero eigenfunction for all Hecke operators $|_{k,\chi}\mathbf{T}$ *with* $\mathbf{T}\in\mathbf{L}_p^n$, *where* $(-1)^k\chi(-1) = \xi(-1)$, *and p is a fixed prime number. Let us set*

$$Q_p(v,F) = \sum_{j=0}^{m} (-1)^j\lambda(\mathbf{q}_j^n(p))v^j, \qquad (m = 2^n),$$

where $\mathbf{q}_j^n(p)\in\mathbf{L}_p^n$ *are the elements* (3.3.77), *and where* $\lambda(\mathbf{T}) = \lambda(\mathbf{T},F)$ *are the eigenvalues corresponding to F. Finally, let* $\alpha_i = \alpha_i(p,F)$ *with* $i = 0, 1, \ldots, n$ *be the parameters of the homomorphism* $\mathbf{T}\to\lambda(\mathbf{T})$ *(see Proposition* 3.3.36). *Prove the following assertions*:

(1) *(Žarkovskaya) For each matrix A of* $\mathbb{A}_n$, *the formal identity holds*

$$\sum_{\delta=0}^{\infty} f(p^\delta A)v^\delta = Q_p(v,F)^{-1}\cdot P(v),$$

where $P(v) = P_p(v,A,F)$ *is a polynomial of degree at most* $2^n - 1$.

[Hint: Use Proposition 3.5.11.]

(2) *The polynomial* $Q_p(v,F)$ *has the following factorization into linear factors*

$$Q_p(v,F) = (1-\alpha_0 v)\prod_{r=1}^{n}\prod_{1\le i_1<\cdots<i_r\le n}(1-\alpha_0\alpha_{i_1}\cdots\alpha_{i_r}v);$$

the parameters $\alpha_0, \alpha_1, \ldots, \alpha_n$ *satisfy*

$$\alpha_0^2\alpha_1\cdots\alpha_n = \chi(p)^n p^{nk-\langle n\rangle}.$$

If all of the roots of $Q_p(v,F)$ *have the same absolute value, then the absolute value is equal to* $p^{-(nk-\langle n\rangle)/2}$, *and the parameters satisfy*

$$|\alpha_0| = p^{(nk-\langle n\rangle)/2}, \qquad |\alpha_1| = \cdots = |\alpha_n| = 1.$$

(3) (*Maass-Žarkovskaya*) *Let us assume that* $n>1$ *and* $k\neq n$, *or* $k=n$ *and* $\chi(p)\neq -1$. *Then the form* $F|\Phi\in\mathfrak{M}_\xi^{n-1}$, *where* Φ *is the Siegel operator, is an eigenfunction for all operators* $|_{k,\chi}\mathbf{T}'$ *with* $\mathbf{T}'\in\mathbf{L}_p^{n-1}$. *If, in addition,* $F|\Phi\neq 0$ *and* $\alpha_0', \alpha_1', \ldots, \alpha_{n-1}'$ *are the parameters corresponding to* $F|\Phi$, *then one can take*

$$(\alpha_0,\alpha_1,\ldots,\alpha_n) = (\alpha_0' p^{k-n}\chi(p), \alpha_1', \ldots, \alpha_{n-1}', p^{n-k}\bar{\chi}(p));$$

in particular,

$$Q_p(v,F) = Q_p(v,F|\Phi)Q_p(p^{k-n}\chi(p)v, F|\Phi).$$

[Hint: Use Theorem 4.2.18, Proposition 4.2.19, and Exercise 4.2.21.]

Exercise 4.3.18. *Let* $F\in\mathfrak{M}_k^2(q,\chi)$ *be a non-zero eigenfunction for all operators* $|_{k,\chi}T$ ($T\in\mathsf{L}^2(q)$) *with the eigenvalues* $\lambda(T)=\lambda(T,F)$, *and let* $\alpha_0(p)$, $\alpha_1(p)$, $\alpha_2(p)$ *with* $p\in\mathbb{P}_{(q)}$ *be the parameters of the homomorphism* $\mathbf{T}\to\lambda(\mathrm{i}_q^{-1}(\mathbf{T}))$ *of* $\mathbf{L}_p^2 = \mathrm{i}_q(\mathsf{L}_p^2(q))$. *Prove the following assertions*:

(1) *The parameters satisfy*

$$\max(|\alpha_0(p)|, |\alpha_0(p)\alpha_1(p)|, |\alpha_0(p)\alpha_2(p)|, |\alpha_0(p)\alpha_1(p),\alpha_2(p)|) \le p^{2\tau k}$$

with the same τ *as in Theorem* 4.3.16;

(2) *If* $F|\Phi\neq 0$, *and* $k\neq 2$, *or* $k=2$ *and* $\chi(p)\neq -1$ *for all* $p\in\mathbb{P}_{(q)}$, *then the form* $F|\Phi\in\mathfrak{M}_k(q,\chi)$ *is an eigenfunction for all operators* $|T'$ *with* $T'\in\mathsf{L}(q)$, *and the zeta-function* (4.3.35) *of* F *satisfies*

$$\zeta(s,\gamma;F) = \zeta(s,\gamma;F|\Phi)\cdot\zeta(s-k+2,\chi\gamma;F|\Phi),$$

where

$$\zeta(s,\gamma;F|\Phi) = \sum_{m\in\mathbb{N}_{(q)}}\frac{\lambda'(m)\gamma(m)}{m^s} = \prod_{p\in\mathbb{P}_{(q)}}(1-\lambda'(p)\gamma(p)p^{-s}+\chi(p)\gamma(p^2)p^{-2s+k-1})^{-1}$$

with $\lambda'(m)=\lambda(T^1(m),F|\Phi)$ *is the similar zeta-function of* $F|\Phi$.

(3) *Assume, in addition, that* $q=1$, $\chi=1$, *and* F *is not a cusp form. Then the zeta-function* $\zeta(s,F)=\zeta(s,1,F)$ *can be meromorphically continued to all of the* s-*plane, has no poles outside the points* $s=0$, $k-2$, k, $2k-2$, *and satisfies the*

functional equation

$$\psi(2k-2-s, F) = (-1)^k \psi(s, F) \quad \textit{with } \psi(s, F) = (2\pi)^{-2s}\Gamma(s)\Gamma(s-k+2)\zeta(s, F)$$

where $\Gamma(s)$ is the gamma-function.

4.3.3. Modular Forms of Arbitrary Genus and Even Zeta-Functions

In this subsection we shall obtain identities analogous to (4.3.29) for modular forms of an arbitrary genus n. The following theory is essentially based on the symmetric factorization of polynomials $\mathbf{R}_p^n(v)$ obtained in §3.5.5. Because coefficients of the polynomials belong to even subrings of corresponding Hecke rings, we start from some remarks concerning the even subrings.

We let

$$\mathsf{E}^n(q) = (\mathsf{L}^n(q))^+ = \mathsf{D}_{\mathbb{Q}}(\Gamma_0^n(q), \mathsf{S}^n(q)^+) \tag{4.3.72}$$

denote the even subring of $\mathsf{L}^n(q)$. The same proof as those of Theorem 3.3.12 shows that $\mathsf{E}^n(q)$ is generated by the even subrings

$$\mathsf{E}_p^n(q) = (\mathsf{L}_p^n(q))^+ = \mathsf{D}_{\mathbb{Q}}(\Gamma_0^n(q), \mathsf{S}_p^n(q)^+) \tag{4.3.73}$$

of $\mathsf{L}_p^n(q)$ for all p in $\mathbb{P}_{(q)}$. The imbedding (3.4.4) maps each of the rings $\mathsf{E}_p^n(q)$ onto the even subring $\mathbf{E}_p^n = (\mathbf{L}_p^n)^+$ of $\mathbf{L}_p^n$. Since the spherical mapping Ω_p^n is monomorphic on $\mathbf{L}_p^n$, it easily follows that the image of $\mathbf{E}_p^n$ under the mapping consists of the polynomials in $\Omega(\mathbf{L}_p^n)$ of even degree in the variable x_0. In particular, all of the coefficients of the polynomial (3.3.52) belong to $\Omega(\mathbf{E}_p^n)$, and therefore the coefficients $\mathbf{r}_a^n(p)$ of $\mathbf{R}_p^n(v)$ belong to $\mathbf{E}_p^n$:

$$\mathbf{R}_p^n(v) \in \mathbf{E}_p^n[v]. \tag{4.3.74}$$

On the other hand, in the notations of Theorem 3.3.30, by (3.3.56) and (3.3.57), we obtain

$$\Omega(\mathbf{E}_p^n) = \mathbb{Q}[(t)^2, \rho_0^{\pm 1}, \rho_1, \ldots, \rho_{n-1}],$$

whence we may write

$$\begin{aligned}\Omega(\mathbf{L}_p^n) &= \mathbb{Q}[t, \rho_0^{\pm 1}, \rho_1, \ldots, \rho_{n-1}] \\ &= \Omega(\mathbf{E}_p^n)[t] \quad \text{with } t^2 \in \Omega(\mathbf{E}_p^n),\end{aligned}$$

which implies that $\mathbf{L}_p^n$ is an extension of $\mathbf{E}_p^n$ of degree 2, more precisely,

$$\mathbf{L}_p^n = \mathbf{E}_p^n \oplus \mathbf{T}\mathbf{E}_p^n, \quad \text{where } \mathbf{T} = \Omega^{-1}(t) \in \mathbf{L}_p^n \quad \text{and } \mathbf{T}^2 \in \mathbf{E}_p^n.$$

Hence we conclude that any non-zero $\mathbb{Q}$-linear homomorphism λ of $\mathbf{E}_p^n$ into $\mathbb{C}$ can be extended to a non-zero $\mathbb{Q}$-linear homomorphism λ' of $\mathbf{L}_p^n$ into $\mathbb{C}$ by setting $\lambda'(\mathbf{T}) = \pm\lambda(\mathbf{T}^2)^{1/2}$, and the extension is unique up to a sign of $\lambda'(\mathbf{T})$. Let $\alpha_0, \alpha_1, \ldots, \alpha_n$ be the parameters of λ'. Since the replacement of $\lambda'(\mathbf{T})$ by $-\lambda'(\mathbf{T})$ only substitutes $-\gamma$ for γ in the last equation in (3.3.83), it easily follows that the replacement replaces the system $\alpha_0, \alpha_1, \ldots, \alpha_n$ by $-\alpha_0, \alpha_1, \ldots, \alpha_n$. The numbers $\pm\alpha_0, \alpha_1, \ldots, \alpha_n$ will be called *the parameters of the homomorphism* $\lambda\colon \mathbf{E}_p^n \to \mathbb{C}$.

We now turn to eigenfunctions of even Hecke operators and the corresponding zeta-functions. As usual, we assume that a character ξ of $\{\pm 1\}$, an integer k, and a Dirichlet character χ satisfy $\xi(-1) = (-1)^k\chi(-1)$. If $F \in \mathfrak{M}_\xi^n$ (resp. $f \in \mathfrak{F}_\xi^n$) is an eigenfunction for all operators $|_{k,\chi}\mathbf{T}$ for $\mathbf{T} \in \mathbf{E}_p^n$ with the eigenvalues $\lambda(\mathbf{T}, F)$ (resp. $\lambda(\mathbf{T}, f)$), then, in view of (4.3.74), we can define the polynomial

$$R_p(v, F) = \sum_{a=0}^{2n} (-1)^a \lambda(\mathbf{r}_a^n(p), F)v^a \tag{4.3.75}$$

(resp.

$$R_p(v, f) = \sum_{a=0}^{2n} (-1)^a \lambda(\mathbf{r}_a^n(p), f)v^a). \tag{4.3.76}$$

Suppose that values of the function f are Fourier coefficients of the form F, and let $\pm\alpha_0(p), \alpha_1(p), \ldots, \alpha_n(p)$ be the parameters of the homomorphism $\mathbf{T} \to \lambda(\mathbf{T}, F) = \lambda(\mathbf{T}, f)$ of the ring $\mathbf{E}_p^n$ into $\mathbb{C}$, then we have the relations

$$R_p(v, F) = R_p(v, f) = \prod_{i=1}^{n} (1 - \alpha_i(p)^{-1}v)(1 - \alpha_i(p)v), \tag{4.3.77}$$

which easily follow from definitions.

Finally, turning to the global situation, we let $F \in \mathfrak{M}_k^n(q, \chi)$ be an eigenfunction for all *even Hecke operators*, i.e. the operators $|_{k,\chi}T$ with $T \in \mathsf{E}^n(q)$. Considering F as a function of $\mathfrak{M}_\xi^n$, we conclude that it is an eigenfunction for all operators $|_{k,\chi}\mathbf{T}$ with $\mathbf{T} \in \mathbf{E}_p^n = i_q(\mathsf{E}^n(q))$ and $p \in \mathbb{P}_{(q)}$. Hence, we can define the polynomials $R_p(v, F)$ for each $p \in \mathbb{P}_{(q)}$. We call the function

$$\zeta_p^+(s, \gamma, F) = R_p(\gamma(p)p^{-s}, F)^{-1} \tag{4.3.78}$$

and the Euler product

$$\zeta^+(s, \gamma, F) = \prod_{p \in \mathbb{P}_{(q)}} \zeta_p^+(s, \gamma, F), \tag{4.3.79}$$

where $\gamma\colon \mathbb{N}_{(q)} \to \mathbb{C}$ is a completely multiplicative function, the even local zeta-function of F with the character γ and the even zeta-function of F with the character γ, respectively. In the notation, the even zeta-functions (4.3.26) and (4.3.25) of § 4.3.1 can be written respectively as $\zeta_p^+(s, \chi, F)$ and $\zeta^+(s, \chi, F)$.

We now can formulate the main result of this subsection.

Theorem 4.3.19. *Let*

$$F(Z) = \sum_{A \in \mathbb{A}_n} f(A)e\{AZ\} \in \mathfrak{M}_k^n(q, \chi)$$

be a non-zero eigenfunction for all of the Hecke operators $|_{k,\chi}T$ *with* $T \in \mathsf{E}^n(q)$. *Then, for each completely multiplicative function* γ *on* $\mathbb{N}_{(q)}$ *and each fixed matrix* A_0 *of* $\mathbb{A}_n$, *the following formal identity holds*

$$\sum_{\substack{M\in\Lambda^n\backslash\mathrm{M}_n,\\ |\det M|\in\mathbb{N}_{(q)}}} \frac{\gamma(|\det M|)\bar{\chi}(\det M)f(MA_0{}^tM)}{|\det M|^{s-1}(\det M)^k}$$
$$= X(s,f,A_0)\left[\prod_{p\in\mathbb{P}_{(q)}} B_p^n(\gamma(p)p^{-s},A_0)\right]\zeta^+(s,\gamma,F), \tag{4.3.80}$$

where in the right side $\zeta^+(s,\gamma,F)$ *is the even zeta-function of* F, $B_p^n(v,A)$ *are the polynomials* (4.2.93) *calculated in Theorem* 4.2.27, *and where*

$$X(s,f,A_0) = \left(f|_{k,\chi}\prod_{\substack{p\in\mathbb{P}_{(q)}\\ p|\det A_0}}\left(\sum_{i=0}^{n}(-1)^i x_{-i}^n(p)p^{-is}\gamma(p^i)\right)\right)(A_0) \tag{4.3.81}$$

with the elements $x_{-i}^n(p)\in\mathcal{L}_p^n$ *defined by* (3.5.70), *the function* $f\colon \mathbb{A}_n\to\mathbb{C}$ *being considered as an element of the corresponding space* $\mathfrak{F}_\xi^n$.

Moreover, if F *is not a singular form, and* γ *satisfies* $|\gamma(m)|\le cm^\sigma$ $(m\in\mathbb{N}_{(q)})$ *with* c *and* σ *independent of* m, *then the Dirichlet series in the left side of* (4.3.80) *and the infinite products in the right side converge absolutely and uniformly in any half-plane of the form* $\operatorname{Re}s>(2\tau-1)k+\sigma+n+1+\varepsilon$ *with* $\varepsilon>0$, *where* $\tau=1/2$, *if* F *is a cusp form, and* $\tau=1$, *otherwise.*

Remarks. (1) By (3.5.20), $x_{-i}^n(p)\in\mathcal{C}_{-p}^n\subset\mathcal{C}_-^n(q)$. The ring $\mathcal{C}_-^n(q)$ is commutative, by Theorem 3.4.7. Hence, the right side in (4.3.81) is independent of the choice of the ordering of primes p.

(2) The action of operators $|_{k,\chi}x_{-i}^n(p)$ on $\mathfrak{F}_\xi^n$ is given by the formula (4.3.84) below.

(3) We shall see that, for non-singular F, there is a matrix $A_0\in\mathbb{A}_n$ such that $X(s,f,A_0)=f(A_0)\neq 0$.

We first prove three lemmas.

Lemma 4.3.20. *Let, for* $a\in\mathbb{N}$,

$$t^+(a)=t_n^+(a)=\sum_{\substack{D\in\Lambda^n\backslash\mathrm{M}_n/\Lambda^n,\\ \det D=\pm a}}(U(D))_{\Gamma_0^n}\in\mathcal{L}^n.$$

Then the following assertions hold:

(1) *The elements* $t^+(a)$ *belong to the subring* $\mathcal{C}_+^n(1)$ *of* $\mathcal{L}^n$; *in particular the elements commute with each other.*

(2) *If* a *and* b *are coprime, then*

$$t^+(a)t^+(b)=t^+(ab). \tag{4.3.82}$$

(3) *For any* $g\in\mathfrak{F}_\xi^n$ *with* $\xi(-1)=(-1)^k\chi(-1)$, $a\in\mathbb{N}_{(q)}$, *and* $A\in\mathbb{A}_n$, *one can write the formula*

$$a\sum_{\substack{M\in\Lambda^n\backslash\mathrm{M}_n,\\ \det M=\pm a}}\bar{\chi}(\det M)(\det M)^{-k}g(A[{}^tM])=(g|_{k,\chi}a^{-n}t^+(a))(A). \tag{4.3.83}$$

Proof. The assertion (1) follows from Proposition 3.4.4 and Theorem 3.4.7.

For $D \in \mathsf{M}_n$ with $|\det D| = a > 0$, by (3.4.22), (3.4.26), and (3.5.78), we have

$$\Phi^n((U(D))_{\Gamma_0}) = a^{n+1}(D)_{\Lambda^n},$$

whence, $\Phi^n(t^+(a)) = a^{n+1}t(a)$, where $t(a) = t_n(a) \in \mathsf{H}^n$ are elements of the form (3.2.10). Then the formula (4.3.82) follows from (3.2.14) and Theorem 3.4.7.

By (4.2.24), we have

$$(g|_{k,\chi}(U(D))_{\Gamma_0})(A) = a^{n+1} \sum_{M \in \Lambda\backslash\Lambda D\Lambda} \bar{\chi}(\det M)(\det M)^{-k} g(A[{}^tM]).$$

Dividing the two sides by a^n and summing over all double cosets $\Lambda D\Lambda \subset \mathsf{M}_n$ with $\det D = \pm a$, we obtain the last formula. □

Lemma 4.3.21. *Let $g \in \mathfrak{F}^n_\xi$ be an eigenfunction for all operators $|\mathbf{T}$ ($\mathbf{T} \in \mathbf{E}^n_p$), for a prime p, and let $R_p(v, g)$ be the polynomial* (4.3.76) *associated to g. Then, in the notation of the theorem, the following formal identity holds for every $A \in \mathbb{A}_n$:*

$$R_p(v, g) \cdot \sum_{\delta=0}^{\infty} (g|p^{-\delta n}t^+(p^\delta))(A)v^\delta = B^n_p(v, A) \sum_{i=0}^{n} (-1)^i (g|x^n_{-i}(p))(A)v^i.$$

Proof. With the notation (4.2.88), we can write the left side in the form

$$\sum_{\delta=0}^{\infty} \sum_{a=0}^{2n} (-1)^a (g|\mathbf{r}^n_a(p)|p^{-\delta n}t^+(p^\delta))(A)v^a v^\delta$$

$$= \left(g \left| \sum_{a=0}^{2n} (-1)^a \mathbf{r}^n_a(p)v^a \right| \sum_{\delta=0}^{\infty} p^{-\delta n}t^+(p^\delta)v^\delta\right)(A)$$

$$= \left(g|\mathbf{R}^n_p(v) \sum_{\delta=0}^{\infty} p^{-\delta n}t^+(p^\delta)v^\delta\right)(A),$$

since $\lambda(\mathbf{r}^n_a(p), g) \cdot g = g|\mathbf{r}^n_a(p)$. In view of the factorization (3.5.69) of $\mathbf{R}^n_p(v)$ and the identity (3.5.74), the last expression is equal to

$$\left(g \left| \left(\sum_{i=0}^{n} (-1)^i x^n_{-i}(p)v^i\right)\left(\sum_{i=0}^{n} (-1)^i b^n_i(p)v^i\right)\right.\right)(A)$$

$$= \sum_{i=0}^{n} (-1)^i ((g|x^n_{-i}(p))|B^n_p(v))(A)v^i$$

with $B^n_p(v) = \sum_{i=0}^{n} (-1)^i b^n_i(p)v^i$. Since each of the functions $g|x^n_{-i}(p)$ along with g is contained in $\mathfrak{F}^n_\xi$, it follows, by Proposition 4.2.24, that

$$((g|x^n_{-i}(p))|B^n_p(v))(A) = B^n_p(v, A)(g|x^n_{-i}(p))(A).$$

The lemma follows from these relations and Theorem 4.2.27. □

Lemma 4.3.22. *The elements $x^n_{-i}(p)$ ($i = 0, 1, \ldots, n$) act on a function g of $\mathfrak{F}^n_\xi$ by the formula*

$$(g|_{k,\chi}x_{-i}^{n}(p))(A) \tag{4.3.84}$$
$$= p^{nk-\langle n\rangle+\langle n-i\rangle}\chi(p)^{n} \sum_{\substack{D\in\Lambda\backslash\Lambda D_{n-i}\Lambda,\\ A[{}^tD]\in p^2\mathbb{A}_n}} \bar{\chi}(\det D)(\det D)^{-k}g(p^{-2}A[{}^tD])$$

where $D_{n-i} = D_{n-i}^{n}(p) = \begin{pmatrix} E_i & 0 \\ 0 & pE_{n-i} \end{pmatrix}$. *In particular,*

$$(g|_{k,\chi}x_{-i}^{n}(p))(A) = 0, \qquad \text{if } p^{2i} \nmid \det A. \tag{4.3.85}$$

Proof. The formula (4.3.84) follows from formulae (3.5.70) and Lemma 4.2.8. If $A[{}^tD]\in p^2\mathbb{A}_n$ with a $D\in\Lambda D_{n-i}\Lambda$, then $p^{2n-2i}\det A\in p^{2n}\mathbb{Z}$, hence $p^{-2i}\det A\in\mathbb{Z}$. □

We now return to the theorem.

Proof of Theorem 4.3.19. For $g\in\mathfrak{F}_{\xi}^{n}$, $A\in\mathbb{A}_n$, and a subset Y of $\mathbb{N}_{(q)}$, we consider formal Dirichlet series of the form

$$D(s,A,g,\mathsf{Y}) = \sum_{\substack{M\in\Lambda^n\backslash\mathsf{M}_n,\\ |\det M|\in\mathsf{Y}}} \frac{\gamma(|\det M|)\bar{\chi}(\det M)g(A[{}^tM])}{|\det M|^{s-1}(\det M)^k}$$

In the notation, the series in the left side of (4.3.80) equals $D(s,A_0,f,\mathbb{N}_{(q)})$. By (4.3.83), we can write

$$D(s,A,g,\mathsf{Y}) = \sum_{a\in\mathsf{Y}} \frac{\gamma(a)}{a^s}\cdot a \sum_{\substack{M\in\Lambda\backslash\mathsf{M}_n\\ \det M=\pm a}} \bar{\chi}(\det M)(\det M)^{-k}g(A[{}^tM])$$
$$= \sum_{a\in\mathsf{Y}} \frac{\gamma(a)(g|a^{-n}t^{+}(a))(A)}{a^s}.$$

Suppose, now, that there is a prime p of $\mathbb{P}_{(q)}$ such that the function g is an eigenfunction for all operators $|\mathbf{T}$, where $\mathbf{T}\in\mathbf{E}_p^n$, with the same eigenvalues as f and the set Y is of the form

$$\mathsf{Y} = \bigcup_{\delta=0}^{\infty} p^{\delta}\mathsf{Y}_1 \qquad \text{with a set } \mathsf{Y}_1\subset\mathbb{N}_{(pq)}. \tag{4.3.86}$$

Then, by (4.3.77) and (4.3.82), we have

$$R_p(v,F)D(s,A,g,\mathsf{Y}) = R_p(v,F)\sum_{a\in\mathsf{Y}_1}\sum_{\delta=0}^{\infty} \frac{\gamma(p^{\delta}a)(g|(p^{\delta}a)^{-n}t^{+}(p^{\delta}a))(A)}{(p^{\delta}a)^s}$$
$$= \sum_{a\in\mathsf{Y}_1} \frac{\gamma(a)}{a^s}R_p(v,g)\sum_{\delta=0}^{\infty}(g|a^{-n}t^{+}(a)|p^{-\delta n}t^{+}(p^{\delta}))(A)v^{\delta},$$

where $v = \gamma(p)p^{-s}$. If $a = \prod_i p_i^{\delta_i}$, where p_i are pairwise distinct prime numbers, then $t^{+}(a) = \prod_i t^{+}(p_i^{\delta_i})$, by (4.3.82). Since $t^{+}(p_i^{\delta_i})\in\mathscr{C}_{p_i}^{n}$, it follows, by Proposition

3.4.11(1), that $t^+(p_i^{\delta_i})$ commutes with each element of $\mathbf{E}_p^n \subset \mathbf{L}_p^n$ if $p_i \neq p$. Hence, $t^+(a)$ commutes with each element of $\mathbf{E}_p^n$ if a is prime to p. Thus, for each a of Y_1, the function $g_a = g|a^{-n}t^+(a)$ together with $g = g_1$ is an eigenfunction for all operators $|\mathbf{T}$, $(\mathbf{T} \in \mathbf{E}_p^n)$ with the same eigenvalues as g. Thus, by Lemma 4.3.21, we obtain

$$R_p(v,g)\cdot \sum_{\delta=0}^{\infty} (g|a^{-n}t^+(a)|p^{-\delta n}t^+(p^\delta))(A)v^\delta = B_p^n(v,A)\sum_{i=0}^{n}(-1)^i(g_a|x_{-i}^n(p))(A)v^i.$$

Since $x_{-i}^n(p) \in \mathscr{C}_{-p}^n$ and $t^+(p_i^{\delta_i}) \in \mathscr{C}_{p_i}^n$, it follows that elements $t^+(a) = \prod_i t^+(p_i^{\delta_i})$ with $a \in \mathbb{N}_{(p)}$ commute with elements $x_{-i}^n(p)$, by Proposition 3.4.11(2). Hence,

$$g_a|x_{-i}^n(p) = g|x_{-i}^n(p)|a^{-n}t^+(a), \qquad (a \in \mathbb{N}_{(p)}). \tag{4.3.87}$$

These formulae imply the following formal identity

$$\begin{aligned} R_p(v,F)D(s,A,g,\mathsf{Y}) &= B_p^n(v,A)\sum_{a\in\mathsf{Y}_1}\frac{\gamma(a)}{a^s}\sum_{i=0}^{h}(-1)^i(g|x_{-i}^n(p)|a^{-n}t^+(a))(A)v^i \\ &= B_p^n(v,A)\sum_{i=0}^{n}(-1)^i D(s,A,g|x_{-i}^n(p),\mathsf{Y}_1). \end{aligned} \tag{4.3.88}$$

In particular, if $p \nmid \det A$, then, by (4.3.87) and (4.3.85), we have

$$\begin{aligned} D(s,A,g|x_{-i}^n(p),\mathsf{Y}_1) &= \sum_{a\in\mathsf{Y}_1}\frac{\gamma(a)}{a^s}(g|a^{-n}t^+(a)|x_{-i}^n(p))(A) \\ &= \begin{cases} 0, & \text{if } i = 1,2,\dots,n \\ D(s,A,g,\mathsf{Y}_1), & \text{if } i = 0 \end{cases}, \end{aligned}$$

and (4.3.88) turns into

$$R_p(v,F)D(s,A,g,\mathsf{Y}) = B_p^n(v,A)D(s,A,g,\mathsf{Y}_1). \tag{4.3.89}$$

The proof of (4.3.80) will proceed in the same way as that of (4.3.64). We let $p_1, \dots, p_b$ be all distinct prime divisors of $\det A_0$ belonging to $\mathbb{N}_{(q)}$, and we set $\mathbb{P}_{(q \det A_0)} = \{p_{b+1}, p_{b+2}, \dots\}$ with $p_i < p_{i+1}$. From (4.3.89), by an obvious induction on $c \in \mathbb{N}$ left to the reader, we get the identities

$$D(s,A_0,f,\mathbb{N}_{(q)}) = D(s,A_0,f,\mathbb{N}_{(q\cdot q(c))})\left[\prod_{p|q(c)}B_p^n(\gamma(p)p^{-s},A_0)\right]\prod_{p|q(c)}R_p(\gamma(p)p^{-s},F)^{-1}$$

where $q(c) = p_{b+1}\cdots p_{b+c}$. Taking the coefficient-wise limit of the Dirichlet series in the two sides as $c \to \infty$, we obtain the identity

$$\begin{aligned} &D(s,A_0,f,\mathbb{N}_{(q)}) \\ &\quad = D(s,A_0,f,\mathsf{Y}(p_1,\dots,p_b))\left[\prod_{p\in\mathbb{P}_{(q\cdot\det A_0)}}B_p(\gamma(p)p^{-s},A_0)\right]\prod_{p\in\mathbb{P}_{(q\cdot\det A_0)}}\zeta_p^+(s,\gamma,F) \end{aligned}$$

where $\mathsf{Y}(p_1,\dots,p_b) = \{a \in \mathbb{N}_{(q)}, a|(p_1\cdots p_b)^\infty\}$. Thus, in order to prove (4.3.80), we only have to establish the identities of the form

$$\left[\prod_{p|p_1\cdots p_r} R_p(\gamma(p)p^{-s},F)\right]D(s,A,g,\mathsf{Y}(p_1\cdots p_r))$$
$$= X_r(s,g,A)\cdot \prod_{p|p_1\cdots p_r} B_p(\gamma(p)p^{-s},A) \tag{4.3.90}$$

with $p_1,\ldots,p_r\in\mathbb{P}_{(q)}$ and $A\in\mathbb{A}^n$, where $g\in\mathfrak{F}_\xi^n$ is an eigenfunction for all operators $|\mathbf{T}$, where $\mathbf{T}\in\mathbf{E}^n_{p_1},\ldots,\mathbf{E}^n_{p_r}$, with the same eigenvalues as f, and where

$$X_r(s,g,A)=\left(g\left|\prod_{p|p_1\cdots p_r}\left(\sum_{i=0}^n(-1)^i x^n_{-i}(p)(\gamma(p)p^{-s})^i\right)\right.\right)(A).$$

We shall apply induction to r. If $r=1$, then (4.3.90) follows from (4.3.88) with $\mathsf{Y}=\mathsf{Y}(p_1)$ and $\mathsf{Y}_1=\{1\}$. Suppose the identity is already proved for some $r\geq 1$. If $\mathsf{p}=p_{r+1}\in\mathbb{P}_{(q)}$ is a prime number different from $p_1,\ldots,p_r$, then, by (4.3.88), we have

$$R_{\mathsf{p}}(\gamma(\mathsf{p})\mathsf{p}^{-s},F)D(s,A,g,\mathsf{Y}(p_1,\ldots,p_r,\mathsf{p}))$$
$$=\left(\sum_{i=0}^n(-1)^iD(s,A,g|x^n_{-i}(\mathsf{p}),\mathsf{Y}(p_1,\ldots,p_r))(\gamma(\mathsf{p})\mathsf{p}^{-s})^i\right)B^n_{\mathsf{p}}(\gamma(\mathsf{p})\mathsf{p}^{-s},A).$$

Since $x^n_{-i}(\mathsf{p})\in\mathscr{C}^n_{-\mathsf{p}}$, it follows from Proposition 3.4.11(1) that the function $g|x^n_{-i}(\mathsf{p})$ together with g is an eigenfunction for all operators $|\mathbf{T}$, where $\mathbf{T}\in\mathbf{E}^n_{p_1},\ldots,\mathbf{E}^n_{p_r}$, with the same eigenvalues as f. Thus, from the last identity, by the inductive hypothesis, we obtain

$$\left[\prod_{p|p_1\cdots p_r} R_b(\gamma(p)p^{-s},F)\right]R_{\mathsf{p}}(\gamma(\mathsf{p})\mathsf{p}^{-s},F)D(s,A,g,Y(p_1,\ldots,p_r,\mathsf{p}))$$
$$=\left(\sum_{i=0}^n(-1)^iX_r(s,g|x^n_{-i}(\mathsf{p}),A)(\gamma(\mathsf{p})\mathsf{p}^{-s})^i\right)$$
$$\times\left[\prod_{p|p_1\cdots p_r} B^n_p(\gamma(p)p^{-s},A)\right]B^n_{\mathsf{p}}(\gamma(\mathsf{p})\mathsf{p}^{-s},A).$$

Since all of the elements $x^n_{-j}(p)$ belong to the commutative ring $\mathscr{C}^n_{-}(1)$, it follows that

$$\sum_{i=0}^n(-1)^iX_r(s,g|x^n_{-i}(\mathsf{p}),A)(\gamma(\mathsf{p})\mathsf{p}^{-s})^i$$
$$=\sum_{i=0}^n(-1)^i\left(g|x^n_{-i}(\mathsf{p})\prod_{p|p_1\cdots p_r}\left(\sum_{j=0}^n(-1)^jx^n_{-j}(p)(\gamma(p)p^{-s})^j\right)\right)(A)(\gamma(\mathsf{p})\mathsf{p}^{-s})^i$$
$$=\left(g\left|\prod_{p|p_1\cdots p_{r+1}}\left(\sum_{i=0}^n(-1)^ix^n_{-i}(p)(\gamma(p)p^{-s})^i\right)\right.\right)(A),$$

which proves (4.3.90).

It remains to prove the assertions concerning convergence. By (2.3.20) and (2.3.43), we conclude that the Dirichlet series in the left side of (4.3.80) is dominated by the series

$$c' \sum_{M \in \Lambda \backslash \mathrm{M}_n, |\det M| \in \mathbb{N}_{(q)}} 1/|\det M|^{\mathrm{Re}\, s-(2\tau-1)k-\sigma-1}$$

with a constant $c' \geq 0$. Let us choose representatives for the left cosets modulo Λ in the form satisfying (3.2.15). Since the number of the representatives with fixed diagonal entries $c_{11}, c_{22}, \ldots, c_{nn}$ is clearly $c_{22}c_{33}^2 \cdots c_{nn}^{n-1}$, it follows that the last series is domineted by the series

$$c' \sum_{c_{11}, c_{22}, \ldots, c_{nn} \in \mathbb{N}_{(q)}} \left[\prod_{i=1}^{n} c_{ii}^{i-1}\right] \Big/ \left[\sum_{i=1}^{n} c_{ii}\right]^t = c' \prod_{i=1}^{n} \sum_{c \in \mathbb{N}_{(q)}} 1/c^{t-i+1}$$

with $t = \mathrm{Re}\, s - (2\tau - 1)k - \sigma - 1$, which has positive terms and converges uniformily in $t \geq n + \varepsilon$, i.e. $\mathrm{Re}\, s \geq (2\tau - 1)k + \sigma + n + 1 + \varepsilon$, with $\varepsilon > 0$. The formulae of Theorem 4.2.27 imply, by standard arguments, that the product of the polynomials $B_p^n(\gamma(p)p^{-s}, A_0)$ in the right side converges absolutely and uniformly in $\mathrm{Re}\, s > 1 + \sigma + \varepsilon$, and the absolute value of the product is bounded from below by a positive constant in each of the half-planes. These arguments along with the identity (4.3.80) imply that the Dirichlet series representing the product $X(s, A_0, f)\zeta^+(s, \gamma, F)$ converges absolutely and uniformly in each of the half-planes indicated in the theorem. Because the zeta-function $\zeta^+(s, \gamma, F)$ is independent of the choice of A_0, in order to complete the proof, it suffices to show that, for some $A_0 \in \mathbb{A}_n$, the term $X(s, A_0, f)$ is equal to a non-zero constant. Since the form F is non-singular, it follows that there are matrices $A \in \mathbb{A}_n^+$ such that $f(A) \neq 0$. Let A_0 be such a matrix whose determinant is minimal. Then the formulae (4.3.81) and (4.3.84) imply that $X(s, A_0, f) = f(A_0) \neq 0$. □

Exercise 4.3.23. *With the notation and assumptions of Theorem* 4.3.19, *show that the roots* $\alpha_i(p)^{\pm 1}$ *of the polynomials* $R_p(v, F)$ *satisfy*

$$|\alpha_i(p)^{\pm 1}| \leq p^{(2\tau-1)k+n} \qquad (i = 1, 2, \ldots, n; p \in \mathbb{P}_{(q)})$$

Exercise 4.3.24. *Let* $F \in \mathfrak{M}_\xi^n$ *be an eigenfunction for all operators* $|_{k,\chi}\mathbf{T}$ *with* $\mathbf{T} \in \mathbf{E}_p^n$, *and let* $F|\Phi \neq 0$. *Show that the form* $F|\Phi \in \mathfrak{M}_\xi^{n-1}$ *is an eigenfunction for all operators* $|_{k,\chi}\mathbf{T}'$ *with* $\mathbf{T}' \in \mathbf{E}_p^{n-1}$, *and the corresponding polynomials* (4.3.75) *satisfy*

$$R_p(v, F) = (1 - p^{n-k}\bar{\chi}(p)v)(1 - p^{k-n}\chi(p)v)R_p(v, F|\Phi).$$

[Hint: See Exercise 4.3.17(3).]

Exercise 4.3.25. *Investigate the convergence of* $\zeta^+(s, \gamma, F)$ *if the form F is singular.*
[Hint: Use Exercise 4.3.24 and Theorem 4.3.19.]

Chapter 5. The Action of Hecke Operators on Theta-Series

In the previous chapter we have exposed the classical scheme of applications of Hecke operators to the study of multiplicative properties of Fourier coefficients of modular forms. The scheme is based on relations between Fourier coefficients of eigenfunctions of Hecke operators and the corresponding eigenvalues. Only in the case of modular forms of one variable this approach reveals the multiplicative properties of individual Fourier coefficients. In other cases we can only establish the multiplicativity of certain sums of the Fourier coefficients, which appear as coefficients of the Dirichlet series in the left side of the identities (4.3.64) and (4.3.80). Perhaps there are other multiplicative sums of the Fourier coefficients, which can be expressed in terms of the eigenvalues, but one can hardly expect that general Fourier coefficients of eigenfunctions by themselves can be expressed in the terms. Note it is still unknown, even for the spaces $\mathfrak{M}_k^2$, whether an eigenfunction for all Hecke operators is determined by the corresponding eigenvalues up to a constant factor. In fact, the formula (4.2.19) shows that any Hecke operator does not relate the Fourier coefficients indexed by the matrices A of $\mathbb{A}_n$ which belong to the distinct classes modulo the equivalence $A \sim \mu^{-1} DA\,{}^tD$ with $\mu \in \mathbb{Q}$ and $D \in \mathrm{GL}_n(\mathbb{Q})$, and one can easily see that the number of the classes equals 2, if $n = 1$, is infinite, if $n > 1$. This suggests that Fourier coefficients of an eigenfunction of genus $n > 1$ can be determined by the corresponding eigenvalues only up to an infinite family of parameters. Relations between the parameters probably are of analytic nature. It looks like the Hecke operators bring far less information on the Fourier coefficients of modular forms of genus $n > 1$, than in the case $n = 1$.

After arriving at such a dim conclusion, we have a good reason to remember that the modular forms by themselves were introduced as a technical tool to study the theta-series of quadratic forms, and that the main object of the theory is the Fourier coefficients of the theta-series, rather than those of arbitrary modular forms. A concrete situation may be more meaningful than the general problem. So it would be just natural to look at the action of Hecke operators on the theta-series themselves.

In this chapter we shall show that, as a rule, Hecke operators map theta-series into linear combinations of theta-series, and the linear combinations will be explicitely determined (see Theorem 5.1.15). It turns out that entries of matrices of Hecke operators acting on certain spaces of theta-series can be expressed in terms of numbers of integral representations of quadratic forms by quadratic forms. This opens a way to study relations between the numbers of representations. The future will show where does the way lead.

§ 5.1. The Action of Hecke Operators on Theta-Series

In this section the explicit formulae will be obtained, which express images under the action of Hecke operators of theta-series of integral positive definite quadratic forms in even number of variables as linear combinations of similar theta-series.

5.1.1. Theta-Series and ε-Series

According to Proposition 1.1.3 and the first formula of Proposition 1.3.14, the theta-series

$$\theta^n(Z, Q) = \sum_{N \in \mathbb{M}_{m,n}} e\{Q[N]Z\} = \sum_{A \in \mathbb{A}_n} r(Q, A)e\{AZ\} \tag{5.1.1}$$

of genus n of a matrix $Q \in \mathbb{A}_m^+$ can be considered as a modular form of the unit character for the triangular group Γ_0^n, i.e.

$$\theta^n(Z, Q) \in \mathfrak{M}_+^n = \mathfrak{M}_{\xi^+}^n, \quad \text{where } \xi^+(-1) = 1 \tag{5.1.2}$$

(see (4.2.8)). The spaces $\mathfrak{M}_+^n$ will serve us as big spaces containing all of the functions under consideration.

The spaces $\mathfrak{M}_+^n$ are related to each other by the Siegel operator Φ and its iterations (see around (4.2.58)). The action of the Siegel operator on theta-series was determined in Proposition 2.3.10; in particular, for the series of the form $\theta^n(Z, Q) = \theta^n(Z, Q|0)$, we have the formula

$$(\theta^n|\Phi)(Z, Q) = \theta^{n-1}(Z, Q), (Z \in \mathbb{H}_{n-1}, Q \in \mathbb{A}_m^+). \tag{5.1.3}$$

This formula along with the Žarkovskaya relations (4.2.69) will allow us to reduce the problem of the action of Hecke operators on theta-series of genus n of matrices of $\mathbb{A}_m^+$ to the special case when $n = m$, which will be refered to as the basic case. The following lemma on linear independence of theta-series play an important role in the reduction.

Lemma 5.1.1. *Let $Q_1, \ldots, Q_h$ be a finite set of matrices of $\mathbb{A}_m^+$ such that the $\mathbb{Z}$-equivalence classes $\{Q_1\}, \ldots, \{Q_h\}$ are pairwise distinct. Then, for any $n \geq m$, the theta-series $\theta^n(Z, Q_1), \ldots, \theta^n(Z, Q_h)$ are linearly independent.*

Proof. If

$$\alpha_1\theta^n(Z, Q_1) + \cdots + \alpha_h\theta^n(Z, Q_h) = 0$$

with some $n \geq m$, then, applying the linear operator $\Phi(n-m)$ times and using each time the formula (5.1.3), we obtain the relation

$$\alpha_1 \theta^m(Z, Q_1) + \cdots + \alpha_h \theta^m(Z, Q_h) = 0$$

which is equivalent with the following system of linear relations for the corresponding Fourier coefficients:

$$\alpha_1 r(Q_1, A) + \cdots + \alpha_h r(Q_h, A) = 0, \ (A \in \mathbb{A}_m). \tag{5.1.4}$$

It easily follows from definitions that, for every $Q, A \in \mathbb{A}_m^+$ with $\det Q \geq \det A$ the following relations hold

$$r(Q, A) = \begin{cases} r(Q, Q) = e(Q) \geq 1, & \text{if } \{Q\} = \{A\} \\ 0, & \text{otherwise.} \end{cases} \tag{5.1.5}$$

Suppose, now, that $\det Q_1 \leq \det Q_2 \leq \cdots \leq Q_h$, and substitute $A = Q_1$ in (5.1.4). Then, by (5.1.5), we get $\alpha_1 e(Q_1) = 0$, and so $\alpha_1 = 0$. Similarly, $\alpha_2 = \cdots = \alpha_h = 0$. □

To explain other reduction of the problem, we need some definitions. Let F be a modular form of $\mathfrak{M}_\xi^n$ with the Fourier coefficients $f(A)$, $(A \in \mathbb{A}_n)$. For $0 \leq r \leq n$, we set

$$F^{(r)}(Z) = \sum_{A \in \mathbb{A}_n \ \operatorname{rank} A = r} f(A) e\{AZ\}.$$

It is clear that each of the functions $F^{(r)}$ is again contained in $\mathfrak{M}_\xi^n$, the mapping $F \to F^{(r)}$ is linear, and $F = F^{(0)} + F^{(1)} + \cdots + F^{(n)}$. The last decomposition and the formulae (4.2.16) imply that any Hecke operator $|_{k,\chi} T$ on $\mathfrak{M}_\xi^n$ with $T \in \mathscr{L}^n$, where k and χ satisfy $(-1)^k \chi(-1) = \xi(-1)$, commutes with the operator $F \to F^{(r)}$, i.e.

$$(F|_{k,\chi} T)^{(r)} = F^{(r)}|_{k,\chi} T, \quad (0 \leq r \leq n). \tag{5.1.6}$$

We call the function $F^{(r)}$ *the r-component of F*, and $F^{(n)}$ is *the leading component of F*. The leading component of a theta-series $\theta^n(Z, Q)$ will be denoted by $\theta^{(n)}(Z, Q)$. In the basic case the leading component of a theta-series is a series of the form

$$\theta^{(m)}(Z, Q) = \sum_{A \in \mathbb{A}_m^+} r(Q, A) e\{AZ\} = \sum_{N \in \mathsf{M}_m, \ \det N \neq 0} e\{Q[N]Z\} \tag{5.1.7}$$

with $Q \in \mathbb{A}_m^+$. The same proof as that of Lemma 5.1.1 shows that the mapping $\theta^m \to \theta^{(m)}$ defines an isomorphism of any linear space spanned by theta-series of genus m of matrices of $\mathbb{A}_m^+$ onto the space spanned by the leading components of the theta-series. Hence, in view of (5.1.6), one may expect that the action of Hecke operators on theta-series of genus m of matrices of $\mathbb{A}_m^+$ can be reduced to their action on the leading components of the series.

For further simplification of the series under consideration, we note that each of the series (5.1.7) is uniquely determined by the partial sum

$$\varepsilon(Z, Q) = \sum_{N \in \mathsf{M}_m, \det N = \pm 1} e\{Q[N]Z\} = \sum_{V \in \Lambda^m} e\{Q[V]Z\}, \tag{5.1.8}$$

because the functions $\varepsilon(Z, Q_1), \ldots, \varepsilon(Z, Q_h)$ with arbitrary pairwise non-equivalent over $\mathbb{Z}$ matrices $Q_1, \ldots, Q_h$ of $\mathbb{A}_m^+$ are clearly linearly independent. The series (5.1.8) will be called *the ε-series of Q*. As we shall see below, all Hecke operators on $\mathfrak{M}_+^n$ carry ε-series to finite linear combinations of ε-series. This reduces the problem of the action of Hecke operators on the leading components (5.1.7) to the question whether the operators commute in some sense with a linear mapping carrying $\varepsilon(Z, Q)$ to $\theta^{(m)}(Z, Q)$. At this point we shall show that such a mapping can be written in terms of suitable Hecke operators. The corresponding elements of the ring $\mathscr{L}^n$ and their action on $\mathfrak{M}_+^n$ are described by the following lemma dual to Lemma 4.3.20.

Lemma 5.1.2. *Let, for $a \in \mathbb{N}$,*

$$t^-(a) = t_n^-(a) = \sum_{\substack{D \in \Lambda^n \backslash \mathsf{M}_n / \Lambda^n, \\ \det D = \pm a}} (U(D^*))_{\Gamma_0^n} \in \mathscr{L}^n.$$

Then the following assertions hold:

(1) *The elements $t^-(a)$ belong to the subring $\mathscr{C}_-^n(1)$ of $\mathscr{L}^n$; in particular, the elements commute with each other.*

(2) *If a and b are coprime*, then

$$t^-(a)t^-(b) = t^-(ab). \tag{5.1.9}$$

(3) *If $F \in \mathfrak{M}_+^n$ and $a \in \mathbb{N}$, then*

$$\sum_{M \in \Lambda^n \backslash \mathsf{M}_n, \det M = \pm a} F(Z[{}^tM]) = \bar{\chi}(a) a^{-k} (F|_{k,\chi} t^-(a))(Z), \tag{5.1.10}$$

where one assumes that $\chi(-1) = (-1)^k$ and $\chi(a) \neq 0$.

Proof. It follows from definitions that the elements $t^-(a)$ are dual to the elements $t^+(a)$ of Lemma 4.3.20 with respect to the antiautomorphism $*$:

$$t^-(a)^* = t^+(a), \; t^+(a)^* = t^-(a). \tag{5.1.11}$$

Thus, the assertions (1) and (2) directly follow from the corresponding assertions of Lemma 4.3.20 and (3.4.18). By (4.2.21), we can write

$$\begin{aligned}
\bar{\chi}(a) a^{-k} F|_{k,\chi} t^-(a) &= \bar{\chi}(a) a^{-k} \sum_{\substack{D \in \Lambda \backslash \mathsf{M}_n / \Lambda \\ \det D = \pm a}} \sum_{N \in \Lambda \backslash \Lambda D^* \Lambda} \bar{\chi}(\det N)(\det N)^{-k} F(Z[N^{-1}]) \\
&= \bar{\chi}(a) a^{-k} \sum_{D} \sum_{M \in \Lambda \backslash \Lambda D \Lambda} \chi(\det M)(\det M)^k F[{}^tM] \\
&= \sum_{M \in \Lambda \backslash \mathsf{M}_n, \det M = \pm a} F(Z[{}^tM]),
\end{aligned}$$

because $\det M = \pm \det D = \pm a$ in each of the terms, and $\chi(\pm a)(\pm a)^k = \chi(a) a^k$. $\square$

According to (5.1.10), the leading component $\theta^{(m)}(Z,Q)$ with $Q\in\mathbb{A}_m^+$ can be reconstructed from $\varepsilon(Z,Q)$ in the form

$$\begin{aligned}\theta^{(m)}(Z,Q) &= \sum_{a\in\mathbb{N}}\ \sum_{M\in\mathrm{M}_m,\ \det M=\pm a} e\{Q[M]Z\}\\ &= \sum_{a\in\mathbb{N}}\ \sum_{\substack{M\in\Lambda\backslash\mathrm{M}_m,\\ \det M=\pm a}}\ \sum_{V\in\Lambda^m} e\{{}^tVQVMZ\,{}^tM\}\\ &= \sum_{a\in\mathbb{N}}(\varepsilon|_{0,1}t^-(a))(Z,Q) = \varepsilon(Z,Q)|_{0,1}\sum_{a=1}^{\infty}t^-(a),\end{aligned}\tag{5.1.12}$$

where 1 is the unit character modulo 1 (for any character modulo $q>1$, it would be necessary to exclude all terms with $a\notin\mathbb{N}_{(q)}$).

We now consider the action of Hecke operators on ε-series.

Proposition 5.1.3. *Let $k\in\mathbb{Z}$, and let χ be a Dirichlet character modulo q with $\chi(-1)=(-1)^k$. Then the image of the ε-series $\varepsilon(Z,Q)$ of a matrix Q of $\mathbb{A}_m^+$ under a Hecke operator $|_{k,\chi}T$ with $T\in\mathscr{L}^m(q)$ is a finite linear combination of ε-series of matrices of $\mathbb{A}_m^+$. More precisely:*

$$\varepsilon(Z,Q)|_{k,\chi}T = \sum_{\{Q'\}\in\mathbb{A}_m^+}\varepsilon(Q,Q';T)\varepsilon(Z,Q'),\tag{5.1.13}$$

where the sum is taken over a set of representatives for the classes of $\mathbb{A}_m^+$ modulo the equivalence over $\mathbb{Z}$, and where the coefficients

$$\varepsilon(Q,Q',T)=\varepsilon_{k,\chi}(Q,Q',T)\tag{5.1.14}$$

are determined by the following conditions:

(1) *If $M_0=\begin{pmatrix}\mu D_0^* & B_0\\ 0 & D_0\end{pmatrix}\in\mathscr{S}^m(q)$ and $Q,\,Q'\in\mathbb{A}_m^+$, then*

$$\varepsilon(Q,Q',(M_0)_{\Gamma_0})=\alpha_{k,\chi}(M_0)\sum_{D\in\Lambda D_0\Lambda/\Lambda,\ \mu Q[D^*]\in\{Q'\}} s(Q,D,M_0),\tag{5.1.15}$$

with

$$\alpha_{k,\chi}(M_0)=\mu^{mk-\langle m\rangle}(\det D_0)^{-k}\chi(\det\mu D_0^*),\tag{5.1.16}$$

$s(Q,D,M_0)$ being the sums (4.2.20), *and $\Lambda=\Lambda^m$;*

(2) *The function $\varepsilon(Q,Q';T)$ is linear in T.*

The function $\varepsilon(Q,Q';T)$ on $\mathbb{A}_m^+\times\mathbb{A}_m^+\times\mathscr{L}^m(q)$ defined by the conditions (1), (2) *has the following properties: the function depends only on the classes $\{Q\}$ and $\{Q'\}$ of the matrices Q and Q'; if the class $\{Q\}$ (resp. $\{Q'\}$) and the element T are fixed, then $\varepsilon(Q,Q',T)=0$ for all but a finite number of the classes $\{Q'\}$ (resp. $\{Q\}$); for every Q,Q'* of *$\mathbb{A}_m^+$ and T,T' of $\mathscr{L}^m(q)$ the relation holds*

$$\varepsilon(Q,Q',TT')=\sum_{\{A\}\in\mathbb{A}_m^+}\varepsilon(Q,A,T)\varepsilon(A,Q',T').\tag{5.1.17}$$

Proof. We start from the action of the operator $|_{k,\chi}(M_0)_{\Gamma_0}$. Because $(-1)^k\chi(-1)=1$, one can write

$$\varepsilon(Z,Q) = \sum_{V \in \Lambda} e\{QVZ{}^tV\} = \sum_{V \in \Lambda} e\{QZ\}|_{k,\chi} U(V^*)$$

$$\overset{V \to V^*}{=} \sum_{V \in \Lambda} e\{QZ\}|_{k,\chi} U(V). \tag{5.1.18}$$

Hence, by definitions, Lemma 3.3.25, and (4.1.4), we have

$$\begin{aligned} \varepsilon(Z,Q)|_{k,\chi}(M_0)_{\Gamma_0} &= \sum_{\substack{D \in \Lambda \backslash \Lambda D_0 \Lambda \\ B \in \mathrm{B}(D,M_0)/\mathrm{mod}\, D}} \varepsilon(Z,Q)|_{k,\chi} \begin{pmatrix} \mu D^* & B \\ 0 & D \end{pmatrix} \\ &= \sum_{V \in \Lambda} \sum_{\substack{D \in \Lambda \backslash \Lambda D_0 \Lambda, \\ B \in \mathrm{B}(D,M_0)/\mathrm{mod}\, D}} e\{QZ\}|_{k,\chi} \begin{pmatrix} V^* & 0 \\ 0 & V \end{pmatrix} \begin{pmatrix} \mu D^* & B \\ 0 & D \end{pmatrix} \\ &= \sum_{\substack{D \in \Lambda D_0 \Lambda \\ B \in \mathrm{B}(D,M_0)/\mathrm{mod}\, D}} e\{QZ\}|_{k,\chi} \begin{pmatrix} \mu D^* & B \\ 0 & D \end{pmatrix}, \end{aligned}$$

where we have also used the obvious equalities of the sets of representatives

$$V^*\{\mathrm{B}(D,M_0)/\mathrm{mod}\, D\}\, V' = \mathrm{B}(VDV',M_0)/\mathrm{mod}\, VDV', (V, V' \in \Lambda) \tag{5.1.19}$$

with $V' = E_m$. We now decompose the double coset $\Lambda D_0 \Lambda$ into the right cosets modulo Λ and use (5.1.19) with $V = E_m$. Then we get

$$\begin{aligned} &\varepsilon(Z,Q)|_{k,\chi}(M_0)_{\Gamma_0} \\ &= \sum_{\substack{D \in \Lambda D_0 \Lambda/\Lambda \\ B \in \mathrm{B}(D,M_0)/\mathrm{mod}\, D}} \sum_{V' \in \Lambda} e\{QZ\} \left| \begin{pmatrix} \mu D^* & B \\ 0 & D \end{pmatrix} \right| U(V') \qquad (5.1.20) \\ &= \sum_{V' \in \Lambda} \sum_{D \in \Lambda D_0 \Lambda/\Lambda} \sum_{B \in \mathrm{B}(D,M_0)/\mathrm{mod}\, D} e\{QZ\} \left| \begin{pmatrix} \mu D^* & B \\ 0 & D \end{pmatrix} \right| U(V') \end{aligned}$$

(all of the series under consideration converge absolutely in $\mathbb{H}_m$). By definitions (see (4.2.2) and (4.2.20)), the inner sum can be represented in the form

$$\begin{aligned} &\sum_{B \in \mathrm{B}(D,M_0)/\mathrm{mod}\, D} \left| \begin{pmatrix} \mu D^* & B \\ 0 & D \end{pmatrix} \right. \\ &= \mu^{mk - \langle m \rangle} (\det D)^{-k} \chi(\det \mu D^*) \sum_{B \in \mathrm{B}(D,M_0)/\mathrm{mod}\, D} e\{Q(\mu D^* Z + B)D^{-1}\} \qquad (5.1.21) \\ &= \alpha_{k,\chi}(M_0) e\{\mu Q[D^*]Z\} s(Q,D,M_0), \end{aligned}$$

because $(\det D)^{-k}\chi(\det \mu D^*)$ is independent of the choice of D in $\Lambda D_0 \Lambda$.

We now show that

$$s(Q,D;M_0) = 0, \quad \text{if } \mu Q[D^*] \notin \mathbb{E}_m. \tag{5.1.22}$$

In fact, if $B_1, \ldots, B_c$ is a set of representatives for the right residue classes of $\mathrm{B}(D,M_0)$ modulo D, then the same is true for the set $B_1 + \mu D^* S, \ldots, B_c + \mu D^* S$

with any $S \in \mathsf{S}_m(\mathbb{Z})$, as easily follows from definitions. Since the sum $s(Q, D, M_0)$ with $Q \in \mathbb{E}_m$ is independent of choice of representatives, it follows that

$$s(Q, D, M_0) = \sum_i e\{QB_iD^{-1}\}$$
$$= \sum_i e\{QB_iD^{-1} + Q\mu D^*SD^{-1}\} = e\{\mu Q[D^*]S\}s(Q, D, M_0).$$

If $\mu Q[D^*] \notin \mathbb{E}_m$, then there is a matrix S of $\mathsf{S}_m(\mathbb{Z})$ such that $\sigma(\mu Q[D^*]S) \notin 2\mathbb{Z}$. For such a matrix, we have $e\{\mu Q[D^*]S\} \neq 1$. Thus, $s(Q, D, M_0) = 0$.

By formulae (5.1.20)–(5.1.22), we obtain

$$\begin{aligned} &\varepsilon(Z, Q)|(M_0)_{\Gamma_0} \\ &= \sum_{V' \in \Lambda} \sum_{\substack{D \in \Lambda D_0 \Lambda / \Lambda \\ \mu Q[D^*] \in \mathbb{A}_m^+}} \alpha_{k,\chi}(M_0) s(Q, D, M_0) e\{\mu Q[D^*]Z\} | U(V') \qquad (5.1.23) \\ &= \alpha_{k,\chi}(M_0) \sum_{\substack{D \in \Lambda D_0 \Lambda / \Lambda \\ \mu Q[D^*] \in \mathbb{A}_m^+}} s(Q, D, M_0) \varepsilon(\mu Q[D^*], Z), \end{aligned}$$

where we have also used (5.1.18) with $\mu Q[D^*]$ in place of Q. Since the number of the right cosets modulo Λ contained in $\Lambda D_0 \Lambda$ is finite, it follows that the last sum is finite. Putting together all of the terms in the sum with the matrix $\mu Q[D^*]$ belonging to a fixed class $\{Q'\} \subset \mathbb{A}_m^+$, we get the formula (5.1.13) with $T = (M_0)_{\Gamma_0}$ and the formula (5.1.15). The general formula (5.1.13) follows by linearity.

The relations (5.1.17) follow from Lemma 4.2.3 and (5.1.13), because ε-series of pairwise non-equivalent matrices of $\mathbb{A}_m^+$ are linearly independent. The rest is clear. □

To close this subsection, we prove further properties of the coefficients $\varepsilon(Q, Q', T)$, which will be essentially used later.

Lemma 5.1.4. *Let Q be a matrix of $\mathbb{A}_m^+$ of a level q, and let $M_0 = \begin{pmatrix} \mu D_0^* & B_0 \\ 0 & D_0 \end{pmatrix}$ be a matrix of $\mathcal{S}^m(q)$ satisfying*

$$h(M_0) = \det D_0 (\det \mu D_0^*)^{-1} = (\det D_0)^2 \mu^{-m} = 1. \qquad (5.1.24)$$

Suppose that $\varepsilon_{k,\chi}(Q, Q'; (M_0)_{\Gamma_0}) \neq 0$ for a Q' of $\mathbb{A}_m^+$, where χ is a character modulo q, $k \in \mathbb{Z}$, and $\chi(-1) = (-1)^k$. Then the matrix Q' has the same determinant, divisor, and level as those of Q: $\det Q' = \det Q$, $\delta(Q') = \delta(Q)$, and $q(Q') = q(Q)$; if, in addition, the number μ is a rational square, then Q' and Q belong to the same genus (see A.2.6).

We first prove the following simple

Lemma 5.1.5. *The determinant d and the level q of a non-singular matrix Q of $\mathbb{E}_m$ have the same prime divisors.*

Proof. The matrix dQ^{-1} is integral and symmetric. If m is even, then $m-1$ is odd, and Proposition 1.4.12 implies that each of the diagonal entries of dQ^{-1} is even. Hence, for even m, $q|d$. If m is odd, then certainly q divides $2d$, but, by Proposition 1.4.12, in this case q and d are both even. Thus, each of the prime divisors of q divides d. Conversely, since $qQ^{-1} \in \mathbb{E}_m$, it follows that $d|q^m$. □

Proof of Lemma 5.1.4. By the assumption, there is a matrix $D \in \Lambda D_0 \Lambda$ with $\mu Q[D^*] = Q'$. Then the relation (5.1.24) implies the equality $\det Q = \det Q'$. Since the matrices $M_0^{\pm 1}$ are both q-integral, it follows that there is a γ in $\mathbb{N}_{(q)}$ such that each of the matrices γD, γD^{-1}, $\gamma\mu D^*$, and $\gamma\mu^{-1}\,{}^tD$ is integral. Since $\gamma^2 Q' = \gamma D^{-1} Q \gamma\mu D^*$ and $(Q')^{-1} = \gamma\mu^{-1}\,{}^tD(\gamma^{-2}Q^{-1})\gamma D$, it follows that $\delta(Q)|\gamma^2\delta(Q')$ and $q(Q')|\gamma^2 q$. Since $\det Q' = \det Q$, it follows, by the previous lemma, that the numbers $\delta(Q)$ and $q(Q')$ divide q^∞ and therefore are both prime to γ. Thus, $\delta(Q)|\delta(Q')$ and $q(Q')|q$. Similarly, $\delta(Q')|\delta(Q)$ and $q|q(Q')$. Hence, $\delta(Q') = \delta(Q)$ and $q(Q') = q$.

If $\mu = t^2$ with $t \in \mathbb{Q}$, then we can write $Q' = Q[V]$ with $V = tD^*$. By (5.1.24), we have $\det V = \pm 1$. Since $M_0 \in \mathscr{S}^m(q)$, it follows that the denomerators of all of the elementary divisors of V are prime to q. Then Lemma A.2.23 complete the proof. □

The sets

$$\mathscr{S}_0^m(q) = \left\{\begin{pmatrix} \mu D^* & B \\ 0 & D \end{pmatrix} \in \mathscr{S}^m(q); (\det D)^2 \mu^{-m} = 1\right\} \tag{5.1.25}$$

and

$$\mathscr{E}_0^m(q) = \{M \in \mathscr{S}_0^m(q); \mu(M) = t^2, t \in \mathbb{Q}\} \tag{5.1.26}$$

are clearly subgroups of $\mathscr{S}^m(q)$, and $\mathscr{E}_0^m(q)$ is the even subgroup of $\mathscr{S}_0^m(q)$ in the sense of (4.2.70). Hence, linear combinations of the double cosets of matrices satisfying (5.1.24) form a subring $\mathscr{L}_0^m(q)$ of $\mathscr{L}^m(q)$, the Hecke ring of the pair $(\Gamma_0^m, \mathscr{S}_0^m(q))$:

$$\mathscr{L}_0^m(q) = \mathrm{D}_{\mathbb{Q}}(\Gamma_0^m, \mathscr{S}_0^m(q)) \subset \mathscr{L}^m(q). \tag{5.1.27}$$

Similarly, the double cosets of matrices of $\mathscr{E}_0^m(q)$ span a ring $\mathscr{L}_0^m(q)^+$, which is the even subring of $\mathscr{L}_0^m(q)$:

$$\mathscr{L}_0^m(q)^+ = \mathrm{D}_{\mathbb{Q}}(\Gamma_0^m, \mathscr{E}_0^m(q)) \subset \mathscr{L}_0^m(q). \tag{5.1.28}$$

With the notation, Lemma 5.1.4 implies the following

Proposition 5.1.6. *Let Q be a matrix of $\mathbb{A}_m^+$ of a level q, and let $T \in \mathscr{L}_0^m(q)$ (resp. $T \in \mathscr{L}_0^m(q)^+$). Suppose that $\varepsilon_{k,\chi}(Q, Q', T) \neq 0$ for a matrix $Q' \in \mathbb{A}_m^+$, where χ is a character modulo q, $k \in \mathbb{Z}$, and $\chi(-1) = (-1)^k$. Then the determinant, the divisor, and the level of Q' are equal to those of Q (resp. Q' belongs to the same genus as Q).*

Exercise 5.1.7. *Show that the leading component of any F of $\mathfrak{M}^n_+$ can be written in the form*

$$F^{(m)}(Z) = \sum_{\{A\} \in \mathbb{A}^+_m} \frac{f(A)}{e(A)} \varepsilon(Z, A),$$

where $f(A)$ are Fourier coefficients of F. For $f \in \mathscr{F}^m_+ = \mathscr{F}^m_{\xi+}$, prove the formula

$$(f|_{k,\chi} T)(A) = e(A) \sum_{\{Q\} \in \mathbb{A}^+_m} e(Q)^{-1} f(Q) \varepsilon(Q, A, T), \; (T \in \mathscr{L}^m(q), A \in \mathbb{A}^+_m)$$

Exercise 5.1.8. *Let $Q \in \mathbb{A}^+_m$, and let $Q_1, \ldots, Q_h$ be a system of representatives for the classes $\{\cdots\}$ belonging to the genus $\mathsf{G}(Q)$. Show that the function*

$$\sum_{i=1}^{h} e(Q_i)^{-1} \varepsilon(Z, Q_i) = \sum_{A \in \mathsf{G}(Q)} e\{AZ\} \in \mathfrak{M}^m_+$$

is an eigenfunction for all operators $|_{k,\chi} T$ with $T \in \mathscr{L}^m_0(q)^+$, where $k \in \mathbb{Z}$, q is the level of Q, and χ is a character modulo q with $\chi(-1) = (-1)^k$.

Exercise 5.1.9. *Let*

$$\mathsf{C}(\mathbb{A}^+_m) = \{f: \mathbb{A}^+_m \to \mathbb{C}; f(A[V]) = f(A), A \in \mathbb{A}^+_m, V \in \Lambda^m\}.$$

For $f \in \mathsf{C}(\mathbb{A}^+_m)$, $T \in \mathscr{L}^m(q)$, and $Q \in \mathbb{A}^+_m$, we set

$$(T \circ f)(Q) = (T \circ_{k,\chi} f)(Q) = \sum_{\{A\} \in \mathbb{A}^+_m} \varepsilon(Q, A, T) f(A)$$

Prove the following assertions:

(1) *The mapping $T \to T\circ$ is a linear representation of $\mathscr{L}^m(q)$ on $\mathsf{C}(\mathbb{A}^+_m)$;*

(2) *For d, δ, $q \in \mathbb{N}$, let*

$$\mathsf{C}_m(d, \delta, q) = \{f \in \mathsf{C}(\mathbb{A}^+_m); f(A) = 0 \textit{ unless } \det A = d, \delta(A) = \delta, q(A) = q\}$$

and, for $Q \in \mathbb{A}^+_m$,

$$\mathsf{C}(\mathsf{G}(Q)) = \{f \in \mathsf{C}(\mathbb{A}^+_m); f(A) = 0 \quad \textit{if } A \notin \mathsf{G}(Q)\}.$$

Then the subspace $\mathsf{C}_m(d, \delta, q)$ (resp. $\mathsf{C}(\mathsf{G}(Q))$) of $\mathsf{C}(\mathbb{A}^+_m)$ is finite-dimentional and invariant under all of the operators $T \circ_{k,\chi}$ with $T \in \mathscr{L}^m_0(q)$ (resp. $T \in \mathscr{L}^m_0(q)^+$).

5.1.2. The Basic Case

In this subsection we shall consider the action of Hecke operators on theta series of genus m of matrices of $\mathbb{A}^+_m$ with an even m.

Theorem 5.1.10. *Let, for an even m, d, δ, and q of $\mathbb{N}$, Q be a matrix of $\mathbb{A}^+_m$ with* $\det Q = d$, *$\delta(Q) = \delta$, and $q(Q) = q$. Then the image of the theta-series $\theta^m(Z, Q)$ under the action of a Hecke operator $|_{k,\chi} T$ with $T \in \mathsf{L}^m(q)$, where $k = m/2$ and*

$\chi = \chi_Q$ is the character of Q, is a linear combination of theta-series given by the formula

$$\theta^m(Z,Q)|_{k,\chi}T = \sum_{\{Q'\}\in \mathbb{A}_m(d,\delta,q)} \varepsilon_{k,\chi}(Q,Q',\eta(\mathbf{T}))\theta^m(Z,Q') \tag{5.1.29}$$

with

$$\mathbb{A}_m(d,\delta,q) = \{A \in \mathbb{A}_m; \det A = d, \delta(A) = \delta, q(A) = q\},$$

where $\varepsilon_{k,\chi}$ are the coefficients (5.1.14), $\mathbf{T} = \mathrm{i}_q(T) \subset \mathbf{L}^m(q)$, *and where, for a* $X = \sum_i c_i\left(\begin{pmatrix} A_i & B_i \\ 0 & D_i \end{pmatrix}\right)_{\Gamma_0}$ *of* $\mathscr{L}^m$,

$$\eta(X) = \eta_{k,\chi}(X) = \sum_{\substack{(a\in\mathbb{N},i) \\ a^2 \det D_i (\det A_i)^{-1}=1}} c_i a^{-k}\chi(a)\left(\begin{pmatrix} A_i & B_i \\ 0 & D_i \end{pmatrix}\right)_{\Gamma_0} t^+(a),$$

$t^+(a) = t_m^+(a)$ being the elements of $\mathscr{L}^m$ defined in Lemma 4.3.20. *If T belongs to the even subring $\mathsf{E}^m(q)$ of $\mathsf{L}^m(q)$, then*

$$\theta^m(Z,Q)|_{k,\chi}T = \sum_{\{Q'\}\in \mathsf{G}(Q)} \varepsilon_{k,\chi}(Q,Q',\eta(\mathbf{T}))\theta^m(Z,Q'). \tag{5.1.30}$$

Proof. By definitions, we have the following inclusions

$$\eta(X) \in \begin{cases} \mathscr{L}_0^m(q), & \text{if } X \in \mathscr{L}^m(q) \\ \mathscr{L}_0^m(q)^+, & \text{if } X \in \mathscr{L}^m(q)^+ \end{cases} \tag{5.1.31}$$

Hence, by Proposition 5.1.16, a coefficient $\varepsilon(Q,Q';\eta(T))$ equals zero unless $Q' \in \mathbb{A}_m(d,\delta,q)$, or $Q' \in \mathsf{G}(Q)$ and $T \in \mathsf{E}^m(q)$; in particular, the formula (5.1.30) is a consequence of (5.1.29), and we may prove the two formulae in the form

$$\theta^m(Z,Q)|_{k,\chi}T = \sum_{\{Q'\}\in \mathbb{A}_m^+} \varepsilon_{k,\chi}(Q,Q',\eta(\mathbf{T}))\theta^m(Z,Q') \tag{5.1.32}$$

with the sum taken over representatives for all classes in $\mathbb{A}_m^+$.

We first consider a local case with $T \in \mathsf{L}_p^m(q)$ for a fixed prime p of $\mathbb{P}_{(q)}$. In the case, $\mathbf{T} \in \mathbf{L}_p^m$. As we have told, the main point of our consideration is the action of $|\mathbf{T}$ on the leading component $\theta^{(m)}(Z,Q)$ written in the form (5.1.12). For this we shall need the following commutation relations

$$|_{0,1}t^-(a)|_{k,\chi}X = |_{k,\chi}X|_{0,1}t^-(a) \quad \text{if } X \in \mathbf{L}_p^m \text{ and } p \nmid qa. \tag{5.1.33}$$

To prove it, we write the X in the form $X = \sum_{\nu,\rho} X_{\nu,\rho}$, where $X_{\nu,\rho} \in \mathscr{L}_p^m$ is a linear combination of the double cosets $\left(\begin{pmatrix} \mu D_i^* & B_i \\ 0 & D_i \end{pmatrix}\right)_{\Gamma_0}$ with $\mu_i = p^\nu$ and $\det D_i = \pm p^\rho$. Since $p \nmid a$, Lemma 5.1.2 and Proposition 3.4.11(1) imply that $t^-(a)X = Xt^-(a)$. Hence, by "homogenity," we have the relations $t^-(a)X_{\nu,\rho} = X_{\nu,\rho}t^-(a)$ for every ν and ρ, because double cosets occurring in $t^-(a)X_{\nu,\rho}$ or $X_{\nu,\rho}t^-(a)$ for distinct triples a, ν, ρ are clearly distinct. From definitions we easily obtain the relations

$$|_{k,\chi}X_{\nu,\rho} = |_{0,1}\chi(p^{n\nu-\rho})p^{(n\nu-\rho)k}X_{\nu,\rho}. \tag{5.1.34}$$

Thus, we can write

$$\begin{aligned} |_{0,1}t^-(a)|_{k,\chi}X &= \sum_{\nu,\rho} |_{0,1}t^-(a)|_{0,1}\chi(p^{n\nu-\rho})p^{(n\nu-\rho)k}X_{\nu,\rho} \\ &= \sum_{\nu,\rho} |_{0,1}\chi(p^{n\nu-\rho})p^{(n\nu-\rho)k}X_{\nu,\rho}|_{0,1}t^-(a) = |_{k,\chi}X|_{0,1}t^-(a), \end{aligned}$$

which proves (5.1.33). In the notation, we can write $t^-(p^\rho) = t^-(p^\rho)_{0,-\rho}$. Hence, by (5.1.34), we have

$$|_{0,1}t^-(p^\rho) = |_{k,\chi}\frac{\chi(p^\rho)}{p^{k\rho}}t^-(p^\rho) \tag{5.1.35}$$

(the character χ is real, and so $\chi(p^{-\rho}) = \chi(p^\rho)$). Now, using (5.1.12), (5.1.9), (5.1.33), and (5.1.35), we obtain

$$\begin{aligned} \theta^{(m)}(Z,Q)|_{k,\chi}\mathbf{T} &= \sum_{a=1}^{\infty} \varepsilon(Z,Q)|_{0,1}t^-(a)|_{k,\chi}\mathbf{T} \\ &= \sum_{\rho=0}^{\infty}\sum_{a\in\mathbb{N}_{(p)}} \varepsilon(Z,Q)|_{0,1}t^-(p^\rho)|_{0,1}t^-(a)|_{k,\chi}\mathbf{T} \qquad (5.1.36) \\ &= \sum_{a\in\mathbb{N}_{(p)}}\sum_{\rho=0}^{\infty} \varepsilon(Z,Q)|_{k,\chi}(\chi(p)p^{-k})^\rho t^-(p^\rho)\mathbf{T}|_{0,1}t^-(a). \end{aligned}$$

To proceed, we have to consider relations of elements of $\mathbf{L}_p^m$ with the elements $t^-(p^\rho)$ or, rather, with their sum $\sum_{\rho=0}^{\infty}(\chi(p)p^{-k})^\rho t^-(p^\rho)$, which proves to be easier. For this we use the notation (4.2.88) and consider functional series as specializations of the corresponding formal series, according to the following definition. We shall say that a formal power series $\sum_{\rho=0}^{\infty}F_\rho v^\rho$ with the coefficients F_ρ of $\mathfrak{M}_+^m$ *is defined at a point* $v = \beta\in\mathbb{C}$ if the functional series $\sum_{\rho=0}^{\infty}F_\rho(Z)\beta^\rho$ converges absolutely and uniformly on each subset of $\mathbb{H}_m$ of the form $\mathbb{H}_m(\varepsilon)$ with $\varepsilon > 0$. In this case, the sum $F(Z) = \sum_{\rho=0}^{\infty}F_\rho(Z)\beta^\rho$ is clearly contained in $\mathfrak{M}_+^m$. We write

$$F(Z) = \sum_{\rho=0}^{\infty} F_\rho v^\rho|_{v=\beta}$$

and call $F(Z)$ *the value of the formal series at the point* $v = \beta$. In the notation, we can rewrite (5.1.36) in the form

$$\begin{aligned} &\theta^{(m)}(Z,Q)|_{k,\chi}\mathbf{T} \\ &= \sum_{a\in\mathbb{N}_{(p)}} (\varepsilon(Z,Q)|_{k,\chi}\left(\sum_{\rho=0}^{\infty} p^{-\rho m}t^-(p^\rho)v^\rho)\mathbf{T}\right)\Bigg|_{v=\chi(p)p^k}\Bigg|_{0,1} t^-(a) \end{aligned} \tag{5.1.37}$$

The following considerations are essentially based on results of § 3.5.5 for the prime p and $n = m$. By (3.5.73), we can write

$$\left(\sum_{\rho=0}^{\infty} p^{-\rho m}t^-(p^\rho)v^\rho\right)\mathbf{T} = X_-(v)^{-1}\mathbf{T}$$

where $X_-(v)$ is the polynomial over $\mathscr{L}_p^m$ defined in (3.5.79). Hence we obtain a formal identity

$$\left(\sum_{\rho=0}^{\infty} p^{-\rho m} t^-(p^\rho) v^\rho\right) \mathbf{T}$$

$$= X_-(v)^{-1} \mathbf{T} X_-(v) \cdot X_-(v)^{-1} \tag{5.1.38}$$

$$= W(v) \sum_{\rho=0}^{\infty} p^{-\rho m} t^-(p^\rho) v^\rho, \quad \text{where} \quad W(v) = X_-(v)^{-1} \mathbf{T} X_-(v).$$

We now make use of the factorization (3.5.69), which enables us to express the polynomial $X_-(v)$ in the terms of other polynomials occurring in the factorization. We obtain

$$X_-(v) = \mathbf{R}_p^m(v) X_+(v)^{-1} B(v)^{-1},$$

where $B(v)$ and $X_+(v)$ are the last two factors in (3.5.69). Substituting this expression, we can write the series $W(v)$ in the form

$$\begin{aligned} W(v) &= B(v) X_+(v) \mathbf{R}(v)^{-1} T \mathbf{R}(v) X_+(v)^{-1} B(v)^{-1} \\ &= B(v) X_+(v) \mathbf{T} X_+(v)^{-1} B(v)^{-1}, \end{aligned} \tag{5.1.39}$$

where we have also used the relation $\mathbf{R}(v)^{-1} \mathbf{T} \mathbf{R}(v) = \mathbf{T}$, which follows from the commutativity of the ring $\mathbf{L}_p^m$ containing all of the coefficients of $\mathbf{R}(v)$ as well as $\mathbf{T}$.

The last formula will allow us to calculate the action of the series $W(v)$ on $\varepsilon(Q, Z)$ in a closed form. For this we shall need one more definition. For an element $X = \sum_i c_i (M_i)_{\Gamma_0} \in \mathscr{L}^m$ and $\sigma \in \mathbb{Z}$, we set

$$X(\sigma) = X(\sigma)_p = \sum_{i, \sigma_p(M_i) = \sigma} c_i (M_i)_{\Gamma_0},$$

where $\sigma_p(M_i)$ is the p-signature (3.5.83), and call $X(\sigma)$ the *p-homogeneous component of X of p-signature σ*. It is clear that

$$X = \sum_{\sigma \in \mathbb{Z}} X(\sigma). \tag{5.1.40}$$

Lemma 5.1.11. *With the above assumptions and notation, the following identity of formal power series over $\mathfrak{M}_+^m$ is valid*

$$\varepsilon(Z, Q)|_{k,\chi} W(v) = \sum_{\rho \geq 0} \varepsilon(Z, Q)|_{k,\chi} p^{-m\rho} \mathbf{T}(-2\rho) t^+(p^\rho) v^\rho + F(Z, v),$$

where $F(Z, v)$ is a polynomial in v, whose coefficients are finite linear combinations of the functions $\varepsilon(Z, A)$ with $A \in \mathbb{A}_m^+$; the function $F(Z, v)$ is identically equal to zero at $v = \chi(p) p^k$:

$$F(Z, \chi(p) p^k) = 0. \tag{5.1.41}$$

Proof of the lemma. We shall consider the action of each of the factors occurring in (5.1.39) on the ε-series. We first apply Proposition 4.2.24 to the function f of

$\mathfrak{F}_+^m$ whose values are the Fourier coefficients of the series $\varepsilon(Z, A)$ with $A \in \mathbb{A}_m^+$. We get the formula

$$\varepsilon(Z, A)|_{k,\chi} B(v) = B_p^m(v, A)\varepsilon(Z, A), \tag{5.1.42}$$

since, by Theorem 4.2.27(1), the polynomials $B_p^m(v, A')$ with $A' \in \{A\}$ are equal to $B_p^m(v, A)$. From this formula we obtain the formula

$$\varepsilon(Z, A)|_{k,\chi} B(v)^{-1} = B_p^m(v, A)^{-1}\varepsilon(Z, A). \tag{5.1.43}$$

Let, now, X be a p-homogeneous elements of $\mathscr{L}_p^m$. Then we prove that

$$\varepsilon(Z, A)|X = 0, \quad \text{if } \sigma_p(X) > 0 \text{ and } p \nmid \det A. \tag{5.1.44}$$

In fact, without loss of generality we may assume that $X = (M_0)_{\Gamma_0}$ with a matrix $M_0 = \begin{pmatrix} p^\nu D_0^* & B_0 \\ 0 & D_0 \end{pmatrix}$ of $\mathscr{L}_p^m$ satisfying $\det D_0 = \pm p^\rho$ and $\sigma_p(X) = 2\rho - m\nu > 0$, but then (5.1.44) follows from (5.1.13) and (5.1.15), because any matrix of the form $p^\nu A[D^*]$ with $D \in \Lambda D_0 \Lambda$ does not belong to $\mathbb{A}_m$, for $\det p^\nu A[D^*] = \det A / p^{2\rho - m\nu} \notin \mathbb{Z}$.

We return to the action of $W(v)$. By Lemma 5.1.5, we have that $p \nmid \det Q$, since $p \nmid q$. By (3.5.85) and (3.5.71), we see that all of the coefficients of $X_+(v)$, except for the unit free coefficient, are p-homogeneous elements of positive p-signature. Hence, using (5.1.42) and (5.1.44), we have

$$\varepsilon(Z, Q)|_{k,\chi} W(v) = B_p^m(v, Q)\varepsilon(Z, Q)|_{k,\chi} \mathbf{T} X_+(v)^{-1} B(v)^{-1}.$$

Substituting $\mathbf{T} = \sum_\sigma \mathbf{T}(\sigma)$ (the decomposition (5.1.40) into the sum of p-homogeneous components) and replacing $X_+(v)^{-1}$ by the formula (3.5.74), we obtain

$$\varepsilon(Z, Q)|_{k,\chi} W(v) = B_p^m(v, Q)\varepsilon(Z, Q)| \sum_{\sigma,\rho} p^{-\rho m} \mathbf{T}(\sigma) t^+(p^\rho) v^\rho | B(v)^{-1}. \tag{5.1.45}$$

By Proposition 5.1.3, for each σ and ρ, one can write an equality

$$\varepsilon(Z, Q)|\mathbf{T}(\sigma) t^+(p^\rho) = \sum_\alpha c(\sigma, \rho, \alpha)\varepsilon(Z, Q_{\sigma,\rho,\alpha}), \tag{5.1.46}$$

where the sum is finite and all of the matrices $Q_{\sigma,\rho,\alpha}$ belong to $\mathbb{A}_m^+$. If the product $\mathbf{T}(\sigma)t^+(p^\rho)$ is not equal to zero, then, according to (3.5.84) and (3.5.85), the product is p-homogeneous of p-signature $\sigma + 2\rho$. Hence, by (5.1.44), we conclude that

$$\varepsilon(Z, Q)|\mathbf{T}(\sigma) t^+(p^\rho) = 0 \qquad \text{if } \sigma + 2\rho > 0.$$

Thus, substituting (5.1.46) in (5.1.45) and using (5.1.43), we have

$$\begin{aligned} &\varepsilon(Z, Q)| W(v) \\ &\quad = B_p^m(v, Q) \sum_{\substack{\sigma,\rho,\alpha \\ \sigma+2\rho=0}} B_p^m(v, Q_{\sigma,\rho,\alpha})^{-1} c(\sigma, \rho, \alpha) p^{-\rho m} \varepsilon(Z, Q_{\sigma,\rho,\alpha}) v^\rho \\ &\qquad + B_p^m(v, Q) \sum_{\substack{\sigma,\rho,\alpha \\ \sigma+2\rho<0}} B_p^m(v, Q_{\sigma,\rho,\alpha})^{-1} c(\sigma, \rho, \alpha) p^{-\rho m} \varepsilon(Z, Q_{\sigma,\rho,\alpha}) v^\rho. \end{aligned} \tag{5.1.47}$$

If $\sigma + 2\rho = 0$, then $\mathbf{T}(\sigma)t^+(p^\rho) \in \mathscr{L}_0^m$, and so, by Proposition 5.1.6, each of the matrices $Q_{\sigma,\rho,\alpha}$ in (5.1.46) has the same level and the same determinant as those of Q; in particular, each of them is non-singular modulo p and has the same character as that of Q. Hence, by Theorem 4.2.27, we see that $B_p^m(v, Q_{\sigma,\rho,\alpha}) = B_p^m(v, Q)$. Thus, using (5.1.46) again, we conclude that the first expression in the right side of (5.1.47) is equal to

$$\sum_{\sigma+2\rho=0} \varepsilon(Z, Q)|\mathbf{T}(\sigma)t^+(p^\rho)p^{-\rho m}v^\rho.$$

But if $\sigma + 2\rho < 0$, then the formulae of Proposition 5.1.3 imply that each of the matrices $Q_{\sigma,\rho,\alpha}$ in (5.1.46) is degenerate modulo p. Thus, according to Theorem 4.2.27, each of the fractions $B_p^m(v, Q)B_p^m(v, Q_{\sigma,\rho,\alpha})^{-1}$ is a polynomial in v, and the polynomial is divisible by $1 - \chi(p)p^{-k}v = 1 - (\chi(p)p^k)^{-1}v$. It follows that the second expression in (5.1.47) is a polynomial in v whose coefficients are finite linear combinations of the series $\varepsilon(Z, Q_{\sigma,\rho,\alpha})$, and the polynomial is identically equal to zero at $v = \chi(p)p^k$. □

We return to the proof of Theorem 5.1.10. By (5.1.38) and Lemma 5.1.11, we obtain

$$\begin{aligned}\varepsilon(Z, Q)&\left|\left(\sum_{\rho=0}^{\infty} p^{-\rho m}t^-(p^\rho)v^\rho\right)\mathbf{T}\right. \\ &= \left(\sum_{\rho\geq 0} \varepsilon(Z, Q)|p^{-m\rho}\mathbf{T}(-2\rho)t^+(p^\rho)v^\rho\right)\left|\sum_{\nu=0}^{\infty} p^{-\nu m}t^-(p^\nu)v^\nu\right. \\ &\quad + F(Z, v)\left|\sum_{\nu=0}^{\infty} p^{-\nu m}t^-(p^\nu)v^\nu,\right.\end{aligned}$$

with $| = |_{k,\chi}$. From formulae (5.1.10) and trivial estimates we conclude that any formal series of the form

$$G(Z)|_{k,\chi} \sum_{\nu=0}^{\infty} p^{-\nu m}t^-(p^\nu)v^\nu$$

where $G(Z)$ is a finite linear combination of the series $\varepsilon(Z, A)$ with $A \in \mathbb{A}_m^+$, is defined at $v = p^k\chi(p)$. Substituting $v = p^k\chi(p)$, and using (5.1.41), (5.1.13), and (5.1.35), we obtain

$$\begin{aligned}\varepsilon(Z, Q)|_{k,\chi}&\left(\sum_{\rho=0}^{\infty} p^{-\rho m}t^-(p^\rho)v^\rho\right)\mathbf{T}\Bigg|_{v=\chi(p)p^k} \\ &= \sum_{\nu=0}^{\infty}\left(\varepsilon(Z, Q)|_{k,\chi}\sum_{\rho\geq 0}\frac{\chi(p^\rho)}{p^{k\rho}}\mathbf{T}(-2\rho)t^+(p^\rho)\right)\Bigg|_{k,\chi}\frac{\chi(p^\nu)}{p^{\nu k}}t^-(p^\nu) \\ &= \sum_{\nu=0}^{\infty}\left(\sum_{\{Q'\}\in\mathbb{A}_m^+}\varepsilon\left(Q, Q', \sum_{\rho\geq 0}\frac{\chi(p^\rho)}{p^{\rho k}}\mathbf{T}(-2\rho)t^+(p^\rho)\right)\varepsilon(Z, Q')\right)\Bigg|_{0,1} t^-(p^\nu).\end{aligned}$$

According to the definition of p-homogeneous components, $\mathbf{T}(-2\rho)$ is the sum of those summands $a_i\left(\begin{pmatrix} A_i & B_i \\ 0 & D_i \end{pmatrix}\right)$ of $\mathbf{T}$ which satisfy $\det D_i(\det A_i)^{-1} = \rho^{-2\rho}$. Hence,

$$\sum_{\rho \geq 0} \chi(p^\rho)p^{-\rho k}\mathbf{T}(-2\rho)t^+(p^\rho) = \eta_{k,\chi}(\mathbf{T}).$$

Thus, returning to (5.1.37), we obtain, by (5.1.9) and (5.1.12), the identity

$$\theta^{(m)}(Z,Q)|_{k,\chi}\mathbf{T} = \sum_{\{Q'\} \in \mathbb{A}_m^+} \varepsilon(Q,Q',\eta(\mathbf{T}))\theta^{(m)}(Z,Q').$$

According to (5.1.31) and Proposition 5.1.6, the last sum can be taken only over classes $\{Q'\}$ satisfying $\det Q' = d$, $\delta(Q') = \delta$, and $q(Q') = q$. Hence and by (5.1.6), the last relation can be written in the form

$$\left(\theta^m(Z,Q)|_{k,\chi}T - \sum_{\{Q'\} \in \mathbb{A}_m(d,\delta,q)} \varepsilon(Q,Q',\eta(\mathbf{T}))\theta^m(Z,Q')\right)^{(m)} = 0.$$

By Theorem 2.2.2, the function $\theta^m(Z,Q)$ belongs to the space $\mathfrak{M}_k(\Gamma_0^m(q),\chi_Q^m) = \mathfrak{M}_k^m(q,\chi)$, and so the function $\theta^m(Z,Q)|T$ does. In view of Theorem 1.4.10, the character $\chi_{Q'}$ of a matrix Q' of $\mathbb{A}_m^+$ depends only on determinant and level of Q'. Hence, each of the theta-series $\theta^m(Z,Q')$ in the last relation belongs to the space $\mathfrak{M}_k^m(q,\chi)$ as well. Thus, the relation has the form $F^{(m)} = 0$ with a function F of $\mathfrak{M}_k^m(q,\chi)$, which means that the form F is singular. We have $k = m/2$, and so, by Theorem 2.3.16, the space $\mathfrak{M}_k^n(q,\chi)$ contains no singular forms except the zero-form. Therefore, $F = 0$, which proves (5.1.29) with $T \in \mathsf{L}_p^m(q)$.

In order to treat the global case, we first prove the following lemma.

Lemma 5.1.12. *Let* $\mathbf{T} \in \mathbf{L}_p^m$, $\mathbf{T}_1 \in \mathbf{L}_{p_1}^m, \ldots, \mathbf{T}_r \in \mathbf{L}_{p_r}^m$, *where* $p, p_1, \ldots, p_r$ *are pairwise distinct prime numbers prime to* q. *Then relation holds*

$$\eta(\mathbf{T}_1 \cdots \mathbf{T}_r)\eta(\mathbf{T}) = \eta(\mathbf{T}_1 \cdots \mathbf{T}_r\mathbf{T}),$$

where $\eta = \eta_{k,\chi}$.

Proof. We write each of the elements $\mathbf{T}_i$ as the sum (5.1.40) of p_i-homogeneous components:

$$\mathbf{T}_i = \sum_\sigma \mathbf{T}_i(\sigma)_{p_i}.$$

If (M) is a double coset occurring in $\mathbf{T}_1(\sigma_1)\cdots\mathbf{T}_r(\sigma_r)$ then clearly $h(M) = p_1^{\sigma_1}\cdots p_r^{\sigma_r}$, where h is the mapping (3.5.82). Hence, a relation $a^2h(M) = 1$ with $a \in \mathbb{N}_{(q)}$ implies that all of the numbers $\sigma_1, \ldots, \sigma_r$ are even and non-positive, say $\sigma_i = -2\alpha_i$, and $a = p_1^{\alpha_1}\cdots p_r^{\alpha_r}$. Thus, by the definition of η, we have

$$\eta(\mathbf{T}_1\cdots\mathbf{T}_r) = \sum_{\alpha_1,\ldots,\alpha_r \geq 0} \frac{\chi(p_1^{\alpha_1}\cdots p_r^{\alpha_r})}{(p_1^{\alpha_1}\cdots p_r^{\alpha_r})^k}\mathbf{T}_1(-2\alpha_1)\cdots\mathbf{T}_r(-2\alpha_r)t^+(p_1^{\alpha_1}\cdots p_r^{\alpha_r}). \tag{5.1.48}$$

Similarly,

$$\eta(\mathbf{T}) = \sum_{\alpha \geq 0} \chi(p^\alpha) p^{-\alpha k} \mathbf{T}(-2\alpha) t^+(p^\alpha) \tag{5.1.49}$$

By Proposition 3.4.11, $\mathbf{T}$ commutes with each of the elements $t^+(p_i^{\alpha_i}) \in \mathscr{C}_{p_i}^m$. Hence, in view of Lemma 4.3.20, $\mathbf{T}$ commutes with $t^+(p_1^{\alpha_1} \cdots p_r^{\alpha_r}) = t^+(p_1^{\alpha_1}) \cdots t^+(p_r^{\alpha_r})$. By comparing p-homogeneous components of the elements $\mathbf{T}t^+(p_1^{\alpha_1} \cdots p_r^{\alpha_r})$ and $t^+(p_1^{\alpha_1} \cdots p_r^{\alpha_r})\mathbf{T}$, we conclude that $t^+(p_1^{\alpha_1} \cdots p_r^{\alpha_r})$ commutes with each of the p-homogeneous components $\mathbf{T}(\sigma)$ of $\mathbf{T}$. Thus,

$$\begin{aligned}\eta(\mathbf{T}_1 \cdots \mathbf{T}_r)\eta(\mathbf{T}) &= \sum_{\alpha_1, \ldots, \alpha_r, \alpha \geq 0} \frac{\chi(p_1^{\alpha_1} \cdots p_r^{\alpha_r})\chi(p^\alpha)}{(p_1^{\alpha_1} \cdots p_r^{\alpha_r})^k p^{\alpha k}} \mathbf{T}_1(-2\alpha_1) \cdots \mathbf{T}_r(-2\alpha_r)\mathbf{T}(-2\alpha) \\ &\quad \times t^+(p_1^{\alpha_1} \cdots p_r^{\alpha_r}) t^+(p^\alpha) \\ &= \eta(\mathbf{T}_1 \cdots \mathbf{T}_r \cdot \mathbf{T}).\end{aligned}$$

□

We complete the proof of the theorem. The ring $\mathsf{L}^m(q)$ is generated by the local subrings $\mathsf{L}_p^m(q)$. Hence, by linearity, it suffices to prove (5.1.29) for elements T of the form $T = T_1 \cdots T_r$ with $T_i \in \mathsf{L}_{p_i}^m(q)$, where $p_1, \ldots, p_r$ are pairwise distinct prime numbers of $\mathbb{P}_{(q)}$. We shall apply induction to r. The case $r = 1$ have been considered. Suppose the formula is already proved for r factors, and set $T_{r+1} = T$. Then, by (5.1.17) and Lemma 5.1.12, we obtain

$$\begin{aligned}\theta^m(Z, Q)|T_1 \cdots T_{r+1} &= \theta^m(Z, Q)|T_1 \cdots T_r|T \\ &= \sum_{\{Q'\}} \varepsilon(Q, Q', \eta(\mathbf{T}_1 \cdots \mathbf{T}_r))\theta^m(Z, Q')|T \\ &= \sum_{\{Q'\}} \varepsilon(Q, Q', \eta(\mathbf{T}_1 \cdots \mathbf{T}_r)) \sum_{\{Q''\}} \varepsilon(Q', Q'', \eta(\mathbf{T}))\theta^m(Z, Q'') \\ &= \sum_{\{Q''\}} \varepsilon(Q, Q'', \eta(\mathbf{T}_1 \cdots \mathbf{T}_r) \cdot \eta(\mathbf{T}))\theta^m(Z, Q'') \\ &= \sum_{\{Q'\}} \varepsilon(Q, Q'', \eta(\mathbf{T}_1 \cdots \mathbf{T}_r \mathbf{T}_{r+1}))\theta^m(Z, Q'').\end{aligned}$$

□

5.1.3. The General Case

In this subsection we shall derive general formulae for the action of Hecke operators on theta-series from the basic case considered above. In a local situation it can be done with the help of techniques of § 4.2.4. To treat the global case, we need first to get a global version of main results of § 4.2.4.

Let

$$X = \sum_i a_i \left(\Gamma_0^n \begin{pmatrix} \mu_i D^* & B_i \\ 0 & D_i \end{pmatrix} \right)$$

be an element of $\mathscr{L}^n$. By Lemma 3.2.7, we may assume that each of the matrices D_i is of the form

$$D_i = \begin{pmatrix} D_i' & * \\ 0 & d_i \end{pmatrix} \quad \text{with } D_i' \in \mathsf{G}^{n-1} \text{ and } d_i > 0.$$

Then we define a global analog of (4.2.64) by setting

$$\Psi(X, \ldots, u_p, \ldots) = \sum_i a_i \left[\prod_{p \in \mathbb{P}} u_p^{-v_p(\mu_i)} (u_p p^{-n})^{v_p(d_i)} \right] \left(\Gamma_0^{n-1} \begin{pmatrix} \mu_i (D_i')^* & B_i' \\ 0 & D_i' \end{pmatrix} \right) \tag{5.1.50}$$

where $\ldots, u_p, \ldots$ are independent variables indexed by prime numbers, $v_p(r)$ is the p-exponent of a rational $r \neq 0$, and where, for a matrix A of order n, A' is the matrix of order $n-1$ such that $A = \begin{pmatrix} A' & * \\ * & * \end{pmatrix}$. If $n = 1$, then $D_i = d_i$, and we set

$$\Psi(X, \ldots, u_p, \ldots) = \sum_i a_i \prod_{p \in \mathbb{P}} u_p^{-v_p(\mu_i)} (u_p p^{-1})^{v_p(d_i)}. \tag{5.1.51}$$

It is easy to see that the mapping $X \to \Psi(X, \ldots, u_p, \ldots)$ does not depend on the choice of representatives in the left cosets and is a homomorphism of $\mathscr{L}^n$ into the ring

$$\mathscr{L}^{n-1}[\ldots, u_p^{\pm 1}, \ldots] = D_{\mathbb{Q}[\ldots, u_p^{\pm 1}, \ldots]}(\Gamma_0^{n-1}, \mathscr{S}^{n-1})$$

(we set $\mathscr{L}^0 = \mathbb{Q}$). The same proof as that of Theorem 4.2.18 gives us the following global variant of Žarkovskaya relations.

Theorem 5.1.13. *Let $k \in \mathbb{Z}$, $q \in \mathbb{N}$, and let χ be a character modulo q. Then, for any function F of $\mathfrak{M}_\xi^n$ with $\xi(-1) = \chi(-1)(-1)^k$, and any element X of the complexification $\tilde{\mathscr{L}}^n(q)$ of $\mathscr{L}^n(q)$, the following relation holds*

$$(F|_{k,\chi} X)|\Phi = (F|\Phi)|_{k,\chi} \Psi_{k,\chi}(X), \tag{5.1.52}$$

where Φ is the Siegel operator,

$$\Psi_{k,\chi}(X) = \Psi(X, \ldots, p^{n-k}\bar{\chi}(p), \ldots), \tag{5.1.53}$$

and where, for $n = 1$, the operator $|_{k,\chi} \Psi$ acts as multiplication by Ψ.

Finally, the global variant of Proposition 4.2.19 looks as follows:

Proposition 5.1.14. *In the notation of* § 4.2.4 *and Theorem* 5.1.13, *the following assertions hold:*

(1) *If X is contained in $\tilde{\mathbf{L}}^n(q)$ (resp. $\tilde{\mathbf{E}}^n(q) = \tilde{\mathbf{L}}^n(q)^+$), then the element $\Psi_{k,\chi}(X)$ is contained in $\tilde{\mathbf{L}}^{n-1}(q)$(resp. $\tilde{\mathbf{E}}^{n-1}(q)$), where we set $\tilde{\mathbf{L}}^0(q) = \tilde{\mathbf{E}}^0(q) = \mathbb{C}$;*

(2) *The mapping*

$$\Psi_{k,\chi} : \tilde{\mathbf{L}}^n(q) \to \tilde{\mathbf{L}}^{n-1}(q) \tag{5.1.54}$$

is an epimorphism, except in the case when $n = k > 1$ and there are primes p in $\mathbb{P}_{(q)}$ with $\chi(p) = -1$. In this case the image of $\tilde{\mathbf{L}}^n(q)$ is generated by the local rings

Exercise 5.1.17. *In the notation of Exercise* 5.1.16, *show that* G *is an eigenfunction for all of the operators* $|_{k,\chi}T^n(a)$ *with* $a \in \mathbb{N}_{(q)}$ *satisfying the conditions:* $v_p(a)$ *is odd or* $\chi_Q(p) = 1$ *for each prime divisor* p *of* a. *Show that, for such an* a *and* $n = 1, 2$, *the corresponding eigenvalues are given by*

$$\lambda(T^1(a) = \sum_{d|a} \chi_Q(d) d^{k-1}$$

$$\lambda(T^2(a)) = \sum_{d|a} \chi_Q(d) d^{k-2} \omega(d) \sigma_{2k-3}(a/d)$$

where $\omega(d) = d \prod_{p|d}(1 + p^{-1})$ *and* $\sigma_l(b) = \sum_{\delta|b} \delta^l$.

[Hint: For the last assertion, use the formulae (3.4.31) and (3.4.32).]

Exercise 5.1.18. (*Siegel*) *Let* $Q_1, \ldots, Q_h$ *be a set of representatives for all of the classes* $\{Q\} \in \mathbb{A}_m$ *with* $\det Q = 1$. *Prove the identity*

$$\left(\sum_{i=1}^{h} \frac{1}{e(Q_i)}\right)^{-1} \sum_{i=1}^{h} \frac{1}{e(Q_i)} \theta^1(z, Q_i) = (2\zeta(m/2))^{-1} G_{m/2}(z), \ (z \in \mathbb{H}_1)$$

where $G_{m/2}(z)$ *is an Eisenstein series of Exercise* 2.5.6, $\zeta(s)$ *being the Riemann zeta. For each* $a \in \mathbb{N}$, *prove the relation*

$$\left(\sum_{i=1}^{h} \frac{1}{e(Q_i)}\right)^{-1} \sum_{i=1}^{h} \frac{r(Q_i, 2a)}{e(Q_i)} = \frac{(2\pi)^{m/2}}{\left(\frac{m}{2} - 1\right)! \zeta(m/2)} \sum_{d|a} d^{(m/2)-1};$$

in particular,

$$\zeta\left(\frac{m}{2}\right) = (2\pi)^{m/2}((m/2 - 1)!)^{-1} \left(\sum_{i=1}^{h} e(Q_i)^{-1}\right)\left(\sum_{i=1}^{h} \frac{r(Q_i, 2)}{e(Q_i)}\right)^{-1}$$

[Hint: Using Exercises 4.3.5 and 5.1.17, show that the difference of the two sides of the identity is a cusp form, which is an eigenfunction for all of the operators $|_k T^1(a)$ with the eigenvalues $\sum_{d|a} d^{(m/2)-1}$. Then, by (4.3.9), prove that the difference equals zero. The relations for the numbers of representations follow from the identity by comparing the corresponding Fourier coefficients.]

§ 5.2. Theta-Matrices of Hecke Operators and Eichler Matrices

Although the formulae of § 5.1 for the images of theta-series under Hecke operators are quite explicit, the formulae do not reveal the arithmetical meaning of the coefficients of the resulting linear combinations of theta-series. In this section we shall show that the coefficients can be expressed in terms of integral representations of the corresponding quadratic forms by each other. This will allow us to express matrices of Hecke operators operating on certain invariant spaces

Note also that, for $T \in \mathrm{E}^n(q)$, an element of the form $\Psi^{n-m}\mathbf{T}$ is always defined, in view of Proposition 5.1.14.

We prove the formula (5.1.60). If $n = m$, then the formula is the same as (5.1.32). If $n > m$, then, by Proposition 5.1.14 and (5.1.32), we can write

$$\theta^m(Z', Q)|\Psi^{n-m}(\mathbf{T}) = \sum_{\{Q'\} \in \mathbb{A}_m^+} \varepsilon(Q, Q', \eta(\Psi^{n-m}\mathbf{T}))\theta^m(Z', Q') \qquad (5.1.61)$$

By (5.1.3), each of the theta-series $\theta^m(Z', Q')$ is the image of the theta-series of genus n of Q' under the action of the $(n-m)$-th iteration Φ^{n-m} of Φ. Hence, applying Φ to $\theta^n(Z, Q)T$, $n - m$ times, and using (5.1.52) each time, we see that the left side in (5.1.61) is equal to the image of $\theta^n(Z, Q)|T$ under the action of Φ^{n-m}. Thus, the relation (5.1.61) can be written in the form $F|\Phi^{n-m} = 0$, where

$$F = \theta^n(Z, Q)|T - \sum_{\{Q'\}} \varepsilon(Q, Q', \eta(\Psi^{n-m}\mathbf{T}))\theta^n(Z, Q').$$

Since $n > m$, it follows that each of the series $\theta^n(Z, Q')$ and $\theta^n(Z, Q)$ is a singular form. By (5.1.6), the same is true for the form $\theta^n(Z, Q)|T$. Thus, the form F is singular. Since $F \in \mathfrak{M}_k^n(q, \chi)$, it follows, by Theorem 2.3.16, that F is a finite linear combination of the theta-series of genus n of some matrices $Q_1, \ldots, Q_h$ of $\mathbb{A}_m^+$. One can obviously assume that the matrices $Q_1, \ldots, Q_h$ are pairwise non-equivalent over $\mathbb{Z}$. Then we have

$$0 = F|\Phi^{n-m} = \left(\sum_{i=1}^{h} c_i\theta^n(Z, Q_i)\right)\Bigg|\Phi^{n-m} = \sum_{i=1}^{h} c_i\theta^m(Z', Q_i),$$

whence $c_1 = c_2 = \cdots = c_h = 0$, by Lemma 5.1.1, and so $F = 0$.

If $n < m$, then we start with the formula (5.1.32) for an inverse image $\Psi^{n-m}\mathbf{T}$ of $\mathbf{T}$ under the mapping (5.1.57) and apply the operator Φ, $m - n$ times. Using (5.1.3) and (5.1.52) each time, we obtain (5.1.60). □

Exercise 5.1.16. *Let $Q \in \mathbb{A}_m^+$, where m is even, and let $Q_1, \ldots, Q_h$ be a set of representatives for the classes contained in the genus $\mathsf{G}(Q)$ of Q. Show that, for any $n \in \mathbb{N}$, the linear combination of theta-series*

$$G(Z) = \sum_{i=1}^{h} e(Q_i)^{-1}\theta^n(Z, Q_i)$$

is an eigenfunction for all of the operators $|_{k,\chi}T$ with $T \in \mathrm{E}^n(q)$ where $k = m/2$, $\chi = \chi_Q$, and q is the level of Q:

$$G|_{k,\chi}T = \lambda_G(T)\cdot G.$$

Let $\mathrm{i}(T) = \sum_j a_j\left(\Gamma_0\begin{pmatrix}\mu_j D_j^* & B_j\\ 0 & D_j\end{pmatrix}\right)$ be the image of T in $\mathscr{L}^n(q)$. Show that

$$\lambda_G(T) = \sum_j a_j\mu_j^{nk-\langle n\rangle}\chi(\det \mu_j D_j^*)(\det D_j)^{-k}.$$

[Hint: Use Exercise 5.1.8; to calculate $\lambda_G(T)$, compare the 0-components of $G|T$ and $\lambda_G(T)\cdot G$].

Exercise 5.1.17. *In the notation of Exercise* 5.1.16, *show that* G *is an eigenfunction for all of the operators* $|_{k,\chi}T^n(a)$ *with* $a \in \mathbb{N}_{(q)}$ *satisfying the conditions:* $v_p(a)$ *is odd or* $\chi_Q(p) = 1$ *for each prime divisor* p *of* a. *Show that, for such an* a *and* $n = 1, 2$, *the corresponding eigenvalues are given by*

$$\lambda(T^1(a) = \sum_{d|a} \chi_Q(d) d^{k-1}$$

$$\lambda(T^2(a)) = \sum_{d|a} \chi_Q(d) d^{k-2} \omega(d) \sigma_{2k-3}(a/d)$$

where $\omega(d) = d \prod_{p|d}(1 + p^{-1})$ *and* $\sigma_l(b) = \sum_{\delta|b} \delta^l$.

[Hint: For the last assertion, use the formulae (3.4.31) and (3.4.32).]

Exercise 5.1.18. (*Siegel*) *Let* $Q_1, \ldots, Q_h$ *be a set of representatives for all of the classes* $\{Q\} \in \mathbb{A}_m$ *with* $\det Q = 1$. *Prove the identity*

$$\left(\sum_{i=1}^{h} \frac{1}{e(Q_i)}\right)^{-1} \sum_{i=1}^{h} \frac{1}{e(Q_i)} \theta^1(z, Q_i) = (2\zeta(m/2))^{-1} G_{m/2}(z), \ (z \in \mathbb{H}_1)$$

where $G_{m/2}(z)$ *is an Eisenstein series of Exercise* 2.5.6, $\zeta(s)$ *being the Riemann zeta. For each* $a \in \mathbb{N}$, *prove the relation*

$$\left(\sum_{i=1}^{h} \frac{1}{e(Q_i)}\right)^{-1} \sum_{i=1}^{h} \frac{r(Q_i, 2a)}{e(Q_i)} = \frac{(2\pi)^{m/2}}{\left(\frac{m}{2} - 1\right)! \zeta(m/2)} \sum_{d|a} d^{(m/2)-1};$$

in particular,

$$\zeta\left(\frac{m}{2}\right) = (2\pi)^{m/2}((m/2 - 1)!)^{-1} \left(\sum_{i=1}^{h} e(Q_i)^{-1}\right) \left(\sum_{i=1}^{h} \frac{r(Q_i, 2)}{e(Q_i)}\right)^{-1}$$

[Hint: Using Exercises 4.3.5 and 5.1.17, show that the difference of the two sides of the identity is a cusp form, which is an eigenfunction for all of the operators $|_k T^1(a)$ with the eigenvalues $\sum_{d|a} d^{(m/2)-1}$. Then, by (4.3.9), prove that the difference equals zero. The relations for the numbers of representations follow from the identity by comparing the corresponding Fourier coefficients.]

§ 5.2. Theta-Matrices of Hecke Operators and Eichler Matrices

Although the formulae of § 5.1 for the images of theta-series under Hecke operators are quite explicit, the formulae do not reveal the arithmetical meaning of the coefficients of the resulting linear combinations of theta-series. In this section we shall show that the coefficients can be expressed in terms of integral representations of the corresponding quadratic forms by each other. This will allow us to express matrices of Hecke operators operating on certain invariant spaces spanned by theta-series through certain matrices analogous to "Anzahlmatrizen" of Eichler.

5.2.1. The Comparison of Fourier Coefficients

When it is known that a relation of the form (5.1.58) holds with some constant coefficients, the coefficients can be determined directly by comparing the corresponding Fourier coefficients of the series in the two sides. This approach proves to be more convinient than the approach based on the initial formulae for the coefficients.

We first fix some notation. Let $Q \in \mathbb{A}_m^+$, and let $d = \det Q$, $\delta = \delta(Q)$, and $q = q(Q)$. By Theorem 2.1.12, the set $\mathbb{A}_m(d, \delta, q)$ is a finite union of the classes modulo the equivalence over $\mathbb{Z}$,

$$\mathbb{A}_m(d, \delta, q) = \bigcup_{\alpha=1}^{H} \{Q_\alpha\} \qquad (\{Q_\alpha\} \neq \{Q_\beta\} \text{ if } \alpha \neq \beta). \tag{5.2.1}$$

According to Propositions A.2.20 and A.2.21, the genus $\mathsf{G}(Q)$ of Q is contained in $\mathbb{A}_m(d, \delta, q)$. Hence, without loss of generality one may assume that

$$\mathsf{G}(Q) = \bigcup_{\alpha=1}^{h} \{Q_\alpha\} \qquad \text{with } h \leq H, \tag{5.2.2}$$

where $Q_1 = Q$. By the end of this section we fix ordered systems of representatives

$$\{Q = Q_1, \dots, Q_h, \dots, Q_H\} = \mathbb{A}_m(d, \delta, q)/\Lambda$$

and (5.2.3)

$$\{Q_1, \dots, Q_h\} = \mathsf{G}(Q)/\Lambda$$

For each $n \in \mathbb{N}$, we define two *theta-columns of genus n of Q* by

$$\mathbf{t}^n = \mathbf{t}^n(Z, Q) = {}^t(e(Q_1)^{-1}\theta^n(Z, Q_1), \dots, e(Q_H)^{-1}\theta^n(Z, Q_H)), \tag{5.2.4}$$

and

$$\tau^n = \tau^n(Z, Q) = {}^t(e(Q_1)^{-1}\theta^n(Z, Q_1), \dots, e(Q_h)^{-1}\theta^n(Z, Q_h)), \tag{5.2.5}$$

$e(Q_\alpha) = r(Q_\alpha, Q_\alpha)$ being the number of units of Q_α. We shall assume that m is even, say $m = 2k$, and denote by $\chi = \chi_Q = \dots = \chi_{Q_H}$ the character of Q. If T is an element of $\mathsf{L}^n(q)$ such that there is an element of the form $\Psi_{k,\chi}^{n-m}(\mathsf{i}(T)) \in \mathbf{L}^m(q)$, then, in view of Theorem 5.1.15, we have the relations

$$e(Q_\alpha)^{-1}\theta^n(Z, Q_\alpha)|_{k,\chi} T = \sum_{j=1}^{H} \vartheta_{\alpha,\beta}(T) e(Q_\beta)^{-1}\theta^n(Z, Q_\beta) \tag{5.2.6}$$

with $\alpha = 1, 2, \dots, H$, where

$$\vartheta_{\alpha\beta}(T) = e(Q_\alpha)^{-1} e(Q_\beta) \varepsilon_{k,\chi}(Q_\alpha, Q_\beta, \eta(\Psi^{n-m}(\mathsf{i}(T)))). \tag{5.2.7}$$

With the help of the column $\mathbf{t}^n$, we can rewrite the relation (5.2.6) as a matrix relation of the form

$$\mathbf{t}^n|_{k,\chi}T = \Theta(T)\mathbf{t}^n, \tag{5.2.8}$$

where the Hecke operator acts componentwise, and where

$$\Theta(T) = \Theta(\Psi^{n-m}(\mathrm{i}(T))) = (\vartheta_{\alpha\beta}(T)) \in \mathsf{M}_H(\mathbb{C}) \tag{5.2.9}$$

is the matrix with the entries (5.2.7). Similarly, for $T \in \mathsf{E}^n(q)$, we have

$$\tau^n|_{k,\chi}T = \Theta_g(T)\tau^n, \tag{5.2.10}$$

where

$$\Theta_g(T) = \Theta_g(\Psi^{n-m}(\mathrm{i}(T))) = (\vartheta_{\alpha\beta}(T))_{\alpha,\beta=1,\ldots,h}. \tag{5.2.11}$$

In the notation, we have obviously that

$$\Theta(T) = \begin{pmatrix} \Theta_g(T) & 0 \\ 0 & * \end{pmatrix} \quad \text{if } T \in \mathsf{E}^n(q). \tag{5.2.12}$$

Any matrices $\Theta(T)$ and $\Theta_g(T)$ satisfying, respectively, (5.2.8) and (5.2.10) will be called *theta-matrices of* T *or* $|_{k,\chi}T$ *(with respect to* Q*)*.

Lemma 5.2.1. *In the above notation, suppose that* $n \geq m$. *Then the following assertions hold*:

(1) *Theta-matrices* $\Theta(T)$ *and* $\Theta_g(T)$ *are uniquely determined by the relations* (5.2.8) *and* (5.2.10) *respectively*;

(2) *The mapping* $T \to \Theta(T)$ (resp. $T \to \Theta_g(T)$) *is a* $\mathbb{Q}$*-linear homomorphism of the ring* $\mathsf{L}^n(q)$ *(resp.* $\mathsf{E}^n(q)$*) into* $\mathsf{M}_H(\mathbb{C})$ *(resp.* $\mathsf{M}_h(\mathbb{C})$*)*.

Proof. By Lemma 5.1.1, the entries of each of the columns $\mathbf{t}^n(Z, Q)$, $\tau^n(Z, Q)$ are linearly independent over $\mathbb{C}$. This implies the first assertion. The second assertion follows from the first and Proposition 4.1.1. □

We now compare Fourier coefficients of the functions occurring in (5.2.8) and (5.2.10). For $A \in \mathbb{A}_n$ and $T \in \mathsf{L}^n(q)$, we set

$$\mathbf{r}(A) = {}^t(r_1(A), \ldots, r_H(A)), (\mathbf{r}|T)(A) = {}^t((r_1|T)(A), \ldots, (r_H|T)(A)) \tag{5.2.13}$$

and

$$\rho(A) = {}^t(r_1(A), \ldots, r_h(A)), (\rho|T)(A) = {}^t((r_1|T)(A), \ldots, (r_h|T)(A)),$$

where $r_\alpha(A) = e(Q_\alpha)^{-1} r(Q_\alpha, A)$ and $(r_\alpha|T)(A)$ are the Fourier coefficients of $e(Q_\alpha)^{-1}\theta^n(Z, Q_\alpha)$ and $e(Q_\alpha)^{-1}\theta^n(Z, Q_\alpha)|T$ respectively. By comparing the coefficients of $e\{AZ\}$ in the two sides of the relation (5.2.8) (resp. (5.2.10)) we see that the relation is equivalent with the set of matrix relations of the form

$$(\mathbf{r}|T)(A) = \Theta(T)\mathbf{r}(A) \quad \text{with } A \in \mathbb{A}_n \tag{5.2.14}$$

(resp.

$$(\rho|T)(A) = \Theta_g(T)\rho(A)). \tag{5.2.15}$$

These sets of relations can be considered as systems of linear equations relative to the entries $\vartheta_{\alpha\beta}(T)$ of the theta-matrices. According to Theorem 5.1.15, the system (5.2.14) (resp. (5.2.15)) has a solution if there exists $\Psi^{n-m}(\mathrm{i}(T)) \in \mathbf{L}^m(q)$ (resp. if $T \in \mathsf{E}^n(q)$). If, in addition, $n \geq m$, then the solution is unique, by Lemma 5.2.1(1).

Before going on to the calculation of theta-matrices, we note that, in view of (5.1.12), it suffices to calculate only the big matrix $\Theta(T)$. Besides, it suffices to calculate only the matrices $\Theta(T)$ with $T \in \mathsf{L}^m(q)$, since, for $T' \in \mathsf{L}^n(q)$, one can take

$$\Theta(T') = \Theta(\Psi^{n-m}(\mathrm{i}(T'))), \tag{5.2.16}$$

by Theorem 5.1.15.

Nevertheless, we shall formulate and prove the results in a reasonably general setting, unless the contrary is essentially more convenient.

Lemma 5.2.2. *In the above notation, suppose that $n \geq m$.* Then the theta-matrices can be written in the form

$$\Theta(T) = \left((r|T)\begin{pmatrix} 0 & 0 \\ 0 & Q_1 \end{pmatrix}, \ldots, (r|T)\begin{pmatrix} 0 & 0 \\ 0 & Q_H \end{pmatrix}\right), (T \in \mathsf{L}^n(q))$$

$$\Theta_g(T) = \left((\rho|T)\begin{pmatrix} 0 & 0 \\ 0 & Q_1 \end{pmatrix}, \ldots, (\rho|T)\begin{pmatrix} 0 & 0 \\ 0 & Q_h \end{pmatrix}\right), (T \in \mathsf{E}^n(q))$$

where each of the matrices $\begin{pmatrix} 0 & 0 \\ 0 & Q_\alpha \end{pmatrix}$ *is of order n.*

Proof. It sufficies to prove the first formula. Substituting $A = Q'_\alpha = \begin{pmatrix} 0 & 0 \\ 0 & Q_\alpha \end{pmatrix}$ with $1 \leq \alpha \leq H$ in (5.2.14), we obtain the relations

$$\Theta(T)r(Q'_\alpha) = (\mathbf{r}|T)(Q'_\alpha).$$

Since, clearly,

$$r(Q_\beta, Q'_\alpha) = r(Q_\beta, Q_\alpha) = \begin{cases} e(Q_\alpha), & \text{if } \beta = \alpha \\ 0, & \text{if } \beta \neq \alpha \end{cases}$$

it follows that the column $\mathbf{r}(Q'_\alpha)$ is equal to the α-th column of the unit matrix E_H. Hence, the column $\Theta(T)\mathbf{r}(Q'_\alpha)$ is equal to the α-th column of $\Theta(T)$. □

The last lemma reduces calculation of theta-matrices to calculation of the action of Hecke operators on Fourier coefficients of corresponding modular forms. For the latter we shall use the techniques of § 4.2 and decompositions of §§ 3.4 and 3.5. We start from general formulae.

Proposition 5.2.3. *In the above notation, suppose that $n \geq m$. Let $T \in \mathsf{L}^n(q)$, and let*

$$\mathsf{i}(T) = \sum_j a_j \left(\Gamma_0^n \begin{pmatrix} \mu_j D_j^* & B_j \\ 0 & D_j \end{pmatrix} \right)$$

be the image of T in $\mathscr{L}^n(q)$. Then the entries $\vartheta_{\alpha\beta}(T)$ of the theta-matrix $\Theta(T)$ are given by the formula

$$\vartheta_{\alpha\beta}(T) = \sum_{j,\, Q'_\beta[{}^tD_j] \in \mu_j \mathbb{A}_n} a_j \mu_j^{nk - \langle n \rangle} \chi(\det(\mu_j D_j^*))(\det D_j)^{-k} e\{\mu_j^{-1} Q'_\beta[{}^tD_j] B_j D_j^{-1}\}$$
$$\times\, e(Q_\alpha)^{-1} r(Q_\alpha, \mu_j^{-1} Q'_\beta[{}^tD_j]),$$

where $Q'_\beta = \begin{pmatrix} 0 & 0 \\ 0 & Q_\beta \end{pmatrix} \in \mathbb{A}_n$.

Proof. By Lemma 5.2.2, we have $\vartheta_{\alpha,\beta}(T) = (r_\alpha | T)(Q'_\beta)$. Since $r_\alpha(A) = e(Q_\alpha)^{-1} r(Q_\alpha, A)$, the formula follows at once from (4.2.16). □

The formulae (5.2.17) are simpler than (5.2.7). However, the formulae are rather complicated. We shall obtain simple formulae for some specific T below.

Exercise 5.2.4. (*Freitag*). *If* $\det Q = 1$ *and* $n > m$, *prove the formula* (5.2.6) *with the coefficients given by* (5.2.17), *using the results of* § 2.3.6 *on singular forms, instead of Theorem* 5.1.15.

[Hint: Show that, in the case, $\theta^n(Z, Q)|T$ is a singular form of $\mathfrak{M}_k^n$ and use Exercise 2.3.18.]

5.2.2. Theta-Matrices of Elements $T^n(p)$

In this subsection we shall express theta-matrices of elements $T^n(p)$ with p of $\mathbb{P}_{(q)}$ satisfying certain conditions in terms of so-called Eichler matrices, which we first define.

In the notation of § 5.1.1, for $\mu \in \mathbb{Q}$ and $t \in \mathsf{H}^m = \mathsf{D}_\mathbb{Q}(\Lambda^m, \mathsf{G}^m)$, we consider the matrices of the form

$$M(\mu) = M(\mu, Q) = (e(Q_\alpha)^{-1} r(Q_\alpha, \mu Q_\beta) \in \mathsf{M}_H(\mathbb{Q}), \tag{5.2.18}$$

$$M(\mu, t) = M(\mu, t, Q) = (e(Q_\alpha)^{-1} r(Q_\alpha, \mu Q_\beta, t)) \in \mathsf{M}_H(\mathbb{Q}), \tag{5.2.19}$$

and similar matrices related to the genus $\mathsf{G}(Q)$ of Q

$$M_g(\mu) = M_g(\mu, Q) = (e(Q_\alpha)^{-1} r(Q_\alpha, \mu Q_\beta)) \in \mathsf{M}_h(\mathbb{Q}), \tag{5.2.20}$$

$$M_g(\mu, t) = M_g(\mu, t, Q) = (e(Q_\alpha)^{-1} r(Q_\alpha, \mu Q_\beta, t)) \in \mathsf{M}_h(\mathbb{Q}), \tag{5.2.21}$$

with the numbers $r(A, B, t)$ defined by (4.2.43), which will be called *Eichler matrices for Q*. According to (4.2.46), we have the relations

$$M(\mu) = \begin{cases} M(\mu, t(\mu^{m/2})), & \text{if } \mu^{m/2} \in \mathbb{N} \\ 0, & \text{if } \mu^{m/2} \notin \mathbb{N} \end{cases} \tag{5.2.22}$$

where $t(a) \in \mathrm{H}^m$ are the elements (3.2.10). It is also clear that

$$M(\mu) = \begin{pmatrix} M_g(\mu) & * \\ * & * \end{pmatrix}, \quad \text{and} \quad M(\mu, t) = \begin{pmatrix} M_g(\mu, t) & * \\ 0 & * \end{pmatrix}. \tag{5.2.23}$$

Theorem 5.2.5. *In the above notation, suppose that a prime number p of* $\mathbb{P}_{(q)}$ *satisfies the condition* $\chi(p) = \chi_Q(p) = 1$, *if* $n < k$. *Then the theta-matrix* (2.5.9) *of the element* $T^n(p)$ *(see* (3.3.19) *or* (3.3.22)) *is given by the formula*

$$\Theta(T^n(p)) = \beta^n(p, Q) M(p) \tag{5.2.24}$$

where

$$\beta^n(p, Q) = \begin{cases} \prod_{i=1}^{n-k} (1 + \chi(p) p^{-i}), & \text{if } n > k \\ 1, & \text{if } n = k \\ \prod_{i=0}^{k-n-1} (1 + p^i)^{-1} & \text{if } 1 \le n < k \text{ and } \chi(p) = 1, \end{cases}$$

and $M(p)$ *is the Eichler matrix* (5.2.18). *In other words, with the above assumption on p, the image of the theta-series of genus n of a matrix* Q_α *contained in* (5.2.3) *under the action of the operator* $|_{k,\chi} T^n(p)$ *is given by the formula*

$$\theta^n(Z, Q_\alpha)|_{k,\chi} T^n(p) = \beta^n(p, Q) \sum_{\beta=1}^{H} r(Q_\alpha, pQ_\beta) e(Q_\beta)^{-1} \theta^n(Z, Q_\beta). \tag{5.2.25}$$

Proof. We first consider the case $n = m$. By (3.3.58) and (4.2.28), for any $f \in \mathfrak{F}_k^m(q, \chi) \in \mathfrak{F}_+^m$ and $A \in \mathbb{A}_m$, we have the formula

$$\begin{aligned}
(f|_{k,\chi} T^m(p))(A) \\
&= (f|_{k,\chi} \mathbf{T}^m(p))(A) = \sum_{j=0}^{m} (f|_{k,\chi} \Pi_j^m(p))(A) \\
&= p^{mk - \langle m \rangle} \chi(p)^m \sum_{j=0}^{m} p^{\langle j \rangle - jk} \chi(p)^{-j} \sum_{D \in \Lambda \backslash \Lambda D_j^m(p) \Lambda,\, A[{}^tD] \in p\mathbb{A}_m} f(p^{-1} A[{}^tD]).
\end{aligned} \tag{5.2.26}$$

Hence, by Lemma 5.2.2, we conclude that the entries $\vartheta_{\alpha\beta}(T^m(p))$ of $\theta(T^m(p))$ are given by

$$\vartheta_{\alpha\beta}(T^m(p)) = p^{mk - \langle m \rangle} \chi(p)^m \sum_{j=0}^{m} p^{\langle j \rangle - jk} \chi(p)^{-j} \sum_{D \in \Lambda \backslash \Lambda D_j^m(p) \Lambda} e(Q_\alpha)^{-1} r(Q_\alpha, p^{-1} Q_\beta[{}^tD]),$$

because, by definitions, $r(Q_\alpha, p^{-1} Q_\beta[{}^tD]) = 0$, if $Q_\beta[{}^tD] \notin p\mathbb{A}_m$. We now note that the condition $Q_\beta[{}^tD] \in p\mathbb{A}_m$ for a matrix D of $\Lambda D_j^m(p) \Lambda$ implies that $p^{-m+2j} \det Q_\beta \in \mathbb{N}$, which means that $2j \ge m$, because $p \in \mathbb{P}_{(q)}$. Since m is even and $\chi(p) = \pm 1$, the last formula can be written in the form

$$\vartheta_{\alpha\beta}(T^m(p)) = p^{mk-\langle m\rangle} \sum_{j=k}^{m} p^{\langle j\rangle - jk} \chi(p)^j \sum_{D \in \Lambda \backslash \Lambda D_j^m(p) \Lambda} e(Q_\alpha)^{-1} r(Q_\alpha, p^{-1} Q_\beta[{}^tD]). \tag{5.2.27}$$

Let us prove the following relations

$$\sum_{D \in \Lambda \backslash \Lambda D_j^m(p) \Lambda} r(Q_\alpha, p^{-1} Q_\beta[{}^tD]) = \frac{\varphi_k(p)}{\varphi_{j-k}(p)\varphi_{m-j}(p)} r(Q_\alpha, pQ_\beta), \tag{5.2.28}$$

where $k \leq j \leq m$, $\varphi_r(x)$ being the polynomials (3.2.29). Let N be a matrix of M_m such that

$$Q_\alpha[N] = pQ_\beta. \tag{5.2.29}$$

Since $\det Q_\alpha = \det Q_\beta$, it follows that $\det N = \pm p^k$. Hence, all of the elementary divisors of N are powers of p, and so the same is true for the elementary divisors of pN^{-1}. On the other hand, since $p \nmid q$, it follows that the matrix $pN^{-1} = q^{-1} \cdot (qQ_\beta^{-1}) \cdot {}^tNQ_\alpha$ is p-integral. Thus, the matrix pN^{-1} is integral, whence each of the elementary divisors of N is equal to 1 or p. Then, by $\det N = \pm p^k$, we conclude that $\mathrm{ed}(N) = D_k^m(p)$, i.e.

$$N \in \Lambda D_k^m(p) \Lambda = \Lambda p D_k^m(p)^{-1} \Lambda. \tag{5.2.30}$$

Let us define a matrix R by $pD_k^m(p)^{-1} = RpD_j^m(p)^{-1}$. Since $j \geq k$, it follows that the matrix $R = D_k^m(p)^{-1} D_j^m(p)$ is integral. Hence the matrix N can be written in the form

$$N = VR \cdot pD_j^m(p)^{-1} V_1^* = N_1 pD^*,$$

where $V, V_1 \in \Lambda$, $N_1 = VR \in \mathsf{M}_m$, and $D = D_j^m(p) V_1 \in \Lambda D_j^m(p) \Lambda$. On replacing of N_1 by $N_1 V'$ with $V' \in \Lambda$, we may assume that D is equal to any given matrix in ΛD, In view of (5.2.29), we have

$$Q_\alpha[N_1] = p^{-1} Q_\beta[{}^tD]. \tag{5.2.31}$$

Conversely, if matrices $N_1 \in \mathsf{M}_m$ and $D \in \Lambda \backslash \Lambda D_j^m(p) \Lambda$ satisfy (5.2.31), then the matrix $N = N_1 pD^*$ is an integral matrix satisfying (5.2.29). By the above, each of the integral matrices N satisfying (5.2.29) can be obtained in such a way. In addition, such a matrix N results $\alpha(N)$ times, where $\alpha(N)$ is the number of D in $\Lambda \backslash \Lambda D_j^m(p) \Lambda$ satisfying $N\,{}^tD \equiv 0 \pmod{p}$. Hence, in order to prove (5.2.28), we only have to check that

$$\alpha(N) = \varphi_k(p)/\varphi_{j-k}(p)\varphi_{m-j}(p),$$

for any integral N satisfying (5.2.29). In view of (5.2.30), we may assume that N belongs to $\Lambda D_k^m(p) \Lambda$. Since clearly $\alpha(N)$ depends only on the double coset $\Lambda N \Lambda = \Lambda D_k^m(p) \Lambda$, we may take $N = D_k^m(p)$. Choosing a set of representatives for the left cosets of $\Lambda D_j^m(p) \Lambda$ modulo Λ in the form (3.2.27) with $n = m$, $i = j$, and $C = D$, we readily obtain that a congruence $D_k^m(p)\,{}^tD \equiv 0 \pmod{p}$ holds if and only if $D = \begin{pmatrix} PE_k & 0 \\ 0 & D' \end{pmatrix}$, where D' belongs to the system (3.2.27) of representatives for the left cosets of $\Lambda^k D^k_{j-k}(p)\Lambda^k$ modulo Λ^k. Hence, $\alpha(D^m_k(p))$ is equal to the number of these left cosets and can be calculated by the formula (3.2.28) with $n = k$ and $i = j - k$. This proves the formula for $\alpha(N)$ and the relations (5.2.28).

We return to our calculation of the entries $\vartheta_{\alpha\beta}(T^m(p))$. By (5.2.27) and (5.2.28), we have

$$\vartheta_{\alpha\beta}(T^m(p)) = p^{mk-\langle m\rangle}\left(\sum_{j=k}^{m} p^{\langle j\rangle - jk}\chi(p)^j \frac{\varphi_k(p)}{\varphi_{j-k}(p)\varphi_{m-j}(p)}\right) e(Q_\alpha)^{-1} r(Q_\alpha, pQ_\beta)$$

$$\overset{j=k+i}{=} p^{-\langle k\rangle}\chi(p)^k\left(\sum_{i=0}^{k} \chi(p)^i \frac{p^{\langle i\rangle}\varphi_k(p)}{\varphi_i(p)\varphi_{k-i}(p)}\right) e(Q_\alpha)^{-1} r(Q_\alpha, pQ_\beta),$$

which, in view of the identity (3.2.34) with $x = p$, $t = \chi(p)$, and $n = k$, equals

$$p^{-\langle k\rangle}\chi(p)^k \prod_{i=1}^{k} (1 + \chi(p)p^i) e(Q_\alpha)^{-1} r(Q_\alpha, pQ_\beta).$$

The formula (5.2.24) with $n = m$ is proved.

To deduce the general case, we make use of the relations (5.2.16). We first consider the elements $\Psi^{n-m}(\mathbf{T}^n(p))$, where $\mathbf{T}^n(p) = \mathrm{i}(T^n(p))$. Since the diagram (4.2.67) is commutative, it follows, by (3.3.70), that

$$\Omega^{n-1}_p(\Psi^n_p(\mathbf{T}^n(p), u)) = \Xi_n \Omega^n_p(\mathbf{T}^n(p)) = \Xi_n\left(x_0 \prod_{i=1}^{n} (1 + x_i)\right)$$

$$= u^{-1}(1 + u)x_0 \sum_{i=1}^{n-1} (1 + x_i) = \Omega^{n-1}_p((1 + u^{-1})\mathbf{T}^{n-1}(p)).$$

Hence,

$$\Psi^n_p(\mathbf{T}^n(p), u) = (1 + u^{-1})\mathbf{T}^{n-1}(p)$$

because the restriction of Ω^{n-1}_p to $\mathbf{L}^{n-1}_p[u^{\pm 1}]$ is clearly monomorphic. The last relation and the Žarkovskaya relations (4.2.69) imply that

$$\Psi(\mathbf{T}^n(p)) = \Psi^n_p(\mathbf{T}^n(p), p^{n-k}\chi(p)^{-1}) = \left(1 + \frac{\chi(p)}{p^{n-k}}\right)\mathbf{T}^{n-1}(p). \tag{5.2.32}$$

If $n > m$, then the $(n - m)$-th iteration of (5.2.32) leads to the relation

$$\Psi^{n-m}(\mathbf{T}^n(p)) = (1 + \chi(p)p^{k-n})\cdots(1 + \chi(p)p^{k-m-1})\mathbf{T}^m(p)$$

$$= \beta^n(p, Q)\beta^m(p, Q)^{-1}\mathbf{T}^m(p),$$

whence, by the above, we have

$$\Theta(T^n(p)) = \Theta(\Psi^{n-m}(\mathbf{T}^n(p)))$$

$$= \beta^n(p, Q)\beta^m(p, Q)^{-1}\Theta(\mathbf{T}^m(p)) = \beta^n(p, Q)M(p).$$

But if $n < m$, then the $(m - n)$-th iteration of (5.2.32) with n replaced by m gives

$$\Psi^{m-n}(\mathbf{T}^m(p)) = (1+\chi(p)p^{k-m})\cdots(1+\chi(p)p^{k-n-1})\mathbf{T}^n(p)$$
$$= \begin{cases} \beta^m(p,Q)\beta^n(p,Q)^{-1}\mathbf{T}^n(p), & \text{if } n \geq k \\ \beta^m(p,Q)\left[\sum_{i=0}^{k-n-1}(1+\chi(p)p^i)\right]\mathbf{T}^n(p), & \text{if } 1 \leq n < k \end{cases}$$

whence we again have

$$\Psi^{n-m}(\mathbf{T}^n(p)) = \beta^n(p,Q)\beta^m(p,Q)^{-1}\mathbf{T}^m(p),$$

where, in the case $1 \leq n < k$, it is assumed that $\chi(p) = 1$. The formula (5.2.24) follows. □

Exercise 5.2.6. *Let $Q \in \mathbb{A}_2^+$ and $a \in \mathbb{N}_{(q)}$, where q is the level of Q. Prove the following formulae*

$$\Theta(T^2(a)) = \sum_{a_1a_2=a} a_2^{-1}\chi(a_2)M(a_1)M\left(a_2, \left(\begin{pmatrix}1 & 0\\ 0 & a_2\end{pmatrix}\right)_{A^2}\right),$$

$$M(a) = \sum_{b^2|a} M\left(a/b^2, \left(\begin{pmatrix}1 & 0\\ 0 & a/b^2\end{pmatrix}\right)_{A}\right),$$

$$M\left(a, \left(\begin{pmatrix}1 & 0\\ 0 & a\end{pmatrix}\right)_{A}\right) = \sum_{b^2|a} \mu(b)M(a/b^2),$$

$$\Psi_{1,\chi}(\mathbf{T}^2(a)) = \sum_{a_1a_2=a} a_2^{-1}\chi(a_2)\mathbf{T}^1(a_1)\sum_{b^2|a_2}\mu(b)\mathbf{T}^1(a_2/b^2),$$

and one can take

$$\Theta(T^1(a)) = M(a),$$

where $\chi = \chi_Q$, μ is the Möbius function, and where all of the theta-matrices and Eichler matrices are taken with respect to Q.

[Hints: For the first formula, use Lemma 5.2.2, the formula of Exercise 4.2.14, and the formulae (4.2.42), (4.2.49); for the fourth formula, with the help of (3.3.73) and (3.3.75), prove the formal identity

$$\sum_{\delta=0}^{\infty} \Psi(\mathbf{T}^2(p^\delta))v^\delta = (1-(p^{-1}\chi(p)v)^2)\sum_{\delta=0}^{\infty}\mathbf{T}^1(p^\delta)v^\delta\sum_{\gamma=0}^{\infty}\mathbf{T}^1(p^\gamma)(p^{-1}\chi(p)v)^\gamma$$

for each $p \in \mathbb{P}_{(q)}$, and then use (3.3.20); the last formula follows from previous formulae and (5.2.16) by an induction on the number of divisors of a.]

Exercise 5.2.7. *Let $Q \in \mathbb{A}_m^+$, and let p be a prime number prime to the level of Q. Show that $M(p,Q) = 0$ if m is odd, or m is even and $\chi_Q(p) = -1$.*

[Hint: For even m, use Corollary A.2.15.]

Exercise 5.2.8 (Eichler). *Let $Q \in \mathbb{A}_{2k}^+$, and let $M(a) = M(a,Q)$ be the Eichler matrices* (5.2.18). *Prove that a matrix Dirichlet series of the form*

$$\sum_{a \in \mathbb{N}_{(q)}} M(a) \cdot a^{-s},$$

where q is the level of Q, converges (entry-wise) absolutely and uniformly in a right half-plane of the complex variable s, and has an Euler product expansion of the form

$$= \prod_{p \in \mathbb{P}_{(q)}} \left(\sum_{\delta=0}^{\infty} M(p^{\delta}) p^{-\delta s} \right).$$

For $p \in \mathbb{P}_{(q)}$, we let $q_i^{2k}(p)$ be the inverse images in $\mathsf{L}_p^{2k}(q)$ of the elements (3.3.77) under the action i_q, and $\Theta(q_i^{2k}(p))$ the theta-matrices of these elements with respect to Q. Prove that the matrix

$$\Theta_p(v) = \sum_{i=0}^{4^k} (-1)^i \Theta(q_i^{2k}(p)) v^i$$

is invertible in the ring of formal power series in v over $\mathsf{M}_H(\mathbb{Q})$, and an identity holds of the form

$$\sum_{\delta=0}^{\infty} M(p^{\delta}) v^{\delta} = \Theta_p(v)^{-1} \cdot N_p(v),$$

where $N_p(v)$ is a polynomial in v over $\mathsf{M}_H(\mathbb{Q})$ of degree $\leq 4^k - 1$.

[Hints: See Exercise 2.5.17 below; as to the p-factors, in the notation (5.2.13), the β-th column of $M(a)$ is $\mathbf{r}(aQ_\beta)$, which, by (4.2.28), can be written in the form $(\mathbf{r}|_{k,\chi} \Pi_+^{2k}(a))(Q_\beta)$ with $\chi = \chi_Q$. Hence, by (5.2.15), one gets

$$\Theta_p(v) \sum_{\delta=0}^{\infty} \mathbf{r}(p^{\delta} Q_\beta) v^{\delta}$$

$$= \left(\mathbf{r}|_{k,\chi} \sum_{i=0}^{4^k} (-1)^i q_i^{2k}(p) v^i \sum_{\delta=0}^{\infty} \Pi_+^{2k}(p)^{\delta} v^{\delta} \right)(Q_\beta).$$

Next use the second factorization of Proposition 3.5.11.]

5.2.3. Theta-Matrices of Coefficients of Even Local Zeta-Functions

A formal power series over $\mathbf{L}_p^n$ defined by

$$\mathbf{Z}_p^+(v) = \mathbf{Z}_p^+(v)^{(n)} = \mathbf{R}_p^n(v)^{-1} = \sum_{\delta=0}^{\infty} \mathbf{z}^n(p^{\delta}) v^{\delta}, \tag{5.2.33}$$

where $\mathbf{R}_p^n(v)$ is the polynomial (3.5.15), will be called *an even local zeta-function of* $\mathbf{L}^n$. Respectively, if $p \in \mathbb{P}_{(q)}$ and

$$R_p^n(v) = \sum_{j=0}^{2n} (-1)^j r_j^n(p) v^j \tag{5.2.34}$$

is the polynomial over $\mathsf{L}_p^n(q)$ with $\mathsf{i}_q(r_j^n(p)) = \mathbf{r}_j^n(p)$ for $0 \leq j \leq 2n$, then a formal power series over $\mathsf{L}_p^n(q)$ defined by

$$Z_p^+(v) = Z_p^+(v)^{(n)} = R_p^n(v)^{-1} = \sum_{\delta=0}^{\infty} z^n(p^\delta)v^\gamma \tag{5.2.35}$$

will be called *an even local zeta-function of* $\mathsf{L}^n(q)$. In this subsection the theta-matrices of the coefficients $z^n(p^\delta)$ of even local zeta-functions will be expressed in terms of Eichler matrices.

Theorem 5.2.9. *For every Q of $\mathbb{A}_m^+$ with even $m = 2k$, prime number p prime to the level q of Q, and $n \in \mathbb{N}$, the generating series of the theta matrices of the elements $z^n(p^\delta)$ with respect to Q is given by the formula*

$$\sum_{\delta=0}^{\infty} \Theta(z^n(p^\delta))v^\delta = \gamma^n(v,Q)B_p^m(v,Q)^{-1} \sum_{\delta=0}^{\infty} M(t(p^\delta)\cdot j(t(p^\delta)))(\chi(p)p^{1-k}v)^\delta \tag{5.2.36}$$

with

$$\gamma^n(v,Q) = \begin{cases} \left\{\prod_{i=1}^{n-m}(1-p^{k+i}\chi(p)v)(1-p^{-k-i}\chi(p)v)\right\}^{-1}, & \text{if } n > m, \\ 1, \quad \text{if } n = m \\ \prod_{i=1}^{m-n}(1-p^{i+n-k}\chi(p)v)(1-p^{k-i-n}\chi(p)v), & \text{if } n < m \end{cases}$$

and $\chi = \chi_Q$, where $B_p^m(v,Q)$ is the polynomial (4.2.93),

$$M(t) = M(1,t,Q) \qquad (t \in \mathsf{H}^n) \tag{5.2.37}$$

are Eichler matrices of the form (5.2.19), *$t(a) = t_m(a) \subset \mathsf{H}^m$ are the elements* (3.2.10), *and where* j *is the anti-automorphism* (3.1.17) *of* H^m.

The formula (5.2.36) *remains true if one replaces the theta-matrices in the left side by $\Theta_g(z^n(p^\delta))$ and the Eichler matrices in the right side by $M_g(t(p^\delta)\mathrm{j}(t(p^\delta)))$, where we set*

$$M_g(t) = M_g(1,t,Q), \ (\mathrm{t} \in \mathsf{H}^m). \tag{5.2.38}$$

Proof. As usual, we first consider the case $n = m$. Using the factorization (3.5.69) of $\mathbf{R}_p^m(v)$ and the identities (3.5.73), (3.5.74), we obtain the identity

$$\mathbf{Z}_p^+(v) = \sum_{\delta=0}^{\infty} \mathbf{z}^m(p^\delta)v^\delta = \left(\sum_{\delta_1=0}^{\infty} p^{-\delta_1 m}t^+(p^{\delta_1})v^{\delta_1}\right)B_p^m(v)^{-1}\left(\sum_{\delta_2=0}^{\infty} p^{-\delta_2 m}t^-(p^{\delta_2})v^{\delta_2}\right).$$

By this identity and Lemma 5.2.2, in the notation (4.2.88), we conclude that the entries $\vartheta_{\alpha\beta}(z^m(p^\delta))$ of $\Theta(z^m(p^\delta))$ satisfy

$$\sum_{\delta=0}^{\infty} \vartheta_{\alpha\beta}(z^m(p^\delta))v^\delta$$

$$= \sum_{\delta=0}^{\infty} (r_\alpha|_{k,\chi}\mathbf{z}^m(p^\delta))(Q_\beta)v^\delta$$

$$= \left(r_\alpha \left| \left(\sum_{\delta_1=0}^{\infty} p^{-\delta_1 m} t^+(p^{\delta_1})v^{\delta_1}\right) B_p^m(v)^{-1} \left(\sum_{\delta_2=0}^{\infty} p^{-\delta_2 m} t^-(p^{\delta_2})v^{\delta_2}\right)\right.\right)(Q_\beta)$$

$$= \sum_{\delta_1,\delta_2=0}^{\infty} p^{-(\delta_1+\delta_2)m} v^{\delta_1+\delta_2}(r_\alpha|t^+(p^{\delta_1})|B_p^m(v)^{-1}|t^-(p^{\delta_2}))(Q_\beta).$$

By (3.5.85), the element $t^-(p^\delta) \in \mathscr{C}_{-p}^m$ is p-homogeneous of p-signature -2δ. Hence, applying the formula (4.2.22) to each of the double cosets occurring in $t^-(p^\delta)$, we see that, for any f of $\mathscr{F}_+^m$,

$$(f|t^-(p^\delta))(Q_\beta) = \begin{cases} 0 & \text{if } \delta > 0 \\ f(Q_\beta) & \text{if } \delta = 0, \end{cases}$$

because the number $\det Q_\beta = \det Q$ along with q is prime to p. Hence, continuing our calculation, we get

$$= \sum_{\delta=0}^{\infty} p^{-\delta m} v^\delta (r_\alpha|t^+(p^\delta)|B_p^m(v)^{-1})(Q_\beta)$$

which, by Proposition 4.2.24, can be written in the form

$$= B_p^m(v, Q_\beta)^{-1} \sum_{\delta=0}^{\infty} p^{-\delta m} v^\delta (r_\alpha|t^+(p^\delta))(Q_\beta).$$

By Theorem 4.2.27, we have $B_p^m(v, Q_\beta) = B_p^m(v, Q)$. Hence, using (4.3.83) and then (4.2.57) and (4.2.44), we see that the last expression is equal to

$$B_p^m(v, Q)^{-1} \sum_{\delta=0}^{\infty} \sum_{\substack{M \in \Lambda^m \backslash \mathrm{M}^m, \\ \det M = \pm p^\delta}} \bar{\chi}(\det M)(\det M)^{-k} r_\alpha(Q_\beta[{}^tM])(pv)^\delta$$

$$= B_p^m(v, Q)^{-1} \sum_{\delta=0}^{\infty} \sum_{\{A\} \in \mathbb{A}_m^+} e(A)^{-1} r(Q_\beta, A, t(p^\delta)) r_\alpha(A)(\chi(p)p^{1-k}v)^\delta.$$

By the definition of r_α, and formulae (4.2.50), (4.2.46), (4.2.47), we have

$$\sum_{\{A\} \in \mathbb{A}_m^+} e(A)^{-1} r(Q_\beta, A, t(p^\delta)) e(Q_\alpha)^{-1} r(Q_\alpha, A)$$

$$= e(Q_\alpha)^{-1} \sum_{\{A\} \in \mathbb{A}_m^+} e(A)^{-1} r(A, Q_\beta, \mathrm{j}(t(p^\delta))) r(Q_\alpha, A, t(p^\delta))$$

$$= e(Q_\alpha)^{-1} r(Q_\alpha, Q_\beta, t(p^\delta)\mathrm{j}(t(p^\delta))).$$

Thus, finally, we obtain

$$\sum_{\delta=0}^{\infty} \vartheta_{\alpha\beta}(z^m(p^\delta))v^\delta = B_p^m(v, Q)^{-1} \sum_{\delta=0}^{\infty} e(Q_\alpha)^{-1} r(Q_\alpha, Q_\beta; t(p^\delta)\mathrm{j}(t(p^\delta))(\chi(p)p^{1-k}v)^\delta,$$

which proves (5.2.36) with $n = m$.

By the formula for $\Psi_p^n(\mathbf{R}_p^n(v), u)$ obtained in the beginning of the proof of Theorem 2.4.27, we have the identity

$$\Psi_{k,\chi}(\mathbf{R}_p^n(v)) = \sum_{j=0}^{2n} (-1)^j \Psi_p^n(\mathbf{r}_j^n(p), p^{n-k}\overline{\chi}(p))v^j$$
$$= (1 - p^{n-k}\overline{\chi(p)}\,v)(1 - p^{k-n}\chi(p)v)\mathbf{R}_p^{n-1}(v),\ (n > 1)$$

whence, since $\Psi_{k,\chi} = \Psi_p^n(\dots, p^{n-k}\overline{\chi}(p))$ is a homomorphism, it follows that

$$\Psi(\mathbf{Z}_p^+(v)) = \sum_{\delta=0}^{\infty} \Psi(\mathbf{z}^n(p^\delta), p^{n-k}\overline{\chi}(p))v^\delta$$
$$= \{(1 - p^{n-k}\overline{\chi}(p)v)(1 - p^{k-n}\chi(p)v)\}^{-1}\mathbf{Z}_p^+(v)^{(n-1)}.$$

If $n > m$, we apply this formula $(n - m)$ times, which gives us the identity

$$\Psi^{n-m}(\mathbf{Z}_p^+(v)^{(n)}) = \gamma^n(v, Q)\mathbf{Z}_p^+(v)^{(m)},$$

in the notation of the theorem. Similarly, if $n < m$, we obtain the identity

$$\Psi^{m-n}(\mathbf{Z}_p^+(v)^{(m)}) = \gamma^n(v, Q)^{-1}\mathbf{Z}_p^+(v)^{(n)},$$

and so one can take

$$\Psi^{n-m}(\mathbf{Z}_p^+(v)^{(n)}) = \gamma^n(v, Q)\mathbf{Z}_p^+(v)^{(m)}.$$

These relations along with (5.2.16) reduce the general formula (5.2.36) to the case $n = m$.

It follows from (4.3.74) that each of the elements $z^n(p^\delta)$ belongs to the even ring $\mathsf{E}^n(q)$. Thus, the last assertion follows from (5.2.12) and the definition of Elichler matrices. □

5.2.4. Theta-Matrices of Generators of Hecke Rings

We have calculated theta-matrices for a number of elements of local Hecke rings. The important thing is that among these elements there are generators for the ring. This implies that any theta-matrix is a polynomial in Eichler matrices of a special form and enables us to prove some general properties of the theta-matrices.

We start from the generators.

Proposition 5.2.10. *Let $n \in \mathbb{N}$, $q \in \mathbb{N}$, and $p \in \mathbb{P}_{(q)}$. Then each of the sets*

$$\{T^n(p), [p]_n^{\pm 1}, r_1^1(p), \dots, r_{n-1}^n(p)\} \subset \mathsf{L}_p^n(q) \tag{5.2.39}$$

and

$$\{T^n(p), [p]_n^{\pm 1}, z^n(p), \dots, z^n(p^{n-1})\} \subset \mathsf{L}_p^n(q) \tag{5.2.40}$$

is a set of generators for the ring $\mathsf{L}_p^n(q)$ over $\mathbb{Q}$; the elements (5.2.39) minus $[p]_n^{-1}$ are algebraically independent over $\mathbb{Q}$; each of the sets

$$\{[p]_n^{\pm 1}, r_1^n(p), \ldots, r_n^n(p)\} \subset \mathsf{E}_p^n(q) \tag{5.2.41}$$

and

$$\{[p]_n^{\pm 1}, z^n(p), \ldots, z^n(p^n)\} \subset \mathsf{E}_p^n(q) \tag{5.2.42}$$

is a set of generators for the ring $\mathsf{E}_p^n(q)$ *over* $\mathbb{Q}$.

Proof. The assertions on the set (5.2.39) follow from Theorem 3.3.30, Lemmas 3.3.26 and 3.3.34 and the formulae (3.3.70) and (3.5.16).

By (5.2.36), we have the identity

$$R_p^n(v) = \left(\sum_{\delta=0}^{\infty} z^n(p^\delta)v^\delta\right)^{-1} = 1 + \sum_{j=1}^{\infty}(-1)^j\left(\sum_{\delta=1}^{\infty} z^n(p^\delta)v^\delta\right)^j.$$

Hence, by comparing the coefficients of v^j in the two sides, we conclude that each of the elements $r_j^n(p)$ with $j = 1, 2, \ldots, 2n$ is a polynomial in $z^n(p), \ldots, z^n(p^j)$:

$$r_j^n(p) = D_j(z(p), \ldots, z(p^j)). \tag{5.2.43}$$

Thus, the set (5.2.40) generates the ring $\mathsf{L}_p^n(q)$, as soon as the set (5.2.39) does.

It follows from the properties of the set (5.2.39) listed above and definitions that the set

$$\{T^n(p)^2, [p]_n^{\pm 1}, r_1^n(p), \ldots, r_{n-1}^n(p)\}$$

generates the ring $\mathsf{E}_p^n(q)$. From (4.2.78) and the last formula of Lemma 3.3.34 we obtain the formula

$$T^n(p)^2 = p^n[p]_n(2 + 2(r_1^n(p) + \cdots + r_{n-1}^n(p)) + r_n^n(p)).$$

Hence, the set (5.2.41) generates $\mathsf{E}_p^n(q)$, and, by (5.2.43), the same is true for the set (5.2.42). □

Let us return to theta-matrices.

Theorem 5.2.11. *Let Q be a matrix of $\mathbb{A}_m^+$ with even $m = 2k$, $q = q(Q)$, and $\chi = \chi_Q$. Let $n \in \mathbb{N}$ and $p \in \mathbb{P}_{(q)}$. Then the following assertions hold:*

(1) *The theta-matrix $\Theta(T)$ (resp. $\Theta_g(T)$) of every element T of $\mathsf{E}_p^n(q)$ with respect to Q is a polynomial in the Eichler matrices*

$$M(t_m(p^\delta)\cdot \mathrm{j}(t_m(p^\delta)), Q) \text{ with } \delta = 0, 1, \ldots, m \tag{5.2.44}$$

(resp.

$$M_g(t_m(p^\delta)\cdot \mathrm{j}(t_m(p^\delta)), Q) \text{ with } \delta = 0, 1, \ldots, m) \tag{5.2.45}$$

with rational coefficients;

(2) *Suppose that $n \geqslant k$ if $\chi(p) = -1$. Then the theta-matrix $\Theta(T)$ of every element T of $\mathsf{L}_p^n(q)$ with respect to Q is a polynomial in the Eichler matrices*

$$M(p, Q) \text{ and } M(t_m(p^\delta)\mathrm{j}(t_m(p^\delta)), Q) \text{ with } \delta = 0, 1, \ldots, m-1 \tag{5.2.46}$$

with rational coefficients.

Proof. According to Proposition 5.2.10, we only have to prove the first assertion for T belonging to the set (5.2.42) and the second assertion for T belonging to the set (5.2.40). By (5.2.12), it suffices to consider only the matrices $\Theta(T)$. The matrix $\Theta([p]_n)$ is clearly proportional to the unit matrix $E_H = M(t(1)\mathrm{j}(t(1))) = M(1)$. By Theorem 5.2.5, the matrix $\Theta(T^n(p))$ is proportional to the matrix $M(p, Q)$. Hence, in order to prove the theorem, it suffices to check that each of the matrices $\Theta(z^n(p^\delta))$ with $\delta \in \mathbb{N}$ is a rational linear combination of the matrices $M(t(p^\alpha)\mathrm{j}(t(p^\alpha)))$ with $\alpha = 0, 1, \ldots, \delta$, but this follows at once by comparing the coefficients of v^δ in the two sides of (5.2.36). □

Corollary 5.2.12. *With the assumptions of the theorem, each of the matrices $D^{-1}\Theta(T)D$ and $D_g^{-1}\Theta_g(T)D_g$, where*

$$D = D(Q) = \mathrm{diag}(e(Q_1)^{-1/2}, \ldots, e(Q_H)^{-1/2})$$

and

$$D_g = D_g(Q) = \mathrm{diag}(e(Q_1)^{-1/2}, \ldots, e(Q_h)^{-1/2}), \tag{5.2.47}$$

is symmetric.

Proof. It suffices to check that the matrices

$$D^{-1}M(p, Q)D^{-1} = ((e(Q_\alpha)e(Q_\beta))^{-1/2} r(Q_\alpha, pQ_\beta))$$

and

$$D^{-1}M(t, Q)D = ((e(Q_\alpha)e(Q_\beta))^{-1/2} r(Q_\alpha, Q_\beta, t)) \qquad \text{with } t = t(p^\delta)\mathrm{j}(t(p^\delta))$$

are symmetric. For the latter matrices, this follows at once from (4.2.50), since

$$\mathrm{j}(t) = \mathrm{j}(t(p^\delta)\mathrm{j}(t(p^\delta))) = \mathrm{j}^2(t(p^\delta)) \cdot \mathrm{j}(t(p^\delta)) = t.$$

As to the first matrix, since (5.2.29) implies (5.2.30), it follows that

$$r(Q_\alpha, pQ_\beta) = r(Q_\alpha, pQ_\beta, (D_k^m(p))_\Lambda). \tag{5.2.48}$$

Hence, by (4.2.50), we obtain

$$\begin{aligned} r(Q_\alpha, pQ_\beta) &= r(pQ_\beta, Q_\alpha, (D_k^m(p)^{-1})_\Lambda) \\ &= r(Q_\beta, pQ_\alpha, (pD_k^m(p)^{-1})_\Lambda) = r(Q_\beta, pQ_\alpha). \end{aligned} \tag{5.2.49}$$

□

Corollary 5.2.13. *With the assumptions of the theorem, all of the matrices $\Theta(T)$ (resp. $\Theta_g(T)$) can be simultaneously diagonalised. Phrased another way: The space*

$$\mathfrak{F}^n(Q) = \left\{ \sum_{\alpha=1}^{H} c_\alpha \theta^n(Z, Q_\alpha); c_\alpha \in \mathbb{C} \right\} \tag{5.2.50}$$

(resp.

$$\mathfrak{F}_g^n(Q) = \left\{ \sum_{\alpha=1}^{h} c_\alpha \theta^n(Z, Q_\alpha); c_\alpha \in \mathbb{C} \right\}) \tag{5.2.51}$$

where $\{Q_1, \ldots, Q_H\}$ *and* $\{Q_1, \ldots, Q_h\}$ *are the sets* (5.2.3), *is invariant under all of the Hecke operators* $|_{k,\chi}T$ *with* $T \in \mathsf{E}_p^n(q)$, *in the general case, and* $T \in \mathsf{L}_p^n(q)$, *unless* $1 \le n < k$ *and* $\chi(p) = -1$, (*resp.* $T \in \mathsf{E}_p^n(q)$) *and has an basis of common eigenfunctions for all of the operators. All of the eigenvalues of the operators on the spaces are real algebraic numbers.*

Proof. We consider, for example, the matrices $\Theta(T)$. By Theorem 5.2.11 and Corollary 5.2.12, since the Hecke rings are commutative, it follows that all matrices $D^{-1}\Theta(T)D$ are polynomials in a finite number of real symmetric matrices commuting with each other. A well-known theorem of linear algebra, which was actually proved in the proof of Theorem 4.1.8, says that any finite set of pairwise commuting Hermitian matrices can be simultaneously diagonalized. Hence, the matrices $\Theta(T)$ can be simultaneously diagonalized. Since characteristic values of Hermitian matrices are real and all entries of $\Theta(T)$ are rational numbers, it follows that all of the characteristic values of $\Theta(T)$ are real algebraic numbers.

□

Since the global rings $\mathsf{L}^n(q)$ and $\mathsf{E}^n(q)$ are generated by the local subrings $\mathsf{L}_p^n(q)$ and $\mathsf{E}_p^n(q)$ with $p \in \mathbb{P}_{(q)}$, the local results obtained above imply the following global

Theorem 5.2.14. *Let* Q *be a matrix of* $\mathbb{A}_m^+$ *with even* $m = 2k$, $q = q(Q)$, *and* $\chi = \chi_Q$. *Then the following assertions hold:*

(1) *The theta-matrix* $\Theta(T)$ (*resp.* $\Theta_g(T)$) *of every element* T *of* $\mathsf{E}^n(q)$ *with respect to* Q *is a polynomial in finite number of the Eichler matrices* (5.2.44) (*resp.* (5.2.45)) *for primes of* $\mathbb{P}_{(q)}$ *with rational coefficients;*

(2) *Suppose that* $n \ge k$ *unless* χ *is trivial. Then the theta-matrix* $\Theta(T)$ *of every element* T *of* $\mathsf{L}^n(q)$ *with respect to* Q *is a polynomial in finite number of the Eichler matrices* (5.2.46) *for primes of* $\mathbb{P}_{(q)}$ *with rational coefficients;*

(3) *In these cases, each of the matrices of the form* $D^{-1}\Theta(T)D$ *and* $D_g^{-1}\Theta_g(T)D_g$, *where* D *and* D_g *are the matrices* (5.2.47), *is symmetric; all of the matrices* $\Theta(T)$ (*resp.* $\Theta_g(T)$) *can be simultaneously diagonalized, and all of their characteristic values are real algebraic numbers; the spaces* $\mathfrak{F}^n(Q)$ *and* $\mathfrak{F}_g^n(Q)$ *are invariant under the operators* $|_{k,\chi}T$ *and are spanned by common eigenfunctions of all of the operators.*

It would be interesting to find simple formulae expressing theta-matrices in term of Eichler matrices, like the formulae (5.2.24) for the elements $T^n(p)$.

5.2.5. Relations, Relations ...

The time has come to say good-bye to the reader and, perhaps, to the subject itself. But the arithmetic of quadratic forms is inexhaustible, and there is no doubt that Hecke operators will further serve as an important technical tool.

In particular, one may hope that the machinery developed above will help to discover and study new relations between the numbers of integral representations of quadratic forms by each other. The general results in this direction are still waiting to be found. Here we shall limit ourselves to a few remarks and related exercises.

In the previous subsections the relations (5.2.14) and (5.2.15) were used to calculate the theta-matrices $\Theta(T)$ and $\Theta_g(T)$. But if the theta-matrices are already known, then the same relations can be considered as relations between the numbers of representations of quadratic forms by quadratic forms appearing in the entries of the corresponding columns and the theta-matrices. For example, the relations (5.2.14) and formula (4.2.16) for $f = r_\alpha = e(Q_\alpha)^{-1} r(Q_\alpha, \ldots)$ lead to the following result: With the assumptions and notation of Theorem 5.2.14, let

$$\mathfrak{i}_q(T) = \sum_j a_j \left(\Gamma_0^n \begin{pmatrix} \mu_j D_j^* & B_j \\ 0 & D_j \end{pmatrix} \right)$$

be the image of T in $\mathscr{L}^n(q)$. Then, for each $\alpha = 1, \ldots, H$ and $A \in \mathbb{A}_n$, the relation holds

$$\begin{aligned} &\sum_j a_j \mu_j^{nk-\langle n\rangle} \chi(\det(\mu_j D_j^*))(\det D_j)^{-k} e\{\mu_j^{-1} A[{}^tD_j] B_j D_j^{-1}\} \\ &\qquad \times e(Q_\alpha)^{-1} r(Q_\alpha, \mu_j^{-1} A[{}^tD_j])(=(r_\alpha | T)(A)) \qquad (5.2.52) \\ &= \sum_\beta \vartheta_{\alpha\beta}(T) e(Q_\beta)^{-1} r(Q_\beta, A), \end{aligned}$$

where $\vartheta_{\alpha\beta}(T)$ are entries of $\Theta(T)$.

In view of Theorem 5.2.14, the entries $\vartheta_{\alpha\beta}(T)$ can be expressed in terms of numbers of representations of the shape $r(Q_\alpha, \mu Q_\beta, t)$ Hence, (5.2.52) can be considered as a relation between these numbers and the numbers of representations of the shape $r(Q_\alpha, B)$ with $B \in \mathbb{A}_n$. Of course, the chances to make use of the relations depend on whether we have good enough formulae for the sums in the left side and for the coefficients $\vartheta_{\alpha\beta}(T)$ on the right.

Another approach is provided by relations between elements of Hecke rings, because the relations are also satisfied by the corresponding theta-matrises and hence by certain polynomials in Eichler matrices.

The following exercises illustrate what has been said.

Exercise 5.2.15. *Let $Q \in \mathbb{A}_m^+$ with even $m = 2k$, $q = q(Q)$, and $p \in \mathbb{P}_{(q)}$. For every $\delta \in \mathbb{N}$ and $\alpha, \beta = 1, \ldots, H$, prove the relations*

$$r(Q_\alpha, p^{\delta+1} Q_\beta, \tau(T^m(p^\delta))) = \beta^m(p, Q) \sum_{\gamma=1}^{H} r(Q_\alpha, pQ_\gamma) e(Q_\gamma)^{-1} r(Q_\gamma, p^\delta Q_\beta),$$

where $Q_1, \ldots, Q_H$ is the set (5.2.3),

$$\tau(T^m(p^\delta)) = \sum_{i=0}^{m} p^{\langle i-1\rangle - ik} \chi_Q(p)^i \pi_i^m(p) t_m(p^{k(\delta+1)-i}) \in \mathrm{H}_p^m$$

with elements $\pi_i^m(p)$ *defined by* (3.2.26), $\beta^m(p,Q)$ *being the numbers of Theorem* 5.2.5. *Deduce the relation*

$$M(p,Q)M(p^\delta,Q) = \beta^m(p,Q)^{-1}M(p^{\delta+1},\tau(T^m(p^\delta)),Q).$$

[Hints: Use (5.2.52) with the left side written in the form (5.2.26) for $A = p^\delta Q_\beta$, and then transform the inner sum in the left side with the help of (4.2.46), the first formula of Exercise 4.2.15, and the obvious relation $\pi_m^m(p)\mathrm{j}(\pi_i^m(p)) = \pi_{m-i}^m(p)$: for the right side use Theorem 5.2.5.]

Exercise 5.2.16. *In the notation of Exercise* 5.2.15, *prove the relation*

$$\tau(T^2(p^\delta)) = (1+\chi(p)p^{-1})(t_2(p^{\delta+1}) + \chi(p)\pi_2^2(p)t_2(p^{\delta-1})).$$

Use this and the last formula of Exercise 5.2.15 *to prove that, for* $m = 2$,

$$M(p,Q)M(p^\delta,Q) = M(p^{\delta+1},Q) + \chi_Q(p)M(p^{\delta-1},Q).$$

[Hint: Use Proposition 3.2.22 or Exercise 3.2.12.]

Exercise 5.2.17. *Let* $Q \in \mathbb{A}_m^+$ *and* $a, b \in \mathbb{N}_{(q)}$, *where q is the level of Q. Prove that*

$$M(a,Q)M(b,Q) = M(ab,Q) \quad \textit{if } \text{g.c.d.}(a,b) = 1.$$

[Hint: Use the formulae (4.2.46), (4.2.47), and (3.2.14).]

Exercise 5.2.18. *Let* $Q \in \mathbb{A}_2^+$ *and* $a, b \in \mathbb{N}_{(q(Q))}$. *Using the formulae of Exercises* 5.2.16 *and* 5.2.17, *prove the formulae*

$$M(a,Q)M(b,Q) = \sum_{d|a,b} \chi_Q(d)M(ab/d^2,Q),$$

$$M(ab,Q) = \sum_{\delta|a,b} \chi_Q(\delta)\mu(\delta)M(a/\delta,Q)M(b/\delta,Q),$$

where μ *is the Möbius function. Assuming that the theta-series* $\theta^1(z,Q_1), \ldots, \theta^1(z,Q_H)$ *are linearly independent, obtain these formulae with the help of the last formula of Exercise* 5.2.6 *and formulae of Exercise* 3.3.18 *or* 4.3.6.

Exercise 5.2.19. *Let Q be a matrix of* $\mathbb{A}_2^+$ *of divisor* $\delta(Q) = 1$ *and determinant* $\det Q = d$. *Denote by* $\hat{H}(-d)$ *the group of characters of the group* $\mathsf{H}(-d)$ *considered as the group of proper classes of primitive matrices in* $\mathbb{A}_2^+$ *of discriminant* $-d$ *with respect to the composition* (A.3.15). *For* $\psi \in \hat{\mathsf{H}}(-d)$, *let us set*

$$\mathbf{r}(\psi) = ((\psi)(Q_1), \ldots, (\psi)(Q_H))$$

where $Q_1, \ldots, Q_H$ *is a set of representatives for the classes* $\{Q'\} \in \mathbb{A}_2(1,d,d)$, *and where* $(\psi)(Q_\alpha) = \psi(Q_\alpha) + \psi(Q'_\alpha)$ *with* $Q'_\alpha = Q_\alpha\left[\begin{pmatrix}1 & 0\\ 0 & -1\end{pmatrix}\right]$. *Prove the following assertions*:

(1) *The row* $\mathbf{r}(\psi)$ *is independent of the choice of representatives* Q_α *in* $\{Q_\alpha\}$;

(2) *Each of the rows* $\mathbf{r}(\psi)$ *is an eigenrow for all of the Eichler matrices* $M(a) = M(a, Q)$ *with* $a \in \mathbb{N}_{(d)}$, *i.e.*

$$\mathbf{r}(\psi)M(a) = \lambda(a, \psi)\mathbf{r}(\psi),$$

where $\lambda(a, \psi)$ *is a scalar;*

(3) *Each linear combination of theta-series of the form*

$$\sum_{\alpha=1}^{H} (\psi)(Q_\alpha)e(Q_\alpha)^{-1}\theta^n(Z, Q_\alpha) \quad \text{with } \psi \in \hat{\mathsf{H}}(-d) \text{ and } n = 1, 2$$

is an eigenfunction for all of the Hecke operators $|T^n(a)$ *with* $a \in \mathbb{N}_{(d)}$ *(and therefore for all operators of* $\mathsf{L}^n(d)$*);*

(4) *The characteristic values* $\lambda(a, \psi)$ *of* $\mathsf{E}(a)$ *in* (2) *are given by the formula*

$$\lambda(a, \psi) = \sum_{\alpha=1}^{H} (\psi)(Q_\alpha)e(Q_\alpha)^{-1}r(Q_\alpha, 2a);$$

(5) *The entries of a matrix* $M(a, Q)$ *with* $a \in \mathbb{N}_{(d)}$ *satisfy*

$$e(Q_\alpha)^{-1}r(Q_\alpha, aQ_\beta) = e(Q_\alpha \times Q_\beta)^{-1}r(Q_\alpha \times Q_\beta, 2a) + e(Q_\alpha \times Q'_\beta)^{-1}r(Q_\alpha \times Q'_\beta, 2a),$$

where $\times$ *is the Gauss composition* (A.3.15).

[Hints. For (1): Check first that $\{Q_\alpha\} = \{Q_\alpha\}_+ \cup \{Q'_\alpha\}_+$, where $\{\cdots\}_+$ stands for the classes modulo the proper equivalence, and that the classes $\{Q_\alpha\}_+$ and $\{Q'_\alpha\}_+$ are inverse to each other with respect to the Gauss composition. For (2): by the formulae of Exercise 5.2.18, it suffices to consider the case $a = p \in \mathbb{P}_{(d)}$, then, by (5.2.49), (5.2.48), (4.2.44), and (4.2.34), one gets

$$\begin{aligned} e(Q_\alpha)^{-1}r(Q_\alpha, pQ_\beta) &= e(Q_\alpha)^{-1}r(Q_\beta, pQ_\alpha) = e(Q_\alpha)^{-1}r(Q_\beta, pQ_\alpha, \pi_1^2(p)) \\ &= r^*(Q_\beta, pQ_\alpha, \pi_1^2(p)) = \sum_{\substack{D \in \Lambda\backslash\Lambda D_1^2(p)\Lambda, \\ p^{-1}Q_\beta[{}^tD] \in \{Q_\alpha\}}} 1, \end{aligned}$$

hence, by (4.3.41), the β-th entry of the row $\mathbf{r}(\psi)M(p)$ can be written in the form

$$\begin{aligned} &\sum_{\alpha=1}^{H} (\psi)(Q_\alpha)e(Q_\alpha)^{-1}r(Q_\alpha, pQ_\beta) \\ &\quad = \sum_{\alpha=1}^{H} (\psi)(Q_\alpha) \sum_{\substack{D \in \Lambda\backslash\Lambda D_1^2(p)\Lambda \\ p^{-1}Q_\beta[{}^tD] \in \{Q_\alpha\}}} 1 = \sum_{D} (\psi)(p^{-1}Q_\beta[{}^tD]) \\ &\quad = \chi(p)^{-1}p((\psi)|_{1,\chi}\Pi(p))(Q_\beta), \end{aligned}$$

where $\chi = \chi_Q$; by Proposition 4.3.14, the last expression depends on $\varepsilon_p(Q_\beta) = \chi_{Q_\beta}(p) = \chi_Q(p) = \chi(p)$: if $\chi(p) = 1$, then, in the notation of the proposition, it is equal to

$$\begin{aligned} &(\psi)(Q_\beta \times \mathfrak{p}) + (\psi)(Q_\beta \times \bar{\mathfrak{p}}) \\ &\quad = \psi(Q_\beta \times \mathfrak{p}) + \psi((Q_\beta \times \mathfrak{p})') + \psi(Q_\beta \times \bar{\mathfrak{p}}) + \psi((Q_\beta \times \bar{\mathfrak{p}})') \\ &\quad = (\psi(\mathfrak{p}) + \psi(\bar{\mathfrak{p}}))(\psi(Q_\beta) + \psi(Q'_\beta)) = (\psi(\mathfrak{p}) + \psi(\bar{\mathfrak{p}}))(\psi)(Q_\beta), \end{aligned}$$

but if $\chi(p) = -1$, then it is equal to $0 = 0 \cdot (\psi)(Q_\beta)$. For (3): use (2) and formulae of Exercise 5.2.6. For (4): since $\Theta(T^1(a)) = M(a)$ (see Exercise 5.2.6), it follows that the function

$$\sum_{\alpha=1}^{H} (\psi)(Q_\alpha)e(Q_\alpha)^{-1}\theta^1(z, Q_\alpha) = \sum_{a=0}^{\infty} \left(\sum_{\alpha=1}^{H} (\psi)(Q_\alpha)e(Q_\alpha)^{-1}r(Q_\alpha, 2a) \right) e^{2\pi i a z}$$

is an eigenfunction for $|T^1(a)$ with the eigenvalue $\lambda(a, \psi)$, then use (4.3.12) with $\kappa = 1$. For (5): since $\mathsf{H}(-d)$ is a group and ψ is a character of the group, it follows, by (2) and (4) that

$$\sum_{\alpha=1}^{H} (\psi)(Q_\alpha)e(Q_\alpha)^{-1}r(Q_\alpha, aQ_\beta)$$

$$= \left(\sum_{\alpha=1}^{H} (\psi)(Q_\alpha)e(Q_\alpha)^{-1}r(Q_\alpha, 2a) \right) (\psi)(Q_\beta)$$

$$= \sum_{\alpha=1}^{H} (\psi(Q_\alpha \times Q_\beta) + \psi(Q_\alpha \times Q'_\beta))e(Q_\alpha)^{-1}r(Q_\alpha, 2a)$$

$$= \sum_{\alpha=1}^{H} \psi(Q_\alpha)(e(Q_\alpha \times Q'_\beta)^{-1}r(Q_\alpha \times Q'_\beta, 2a) + e(Q_\alpha \times Q_\beta)^{-1}r(Q_\alpha \times Q_\beta, 2a)),$$

next use the orthogonality relations for the characters.]

Exercise 5.2.20. *With the assumptions and notation of Theorem 5.2.5, prove the relations*

$$r(Q_\alpha, 2ap) + p^{k-1}\chi(p)r(Q_\alpha, 2a/p) = \beta^1(p, Q) \sum_{\beta=1}^{H} r(Q_\alpha, pQ_\beta)e(Q_\beta)^{-1}r(Q_\beta, 2a),$$

for $\alpha = 1, \ldots, H$, *and* $a = 0, 1, \ldots$.

Appendix 1. Symmetric Matrices over Fields

A.1.1. Arbitrary Fields

We let $\mathsf{S}_n(\mathsf{F})$ be the set of all symmetric matrices of order n over a field F. Two matrices A, A' of $\mathsf{S}_n(\mathsf{F})$ are said to be equivalent over F if

$$A' = {}^tCAC = A[C] \qquad \text{with } C \in \mathsf{GL}_n(\mathsf{F}). \tag{A.1.1}$$

The following simple identity often helps to simplify a matrix by passing to an equivalent matrix. If $A = \begin{pmatrix} A_1 & A_2 \\ A_3 & A_4 \end{pmatrix} \in \mathsf{S}_n(\mathsf{F})$, where the block A_1 of dimension $r \times r$ with $0 < r < n$ is non-singular, then

$$A = \begin{pmatrix} A_1 & A_2 \\ {}^tA_2 & A_4 \end{pmatrix} = \begin{pmatrix} A_1 & 0 \\ 0 & A_4 - A_1^{-1}[A_2] \end{pmatrix} \left[\begin{pmatrix} E_r & -A_1^{-1}A_2 \\ 0 & E_{n-r} \end{pmatrix} \right]. \tag{A.1.2}$$

Theorem A.1.1. *Let $A \in \mathsf{S}_n(\mathsf{F})$. Suppose that there is a column $\mathbf{c}$ of $\mathsf{M}_{n,1}(\mathsf{F})$ such that $a = A[\mathbf{c}] \neq 0$. Then A is equivalent over F to a matrix of the form $\begin{pmatrix} a & 0 \\ 0 & A_1 \end{pmatrix}$ with $A_1 \in \mathsf{S}_{n-1}(\mathsf{F})$.*

Proof. Replacing A by $A[C]$ with a non-singular C of the form $C = (\mathbf{c}, *)$, we may assume that $a_{11} = a \neq 0$. Then the theorem follows from (A.1.2). □

If the characteristic of F is not 2, then the condition of Theorem A.1.1 holds for each non-zero A of $\mathsf{S}_n(\mathsf{F})$. Hence, by induction on n, we have

Theorem A.1.2. *If the characteristic of F is not equal to 2, then any matrix of $\mathsf{S}_n(\mathsf{F})$ is equivalent over F to a diagonal matrix.*

Theorem A.1.3. *Let A be a matrix of $\mathsf{S}_n(\mathsf{F})$ of rank r with $0 < r < n$. Then A is equivalent over F to a matrix of the form*

$$\begin{pmatrix} A_1 & 0 \\ 0 & 0 \end{pmatrix} \qquad \text{with } A_1 \in \mathsf{S}_r(\mathsf{F}) \text{ and } \det A_1 \neq 0. \tag{A.1.3}$$

roof. Since rank $A = r$, it follows that there are $n - r$ linearly independent olumns, say $\mathbf{c}_{r+1}, \ldots, \mathbf{c}_n$ in $\mathsf{M}_{n,1}(\mathsf{F})$ satisfying the equation $A\mathbf{x} = \mathbf{0}$. Let C be a on-singular matrix whose last $n - r$ columns are $\mathbf{c}_{r+1}, \ldots, \mathbf{c}_n$. Then the matrix $[C]$ is of the form (A.1.3). □

.1.2. The Field $\mathbb{R}$

heorem A.1.4. *For any matrix A of $\mathsf{S}_n(\mathbb{R})$, there is a real orthogonal matrix C ch that*

$$A[C] = \operatorname{diag}(\lambda_1, \ldots, \lambda_n), \tag{A.1.4}$$

here $\lambda_1, \ldots, \lambda_n$ are the characteristic values of A.

The proof of the theorem is well-known and can be found in any manual on ear algebra.

A matrix A of $\mathsf{S}_n(\mathbb{R})$ is said to be *positive semi-definite* if $A[\mathbf{l}] \geq 0$ for each olumn $\mathbf{l} \in \mathsf{M}_{n,1}(\mathbb{R})$, we write $A \geq 0$. If $A \geq 0$ and the relation $A[\mathbf{l}] = 0$ holds ly for $\mathbf{l} = \mathbf{0}$, then the matrix A is called *positive definite*, we write $A > 0$. It llows from the definitions that A is positive semi-definite (resp. positive definite) and only if each of the matrices equivalent to A is. A diagonal matrix is positive mi-definite (resp. positive definite) if and only if all its diagonal entries are on-negative (resp. positive). Hence, by Theorem A.1.4, a matrix of $\mathsf{S}_n(\mathbb{R})$ is ositive semi-definite (resp. positive definite) if and only if all its characteristic lues are non-negative (resp. positive). The entries of definite matrices satisfy a umber of inequalities used in this book. Let $A = (a_{ij})$ be a positive semi-definite atrix of $\mathsf{S}_n(\mathbb{R})$. Since

$$A[\mathbf{e}_\alpha \pm \mathbf{e}_\beta] = a_{\alpha\alpha} \pm 2a_{\alpha\beta} + a_{\beta\beta} \geq 0,$$

here $\mathbf{e}_1, \ldots, \mathbf{e}_n$ are the columns of the unit matrix E_n, it follows that $|a_{\alpha\beta}| \leq$ $_{\alpha\alpha} + a_{\beta\beta})/2$, whence

$$|a_{\alpha\beta}| \leq \sigma(A) = a_{11} + a_{22} + \cdots + a_{nn}. \tag{A.1.5}$$

urther, if $A - B \geq 0$ and $R \geq 0$, then $R = D[C]$, where C is a non-singular atrix and D is a diagonal matrix with non-negative entries, whence

$$\sigma((A - B)R) = \sigma((A - B)[{}^tC] \cdot D) \geq 0$$

cause all of the diagonal entries of $(A - B)[{}^tC]$ are non-negative. Thus,

$$\sigma(AR) \geq \sigma(BR) \qquad \text{if } A - B \geq 0 \text{ and } R \geq 0. \tag{A.1.6}$$

nally, it is clear that

$$\begin{aligned} A + B &\geq 0 \qquad \text{if } A \geq 0 \text{ and } B \geq 0; \\ A + B &> 0 \qquad \text{if } A > 0 \text{ and } B \geq 0. \end{aligned} \tag{A.1.7}$$

e write $A \geq B$ (resp. $A > B$), if $A - B \geq 0$ (resp. $A - B > 0$).

Theorem A.1.5. *The following three conditions on a matrix $A = (a_{ij}) \in \mathsf{S}_n(\mathbb{R})$ are equivalent*:

(1) $A > 0$;

(2) $\det A^{(r)} = \det \begin{pmatrix} a_{11} & \cdots & a_{1r} \\ \cdots & \cdots & \cdots \\ a_{r1} & \cdots & a_{rr} \end{pmatrix} > 0$ *for* $r = 1, \ldots, n$;

(3) $A = {}^tLL$ *with a* $L = \begin{pmatrix} l_{11} & l_{12} & \cdots & l_{1n} \\ 0 & l_{22} & \cdots & l_{2n} \\ \cdots & \cdots & \cdots & \cdots \\ 0 & 0 & \cdots & l_{nn} \end{pmatrix} \in \mathsf{GL}_n(\mathbb{R})$.

Proof. (1) → (2). If $A > 0$, then each of the matrices $A^{(1)} = (a_{11})$, $A^{(2)} = \begin{pmatrix} a_{11} & a_{12} \\ a_{21} & a_{22} \end{pmatrix}$, $\ldots$, $A^{(n)} = A$ is clearly positive definite. Hence, $\det A^{(r)}$ is the product of positive characteristic values and so is positive.

(2) → (3). The induction on n. The case $n = 1$ is clear. Suppose it is true for the matrices of order $n - 1 \geq 1$. If $A \in \mathsf{S}_n(\mathbb{R})$ satisfies the condition (2), then, by the inductive hypothesis, we have

$$A^{(n-1)} = {}^tL'L' \text{ with a } L' = \begin{pmatrix} l'_{11} & \cdots & l'_{1,n-1} \\ 0 & \cdots & l'_{2,n-1} \\ 0 & \cdots & l'_{n-1,n-1} \end{pmatrix} \in \mathsf{GL}_{n-1}(\mathbb{R}).$$

Since $\det A^{(n-1)} \neq 0$, it follows, by (A.1.2), that A can be written in the form

$$A = \begin{pmatrix} A^{(n-1)} & 0 \\ 0 & d \end{pmatrix}\left[\begin{pmatrix} E_{n-1} & \mathbf{h} \\ 0 & 1 \end{pmatrix}\right] = \begin{pmatrix} A^{(n-1)} & 0 \\ 0 & 1 \end{pmatrix}\left[\begin{pmatrix} E_{n-1} & \mathbf{h} \\ 0 & d^{1/2} \end{pmatrix}\right]$$

(note that $d = \det A / \det A^{(n-1)} > 0$). Then $A = {}^tLL$, where $L = \begin{pmatrix} L' & 0 \\ 0 & 1 \end{pmatrix} \times \begin{pmatrix} E_{n-1} & \mathbf{h} \\ 0 & d^{1/2} \end{pmatrix}$.

The implication (3) → (1) is clear, since (3) implies that A is equivalent to the positive definite unit matrix. □

In particular, Theorem A.1.5 implies that the set P_n of all positive definite matrices of order n is an open subset of $\mathsf{S}_n(\mathbb{R})$.

Theorem A.1.6. *Let $A, B \in \mathsf{S}_n(\mathbb{R})$, where $A > 0$. Then there is a matrix $C \in \mathsf{GL}_n(\mathbb{R})$ such that $A[C] = E_n$ and the matrix $B[C]$ is diagonal.*

Proof. By Theorem A.1.5, there is a C_1 with $A[C_1] = E_n$. By Theorem A.1.4, there is an orthogonal C_2 such that the matrix $B[C_1][C_2] = B[C_1C_2]$ is diagonal. Since $A[C_1C_2] = E_n[C_2] = {}^tC_2C_2 = E_n$, it follows that the matrix $C = C_1C_2$ satisfies the theorem. □

Finally, we prove several useful inequalities on determinants.

Theorem A.1.7. *The following inequalities hold*:

$$\det A \leq a_{11}a_{22}\cdots a_{nn}, \qquad \text{if } A = (a_{ij}) \in \mathsf{S}_n(\mathbb{R}),\ A \geq 0; \tag{A.1.8}$$

$$\det(A + B) \geq \det A + \det B, \qquad \text{if } A, B \in \mathsf{S}_n(\mathbb{R}),\ A \geq 0,\ B \geq 0; \tag{A.1.9}$$

$$|\det(A + iB)| \geq \det B, \qquad \text{if } A, B \in \mathsf{S}_n(\mathbb{R}),\ B > 0. \tag{A.1.10}$$

Proof. If A does not satisfies $A > 0$, but satisfies $A \geq 0$, then $\det A = 0$, and (A.1.8) is obvious. (A.1.8) is true for $n = 1$. Suppose it is already proved for positive definite matrices of order $n - 1 \geq 1$. If A is a positive definite matrix of order n, then $A^{(n-1)} > 0$. Hence, by (A.1.2) and the inductive hypothesis, we have

$$\begin{aligned} \det A &= \det A^{(n-1)}(a_{nn} - (A^{(n-1)})^{-1}[A_2]) \\ &\leq a_{11}a_{22}\cdots a_{nn}, \end{aligned}$$

which proves (A.1.8).

(A.1.9) is certainly true, if $\det A = \det B = \det(A + B) = 0$. If, for example, $\det(A + B) \neq 0$, then $A + B > 0$ and, by Theorem A.1.6, the matrices $A + B$ and B, and therefore A and B, can be simultaneously diagonalized. Hence, it suffices to prove (A.1.9) for diagonal positive semi-definite matrices, which is obvious.

The proof of (A.1.10), by Theorem A.1.6, can be reduced to the case then $B = E_n$ and A is a diagonal matrix, where it is obvious. □

Appendix 2. Quadratic Spaces

In some questions of the theory of quadratic forms, especially when the basic ring is a finite field, it turns out to be convenient to use the geometric language. In this appendix we shall present the basic notions, result and applications of the theory of quadratic spaces used in this book.

A.2.1. The Geometric Language

Let K be a commutative ring, and let V be a free K-module of finite rank. A function $f\colon \mathsf{V} \to \mathsf{K}$ is said to be *quadratic*, if there is a basic $\mathbf{e}_1, \ldots, \mathbf{e}_n$ for V over K such that f is given by a quadratic form in coordinates; i.e.

$$f\left(\sum_{i=1}^{n} u_i \mathbf{e}_i\right) = \mathsf{q}(u_1, \ldots, u_n), \qquad (u_i \in \mathsf{K}), \tag{A.2.1}$$

where

$$\mathsf{q}(x_1, \ldots, x_n) = \sum_{1 \le i \le j \le n} q_{ij} x_i x_j, \qquad (q_{ij} \in \mathsf{K}).$$

The definition is clearly independent of the choice of basis. A pair (V, f), where V is a free K-module of finite rank and f is a quadratic function on V will be called *a quadratic space* (over K). The quadratic form q is called *the form of* (V, f) *relative to the basis* $\mathbf{e}_1, \ldots, \mathbf{e}_n$. It is clear, that another basis gives us a form, which is equivalent to q over K, and, conversely every two quadratic forms, which are equivalent over K, can be considered as forms of the same quadratic space over K relative to certain basises. Thus, the class $\{\mathsf{q}\}_{\mathsf{K}}$ of q over K is uniquely determined by the space (V, f). We call the class $\{\mathsf{q}\}_{\mathsf{K}}$ *the type of* (V, f).

A morphism $\varphi\colon (\mathsf{V}, f) \to (\mathsf{V}_1, f_1)$ *of quadratic spaces over* K is a K-linear mapping $\varphi\colon \mathsf{V} \to \mathsf{V}_1$ satisfying $f_1(\varphi(\mathbf{v})) = f(\mathbf{v})$ for all $\mathbf{v} \in \mathsf{V}$. A one-to-one morphism is called *an isomorphism.* If there is an isomorphism $\varphi\colon (\mathsf{V}, f) \to (\mathsf{V}_1, f_1)$, then the spaces (V, f) and (V_1, f_1) are called *isomorphic.* It is clear, that two spaces are isomorphic if and only if they have the same type.

Let (V, f) be a quadratic space over K. For each pair of vectors $\mathbf{u}$, $\mathbf{v} \in \mathsf{V}$ we define *the scalar product* $\mathbf{u} \cdot \mathbf{v} \in \mathsf{K}$ by

$$\mathbf{u} \cdot \mathbf{v} = f(\mathbf{u} + \mathbf{v}) - f(\mathbf{u}) - f(\mathbf{v}). \tag{A.2.2}$$

Vectors $\mathbf{u}, \mathbf{v} \in \mathsf{V}$ with $\mathbf{u}\cdot\mathbf{v} = 0$ are called *orthogonal*. Writing the f in the form (A.2.1), we express the scalar product in terms of the coordinates with respect to the basic $\mathbf{e}_1, \ldots, \mathbf{e}_n$:

$$\left(\sum_{i=1}^{n} u_i\mathbf{e}_i\right)\left(\sum_{j=1}^{n} v_j\mathbf{e}_j\right) = \sum_{1\le i\le j\le n} q_{ij}[(u_i + v_i)(u_j + v_j) - u_iu_j - v_iv_j]$$
$$= \sum_{1\le i\le j\le n} q_{ij}(u_iv_j + v_iu_j) = \sum_{i,j=1}^{n} Q_{ij}u_iv_j, \tag{A.2.3}$$

where $Q_{ij} = Q_{ji} = q_{ij}$, if $1 \le i < j \le n$, and $Q_{ii} = 2q_{ii}$ for $1 \le i \le n$; in particular, the scalar product is linear in each of the factors. The matrix

$$Q = (Q_{ij}) = (\mathbf{e}_i\cdot\mathbf{e}_j) \in \mathsf{S}_n(\mathsf{K})$$

is called *the matrix of f* (or *of the scalar product*) *relative to the basis* $\mathbf{e}_1, \ldots, \mathbf{e}_n$; Q is equal to the matrix (1.1.3) of the form q. It is easy to see that the replacement of the basic $\mathbf{e}_1, \ldots, \mathbf{e}_n$ by a basic $\mathbf{e}_i' = \sum_{j=1}^{n} a_{ij}\mathbf{e}_j$ replaces the matrix Q by

$$Q' = (a_{ij})Q^t(a_{ij}). \tag{A.2.4}$$

This implies that the coset $(\mathsf{K}^*)^2 \cdot \det Q$ of $\det Q$ module the group $(\mathsf{K}^*)^2$ of squares of units in K does not depend on the choice of basis. The coset is called *the determinant of* (V, f). It will be denoted by $d(\mathsf{V}) = d(\mathsf{V}, f)$. If $d(\mathsf{V})$ is a unit in K then the space (V, f) is said to be *non-singular.*

A quadratic space (V_1, f_1) is called *a subspace of* (V, f), if $\mathsf{V}_1 \subset \mathsf{V}$ and the restriction of f to V_1 is equal to f_1. A space (V, f) is said to be *the direct sum of subspaces* (V_i, f_i) *with* $1 \le i \le t$, we write

$$(\mathsf{V}, f) = \bigoplus_{i=1}^{t}(\mathsf{V}_i, f_i),$$

if each $\mathbf{v}$ of V is uniquelly of the shape $\mathbf{v} = \mathbf{v}_1 + \cdots + \mathbf{v}_t$ with $\mathbf{v}_i \in \mathsf{V}_i$, and $f(\mathbf{v}) = f_1(\mathbf{v}_1) + \cdots + f_t(\mathbf{v}_t)$. In this case, clearly,

$$d(\mathsf{V}, f) = \prod_{i=1}^{t} d(\mathsf{V}_i, f_i). \tag{A.2.5}$$

Let $\mathsf{K} = \mathsf{F}$ be a field. If (V_1, f_1) is a subspace of (V, f), then the pair (V_2, f_2), where

$$\mathsf{V}_2 = \{\mathbf{u} \in \mathsf{V}; \mathbf{u}\cdot\mathbf{v} = 0 \text{ for all } \mathbf{v} \in \mathsf{V}_1\}$$

and f_2 is the restriction of f to V_2 is also a subspace of (V, f). The subspace (V_2, f_2) is called *the orthogonal complement of* (V_1, f_1) (*in* (V, f)) we write $(\mathsf{V}_2, f_2) = (\mathsf{V}_1, f_1)^\perp = (\mathsf{V}_1^\perp, f_1^\perp)$ The set $\mathsf{R}(\mathsf{V}) = \mathsf{V}^\perp$ is called *the radical of* (V, f).

Lemma A.2.1. *If* (V, f) *is a quadratic space over a field, then the conditions* $\mathsf{R}(\mathsf{V}) = \{0\}$ *and* $d(\mathsf{V}, f) \neq 0$ *are equivalent.*

Proof. Writing the scalar product in V in the form (A.2.3), we see that a vector $\mathbf{u} = \sum_i u_i \mathbf{e}_i$ belongs to R(V) if and only if $\sum_{i=0}^{n} Q_{ij} u_i = 0$ for $j = 1, 2, \ldots, n$. A non-zero solution of this system of linear homogeneous equations exists if and only if $\det(Q_{ij}) = 0$. □

Lemma A.2.2. *Let* (V_1, f_1) *be a subspace of a quadratic space* (V, f) *over a field.* F. *Suppose that* $V_1 \neq \{0\}$ *and* $V_1 \subset R(V)$. *Then there is a subspace* (V_2, f_2) *of* (V, f) *such that*

$$(V, f) = (V_1, f_1) \oplus (V_2, f_2). \tag{A.2.6}$$

Proof. Let $\mathbf{e}_1, \ldots, \mathbf{e}_r$ be a basis for V_1. We choose $\mathbf{e}_{r+1}, \ldots, \mathbf{e}_n \in V$ is such a way that the elements $\mathbf{e}_1, \ldots, \mathbf{e}_r, \mathbf{e}_{r+1}, \ldots, \mathbf{e}_n$ form a basis for V. If $V_2 = \{F\mathbf{e}_{r+1} + \cdots + F\mathbf{e}_n\}$ and f_2 is the restriction of f to V_2, then clearly (A.2.6) holds. □

Theorem A.2.3. *Let* (V_1, f_1) *be a non-singular subspace of a quadratic space* (V, f) *over a field. Then the following decomposition holds*:

$$(V, f) = (V_1, f_1) \oplus (V_1^{\perp}, f_1^{\perp}). \tag{A.2.7}$$

Proof. We choose a basis $\mathbf{e}_1, \ldots, \mathbf{e}_n$ for V in such a way that first $r = \dim V_1$ of the basis vectors form a basis for V_1. Then the condition $\mathbf{u} = \sum_{1 \le j \le n} u_j \mathbf{e}_j \in V_1^{\perp}$ is equivalent with the following system of linear equations

$$\mathbf{e}_i \mathbf{u} = \sum_{j=1}^{n} u_j \mathbf{e}_i \mathbf{e}_j = 0 \qquad (i = 1, \ldots, r).$$

Since $\det(\mathbf{e}_i \mathbf{e}_j)_{i,j=1,\ldots,r} = d(V_1) \neq 0$, it follows that the rank of the matrix of the system is equal to r. Hence, $\dim V_1^{\perp} = n - r$. On the other hand, $V_1 \cap V_1^{\perp} = \{0\}$, since $V_1 \cap V_1^{\perp} \subset R(V_1)$, and $R(V_1) = \{0\}$, by Lemma A.2.1. These considerations imply that each $\mathbf{u}$ of V is uniquelly of the shape $\mathbf{u} = \mathbf{v}_1 + \mathbf{v}_2$ with $\mathbf{v}_1 \in V_1$ and $\mathbf{v}_2 \in V_1$. Since

$$f(\mathbf{u}) = f(\mathbf{v}_1 + \mathbf{v}_2) = \mathbf{v}_1 \cdot \mathbf{v}_2 + f(\mathbf{v}_1) + f(\mathbf{v}_2) = f(\mathbf{v}_1) + f(\mathbf{v}_2),$$

the theorem follows. □

A quadratic space is said to be *reducible*, if it is a direct sum of two proper subspaces, otherwise the space is called *irreducible*. It follows from the definition that each quadratic space is a direct sum of irreducible subspaces.

Theorem A.2.4. *Let* (V, f) *be a irreducible quadratic space over a field* F. *Then* $\dim V \le 2$. *If the characteristic of* F *is not equal to* 2, *then* $\dim V \le 1$.

Proof. We first assume that the characteristic is not 2. If f is identically zero on V, then any one-dimensional subspace of (V, f) is a direct summand of (V, f), and so $\dim V \leq 1$. Let us now assume that there is $\mathbf{e}_1 \in V$ with $f(\mathbf{e}_1) \neq 0$. Then the determinant of the one-dimensional subspace $V_1 = \{Fe_1\}$ of V, which is equal to $\mathbf{e}_1 \cdot \mathbf{e}_1 = 2f(\mathbf{e}_1)$, is not zero. Hence, by Theorem A.2.3, V is the direct sum of V_1 and $V_1^\perp$. Thus, $V = V_1$ and $\dim V = 1$.

Now let the characteristic of F is 2. If the scalar product is trivial on V, then any one-dimensional subspace is a direct summand, and so $\dim V \leq 1$. Let us assume that there are vectors $\mathbf{u}, \mathbf{v} \in V$ with $\mathbf{u} \cdot \mathbf{v} = a \neq 0$. Then $\mathbf{u}$ and $\mathbf{v}$ are clearly linearly independent and the determinant of the two-dimensional space $V_1 = \{F\mathbf{u} + F\mathbf{v}\}$, which equals $2f(\mathbf{u}) \cdot 2f(\mathbf{v}) - (\mathbf{u} \cdot \mathbf{v})^2 = -a^2$, is not zero. Using Theorem A.2.3, we see that $V = V_1$ and $\dim V = 2$. □

If (V, f) is an one-dimensional quadratic space, then, clearly

$$\text{the type of } (V, f) \text{ is } \{ax^2\}, \quad \text{and } d(V, f) = 2a \quad \text{with } a \in K, \tag{A.2.8}$$

and two such spaces of type $\{ax^2\}$ and $\{bx^2\}$ are isomorphic if and only if $b = at^2$, where t is a unit in K.

Proposition A.2.5. *Each two-dimensional irreducible quadratic space over the field $F = \mathbb{F}_2$ has either the type $\{q_+^2\}$ or $\{q_-^2\}$, where*

$$q_+^2(x_1, x_2) = x_1 x_2, \quad q_-^2(x_1, x_2) = x_1^2 + x_1 x_2 + x_2^2. \tag{A.2.9}$$

The forms q_+^2 and q_-^2 are not equivalent over $\mathbb{F}_2$. Any space of type $\{q_+^2\}_{\mathbb{F}_2}$ or $\{q_-^2\}_{\mathbb{F}_2}$ is irreducible.

Proof. If (V, f) is a two-dimensional irreducible space over $\mathbb{F}_2$ then, as we have seen above, there is a basis $\mathbf{e}_1, \mathbf{e}_2$ for V with $\mathbf{e}_1 \cdot \mathbf{e}_2 = 1$. If $f(\mathbf{e}_1) = f(\mathbf{e}_2) = 1$, then V has the type $\{q_-^2\}$ If $f(\mathbf{e}_1) = 1$ and $f(\mathbf{e}_2) = 0$, then the replacement of the basis $\mathbf{e}_1, \mathbf{e}_2$ by $\mathbf{e}_1, \mathbf{e}_1 + \mathbf{e}_2$ reduces the case to the previous one. Similarly, if $f(\mathbf{e}_1) = 0$ and $f(\mathbf{e}_2) = 1$, we replace $\mathbf{e}_1, \mathbf{e}_2$ by $\mathbf{e}_1 + \mathbf{e}_2, \mathbf{e}_2$ and obtain that the type is again $\{q_-^2\}$. Finally, if $f(\mathbf{e}_1) = f(\mathbf{e}_2) = 0$, then the type of the space is $\{q_+^2\}$.

Since the number of solutions of the equation $q_+^2(x_1, x_2) = 0$ (resp. $q_-^2(x_1, x_2) = 0$) over $\mathbb{F}_2$ is clearly 3 (resp. 1), it follows that the forms q_+^2 and q_-^2 are not equivalent over $\mathbb{F}_2$.

The determinant of any space of type $\{q_+^2\}$ or $\{q_-^2\}$ equals 1, and so the space can not be reducible, because the determinant of any one-dimensional space over $\mathbb{F}_2$ equals zero. □

If 2 is a unit in K, then a quadratic function $f \colon V \to K$ can be reconstructed from the corresponding scalar product $\mathbf{u} \cdot \mathbf{v}$ in the form $f(\mathbf{u}) = 2^{-1} \mathbf{u} \cdot \mathbf{u}$, and therefore a quadratic space can be considered just as a linear space with a bilinear scalar product. In the oppositive case it quite can happen that a space with a non-zero quadratic function has the zero scalar product. Thus, in the general case, it is more convenient to start from quadratic functions.

A.2.2. Non-Degenerate Spaces

A quadratic space (V, f) is said to be *degenerate*, if it is a direct sum of the form

$$(\mathsf{V}, f) = (\mathsf{V}_1, 0) \oplus (\mathsf{V}_2, f_2),$$

where $\dim \mathsf{V}_1 \geq 1$ and 0 denotes the zero-function on V_1; otherwise the space is called *non-degenerate.*

Lemma A.2.6. *A non-singular quadratic space is non-degenerate.*

Proof. Since $d(\mathsf{V}_1, 0) = 0$, the lemma follows from (A.2.5). □

Proposition A.2.7. *A non-degenerate quadratic space over a field of characteristic different from* 2 *is non-singular.*

Proof. Suppose that $d(\mathsf{V}, f) = 0$. Then $\mathsf{R}(\mathsf{V}) \neq \{0\}$, by Lemma A.2.1. Let f_1 be the restriction of f to $\mathsf{R}(\mathsf{V})$. By Lemma A.2.2, the subspace $(\mathsf{R}(\mathsf{V}), f_1)$ is a direct summand of (V, f). On the other hand, if $\mathbf{u} \in \mathsf{R}(\mathsf{V})$, then $f(\mathbf{u}) = 2^{-1}\mathbf{u}\cdot\mathbf{u} = 0$, whence $f_1 = 0$. Thus, (V, f) is degenerate. □

We now consider the case, when K is a field of characteristic 2.

Lemma A.2.8. *Determinant $d(\mathsf{V}, f)$ of any quadratic space (V, f) of odd dimension over a field of characteristic* 2 *equals zero.*

Proof. Let $Q = (Q_{ij})$ be the matrix of f in a basis $\mathbf{e}_1, \ldots, \mathbf{e}_n$, and let A_{ij} be the cofactors of the entries of Q. For $i = 1, \ldots, n$, we have $\det Q = Q_{i1}A_{i1} + \cdots + Q_{in}A_{in}$. Summing up over all $i = 1, \ldots, n$, we get

$$n \det Q = \sum_{1 \leq i, j \leq n} Q_{ij}A_{ij} = \sum_{i=1}^{n} Q_{ii}A_{ii} = 0,$$

because $Q_{ij} = Q_{ji}$, $A_{ij} = A_{ji}$, and $Q_{ii} = 2f(\mathbf{e}_i) = 0$. Since n is odd, it follows that $\det Q = 0$. □

Theorem A.2.9. *Let (V, f) be a non-degenerate quadratic space over the field $\mathsf{F} = \mathbb{F}_2$. If $n = \dim \mathsf{V}$ is even, then (V, f) is nonsingular and has one of the two types $\{\mathsf{q}_+^n\}$, $\{\mathsf{q}_-^n\}$, where*

$$\mathsf{q}_+^n(x_1, \ldots, x_n) = x_1x_2 + x_3x_4 + \cdots + x_{n-1}x_n, \tag{A.2.10}$$

$$\mathsf{q}_-^n(x_1, \ldots, x_n) = x_1x_2 + \cdots + x_{n-3}x_{n-2} + x_{n-1}^2 + x_{n-1}x_n + x_n^2; \tag{A.2.11}$$

but if n is odd, then the type of (V, f) is $\{\mathsf{q}^n\}$, where

$$\mathsf{q}^n(x_1, \ldots, x_n) = x_1x_2 + \cdots + x_{n-2}x_{n-1} + x_n^2. \tag{A.2.12}$$

Proof. We first consider the case of even n. If $d(\mathsf{V}, f) = 0$, then, by Lemma A.2.1, there is a non-zero vector $\mathbf{e}_1$ such that $\mathbf{e}_1 \cdot \mathbf{v} = 0$ for all $\mathbf{v} \in \mathsf{V}$. Let $\mathbf{e}_1, \mathbf{e}_2, \ldots, \mathbf{e}_n$ be a basic for V with the first vector $\mathbf{e}_1$, $\mathsf{V}_2 = \{F\mathbf{e}_2 + \cdots + F\mathbf{e}_n\}$, and let f_2 be the restriction of f to V_2. Since $n-1$ is odd, it follows, by Lemma A.2.8, that $d(\mathsf{V}_2, f_2) = 0$. Again by Lemma A.2.1, we see that the space V_2 contains a non-zero vector $\mathbf{e}'_1$ orthogonal to all of V_2. By the choice of $\mathbf{e}_1$ and $\mathbf{e}'_1$, it follows that the dimension of $\mathsf{V}_1 = \{\mathsf{F}\mathbf{e}_1 + \mathsf{F}\mathbf{e}'_1\}$ is two, the space V_1 is a direct summand in V, and $\mathbf{e}_1 \cdot \mathbf{e}'_1 = 0$. Further, the space (V_1, f_1), where f_1 is the restriction of f to V_1, is degenerate, because it always contains an one-dimensional direct summand of the form $(\mathsf{V}_0, 0)$. In fact, if for example, $f(\mathbf{e}_1) = 0$, then one can take $\mathsf{V}_0 = \{\mathsf{F}\mathbf{e}_1\}$; but if $f(\mathbf{e}_1) = f(\mathbf{e}'_1) = 1$, then we take $\mathsf{V}_0 = \{\mathsf{F}(\mathbf{e}_1 + \mathbf{e}'_1)\}$. Hence, the space (V, f) is degenerate. We have proved that $d(\mathsf{V}, f) \neq 0$. By Theorem A.2.4, the space (V, f) is a direct sum of irreducible subspaces $(\mathsf{V}_1, f_1), \ldots, (\mathsf{V}_k, f_k)$ with $\dim \mathsf{V}_i \leq 2$. If there is i with $\dim \mathsf{V}_i = 1$, then $d(\mathsf{V}_i, f_i) = 0$, and so $d(\mathsf{V}, f) = \prod_j d(\mathsf{V}_j, f_j) = 0$. Hence, $\dim \mathsf{V}_i = 2$ for all i. Then, by Proposition A.2.5, each of the space (V_i, f_i) has either the type $\{\mathsf{q}_+^2\}$ or $\{\mathsf{q}_-^2\}$. Hence, in order to complete the consideration of even-dimensional spaces, we only have to check that the direct sum of two spaces of type $\{\mathsf{q}_-^2\}$ is isomorphic to the direct sum of two spaces of type $\{\mathsf{q}_+^2\}$. The direct sum (V', f') of two spaces of type $\{\mathsf{q}_-^2\}$ has a basis $\mathbf{e}_1$, $\mathbf{e}_2$, $\mathbf{e}'_1$, $\mathbf{e}'_2$ such that $f'(\mathbf{e}_1) = f'(\mathbf{e}_2) = f(\mathbf{e}'_1) = f(\mathbf{e}'_2) = 1$, $\mathbf{e}_1 \cdot \mathbf{e}_2 = \mathbf{e}'_1 \cdot \mathbf{e}'_2 = 0$, and $\mathbf{e}_i \cdot \mathbf{e}'_j = 0$ for $i, j = 1, 2$. Then the subspaces $\mathsf{V}'_1 = \{\mathsf{F}(\mathbf{e}_1 + \mathbf{e}'_1) + \mathsf{F}(\mathbf{e}_2 + \mathbf{e}'_1)\}$ and $\mathsf{V}'_2 = \{\mathsf{F}(\mathbf{e}_1 + \mathbf{e}_2 + \mathbf{e}'_1 + \mathbf{e}'_2) + \mathsf{F}(\mathbf{e}_1 + \mathbf{e}_2 + \mathbf{e}'_2)\}$ are both of type $\{\mathsf{q}_+^2\}$ and (V', f') is the direct sum of the subspaces.

We now assume that n is odd. It follows from Lemmas A.2.8 and A.2.1 that there is a non-zero vector $\mathbf{e} = \mathbf{e}_n \in \mathsf{R}(\mathsf{V})$. We set $\mathsf{V}_2 = \{\mathsf{F}\mathbf{e}_n\}$ and denote by f_2 the restriction of f to V_2. Then, by Lemma A.2.2, we have $(\mathsf{V}, f) = (\mathsf{V}_1, f_1) \oplus (\mathsf{V}_2, f_2)$ with a subspace (V_1, f_1). Since (V, f) is non-degenerate, it follows that (V_1, f_1) and (V_2, f_2) are non-degenerate as well. Then, by the above, $f(\mathbf{e}_n) = 1$, and the space (V_1, f_1) has either the type $\{\mathsf{q}_+^{n-1}\}$ or $\{\mathsf{q}_-^{n-1}\}$. In the first case the type of (V, f) is clearly $\{\mathsf{q}_+^{n-1} + x_n^2\} = \{\mathsf{q}^n\}$. In the second case the space (V_1, f_1) can be represented in the form $(\mathsf{V}_3, f_3) \oplus (\mathsf{V}_4, f_4)$ where (V_3, f_3) has the type $\{\mathsf{q}_+^{n-3}\}$, and (V_4, f_4) has the type $\{\mathsf{q}_-^2\}$. Then the sum $(\mathsf{V}_4, f_4) \oplus (\mathsf{V}_2, f_2)$ has a basis $\mathbf{v}_1$, $\mathbf{v}_2$, $\mathbf{e}_n$ such that $f(\mathbf{v}_1) = f(\mathbf{v}_2) = f(\mathbf{e}_n) = 1$, $\mathbf{v}_1 \cdot \mathbf{e}_n = \mathbf{v}_2 \cdot \mathbf{e}_n = 0$, and $\mathbf{v}_1 \mathbf{v}_2 = 1$. On replacing of the basis by $\mathbf{v}_1 + \mathbf{e}_n$, $\mathbf{v}_2 + \mathbf{e}_n$, $\mathbf{e}_n$ we see that the type of the sum is $\{x_1 x_2 + x_3^2\} = \{\mathsf{q}^3\}$. Hence, the type of $(\mathsf{V}, f) = (\mathsf{V}_3, f_3) \oplus (\mathsf{V}_4, f_4) \oplus (\mathsf{V}_2, f_2)$ is $\{\mathsf{q}_+^{n-3} + x_{n-2} x_{n-1} + x_n^2\} = \{\mathsf{q}^n\}$. □

A.2.3. Gaussian Sums

A gaussian sum of a quadratic space (V, f) *over* $\mathsf{F} = \mathbb{F}_p$ is the sum of p-th roots of unity of the form

$$G_p(\alpha f) = \sum_{\mathbf{v} \in \mathsf{V}} \exp\left(\frac{2\pi i \alpha f(\mathbf{v})}{p}\right), \tag{A.2.13}$$

where each of the elements α, $\alpha f(\mathbf{v})$ of $\mathbb{F}_p = \mathbb{Z}/p\mathbb{Z}$ is considered as an integer of the corresponding residue class modulo p. If the type of (V, f) is $\{\mathsf{q}(x_1, \ldots, x_n)\}$, then the sum $G_p(\alpha f)$ is obviously equal to the gaussian sum of the quadratic form $\alpha\mathsf{q}$:

$$G_p(\alpha f) = G_p(\alpha\mathsf{q}) = \sum_{a_1, \ldots, a_n \in \mathbb{Z}/p\mathbb{Z}} \exp(2\pi i \alpha \mathsf{q}(a_1, \ldots, a_n)/p).$$

If α is fixed, then the sum (A.2.13) depends only on the values of f on V. Hence, isomorphic spaces (resp. equivalent quadratic forms) have equal gaussian sums. It is clear that

$$G_p(\alpha f) = G_p(\alpha\mathsf{q}) = p^{\dim \mathsf{V}} \quad \text{if } \alpha = 0 \text{ or } f = 0. \tag{A.2.14}$$

Further, if $(\mathsf{V}, f) = (\mathsf{V}_1, f_1) \oplus \cdots \oplus (\mathsf{V}_t, f_t)$, then

$$\begin{aligned} G_p(\alpha f) &= \sum_{\mathbf{v}_j \in \mathsf{V}_j (1 \le j \le t)} \exp(2\pi i \alpha (f_1(\mathbf{v}_1) + \cdots + f_t(\mathbf{v}_t))/p) \\ &= \prod_{j=1}^{t} \sum_{\mathbf{v}_j \in \mathsf{V}_j} \exp(2\pi i \alpha f_j(\mathbf{v}_j)/p) = \prod_{j=1}^{t} G_p(\alpha f_j), \end{aligned} \tag{A.2.15}$$

which reduces gaussian sums of arbitrary spaces to those of irreducible spaces.

Before calculating the gaussian sums, we remind the definition and the simplest properties of the Legendre symbol. Let p be an odd prime number, $\mathsf{F} = \mathbb{F}_p$, F^* the multiplicable group of F, and $\mathsf{F}^{*2} = \{\alpha^2, \alpha \in \mathsf{F}^*\}$ the subgroup of squares. Since the kernel of the homomorphism $\alpha \to \alpha^2$ of F^* onto F^{*2} consists of 1 and -1, and $-1 \ne 1$, it follows that the index of F^{*2} in F^* is equal to 2. Let $\alpha \to (\alpha/p)$ denotes the only non-trivial character of the factor group $\mathsf{F}^*/\mathsf{F}^{*2}$ considered as a function on F^*. In other words, $(\alpha/p) = 1$, if α is a square, and $(\alpha/p) = -1$, otherwise. For an integer a prime to p, *the Legendre symbol* (a/p) is (α/p) where α is the residue class of a in $\mathbb{Z}/p\mathbb{Z}$. The definition implies that the Legendre symbol is multiplicable i.e. it satisfies

$$\left(\frac{ab}{p}\right) = \left(\frac{a}{p}\right)\left(\frac{b}{p}\right) \quad \text{if } p \nmid ab. \tag{A.2.16}$$

The multicativity holds for all integers if we set $\left(\frac{a}{p}\right) = 0$ for $a \in p\mathbb{Z}$. Since $\left(\frac{\alpha}{p}\right)$ is a non-trivial character of the finite group F^*, it follows that

$$\sum_{1 \le a \le p-1} \left(\frac{a}{p}\right) = \sum_{\alpha \in \mathsf{F}^*} \left(\frac{\alpha}{p}\right) = 0. \tag{A.2.17}$$

Lemma A.2.10. *The following formulae hold:*

$$G_2(x^2) = 0, \quad G_2(\mathsf{q}_+^2) = 2, \quad G_2(\mathsf{q}_-^2) = -2, \tag{A.2.18}$$

where $\mathsf{q}_\pm^2$ *are the forms* (A.2.9);

$$G_p(ax^2) = \left(\frac{a}{p}\right) G_p(x^2), \quad G_p(x^2)^2 = \left(\frac{-1}{p}\right) p, \tag{A.2.19}$$

where p is an odd prime number and a is prime to p.

Proof. The formulae (A.2.18) follow from definitions. The formulae (A.2.19) have been proved in the proof of Proposition 1.4.9. □

Proposition A.2.11. *Let* (V, f) *be a non-degenerate quadratic space over a finite field* $\mathsf{F} = \mathbb{F}_p$. *Then the gaussian sum* $G_p(\alpha f)$ *with* $\alpha \in \mathsf{F}^*$ *is given by the following formulae: if* $\dim \mathsf{V}$ *is even, say* $\dim \mathsf{V} = 2k$, *then*

$$G_p(\alpha f) = \varepsilon(\mathsf{V}, f) p^k \tag{A.2.20}$$

where

$$\varepsilon(\mathsf{V}, f) = \begin{cases} \left(\dfrac{(-1)^k d(\mathsf{V}, f)}{p}\right) & \text{if } p \neq 2, \\ 1 & \text{if } p = 2 \text{ and } (\mathsf{V}, f) \text{ is of type } \{\mathsf{q}_+^n\} \\ -1 & \text{if } p = 2 \text{ and } (\mathsf{V}, f) \text{ is of type } \{\mathsf{q}_-^n\}, \end{cases} \tag{A.2.21}$$

but if $\dim \mathsf{V}$ *is odd, say* $\dim \mathsf{V} = 2k + 1$, *then*

$$G_p(\alpha f) = \left(\frac{\alpha}{p}\right) G_p(x^2) \left(\frac{(-1)^k 2d(V, f)}{p}\right) p^k, \quad \text{for } p \neq 2, \tag{A.2.22}$$

and

$$G_2(f) = 0.$$

Proof. If $p \neq 2$, then it follows from Theorem A.2.4 that (V, f) is the direct sum of n one-dimensional subspaces $(\mathsf{V}_1, f_1), \ldots, (\mathsf{V}_n, f_n)$. Since (V, f) is non-degenerate, it follows that each of the (V_i, f_i) is non-degenerate as well. Then the type of (V_i, f_i) is $\{\alpha_i x^2\}$ with $\alpha_i \neq 0$, and, by (A.2.15), (A.2.16) and (A.2.19), we have

$$\begin{aligned} G_p(\alpha f) &= \prod_{i=1}^n G_p(\alpha f_i) = \prod_{i=1}^n G_p(\alpha \alpha_i x^2) \\ &= \left[\prod_{i=1}^n \left(\frac{\alpha \alpha_i}{p}\right)\right] G_p(x^2)^n = \left(\frac{\alpha}{p}\right)^n \left(\frac{2}{p}\right)^n \left(\frac{2^n \alpha_1 \cdots \alpha_n}{p}\right) G_p(x^2)^n \\ &= \left(\frac{d(\mathsf{V}, f)}{p}\right) G_p(x^2)^n \end{aligned}$$

(by (A.2.5), $d(\mathsf{V}, f) = d(\mathsf{V}_1, f_1) \cdots d(\mathsf{V}_n, f_n) = 2\alpha_1 \cdots 2\alpha_n$). Substituting $G_p(x^2)^2 = (-1/p)p$, we get the formulae for odd primes.

The formulae for $G_2(f)$ follow from (A.2.15) and (A.2.18), since a space of type $\{\mathsf{q}_+^{2k}\}$ is the direct sum of k spaces of type $\{\mathsf{q}_+^2\}$, a space of type $\{\mathsf{q}_-^{2k}\}$ is the direct sum of $k - 1$ spaces of type $\{\mathsf{q}_+^2\}$ and one space of type $\{\mathsf{q}_-^2\}$, and a space of type

$\{q^{2k+1}\}$ contains as a direct summand a subspace of type $\{x^2\}$ whose gaussian sum is zero. □

The number $\varepsilon(V, f)$ will be called *the sign of a non-degenerate space* (V, f) *of even dimension over* $\mathbb{F}_p$. It follows from (A.2.15) and (A.2.20) that the sign is multiplicative in the sense that

$$\varepsilon(V_1, f_1)\varepsilon(V_2, f_2) = \varepsilon((V_1, f_1) \oplus (V_2, f_2)) \tag{A.2.23}$$

if (V_i, f_i) are non-degenerate spaces of even dimension.

A.2.4. Isotropic Subspaces of Non-Degenerate Spaces over the Fields $\mathbb{F}_p$

A quadratic space of positive dimension with the zero quadratic function (or form) is called *isotropic*.

Proposition A.2.12. *Let* (V, f) *be a non-degenerate quadratic space over a finite field* $F = \mathbb{F}_p$. *Let* $(V_1, 0)$ *be an isotropic subspace of* (V, f), *and let* $\mathbf{e}_1, \ldots, \mathbf{e}_r$ *be a basis for* V_1. *Then there are vectors* $\mathbf{e}_1^*, \ldots, \mathbf{e}_r^*$ *in* V *satisfying the following conditions*:

(1) $\mathbf{e}_1 \cdot \mathbf{e}_1^* = \cdots = \mathbf{e}_r \cdot \mathbf{e}_r^* = 1, f(\mathbf{e}_1^*) = \cdots = f(\mathbf{e}_r^*) = 0$;

(2) *the subspaces* $(P_1, f_1), \ldots, (P_r, f_r)$, *where* $P_i = \{F\mathbf{e}_i + F\mathbf{e}_i^*\}$ *and* f_i *is the restriction of* f *to* P_i, *are pairwise orthogonal.*

We first prove a lemma.

Lemma A.2.13. *If* (V, f) *is a quadratic space over a field with* $d(V, f) \neq 0$, *then, for any subspace* (V_1, f_1) *of* (V, f), *the following relations hold*

$$\dim V_1^\perp = \dim V - \dim V_1, \quad (V_1^\perp)^\perp = V_1$$

Proof of the lemma. Let $\mathbf{e}_1, \ldots, \mathbf{e}_n$ be a basis for V such that $\mathbf{e}_1, \ldots, \mathbf{e}_r$ form a basis for V_1. Since $d(V, f) = \det(\mathbf{e}_i \cdot \mathbf{e}_j) \neq 0$, it follows that rows of the matrix $(\mathbf{e}_i \cdot \mathbf{e}_j)$ are linearly independent, in particular, the first r rows are linearly independent. This implies that the system of r linear equations $u\mathbf{e}_1 = 0, \ldots, u\mathbf{e}_r = 0$ which determines vectors $\mathbf{u} = \sum_i u_i \mathbf{e}_i$ of $V_1^\perp$, has $n - r$ linearly independent solutions. This proves the formulae for dimensions; in particular,

$$\dim(V_1^\perp)^\perp = \dim V - (\dim V - \dim V_1) = \dim V_1.$$

On the other hand, it is clear that $V_1 \subset (V_1^\perp)^\perp$. □

Proof of the proposition. We first assume that either $n = \dim V$ is even or p is odd, we shall apply an indunction to r. Since $f(\mathbf{e}_1) = 0$ and the space (V, f) is non-degenerate, it follows that there is $\mathbf{v} \in V$ with $\mathbf{e}_1 \mathbf{v} = \alpha \neq 0$ (see Lemma A.2.2). We set $\mathbf{v}_1 = \alpha^{-1}\mathbf{v}$ and $\mathbf{e}_1^* = \mathbf{v}_1 - f(\mathbf{v}_1)e_1$. Then $\mathbf{e}_1 \mathbf{e}_1^* = \mathbf{e}_1 \mathbf{v}_1 = 1$ and $f(\mathbf{e}_1^*) =$

$f(\mathbf{v}_1) - \mathbf{v}_1 \cdot (f(\mathbf{v}_1)\mathbf{e}_1) = 0$, which proves the assertion for $r = 1$. Suppose that the assertion is already proved for subspaces of dimension $r - 1$ with $r > 1$. Let us set $\mathsf{V}_0 = \{\mathsf{Fe}_1 + \cdots + \mathsf{Fe}_{r-1}\}$. According to Proposition A.2.7 and Theorem A.2.9, $d(V, f) \neq 0$. Hence, by Lemma A.2.13, $(\mathsf{V}_0^\perp)^\perp = \mathsf{V}_0$. The vector $\mathbf{e}_r$ clearly belongs to $\mathsf{V}_0^\perp$ but does not belong to V_0. Thus, $\mathbf{e}_r$ does not belong to $(V_0^\perp)^\perp$, and so there is a vector $\mathbf{u} \in \mathsf{V}_0^\perp$ with $\mathbf{e}_r \cdot \mathbf{u} = \beta \neq 0$. On replacing $\mathbf{u}$ by $\mathbf{e}_r^* = \mathbf{u}_1 - f(\mathbf{u}_1)\mathbf{e}_r$, where $\mathbf{u}_1 = \beta^{-1}\mathbf{u}$, we get the vector $\mathbf{e}_r^*$ in $\mathsf{V}_0^\perp$ satisfying $\mathbf{e}_r \cdot \mathbf{e}_r^* = \mathbf{e}_r \cdot \mathbf{u}_1 = 1$ and $f(\mathbf{e}_r^*) = \mathbf{u}_i(-f(\mathbf{u}_1)\mathbf{e}_r) + f(\mathbf{u}_1) = 0$. Let us set $\mathsf{P}_r = \{F\mathbf{e}_r + F\mathbf{e}_r^*\}$. Since $\mathsf{P}_r \subset \mathsf{V}_0^\perp$, it follows that $\mathsf{V}_0 \subset \mathsf{P}_r^\perp$. Since the determinant of P_r is equal to 1, it follows from Theorem A.2.3, (A.2.5), and Lemma A.2.6 that $\mathsf{P}_r^\perp$ is a non-degenerate space of dimension $n - 2$. By the inductive hypothesis there are vectors $\mathbf{e}_1^*, \ldots, \mathbf{e}_{r-1}^* \in \mathsf{P}_r^\perp$ satisfying the conditions (1), (2) with respect to the basis $\mathbf{e}_1, \ldots, \mathbf{e}_{r-1}$, of the isotropic subspace $\mathsf{V}_0 \subset \mathsf{P}_r^\perp$. Then the vectors $\mathbf{e}_1^*, \ldots, \mathbf{e}_{r-1}^*, \mathbf{e}_r^*$ satisfies the conditions of the proposition.

Let us now suppose that $n = \dim \mathsf{V}$ is odd and $p = 2$. Since the rank of the matrix of the quadratic form (A.2.12) over $\mathbb{F}_2$ is clearly equal to $n - 1$, it follows that the radical $\mathsf{R}(\mathsf{V}) = \mathsf{V}^\perp$ of V is one-dimensional. Since V is non-degenerate, it follows that $\mathsf{R}(\mathsf{V})$ contains the unique vector $\mathbf{e}_0$ with $f(\mathbf{e}_0) = 1$. Then $\mathbf{e}_0 \notin \mathsf{V}_1$. Hence, the vectors $\mathbf{e}_0, \mathbf{e}_1, \ldots, \mathbf{e}_r$ are linearly independent and there is a basis for V of the form $\mathbf{e}_0, \mathbf{e}_1, \ldots, \mathbf{e}_r, \ldots, \mathbf{e}_{n-1}$. Let us set $\mathsf{V}' = \{\mathsf{Fe}_1 + \cdots + \mathsf{Fe}_{n-1}\}$. Then

$$(\mathsf{V}, f) = (\{\mathsf{Fe}_0\}, f_0) \oplus (\mathsf{V}', f'),$$

where f_0 and f' are the restrictions of f to the corresponding subspaces. Since (V, f) is non-degenerate, it follows from the decomposition that the space (V', f') is non-degenerate too. Since $\mathsf{V}_1 \subset \mathsf{V}'$ and $\dim \mathsf{V}'$ is even, the proposition follows from the case considered above. □

A set of vectors $\mathbf{e}_1, \ldots, \mathbf{e}_r$ of a quadratic space (V, f) is said to be *isotropic*, if the vectors are linearly independent and span an isotropic subspace.

Proposition A.2.14. *Let* (V, f) *be a non-degenerate quadratic space over* $\mathsf{F} = \mathbb{F}_p$. *Then the number* $i(\mathsf{V}, f, r)$ *of isotropic sets of vectors of* (V, f) *consisting of* r *vectors with* $r = 1, 2, \ldots, \dim \mathsf{V}$ *is equal to*

$$p^{\langle r-1 \rangle}(p^k - \varepsilon)(p^{k-r} + \varepsilon)(p^{2(k-1)} - 1) \cdots (p^{2(k-r+1)} - 1) \qquad \text{(A.2.24)}$$

if $\dim \mathsf{V} = 2k$, *where* $\varepsilon = \varepsilon(\mathsf{V}, f)$ *is the sign* (A.2.21) *of* (V, f), *and it is equal to*

$$p^{\langle r-1 \rangle}(p^{2k} - 1)(p^{2(k-1)} - 1) \cdots (p^{2(k-r+1)} - 1) \qquad \text{(A.2.25)}$$

if $\dim \mathsf{V} = 2k + 1$.

Proof. An induction on r. By the definition, $i(\mathsf{V}, f, 1)$ is the number of non-zero vectors $\mathbf{v}$ in V satisfying $f(\mathbf{v}) = 0$. The number of all solutions of $f(\mathbf{v}) = 0$ is clearly

$$i(\mathsf{V}, f, 1) + 1 = p^{-1} \sum_{0 \le \alpha \le p-1} G_p(\alpha f) = p^{n-1} + p^{-1} \sum_{1 \le \alpha \le p-1} G_p(\alpha f),$$

where $n = \dim V$. Hence, if $n = 2k$, then by (A.2.20) we have

$$i(V, f, 1) = p^{2k-1} - 1 + \varepsilon p^{k-1}(p-1) = (p^k - \varepsilon)(p^{k-1} + \varepsilon),$$

but if $n = 2k+1$ then, by (A.2.22) and (A.2.17), we obtain

$$i(V, f, 1) = p^{n-1} - 1 = p^{2k} - 1,$$

which proves the proposition for $r = 1$. Suppose the proposition is already proved for sets of $r-1$ vectors with $r > 1$. If $i(V, f, r-1) = 0$, then $i(V, f, r) = 0$, and the proposition easily follows from the inductive hypothesis. Let us now assume that $i(V, f, r-1) \neq 0$, and let $\mathbf{e}_1, \ldots, \mathbf{e}_{r-1}$ be a isotropic set of $r-1$ vectors. Since the space spanned by $\mathbf{e}_1, \ldots, \mathbf{e}_{r-1}$ is isotropic, then, for the basis $\mathbf{e}_1, \ldots, \mathbf{e}_{r-1}$ of the space, there are vectors $\mathbf{e}_1^*, \ldots, \mathbf{e}_{r-1}^*$ satisfying the relations of Proposition A.2.12. The determinant of each of the subspaces $(P_1, f_1), \ldots, (P_{r-1}, f_{r-1})$ defined in Proposition A.2.12 is equal to -1. Thus, in view of Theorem A.2.3, the space (V, f) has a decomposition of the form

$$(V, f) = (P_1, f_1) \oplus \cdots \oplus (P_{r-1}, f_{r-1}) \oplus (V', f'). \tag{A.2.26}$$

Hence, each vector $\mathbf{v}$ of V is uniquely of the shape

$$\mathbf{v} = \sum_{i=1}^{r-1} \alpha_i \mathbf{e}_i + \sum_{j=1}^{r-1} \beta_j \mathbf{e}_j^* + \mathbf{v}',$$

where $\alpha_i, \beta_j \in F$ and $\mathbf{v}' \in V'$. Since $\mathbf{v} \cdot \mathbf{e}_i = \beta_i$, it follows that $\mathbf{v}$ is orthogonal to all of the vectors $\mathbf{e}_1, \ldots, \mathbf{e}_{r-1}$ if and only if $\beta_1 = \cdots = \beta_{r-1} = 0$. Since the vectors $\mathbf{e}_1, \ldots, \mathbf{e}_{r-1}, \mathbf{v}'$ are pairwise orthogonal, it follows that such a $\mathbf{v}$ satisfies $f(\mathbf{v}) = f(\alpha_1 \mathbf{e}_1) + \cdots + f(\alpha_{r-1}\mathbf{e}_{r-1}) + f(\mathbf{v}') = f(\mathbf{v}')$, and so the condition $f(\mathbf{v}) = 0$ is equivalent with $f(\mathbf{v}') = 0$. Finally, the vectors $\mathbf{e}_1, \ldots, \mathbf{e}_{r-1}, \mathbf{v}$ are linearly independent if and only if $\mathbf{v}' \neq 0$. Thus, the number of isotropic sets of the form $\mathbf{e}_1, \ldots, \mathbf{e}_{r-1}, \mathbf{v}$ is equal to $p^{r-1} \cdot i(V', f', 1)$, whence

$$i(V, f, r) = p^{r-1} i(V', f', 1) i(V, f, r-1). \tag{A.2.27}$$

The decomposition (A.2.26) implies that the subspace (V', f') is non-degenerate and $\dim V' = n - 2(r-1)$. Since the sign of each of the (P_i, f_i) is clearly 1, it follows, by (A.2.26) and (A.2.23), that for even n with $n > 2(r-1)$, the sign $\varepsilon(V', f') = \varepsilon(V, f)$. If $n = 2(r-1)$, then $i(V, f, r) = 0$ and the expression in (A.2.24) is also equal to zero. If $n > 2(r-1)$, then using formulae for $i(V', f', 1)$ obtained above and the inductive hypothesis, we get the formula for $i(V, f, r)$. □

Corollary A.2.15. *Let* (V, f) *be a non-degenerate quadratic space over* $F = \mathbb{F}_p$ *and let* $V_0 = (V_0, 0)$ *be a maximal isotropic subspace of* (V, f). *Then*

$$\lambda = \dim V_0 = \begin{cases} k & \text{if } \dim V = 2k \text{ and } \varepsilon(V, f) = 1 \\ k-1 & \text{if } \dim V = 2k \text{ and } \varepsilon(V, f) = -1. \\ k & \text{if } \dim V = 2k+1 \end{cases}$$

A.2.5. Non-Singular Spaces over Residue Class Rings

In this subsection we shall describe non-singular quadratic spaces over the residue-class rings $\mathbb{Z}/a\mathbb{Z}$ with $a \in \mathbb{N}$. Each quadratic space over $\mathbb{Z}/a\mathbb{Z}$ can be obtained as a result of "the reduction modulo a" of a quadratic space (V_0, f_0) over $\mathbb{Z}$. Then we write $(\mathsf{V}, f) = (\mathsf{V}_0, f_0)/\text{mod}\, a$.

If (V, f) is a space over $\mathbb{Z}/a\mathbb{Z}$ and $a_1 | a$, then the canonical epimorphism $\mathbb{Z}/a\mathbb{Z} \to \mathbb{Z}/a_1\mathbb{Z}$ allows us to consider (V, f) as a space over $\mathbb{Z}/a_1\mathbb{Z}$. Definitions and standard properties of congruences (the Chinese reminder theorem) imply the following

Lemma A.2.16. *Let $a = a_1 \cdots a_m$, where the numbers $a_1, \dots, a_m$ of $\mathbb{N}$ are pairwise coprime. Then quadratic spaces over the ring $\mathsf{K} = \mathbb{Z}/a\mathbb{Z}$ are isomorphic if and only if they are isomorphic over each of the rings $\mathbb{Z}/a_1\mathbb{Z}, \dots, \mathbb{Z}/a_m\mathbb{Z}$.*

This lemma reduces the classification of quadratic spaces over $\mathbb{Z}/a\mathbb{Z}$ to the case $a = p^\delta$, where p is a prime number.

Proposition A.2.17. *Let (V, f) be a non-singular quadratic space over the ring $\mathsf{K} = \mathbb{Z}/p^\delta\mathbb{Z}$, where p is an odd prime number and $\delta \geq 1$. Then the type of (V, f) is equal to $\{\mathsf{q}_d^n\}_\mathsf{K}$, where $n = \dim \mathsf{V}$, $d = d(\mathsf{V}, f)$, and where*

$$\mathsf{q}_d^n = x_1^2 + \cdots + x_{n-1}^2 + 2^{-n} d x_n^2. \tag{A.2.28}$$

Proof. If $n = 1$, then the assertion is obvious. Suppose that $n > 1$ and the assertion is already proved for spaces of dimension $< n$.

We first prove by induction on δ, that V contains a vector $\mathbf{v}$ with $f(\mathbf{v}) = 1$. If $\delta = 1$, then the number of such vectors is clearly equal to

$$\begin{aligned} N &= p^{-1} \sum_{0 \leq \alpha \leq p-1} \sum_{\mathbf{v} \in \mathsf{V}} \exp(2\pi i \alpha (f(\mathbf{v}) - 1)/p) \\ &= p^{n-1} + p^{-1} \sum_{1 \leq \alpha \leq p-1} \exp(-2\pi i \alpha/p) G_p(\alpha f), \end{aligned}$$

where $G_p(\alpha f)$ are the gaussian sums (A.2.13). By Lemma A.2.6, the space (V, f) is non-degenerate. Then the formulae of Proposition A.2.11 and the formulae (A.2.19) imply that $|G_p(\alpha f)| = p^{n/2}$ if $\alpha = 1, \dots, p-1$. Hence,

$$\left| p^{-1} \sum_{1 \leq \alpha \leq p-1} \exp(-2\pi i \alpha/p) \right| \leq (p-1) p^{(n/2)-1}$$

and so $N \geq 1$, for $n > 1$. Suppose, now, that $\delta > 1$, and it is already known that the space (V, f) considered over $\mathbb{Z}/p^{\delta-1}\mathbb{Z}$ contains a vector $\mathbf{v}'$ with $f(\mathbf{v}') = 1$. For the initial space (V, f) over K this means that there is $\mathbf{v}' \in \mathsf{V}$ with $f(\mathbf{v}') \equiv 1 \pmod{p^{\delta-1}}$. We shall seek the vector $\mathbf{v}$ in the form $\mathbf{v} = \mathbf{v}' + p^{\delta-1}\mathbf{u}$ with $\mathbf{u} \in \mathsf{V}$. Since $\delta > 1$, it follows, by (A.2.2), that

$$\begin{aligned} f(\mathbf{v}) = f(\mathbf{v}' + p^{\delta-1}\mathbf{u}) &= f(\mathbf{v}') + p^{\delta-1}\mathbf{v}' \cdot \mathbf{u} + p^{2\delta-2} f(\mathbf{u}) \\ &\equiv f(\mathbf{v}') + p^{\delta-1}\mathbf{v}' \cdot \mathbf{u} \pmod{p^\delta}. \end{aligned}$$

Hence, the condition $f(\mathbf{v}) \equiv 1 \pmod{p^\delta}$ is equivalent with the congruence

$$\mathbf{v}' \cdot \mathbf{u} \equiv (1 - f(\mathbf{v}'))/p^{\delta-1} \pmod{p}. \tag{A.2.29}$$

Let $\mathbf{v}' = v_1 \mathbf{e}_1 + \cdots + v_n \mathbf{e}_n$ and $\mathbf{u} = u_1 \mathbf{e}_1 + \cdots + u_n \mathbf{e}_n$, where $\mathbf{e}_1, \ldots, \mathbf{e}_n$ is a basis for V. Then we can write

$$\mathbf{v}' \cdot \mathbf{u} = \sum_{i,j} v_i u_j \mathbf{e}_i \cdot \mathbf{e}_j = \sum_j u_j \sum_i v_i \mathbf{e}_i \cdot \mathbf{e}_j.$$

By the assumption, the determinant $d = \det(\mathbf{e}_i \cdot \mathbf{e}_j)$ is not divisible by p. Since $f(\mathbf{v}') \equiv 1 \pmod{p}$, it follows that not all of the coordinates v_i of $\mathbf{v}'$ are divisible by p. Hence, there is an index t with $1 \leq t \leq n$ such that

$$\sum_i v_i \mathbf{e}_i \cdot \mathbf{e}_t \not\equiv 0 \pmod{p}. \tag{A.2.30}$$

This implies that the product $\mathbf{v}' \cdot \mathbf{u}$ takes arbitrary values modulo p when $u_1, \ldots, u_n$ vary modulo p; in particular, the congruence (A.2.29) has a solution with respect to $\mathbf{u}$.

Let us fix a vector $\mathbf{v} \in \mathsf{V}$ with $f(\mathbf{v}) = 1$ and set $\mathsf{V}_2 = \{\mathbf{u} \in \mathsf{V}, \mathbf{u} \cdot \mathbf{v} = 0\}$. We may assume without loss of generality that the condition (A.2.30) for the coordinates $v_1, \ldots, v_n$ of $\mathbf{v}$ is satisfied by $t = n$. Let us set, for $j = 1, \ldots, n$, $\alpha_j = \sum_i v_i \mathbf{e}_i \cdot \mathbf{e}_j$. Since α_n is invertible in K, it follows that

$$\begin{aligned} \mathsf{V}_2 &= \Big\{\mathbf{u} = u_1 \mathbf{e}_1 + \cdots + u_n \mathbf{e}_n; u_i \in \mathsf{K}, \sum_j \alpha_j u_j = 0\Big\} \\ &= \left\{u_1\left(\mathbf{e}_1 - \frac{\alpha_1}{\alpha_n}\mathbf{e}_n\right) + \cdots + u_{n-1}\left(\mathbf{e}_{n-1} - \frac{\alpha_{n-1}}{\alpha_n}\mathbf{e}_n\right); u_1, \ldots, u_{n-1} \in \mathsf{K}\right\}. \end{aligned}$$

Thus V_2 is a free K-module of dimension $n - 1$, and the pair (V_2, f_2), where f_2 is the restriction of f to V_2, is a quadratic space over K. Let us set $\mathsf{V}_1 = \mathsf{K} \cdot \mathbf{v}$, and let f_1 be the restriction of f to V_1. Then it is easy to see that (V, f) is the direct sum of (V_1, f_1) and (V_2, f_2). By (A.2.5), we have $d(\mathsf{V}_2, f_2) = d(\mathsf{V}, f)/d(\mathsf{V}_1, f_1) = d/2$. Hence, the space (V_2, f_2) along with (V, f) is non-singular and we may apply the inductive hypothesis to (V_2, f_2). Since the type of (V_1, f_1) is $\{x_1^2\}_\mathsf{K}$, it follows that the type of (V, f) is

$$\{x_1^2 + x_2^2 + \cdots + x_{n-1}^2 + 2^{-(n-1)} \cdot 2^{-1} d x_n^2\}_\mathsf{K} = \{\mathsf{q}_d^n\}_\mathsf{K}. \qquad \square$$

Proposition A.2.18. *Let* (V, f) *be a non-singular space over the ring* $\mathsf{K} = \mathbb{Z}/2^\delta\mathbb{Z}$ *with* $\delta \geq 3$. *Then the following assertion holds:*

(1) *The dimension of* V *is even,* $n = \dim V = 2k$;

(2) *The determinant* $d = d(\mathsf{V}, f)$ *of* (V, f) *satisfies one of the congruences*

$$d \equiv (-1)^k \quad \textit{or} \quad d \equiv 3 \cdot (-1)^{k-1} \pmod{8}; \tag{A.2.31}$$

(3) *If* $d \equiv (-1)^k \pmod{8}$ *(resp.* $d \equiv 3 \cdot (-1)^{k-1} \pmod{8}$*) then the type of* (V, f) *is equal to* $\{\mathsf{q}_+^n\}_\mathsf{K}$ *(resp.* $\{\mathsf{q}_-^n\}_\mathsf{K}$*) where* q_+^n *and* q_-^n *are the quadratic forms* (A.2.10) *and* (A.2.11) *respectively.*

Proof. The first assertions follows from Lemma A.2.8. The other assertions will be proved by induction on $n = \dim V$.

If $n = 2$, then it follows from Lemma A.2.6 and Theorem A.2.9 that there is a basis $\mathbf{e}'_1, \mathbf{e}'_2$ for V such that the function f is given by a form $ax_1^2 + bx_1x_2 + cy_1^2$, where either $(a, b, c) \equiv (0, 1, 0) \pmod 2$ or $(a, b, c) \equiv (1, 1, 1) \pmod 2$. In the first case we have

$$d = 4ac - b^2 \equiv -b^2 = -(2b_1 + 1)^2 = -4(b_1^2 + b_1) - 1 \equiv -1 \pmod 8.$$

Similarly, in the second case, $d \equiv 4 - b^2 \equiv 3 \pmod 8$.

Suppose that $(a, b, c) \equiv (0, 1, 0) \pmod 2$ and show that the congruence

$$a + 2bx + 4cx^2 \equiv 0 \pmod{2^t} \tag{A.2.32}$$

has a solution for every $t \in \mathbb{N}$. Since $a = 2a_1$ and $c = 2c_1$, the congruence can be written in the form $a_1 + bx + 4c_1x^2 \equiv 0 \pmod{2^{t-1}}$. Hence, since b is odd, it follows, that the congruence has a solution for $t = 3$. Further, if β is a solution of (A.2.32) for some $t \geq 3$, then, setting $\alpha = \beta + 2^{t-1}\gamma$, we see that

$$a + 2b\alpha + 4c\alpha^2 \equiv a + 2b\beta + 4c\beta^2 + 2^t b\gamma \equiv 0 \pmod{2^{t+1}}$$

provided that $\gamma \equiv a + 2b\beta + 4c\beta^2/2^t \pmod 2$, and so there is a solution modulo 2^{t+1}. We let α be a solution of (A.2.32) for $t = \delta$, and μ a solution of $4ay^2 + 2by + c \equiv 0 \pmod{2^\delta}$, which exists by the same reasons. We now set $\mathbf{e}_1 = \mathbf{e}'_1 + 2\alpha\mathbf{e}'_2$ and $\mathbf{e}_2 = \nu(2\mu\mathbf{e}'_1 + \mathbf{e}'_2)$, where ν is chosen so that

$$\mathbf{e}_1 \cdot \mathbf{e}_2 = \nu(4\mu a + b + 4\alpha\mu b + 2\alpha c) \equiv 1 \pmod{2^\delta}.$$

Then the elements $\mathbf{e}_1$, $\mathbf{e}_2$ form a basis for V and satisfy $f(\mathbf{e}_1) = f(\mathbf{e}_2) = 0$ and $\mathbf{e}_1 \cdot \mathbf{e}_2 = 1$. Hence, the quadratic form of the space (V, f) in the basis $\mathbf{e}_1$, $\mathbf{e}_2$ is $x_1x_2 = q_+^2$.

Suppose now that $(a, b, c) \equiv (1, 1, 1) \pmod 2$. Then the congruences

$$a + 2bx + 4cx^2 \equiv 1 \pmod{2^t} \quad \text{and} \quad 4ay^2 + 2by + c \equiv 1 \pmod{2^t}$$

are clearly solvable for $t = 2$. Hence, in complete analogy with the case of the congruence (A.2.32), it follows that the congruences are solvable for every t. Let α and β be respectively the solutions of the congruences for $t = \delta$. Then the vectors $\mathbf{u} = \mathbf{e}'_1 + 2\alpha\mathbf{e}'_2$ and $\mathbf{v} = 2\beta\mathbf{e}'_1 + \mathbf{e}'_2$ form a basis for V and satisfy the conditions $f(\mathbf{u}) = f(\mathbf{v}) = 1$. The assumptions imply that the number $b_1 = \mathbf{u} \cdot \mathbf{v}$ is odd, although, generally speaking, b_1 is not necessarily equal to 1. We shall seek a basis $\mathbf{e}_1, \mathbf{e}_2$ for V satisfying $f(\mathbf{e}_1) = f(\mathbf{e}_2) = 1$ and $\mathbf{e}_1 \cdot \mathbf{e}_2 = 1$ in the form $\mathbf{e}_1 = \mathbf{u}$, $\mathbf{e}_2 = \gamma\mathbf{u} + \xi\mathbf{v}$, with an odd ξ. Since we already have $f(\mathbf{e}_1) = f(\mathbf{u}) = 1$, it follows that the conditions for $\mathbf{e}_1, \mathbf{e}_2$ are equivalent to the relations

$$f(\mathbf{e}_2) = f(\gamma\mathbf{u}) + f(\xi\mathbf{v}) + (\gamma\mathbf{u}) \cdot (\xi\mathbf{v}) = \gamma^2 + \xi^2 + \gamma\xi b_1 = 1,$$

$$\mathbf{e}_1 \cdot \mathbf{e}_2 = \gamma\mathbf{u} \cdot \mathbf{u} + \xi\mathbf{v} \cdot \mathbf{u} = 2\gamma + \xi b_1 = 1.$$

Since b_1 is odd, the second relation implies that $\xi = b_1^{-1}(1 - 2\gamma)$. Substituting it in the first relation, we see that γ must satisfy $\gamma^2 - \gamma = l$ with $l =$

$(1 - b_1^{-2})(4b_1^2 - 1)^{-1} \equiv 0 \pmod 8$. The equation $x^2 - x = l$ has a solution in K, since the congruence $y^2 - y \equiv l \pmod{2^\delta}$ is equivalent with the congruence $(2y - 1)^2 \equiv 1 + 4l \pmod{2^{\delta+2}}$ solvable for every even l. The quadratic form of (V, f) in the basis $\mathbf{e}_1, \mathbf{e}_2$ is $x_1^2 + x_1x_2 + x_2^2 = q_-^2$.

Suppose the assertions (2) and (3) are already proved for spaces of dimension less than $2k$ with some $k > 1$, and let (V, f) be a non-singular space of dimension $2k$ over $K = \mathbb{Z}/2^\delta\mathbb{Z}$. Since the space (V, f) is also non-singular over the field $\mathbb{Z}/2\mathbb{Z}$, it follows from Theorem A.2.9 that there is a basis $e_1', \ldots, e_n'$ for V such that the quadratic form is congruent modulo 2 to one of the forms q_-^n, q_+^n. In particular, by $2k > 2$, we have $f(\mathbf{e}_1') = f(\mathbf{e}_2') \equiv 0 \pmod 2$ and $\mathbf{e}_1' \cdot \mathbf{e}_2' \equiv 1 \pmod 2$. The above considerations show that the space (V_1, f_1), where $V_1 = K\mathbf{e}_1' + K\mathbf{e}_2'$ and f_1 is the restriction of f to V_1, has a basis $\mathbf{e}_1, \mathbf{e}_2$ with $f_1(\mathbf{e}_1) = f_1(\mathbf{e}_2) = 0$ and $\mathbf{e}_1 \cdot \mathbf{e}_2 = 1$. Let us set $V_2 = \{\mathbf{u} \in V; \mathbf{u} \cdot \mathbf{e}_1 = \mathbf{u} \cdot \mathbf{e}_2 = 0\}$. It is easy to see that V_2 is a free K-module with the basis $\mathbf{e}_i' - (\mathbf{e}_2 \cdot \mathbf{e}_i')\mathbf{e}_1 - (\mathbf{e}_1 \cdot \mathbf{e}_i')\mathbf{e}_2$ $(i = 3, 4, \ldots, 2k)$, and (V, f) is the direct sum of (V_1, f_1) and (V_2, f_2), where f_2 is the restriction of f to V_2. By (A.2.5), we have $d(V_2, f_2) = d(V, f)/d(V_1, f_1) = -d(V, f)$. Applying the inductive hypothesis to (V_2, f_2), we get the assertions. □

By Lemma A.2.16, Proposition A.2.17 and Proposition A.2.18, we get

Theorem A.2.19. *Suppose that a number a of $\mathbb{N}$ is either odd or divisible by 8. Then two non-singular quadratic spaces over $K = \mathbb{Z}/a\mathbb{Z}$ are isomorphic if and only if they have equal dimensions and equal (modulo K^{*2}) determinants.*

A.2.6. The Genus of Quadratic Spaces over $\mathbb{Z}$

Each quadratic space over $\mathbb{Z}$ can be considered as a space over any commutative ring containing $\mathbb{Z}$ and any residue-class ring $\mathbb{Z}/a\mathbb{Z}$. We say that two quadratic spaces over $\mathbb{Z}$ *belong to the same genus* if the corresponding spaces over the field $\mathbb{R}$ and over each of the rings $\mathbb{Z}/a\mathbb{Z}$ with $a \in \mathbb{N}$ are respectively equivalent. *The genus of a space* (V, f) will be denoted by $G(V, f)$. If the type of (V, f) is $\{q\} = \{q\}_\mathbb{Z}$, then a space (V_1, f_1) of type $\{q_1\}$ belongs to $G(V, f)$ if and only if the forms q and q_1 are equivalent over $\mathbb{R}$ and each of the rings $\mathbb{Z}/a\mathbb{Z}$ respectively. In this case we say that q and q_1 *belong to the same genus* and denote by $G(q)$ *the genus* of q. If Q and Q_1 are the matrices of q and q_1, then the condition $q_1 \in G(q)$ means that Q and Q_1 have the same order, say n, and there are matrices $U \in GL_n(\mathbb{R})$ and $U_a \in GL_n(\mathbb{Z}/a\mathbb{Z})$ such that

$$Q[U] = Q_1 \quad \text{and} \quad Q[U_a] \equiv Q_1 \pmod a \quad \text{for all } a \in \mathbb{N}. \qquad \text{(A.2.33)}$$

In this case we say that Q and Q_1 *belong to the same genus and* denote by $G(Q)$ *the genus of* Q.

If matrices $Q, Q_1 \in \mathbb{E}_n$ belong to the same genus and Q is non-singular, then, by (A.2.33), the rational number $\det Q_1/\det Q$ is a squared unit in each of the rings $\mathbb{R}$ and $\mathbb{Z}/a\mathbb{Z}$, which implies

Proposition A.2.20. *Even symmetric matrices belonging to the same genus have equal determinants.*

We remind that *the divisor* of a non-zero matrix $Q \in \mathbb{E}_n$ is the greatest integer $\delta = \delta(Q)$ such that $\delta^{-1}Q \in \mathbb{E}_n$, and *the level* of a non-singular matrix $Q \in \mathbb{E}_n$ is the least natural number $q = q(Q)$ such that $qQ^{-1} \in \mathbb{E}_n$. It is clear that

$$\delta(Q)^n | \det Q \quad \text{and} \quad \delta(Q) | q(Q). \tag{A.2.34}$$

If Q and Q_1 are two non-zero (resp. non-singular) matrices of $\mathbb{E}_n$ belonging to the same genus, then, for each $a \in \mathbb{N}$ the matrix $\delta(Q)^{-1}Q_1 \equiv (\delta(Q)^{-1} \cdot Q)[U_a] \pmod a$ (resp. $q(Q)Q_1^{-1} \equiv (q(Q)Q^{-1})[U_a^*] \pmod a$) is congruent modulo a to a matrix of $\mathbb{E}_n$ and therefore it belongs to $\mathbb{E}_n$. Hence $\delta(Q)|\delta(Q_1)$ (resp. $q(Q_1)|q(Q)$) Similarly, $\delta(Q_1)|\delta(Q)$ (resp. $q(Q)|q(Q_1)$). We have proved

Proposition A.2.21. *Non-zero (resp. non-singular) matrices of the same genus have equal divisors (resp. levels).*

By Proposition A.2.20 and Theorem A.2.12, we get

Theorem A.2.22. *The genus of any matrix of $\mathbb{A}_n^+$ with $n \in \mathbb{N}$ is a finite union of classes modulo the integral equivalence.*

The reader will easily translate Proposition A.2.20 and Theorem A.2.22 into the languange of quadratic spaces.

The following lemma gives a convenient sufficient condition for two matrices to be in the same genus.

Lemma A.2.23. *Let Q be a non-singular matrix of $\mathbb{E}_n$. Suppose that a matrix U of* $\mathsf{M}_n(\mathbb{Q})$ *with* $\det U = \pm 1$ *satisfies the following conditions:*

(1) *The denominators of all of the elementary divisors of U are prime to the level q of Q;*

(2) *The matrix $Q_1 = Q[U]$ belongs to $\mathbb{E}_n$.*

Then the matrices Q_1 and Q belong to the same genus.

Proof. Let (V, f) (resp. (V_1, f_1)) be a quadratic space over $\mathbb{Z}$ whose quadratic form in some basis has the matrix Q (resp. Q_1). Since the first condition in (A.2.33) is satisfied by the assumption, we only have to check that the spaces (V, f) and (V_1, f_1) are isomorphic over each of the rings $\mathbb{Z}/a\mathbb{Z}$. Let $a \in \mathbb{N}$. Let us write the number $8a$ in the form $8a = a_1 \cdot a_2$, where a_1 is prime to q and each of the prime divisors of a_2 divides q. By Lemma 5.1.5, the number a_1 is prime to $\det Q = \det Q_1$. It is also clear that a_1 is either odd or divisible by 8. Hence the spaces (V, f) and (V_1, f_1) considered over $\mathbb{Z}/a_1\mathbb{Z}$ satisfy the assumptions of Theorem A.2.19 and therefore are isomorphic over $\mathbb{Z}/a_1\mathbb{Z}$. According to the condition (1) and the choice of a_2, the matrix U is a_2-integral, and so the spaces (V, f) and (V_1, f_1) are isomorphic over $\mathbb{Z}/a_2\mathbb{Z}$. Then, by Lemma A.2.16, the spaces are isomorphic over $\mathbb{Z}/a_1a_2\mathbb{Z} = \mathbb{Z}/8a\mathbb{Z}$ and therefore over $\mathbb{Z}/a\mathbb{Z}$. □

Appendix 3. Modules in Quadratic Fields and Binary Quadratic Forms

Integral binary quadratic forms of a non-square discriminant Δ are naturally related to modules in the quadratic field $\mathbb{Q}(\Delta^{1/2})$. This allows us to use the natural multiplication of the modules in order to define a multiplication of the forms (the Gauss composition of quadratic forms) and a multiplication of forms by modules.

Although we give all necessary definitions and practically all of the proofs, some knowledge of the algebraic number theory would be helpful for the reader. We may recommed, for instance, §§ 2, 6, 7 of Chapter II of the book [15].

A.3.1. Modules of Algebraic Number Fields

Here we shall remind the simplest definitions and facts on modules in finite extensions of the field $\mathbb{Q}$. Proofs of all of the assertions of this subsection easily follow from definitions and are left to the reader as an exercise.

Let F be a field and $\mathsf{F}_0 \subset \mathsf{F}$ a subfield of F. The field F can be considered as a linear space over F_0. If the space has a finite dimension, say n, then F is called *a finite extension of* F_0 *of degree* $[\mathsf{F}:\mathsf{F}_0] = n$. Let $[\mathsf{F}:\mathsf{F}_0] = n$, and let $\omega_1, \dots, \omega_n$ be a basis for F over F_0. Then, for each $\alpha \in \mathsf{F}$ and $i = 1, \dots, n$ one can write $\alpha\omega_i = \sum_{1\le j\le n} a_{ij}\omega_j$ with $a_{ij} \in \mathsf{F}_0$. The matrix (a_{ij}) is called the matrix of α with respect to $\omega_1, \dots, \omega_n$; its characteristic polynomial

$$\mathrm{ch}(v, \alpha) = \det((a_{ij}) - vE_n) = (-v)^n + s(\alpha)(-v)^{n-1} + \cdots + N(\alpha)$$

is independent of the choice of basis $\{\omega_i\}$. The polynomial $\mathrm{ch}(v, \alpha)$ is called *the characteristic polynomial of* α (over F_0), its coefficients

$$s(\alpha) = \sum_{i=1}^{n} a_{ii} \quad \text{and} \quad N(\alpha) = \det(a_{ij})$$

are called respectively *the trace* and *the norm of* α (from F to F_0). It is clear that

$$s(\alpha + \beta) = s(\alpha) + s(\beta) \quad \text{and} \quad N(\alpha\beta) = N(\alpha)N(\beta) \quad (\alpha, \mathrm{B} \in \mathsf{F}).$$

A finite extension of the field $\mathbb{Q}$ is called *an algebraic number field.* A number of an algebraic number field is said to be *integral* if all of the coefficients of its characteristic polynomial (over $\mathbb{Q}$) are rational integers.

Let F be an algebraic number field. Any $\mathbb{Z}$-submodule $\mathsf{M} \subset \mathsf{F}$ with a finite number of generators is called *a module in* F. Each module M in F is a free abelian group with a finite number of generators and therefore has a basis (over $\mathbb{Z}$), i.e. a finite set $\omega_1, \ldots, \omega_m \in$ M such that every element α of M is uniquely of the form $\alpha = a_1\omega_1 + \cdots + a_m\omega_m$ with $a_i \in \mathbb{Z}$. The number m is independent of the choice of basis and is called *the rank of* M. The rank of any module in F is not greater than the degree of the extension $n = [\mathsf{F} : \mathbb{Q}]$. Modules of rank n are called *complete*. Complete modules M in F can be characterized by the condition $\mathbb{Q}\mathsf{M} = \mathsf{F}$.

If M_1 and M_2 are two complete modules with the bases $\omega_1, \ldots, \omega_n$ and $\eta_1, \ldots, \eta_n$ respectively, then the set of all integral linear combinations of the elements $\omega_i\eta_j$ $(1 \le i,j \le n)$ is again a module, which is called *the product of* M_1 *and* M_2 and is denoted by $\mathsf{M}_1\mathsf{M}_2$. The product of complete modules is a complete module.

A complete module in an algebraic number field F which is a ring with the unity is called *an order in* F. Matrices of elements of an order with respect to a fixed basis of the order are clearly integral. Hence, each order in F is contained in the set $\mathsf{O} = \mathsf{O}_F$ of all integral numbers of F.

Let M be a module in F, then the set

$$\mathsf{O}(\mathsf{M}) = \{\alpha \in \mathsf{F}; \alpha\mathsf{M} \subset \mathsf{M}\}$$

is clearly a ring. The ring O(M) is called *the ring of multipliers of* M. The ring of multipliers of a complete module is an order in F. Two modules M_1, M_2 in F are said to be *equivalent* if $\mathsf{M}_2 = \alpha\mathsf{M}_1$ with a non-zero α of F. Equivalent modules have equal rings of multipliers. For each complete module, there is an equivalent module contained in its ring of multipliers.

Let M be a complete module with the ring of multipliers O(M). Take a basis $\alpha_1, \ldots, \alpha_n$ for O(M) and a basis $\omega_1, \ldots, \omega_n$ for M. Then the number

$$N(\mathsf{M}) = |\det(a_{ij})|, \quad \text{where} \quad \omega_i = \sum_{j=1}^{n} a_{ij}\alpha_j \quad \text{with } a_{ij} \in \mathbb{Q},$$

does not depend on the choice of the bases and is called *the norm of* M. If α is a non-zero number of F, then

$$N(\alpha\mathsf{M}) = |N(\alpha)|\, N(\mathsf{M}), \tag{A.3.1}$$

where $N(\alpha)$ is the norm of α from F to $\mathbb{Q}$. If M is contained in O(M), then $N(\mathsf{M})$ is equal to the index of M in O(M).

Let M be a complete module with a basis $\omega_1, \ldots, \omega_n$. The number $\Delta(\mathsf{M}) = \det(s(\omega_i\omega_j))$, where s means the trace from F to $\mathbb{Q}$, is independent of the choice of the basis and is called the *discriminant of* M.

A.3.2. Modules in Quadratic Fields and Prime Numbers

Let us now assume that F is an extension of $\mathbb{Q}$ of degree 2. It is easy to see that any such F is of the form $\mathsf{F} = \mathbb{Q} + \mathbb{Q}\Delta_0^{1/2} = \mathbb{Q}(\Delta_0^{1/2})$, where $\Delta_0 \neq 0, 1$ is a square free rational integer. Each α of F is uniquely of the shape $\alpha = a + b\Delta_0^{1/2}$ with a, b in $\mathbb{Q}$. The number $\alpha' = a - b\Delta_0^{1/2}$ is called *conjugate to* α *(over* $\mathbb{Q}$*)*. Computing the matrix of $\alpha = a + b\Delta_0^{1/2}$ with respect to the basis $1, \Delta_0^{1/2}$, we find the characteristic polynomial of α:

$$\mathrm{ch}(v, \alpha) = \det\left(\begin{pmatrix} a & b \\ b\Delta_0 & a \end{pmatrix} - \begin{pmatrix} v & 0 \\ 0 & v \end{pmatrix}\right) = v^2 - 2av + (a^2 - \Delta_0 b^2)$$
$$= (v - \alpha)(v - \alpha').$$

In particular, the trace and the norm of α are given by $s(\alpha) = \alpha + \alpha' = 2a$ and $N(\alpha) = \alpha\alpha' = a^2 - \Delta_0 b^2$, and the α is an integral number if $s(\alpha)$ and $N(\alpha)$ are rational integers.

Theorem A.3.1. *The set* O *of integral numbers of a field* $\mathsf{F} = \mathbb{Q}(\Delta_0^{1/2})$, *where* $\Delta_0 \neq 0$, 1 *is a square free rational integer, is an order in* F. *The numbers* $1, \omega$, *where*

$$\omega = \begin{cases} (1 + \Delta_0^{1/2})/2 & \text{if } \Delta_0 \equiv 1 \pmod 4 \\ \Delta_0^{1/2} & \text{if } \Delta_0 \equiv 2, 3 \pmod 4, \end{cases} \tag{A.3.2}$$

form a basis for O. *The discriminant* $\Delta = \Delta(\mathsf{O})$ *of* O *is equal to* Δ_0, *if* $\Delta_0 \equiv 1 \pmod 4$, *and* $4\Delta_0$, *if* $\Delta_0 \equiv 2$ *or* $3 \pmod 4$.

An arbitrary order in F *has the form*

$$\mathsf{O}_l = \mathbb{Z} + \mathbb{Z}l\omega, \tag{A.3.3}$$

where $l \in \mathbb{N}$ *is the index of* O_l *in* O. *The discriminant of* O_l *is equal to* Δl^2.

Proof. Let $\alpha = a + b\Delta_0^{1/2} \in \mathsf{F}$. By definitions, the condition $\alpha \in \mathsf{O}$ is equivalent to the conditions $2a \in \mathbb{Z}$, $a^2 - \Delta_0 b^2 \in \mathbb{Z}$. Let us set $a_1 = 2a$ and $b_1 = 2b$. Then the conditions take the shape $a_1 \in \mathbb{Z}, a_1^2 - \Delta_0 b_1^2 \in 4\mathbb{Z}$. Since Δ_0 is square free, it follows that $b_1 \in \mathbb{Z}$. Hence the condition $\alpha \in O$ can be written in the form a_1, $b_1 \in \mathbb{Z}$, $a_1^2 - \Delta_0 b_1^2 \equiv 0 \pmod 4$. If $\Delta_0 \equiv 1 \pmod 4$, the last congruence means that $a_1 \equiv b_1 \pmod 2$, but if $\Delta_0 = 2$ or $3 \pmod 4$, then it is equivalent to the congruences $a_1 \equiv b_1 \equiv 0 \pmod 2$. Hence, in the first case, we have

$$\mathsf{O} = \left\{\frac{a_1 + b_1\Delta_0^{1/2}}{2}; a, b_1 \in \mathbb{Z}, a_1 \equiv b_1 \pmod 2\right\}$$
$$= \left\{\frac{a_1 - b_1}{2} + b_1\frac{1 + \Delta_0^{1/2}}{2}; \frac{a_1 - b_1}{2} \in \mathbb{Z}, b_1 \in \mathbb{Z}\right\}$$

and, in the second case,

$$\mathsf{O} = \{a + b\Delta_0^{1/2}; a, b \in \mathbb{Z}\}.$$

Thus, in each of the cases we have $\mathsf{O} = \mathbb{Z} + \mathbb{Z}\omega$, and O is a complete module with the basis $1, \omega$. Since $\omega^2 \in \mathsf{O}$, it follows that O is a ring. The formulae for the discriminant $\Delta = \Delta(\mathsf{O})$ follow directly from definitions.

Let O' be an order in F. Then $\mathsf{O}' \subset \mathsf{O}$ and the index $l = [\mathsf{O} : \mathsf{O}']$ is clearly finite. Since $1 \in \mathsf{O}'$ and $l\omega \in \mathsf{O}'$, it follows that $\mathsf{O}' \subset \mathsf{O}_l = \mathbb{Z} + \mathbb{Z}l\omega$. Since $[\mathsf{O} : \mathsf{O}_l] = l$, we have $\mathsf{O}' = \mathsf{O}_l$. On the other hand, by the obvious inclusion $(l\omega)^2 \in \mathsf{O}_l$, we conclude that the module O_l is a ring for any $l \in \mathbb{N}$. The formula for $\Delta(\mathsf{O}_l)$ follows directly from definitions. □

The number $\Delta = \Delta(\mathsf{O})$ is called *the discriminant of the field F*. By Theorem A.3.1, we can write $\mathsf{F} = \mathbb{Q}(\Delta^{1/2})$.

Let us now turn to modules in $\mathsf{F} = \mathbb{Q}(\Delta_0^{1/2})$. For $\alpha_1, \ldots, \alpha_n \in \mathsf{F}$ we shall denote by $\{\alpha_1, \ldots, \alpha_n\}$ the set of all linear combinations of $\alpha_1, \ldots, \alpha_n$ with coefficients in $\mathbb{Z}$. In the notation, every complete module in F is of the shape $\{\alpha, \beta\}$, where $\alpha \neq 0$ and $\gamma = \beta/\alpha \notin \mathbb{Q}$. Since the module $\{\alpha, \beta\}$ is equivalent to the module $\{1, \beta/\alpha\}$, we first consider modules of the form $\{1, \gamma\}$ with $\gamma \notin \mathbb{Q}$. Every γ of F is a root of a polynomial $av^2 + bv + c$ of degree 2 with coefficients in $\mathbb{Z}$ (one can take the characteristic polynomial of γ and multiply it by the common denominator of the coefficients). One can assume that $a > 0$ and g.c.d.$(a, b, c) = 1$. It is clear that the polynomial $av^2 + bv + c$ with these properties is uniquely determined by γ, provided that $\gamma \notin \mathbb{Q}$. In this case the polynomial will be denoted by $\overline{\mathrm{ch}}(v, \gamma)$.

Lemma A.3.2. *Let* $\gamma \in \mathsf{F}$, $\gamma \notin \mathbb{Q}$, *and let* $\overline{\mathrm{ch}}(v, \gamma) = av^2 + bv + c$. *Set* $\mathsf{M} = \{1, \gamma\}$. *Then*

$$\mathsf{O}(\mathsf{M}) = \{1, a\gamma\}, \qquad \Delta(\mathsf{O}(\mathsf{M})) = b^2 - 4ac, \qquad \text{and} \qquad N(\mathsf{M}) = 1/a.$$

Proof. Let $\alpha = x + y\gamma$ with $x, y \in \mathbb{Q}$ be a number of F. Since the inclusion $\alpha\mathsf{M} \subset \mathsf{M}$ is equivalent to the conditions $\alpha \cdot 1 = x + y\gamma \in \mathsf{M}$, $\alpha \cdot \gamma = -cy/a + (x - by/a)\gamma \in \mathsf{M}$, it follows that the α is contained in $\mathsf{O}(\mathsf{M})$, if and only if the numbers x, y, cy/a, and by/a are contained in $\mathbb{Z}$. Since g.c.d.$(a, b, c) = 1$ this is equivalent to the inclusions $x \in \mathbb{Z}$, $y \in a\mathbb{Z}$, whence $\mathsf{O}(\mathsf{M}) = \{1, a\gamma\}$. The formula for $\Delta(\mathsf{O}(\mathsf{M}))$ follows from definitions. If we use the basis 1, $a\gamma$ for $\mathsf{O}(\mathsf{M})$ and the basis 1, γ for M, then we get $N(\mathsf{M}) = \det\begin{pmatrix} 1 & 0 \\ 0 & 1/a \end{pmatrix} = 1/a$. □

For a module M we set $\mathsf{M}' = \{\alpha'; \alpha \in \mathsf{M}\}$. M' is clearly a module. The module M' is called *conjugate to* M. If M is a complete module, then M' is clearly a complete module too. The number ω of Theorem A.3.1 satisfies respectively the relations $\omega' = -\omega + 1$ and $\omega' = -\omega$, which implies that any order O in F satisfies $\mathsf{O}' = \mathsf{O}$. Hence, for every module M, we have $\mathsf{O}(\mathsf{M}') = \mathsf{O}(\mathsf{M})' = \mathsf{O}(\mathsf{M})$, i.e. the rings of multipliers of conjugate modules are equal.

Proposition A.3.3. *For every complete module* M *in* F *the following formula holds*

$$\mathsf{M}\mathsf{M}' = N(\mathsf{M})\mathsf{O}(\mathsf{M}). \tag{A.3.4}$$

Proof. We first assume that M is of the form $\{1, \gamma\}$. Let $\overline{\mathrm{ch}}(v, \gamma) = av^2 + bv + c$. Then

$$\mathsf{MM}' = \{1,\gamma\}\cdot\{1,\gamma'\} = \{1,\gamma,\gamma',\gamma\gamma'\}$$
$$= \left\{1,\gamma,-\gamma-\frac{b}{a},\frac{c}{a}\right\} = \{1,\gamma,-b/a,c/a\} = \frac{1}{a}\{a,b,c,a\gamma\}$$

Since g.c.d.$(a,b,c) = 1$, it follows that $a\mathbb{Z}+b\mathbb{Z}+c\mathbb{Z}=\mathbb{Z}$. Hence, by Lemma A.3.2, we have

$$\mathsf{MM}' = a^{-1}\{1,a\gamma\} = N(\mathsf{M})\mathsf{O}(\mathsf{M}).$$

For the general case, we can write $\mathsf{M} = \alpha\mathsf{M}_1$ with $\mathsf{M}_1 = \{1,\gamma\}$, and so, by the above and (A.3.1), we get

$$\mathsf{MM}' = \alpha\alpha'\mathsf{M}_1\mathsf{M}_1' = N(\alpha)N(\mathsf{M}_1)\mathsf{O}(\mathsf{M}_1) = N(\mathsf{M})\mathsf{O}(\mathsf{M}).$$ □

Proposition A.3.4. *Let* M *and* M_1 *be complete modules in* F. *Then*

$$N(\mathsf{M}\cdot\mathsf{M}_1) = N(\mathsf{M})N(\mathsf{M}_1). \tag{A.3.5}$$

If $\mathsf{O}(\mathsf{M}) = \mathsf{O}_l$ *and* $\mathsf{O}(\mathsf{M}_1) = \mathsf{O}_{l_1}$, *then*

$$\mathsf{O}(\mathsf{MM}_1) = \mathsf{O}_t, \tag{A.3.6}$$

where $t =$ g.c.d.(l,l_1)

Proof. By Proposition A.3.3, we have

$$(\mathsf{M}\cdot\mathsf{M}_1)\cdot(\mathsf{M}\cdot\mathsf{M}_1)' = N(\mathsf{MM}_1)\mathsf{O}(\mathsf{MM}_1).$$

On the other hand,

$$(\mathsf{MM}_1)\cdot(\mathsf{MM}_1)' = \mathsf{MM}'\cdot\mathsf{M}_1\mathsf{M}_1' = N(\mathsf{M})\cdot N(\mathsf{M}_1)\mathsf{O}_l\cdot\mathsf{O}_{l_1},$$

whence we get the relation

$$N(\mathsf{MM}_1)\mathsf{O}(\mathsf{MM}_1) = N(\mathsf{M})N(\mathsf{M}_1)\mathsf{O}_l\cdot\mathsf{O}_{l_1}.$$

By comparing the least positive numbers contained in the two sides of the relation, we get (A.3.5), whence

$$\mathsf{O}(\mathsf{MM}_1) = \mathsf{O}_l\mathsf{O}_{l_1} = \{1,l\omega\}\{1,l_1\omega\}$$
$$= \{1,l\omega,l_1\omega,ll_1\omega^2\} = \{1,t\omega\} = \mathsf{O}_t,$$

for $\mathbb{Z}l\omega + \mathbb{Z}l_1\omega = \mathbb{Z}t\omega$ and $ll_1\omega^2 \in \mathbb{Z}+\mathbb{Z}t\omega$, by $\omega^2 = \omega + (\Delta_0 - 1)/4$ or $\omega^2 = \Delta_0$. □

Let O_l be a fixed order in F. It follows from (A.3.6) and (A.3.4) that all complete modules in F with the ring of multipliers O_l form an abelian group with respect to the multiplication of modules, where the module O_l serves as the unity, and the inverse of a module M is $N(\mathsf{M})^{-1}\mathsf{M}'$. The modules equivalent to O_l form clearly a subgroup of the group of all moduels with the ring of multipliers O_l. The factor group of the group by the subgroup is called *the class group of* O_l and is denoted by

$$\mathsf{H}(O_l) = \mathsf{H}(\Delta l^2),$$

where $\Delta l^2 = \Delta(\mathsf{O}_l)$ is the discriminant of O_l. If $l_1 | l$, then, by (A.3.6), we conclude that the mapping $\mathsf{M} \to \mathsf{O}_{l_1}\mathsf{M}$ induces a homomorphism of the class groups

$$\xi(l, l_1)\colon \mathsf{H}(\mathsf{O}_l) \to \mathsf{H}(\mathsf{O}_{l_1}). \tag{A.3.7}$$

We now consider the relations between modules in quadratic fields and prime numbers.

Proposition A.3.5. *Let $\mathsf{F} = \mathbb{Q}(\Delta^{1/2})$ be a quadratic field with the discriminant Δ, and let O_l be the order in F of discriminant $\Delta(\mathsf{O}_l) = \Delta l^2$ with $l \in \mathbb{N}$. Let p be a prime number prime to l. Then a complete module M in F satisfying the conditions*

$$\mathsf{O}(\mathsf{M}) = \mathsf{O}_l, \quad \mathsf{M} \subset \mathsf{O}_l, \quad \text{and} \quad N(\mathsf{M}) = p \tag{A.3.8}$$

exists if and only if the congruence $x^2 \equiv \Delta \pmod{4p}$ is solvable. In this case, if $p \nmid \Delta$, then there are exactly two modules $\mathfrak{p}$, $\mathfrak{p}_1$ satisfying (A.3.8), *and $\mathfrak{p}_1 = \mathfrak{p}'$. But if the congruence has a solution and $p | \Delta$ then there is only one module $\mathfrak{p}$ satisfying* (A.3.8), *and $\mathfrak{p}' = \mathfrak{p}$. Finally, if the congruence has no solutions, then F contains the only complete module $\mathsf{M} = \mathfrak{p}$ satisfying*

$$\mathsf{O}(\mathsf{M}) = \mathsf{O}_l, \quad \mathsf{M} \subset \mathsf{O}_l, \quad \text{and} \quad N(\mathsf{M}) = p^2,$$

and $\mathfrak{p} = p\mathsf{O}_l$. All of the modules $\mathsf{M} = \mathfrak{p}$, $\mathfrak{p}_1$ are prime ideals of the ring O_l.

Proof. Suppose M satisfies (A.3.8). Since the index of M in O_l is p, it follows that the least positive integer contained in M is p. Hence, M has a basis of the form $p, p\gamma$, $\mathsf{M} = \{p, p\gamma\}$, where $\gamma \in F$, $\gamma \notin \mathbb{Q}$. Let $\overline{\mathrm{ch}}(v, \gamma) = av^2 + bv + c$. Then, by Lemma A.3.2, we have $p = N(\mathsf{M}) = N(p)N\{1, \gamma\} = \dfrac{p^2}{a}$, whence $a = p$. On the other hand, $b^2 - 4ac = \Delta(\mathsf{O}(\mathsf{M})) = \Delta l^2$. Thus, $b^2 \equiv \Delta l^2 \pmod{4p}$. Since $\Delta = 0$ or $1 \pmod 4$ and $p \nmid l$, it follows that the congruence $x^2 \equiv \Delta \pmod{4p}$ has a solution. Conversely, if the congruence has a solution, then there are $b, c \in \mathbb{Z}$ such that $b^2 - 4pc = \Delta$. Since Δ is a discriminant, it follows that g.c.d.$(p, b, c) = 1$, and so g.c.d.$(p, bl, cl^2) = 1$. Let γ be a root of the equation $pv^2 + blv + cl^2 = 0$. Then Lemma A.3.2 implies that the module $\mathsf{M} = \{p, p\gamma\}$ satisfies (A.3.8).

Let $\mathsf{M}_i = \{p, p\gamma_i\}$ with $i = 1, 2$ be two modules satisfying (A.3.8). As we have seen, each of the γ_i satisfies a relation of the form $p\gamma_i^2 + b_i\gamma_i + c_i = 0$, where g.c.d.$(p, b_i, c_i) = 1$ and $b_i^2 - 4pc_i = \Delta l^2$. Since $b_i^2 \equiv \Delta l^2 \pmod{4p}$, it follows that $b_1 \equiv b_2 \pmod{2p}$ or $b_1 \equiv b_2 \pmod{2p}$. In the first case we have $b_2 = b_1 + 2tp$, and

$$\mathsf{M}_2 = \{p, (-b_2 \pm l\Delta^{1/2})/2\} = \{p, (-b_1 \pm l\Delta^{1/2})/2 - tp\} = \{p, (-b_1 \pm l\Delta^{1/2})/2\}.$$

The last module is equal to M_1 or M_1'. Similarly, in the second case, $b_2 = -b_1 + 2tp$, and

$$\mathsf{M}_2 = \{p, (b_1 \pm l\Delta^{1/2})/2 - tp\} = \{p, (-b_1 \mp l\Delta^{1/2})/2\} = \mathsf{M}_1 \text{ or } \mathsf{M}_1'$$

If $p|\Delta$, then $b_1 \equiv -b_1 \pmod{2p}$, whence $\mathsf{M}_1' = \mathsf{M}_1$.

Let us now assume that the congruence $x^2 \equiv \Delta \pmod{4p}$ has no solutions, and let M be a complete module satisfying $\mathsf{O}(\mathsf{M}) = \mathsf{O}_l$, $\mathsf{M} \subset \mathsf{O}_l$, and $N(\mathsf{M}) = p^2$. Then the least positive integer contained in M is equal to p^δ with $\delta = 1$ or 2, and one may assume that $\mathsf{M} = \{p^\delta, p^\delta\gamma\}$. Let $\overline{\mathrm{ch}}(v, \gamma) = av^2 + bv + c$. By Lemma A.3.2, we have $p^2 = N(\mathsf{M}) = N(p^\delta)N\{1, \delta\} = p^{2\delta}/a$, $\mathsf{O}(\mathsf{M}) = \{1, a\gamma\}$, and $\Delta(\mathsf{O}(\mathsf{M})) = b^2 - 4ac$. If $\delta = 2$, then $a = p^2$ and $\Delta l^2 = b^2 - 4ac \equiv b^2 \pmod{4p^2}$, which, by $p \nmid l$, contradicts the assumption. Thus, $\delta = 1$, $a = 1$, $\mathsf{O}(\mathsf{M}) = \mathsf{O}_l = \{1, \gamma\}$, and $\mathsf{M} = p\{1, \gamma\} = p\mathsf{O}_l$. It is also clear that $p\mathsf{O}_l$ satisfies the conditions.

If $N(\mathsf{M}) = p$, then the last assertion is clear. Let $N(\mathsf{M}) = p^2$. Suppose that $\alpha \in \mathsf{O}(\mathsf{M}) = \mathsf{O}_l$, but $\alpha \notin M$. Then the module $\mathsf{M}_1 = \alpha\mathsf{O}_l + \mathsf{M}$ is clearly complete, and $N(\mathsf{M}_1) = p$ or 1, by $\mathsf{M}_1 \neq \mathsf{M}$. As we have seen, in this case the equality $N(\mathsf{M}_1) = p$ is impossible, whence $N(\mathsf{M}_1) = 1$, $\mathsf{M}_1 = \mathsf{O}_l$, and so α is invertible modulo M. Thus, the quotient ring O_l/M is a field, and the ideal M is prime. □

A.3.3. Modules in Imaginary Quadratic Fields and Quadratic Forms

Let

$$\mathsf{q}(x, y) = ax^2 + bxy + cy^2$$

be a binary quadratic form. We remind that the matrix $Q = \begin{pmatrix} 2a & b \\ b & 2c \end{pmatrix}$ is called the matrix of q, the number $\Delta(\mathsf{q}) = -\det Q = b^2 - 4ac$ is the discriminant of q, and the form is called *integral* if $a, b, c \in \mathbb{Z}$. The number $\delta = \delta(\mathsf{q}) = \text{g.c.d.}(a, b, c)$ is called the divisor of q. The forms of divisor 1 are called *primitive*. Two binary forms q and q_1 are said to be *properly equivalent* if the corresponding matrices Q and Q_1 satisfy

$$Q_1 = Q[U] \quad \text{with } U \in \mathsf{SL}_2(\mathbb{Z}). \tag{A.3.9}$$

If q and q_1 are properly equivalent, then clearly $\Delta(q) = \Delta(q_1)$ and $\delta(q) = \delta(q_1)$. Hence, the set of all integral forms with fixed discriminant and divisor is divided into the classes $\{\cdots\}_+$ of properly equivalent forms. If q_1 is equivalent to q over $\mathbb{Z}$ in the sense of § 1.1.1, then q_1 is properly equivalent to one of the forms $\mathsf{q}(x, y)$ or $\mathsf{q}'(x, y) = \mathsf{q}(x, -y)$. Hence, the each class of forms equivalent over $\mathbb{Z}$ consists of at most two classes of properly equivalent forms. Thus, by Theorem 2.1.12, we obtain

Proposition A.3.6. *The number of classes of properly equivalent integral positive definite binary quadratic forms of a fixed discriminant is finite.*

We shall limit ourselves to the consideration of integral primitive positive definite binary quadratic forms $\mathsf{q}(x, y) = ax^2 + bxy + cy^2$. For such a form, $a > 0$ and $\Delta(\mathsf{q}) = b^2 - 4ac < 0$. Hence, the field $\mathsf{F} = \mathsf{F}_\mathsf{q} = \mathbb{Q}(\Delta(\mathsf{q})^{1/2})$ is an imaginary quadratic field. If Δ is the discriminant of F, then clearly $\Delta(q) = \Delta l^2$ with

$l \in \mathbb{N}$. We let $\gamma = \gamma_{\mathsf{q}}$ be the root of $av^2 + bv + c = 0$ belonging to the upper half-plane $\mathbb{H}_1$:

$$\gamma = \gamma_{\mathsf{q}} = (-b + \Delta(\mathsf{q})^{1/2})/2a = (-b + l\Delta^{1/2})/2a. \tag{A.3.10}$$

Then the form q can clearly be written in the form

$$\mathsf{q}(x, y) = a(x - \gamma y)(x - \bar{\gamma} y), \tag{A.3.11}$$

where $\bar{\gamma} = \gamma'$ is the complex conjugate to γ. Let us assign to the form q the module $\mathsf{M} = \mathsf{M}(\mathsf{q})$ in F of the shape $\{a, a\gamma_{\mathsf{q}}\}$:

$$\mathsf{q} \to \mathsf{M}(\mathsf{q}) = \{a, a\gamma_{\mathsf{q}}\} = \{a, (-b + l\Delta^{1/2})/2\}. \tag{A.3.12}$$

It is clear that $\mathsf{M}(\mathsf{q})$ is a complete module. By Lemma A.3.2 the ring of multipliers of $\mathsf{M}(\mathsf{q})$ is the subring O_l of index l in the ring of all integral numbers of F.

Lemma A.3.7. *Integral primitive positive definite binary quadratic forms* q *and* q_1 *are properly equivalent if and only if the modules* $\mathsf{M}(\mathsf{q})$ *and* $\mathsf{M}(\mathsf{q}_1)$ *belong to the same quadratic field and are equivalent.*

Proof. If q_1 is properly equivalent to q, i.e.

$$\mathsf{q}_1(x, y) = \mathsf{q}(sx + ty, ux + vy), \qquad \text{where} \quad \begin{pmatrix} s & t \\ u & v \end{pmatrix} \in \mathsf{SL}_2(\mathbb{Z}),$$

then, by (A.3.11), we have

$$\begin{aligned} \mathsf{q}_1(x, y) &= a(sx + ty - \gamma(ux + vy))(sx + ty - \bar{\gamma}(ux + vy)) \\ &= aN(s - \gamma u)\left(x - \frac{\gamma v - t}{-\gamma u + s} y\right)\left(x - \frac{\bar{\gamma} v - t}{-\bar{\gamma} u + s} y\right). \end{aligned}$$

This factorization along with the definition of numbers $\gamma = \gamma_{\mathsf{q}}$ and $\gamma_1 = \gamma_{\mathsf{q}_1}$ implies the relation $\gamma_1 = (v\gamma - t)(-u\gamma + s)^{-1}$ (note that $\gamma_1 \in \mathbb{H}_1$, since $vs - ut = 1$). The module $\mathsf{M}(\mathsf{q}_1)$ is equivalent to the module

$$\begin{aligned} \{1, \gamma_1\} &= \left\{1, \frac{v\gamma - t}{-u\gamma + s}\right\} = (-u\gamma + s)\{-u\gamma + s, v\gamma - t\} \\ &= (-u\gamma + s)\{1, \gamma\}, \end{aligned}$$

because the inclusion $\begin{pmatrix} -u & s \\ v & -t \end{pmatrix} \in \mathsf{GL}_2(\mathbb{Z})$ implies that $\{1, \gamma\} = \{-u\gamma + s, v\gamma + t\}$. The last module is equivalent to $\mathsf{M}(\mathsf{q})$.

Conversely, if the modules $\mathsf{M}(\mathsf{q})$ and $\mathsf{M}(\mathsf{q}_1)$ are equivalent, then the same is true for the modules $\{1, \gamma\}$ and $\{1, \gamma_1\}$ with $\gamma = \gamma_{\mathsf{q}}$ and $\gamma_1 = \gamma_{\mathsf{q}_1}$. Thus, $\alpha\{1, \gamma_1\} = \{1, \gamma\}$, whence $\alpha = s + t\gamma$ and $\alpha\gamma_1 = u + v\gamma$. Since α, $\alpha\gamma_1$ and $1, \gamma$ are two bases of the same module, it follows that $\begin{pmatrix} s & t \\ u & v \end{pmatrix} \in \mathsf{GL}_2(\mathbb{Z})$. Expressing γ_1, we find that $\gamma_1 = (v\gamma + u)(t\gamma + s)^{-1}$, whence

$$\begin{aligned}\mathsf{q}_1(x,y) &= a_1(x-\gamma_1 y)(x-\bar{\gamma}_1 y)\\ &= a_1\left(x-\frac{v\gamma+u}{t\gamma+s}y\right)\left(x-\frac{v\bar{\gamma}+u}{t\bar{\gamma}+s}y\right)\\ &= \frac{a_1}{aN(t\gamma+s)}a((sx-uy)-\gamma(-tx+vy))((sx-uy)-\bar{\gamma}(-tx+vy))\\ &= \frac{a_1}{aN(t\gamma+s)}\mathsf{q}(sx-uy,-tx+vy),\end{aligned}$$

where a_1 and a are the coefficients of x^2 in q_1 and q respectively. The form $\mathsf{q}'(x,y) = \mathsf{q}(sx-uy,-tx+vy)$ is equivalent to q over $\mathbb{Z}$ and therefore is integral, primitive and positive definite. A relation $\mathsf{q}_1 = r\mathsf{q}'$ with $r \in \mathbb{Q}$, where q_1 and q' are integral primitive positive definite forms, is possible only if $r = 1$. Hence $\mathsf{q}_1 = \mathsf{q}'$. Finally, since $\gamma, \gamma_1 \in \mathbb{H}_1$, it follows that $vs - ut = 1$, and so q' is properly equivalent to q. □

Now let $\mathsf{F} = \mathbb{Q}(\Delta^{1/2})$ be a quadratic field of negative discriminant Δ, and let $\mathsf{M} = \{\alpha, \beta\}$ be a complete module in F with the ring of multipliers O_l. We fix the ordering of the basis α, β for M so that $\gamma = -\beta/\alpha \in \mathbb{H}_1$ and assign to the module M the quadratic form

$$\begin{aligned}\mathsf{q} = \mathsf{q}(\mathsf{M}) &= N(\mathsf{M})^{-1}(\alpha x + \beta y)(\bar{\alpha}x + \bar{\beta}y)\\ &= N(\alpha)N(\mathsf{M})^{-1}(x-\gamma y)(x-\bar{\gamma}y).\end{aligned} \tag{A.3.13}$$

Lemma A.3.8. *The form* $\mathsf{q}(\mathsf{M})$ *is an integral primitive positive definite quadratic form of discriminant* Δl^2. *The module* $\mathsf{M}(\mathsf{q})$ *corresponding to* q *is equivalent to the module* M.

Proof. Let $\overline{\mathrm{ch}}(v,\gamma) = av^2 + bv + c$. Then we have $(v-\gamma)(v-\bar{\gamma}) = a^{-1}(av^2+bv+c)$, whence

$$\mathsf{q}(x,y) = \frac{N(\alpha)}{N(\mathsf{M})}y^2\left(\frac{x}{y}-\gamma\right)\left(\frac{x}{y}-\bar{\gamma}\right) = \frac{N(\alpha)}{aN(\mathsf{M})}(ax^2+bxy+cy^2).$$

It follows from (A.3.1) and Lemma A.3.2 that $N(\mathsf{M}) = N(\alpha\{1,-\gamma\}) = N(\alpha)N\{1,\gamma\} = \dfrac{N(\alpha)}{a}$. Hence, q is equal to the integral primitive positive definite quadratic form $ax^2+bxy+cy^2$. By Lemma A.3.2, we have $\Delta(\mathsf{q}) = b^2 - 4ac = \Delta(\mathsf{O}(\mathsf{M})) = \Delta l^2$. Finally, $\mathsf{M}(\mathsf{q}) = a\{1,\gamma\} = a\{1,-\gamma\} = a\alpha^{-1}\{\alpha,\beta\} = a\alpha^{-1}\mathsf{M}$. □

Lemmas A.3.7 and A.3.8 and Proposition A.3.6 imply the following

Theorem A.3.9. *For every negative rational integer* Δ *the mappings* (A.2.12) *and* (A.3.13) *define one-to-one correspondences inverse to each other between the set*

of proper classes of integral primitive positive definite binary quadratic forms of discriminant Δ and the group $\mathsf{H}(\Delta)$ *of classes of equivalent modules in* $\mathbb{Q}(\Delta^{1/2})$ *with the ring of multiplies of discriminant* Δ; *in particular, the group* $\mathsf{H}(\Delta)$ *is finite.*

Let q and q_1 be two integral primitive positive definite binary quadratic forms, let $Q = Q(\mathsf{q})$ and $Q_1 = Q(\mathsf{q}_1)$ be the matrices of q and q_1, and let $\Delta = -\det Q$, $\Delta_1 = -\det Q_1$ be their discriminants. Let us assume that the quotient Δ/Δ_1 is a rational square. Then $\mathbb{Q}(\Delta^{1/2}) = \mathbb{Q}(\Delta_1^{1/2}) = \mathsf{F}$. Hence, the modules $\mathsf{M}(\mathsf{q})$ and $\mathsf{M}(\mathsf{q}_1)$ are complete modules in the same imaginary quadratic field, and we may multiply them by each other and by arbitrary complete modules in the field. We define the product $\mathsf{q} \times \mathsf{q}_1$ of two forms q and q_1 by

$$\mathsf{q} \times \mathsf{q}_1 = \mathsf{q}(\mathsf{M}(\mathsf{q}) \cdot \mathsf{M}(\mathsf{q}_1)) \tag{A.3.14}$$

and denote by $Q \times Q_1$ the matrix of $\mathsf{q} \times \mathsf{q}_1$:

$$Q \times Q_1 = Q(\mathsf{q} \times \mathsf{q}_1). \tag{A.3.15}$$

Further, for any complete module M in F, we set

$$\mathsf{q} \times \mathsf{M} = \mathsf{M} \times \mathsf{q} = \mathsf{q}(\mathsf{M}(\mathsf{q}) \cdot \mathsf{M}), \tag{A.3.16}$$

and denote by $Q \times \mathsf{M}$ the matrix of $\mathsf{q} \times \mathsf{M}$:

$$Q \times \mathsf{M} = \mathsf{M} \times Q = Q(\mathsf{q} \cdot \mathsf{M}). \tag{A.3.17}$$

The multiplication (A.3.14) is called *the Gauss composition of quadratic forms.* According to Lemmas A.3.7 and A.3.8, the proper class of the product $\mathsf{q} \times \mathsf{q}_1$ depends only on proper classes of q and q_1, and the proper class of $\mathsf{q} \times \mathsf{M}$ depends only on the proper class of q and the class of M. Thus, the formulae (A.3.14)–(A.3.17) define the product of the corresponding classes.

According to Theorem A.3.9, the set of proper classes of integral primitive positive definite binary quadratic form of a fixed negative discriminant Δ (resp. the set of the classes under (A.3.9) of matrices of these form) form a finite abelian group which is isomorphic to the group $\mathsf{H}(\Delta)$ of classes of equivalent modules in $\mathbb{Q}(\Delta^{1/2})$ with the ring of multipliers of discriminant Δ. We denote by $\mathsf{H}(\Delta)$ each of these three realization of the same abstract group.

Notes

On Chapter 1

§§ 1.3.1, 1.3.2. Our presentation of the properties of theta-functions and theta-groups follows Eichler's approach in [17], Appendix to Chapter 1.

§ 1.3.3. The idea to interpret all theta-series as specializations of theta-functions and realize transformation groups for the theta-series as subgroups of those for the theta-functions was first used by Eichler in the case of one variable in [17], Appendix to Chapter 1; the general case was considered in [14].

§ 1.4.2. The automorphic properties of theta-series of integral positive definite quadratic forms of determinant 1 were first considered by Witt in [52].

§§ 1.4.3–1.4.5. For theta-series of one variable, the calculation of multipliers is all standard (see, for example, [40], [36], [17]); the general case was treated in [14].

On Chapter 2

In more details the theory of Siegel modular forms may be found in the books [31], [33], [25]. Old N. Bourbaki fans may use the notes [41]. For modular forms of one variable see the books [34] and [32]: an initial presentation can be found in [42]; the relations to algebraic geometry are presented in [44]. We have drawn largely on the account in [33] and [29].

§ 2.1.2. For more of the reduction theory see e.g. [16].

§ 2.2.3. In the case of one variable, our definition of modular forms of half-integral weight for $\Gamma_0^n(q)$ is the same as that of Shimura in [45].

§ 2.3.3. The notation (2.3.14) is attributed to H. Petersson.

§ 2.3.6. We follow Freitag's papers [22], [23], [24].

§ 2.4.1. The proof of Proposition 2.4.1 follows [42].

§ 2.5.1 The scalar product of modular forms of one variable was introduced by Petersson in [35]; the consideration of the general case is based on the same idea and was started by Maass.

§ 2.5.2. For the Eisenstein series of one variable see [26]. There are a number of papers on asymptotic formulae for the Fourier coefficients of modular forms and theta-series; for the beginning see, for example, [28], [34], [37].

On Chapter 3

§ 3.1.1. Hecke has introduced his operators in [27].

§ 3.1.2. We do not know who has initially introduced the abstract Hecke rings. In our early papers we used the definition from Shimura's paper [43]. The definition used in this book was introduced in [6]; over $\mathbb{Z}$ our definition is equivalent to that of Shimura, but turns out to be more convenient.

§ 3.1.4. The properties of the anti-isomorphism j were considered in [8].

§ 3.2. Hecke rings of the general linear group are considered in the spirit of Tamagawa's paper [51].

§ 3.2.2. The formulae of Lemma 3.2.18 is a particular case of the formulae of Lemmas 6 and 9 of [2].

§ 3.2.3. For explicit formulae for the spherical mapping see [2].

§ 3.3. The structure of Hecke ring for Γ^n was indicated in [39] and [43] (Who was the first?). The passage to congruence subgroups goes back to Hecke's ideas in [27]. Our presentation uses a common sense and an analogy with the case of the general linear group rather than either historical source.

§ 3.3.3. The theory of spherical mappings in the form of the theory of zonal spherical functions on reductive algebraic groups over $\mathfrak{p}$-adic fields is due to Satake in [39]. We use an elementary approach based on explicit formulae. For $n = 1, 2$ the series (3.3.71) were summed up in [43], the general case was considered in [1] and [2].

§ 3.4. The imbeddings of local Hecke rings of the symplectic group into Hecke rings of the triangular subgroup were first used in [6] to considered factorizations of symplectic polynomials. The centralizer of $\Pi^n_+(p)$ was considered in [6].

§ 3.4.3. For the corresponding operators acting on Fourier coefficients of modular forms, the decomposition (3.4.32) was obtained and used in [4].

§ 3.5. The theory of factorization of symplectic polynomials has arisen as a result of attempts to understand and generalize the factorizations of the operational polynomials corresponding to $Q^2_p(v)$ and $R^2_p(v)$, which were obtained in [4] and [5] respectively. A preliminary variant of the theory was presented in [6]; the formulae (3.5.56) were obtained in [7]; in [8] the duality arguments were added, which has led to the symmetric factorization of $R^n_p(v)$.

On Chapter 4

§4.1.1. Hecke operators for the group $\Gamma^1 = \mathrm{SL}_2(\mathbb{Z})$ and some congruence subgroups were defined in [27]. The Hecke operators on Siegel modular forms were first introduced by Sugawara in [49] and [50] and then, after the war, were studied by Maass in [30].

§4.1.2. The existence of a basis of eigenfunctions for all Hecke operators acting on cusp forms of one variable was proved by Petersson in [35]. In the situation of Siegel modular forms Petersson's ideas were used by Maass in [30]. In [20] Evdokimov proves that a basis of the eigenfunctions exists for modular forms for a wide class of congruence subgroups, but the paper contains quite a few of inaccurucies, partly indicated in [21], and should be corrected before use. In this book we prove the existence of a basis of eigenfunction for all Hecke operators in the following three cases: for spaces of cusp forms for a q-regular pair (see Theorem 4.1.8), for invariant subspaces of the spaces $\mathfrak{M}_k^n$ (see Theorem 4.2.20), and for invariant spaces spanned by theta-series (see Corollary 5.2.13).

§4.2.3. Here we follow the paper [12].

§4.2.4. The relations (4.2.69) for the group Γ^n and the unit character are due to Žarkovskaya: see [53], our consideration follows the same idea. The paper [53] has arisen as a result of attempts to generalize Maass' commutation relations for the Siegel operator and the Hecke operators $|T^2(a)$ obtained in [30].

§4.2.5. The exposed results were published in [7] and [8].

§4.3.1. The results for the series $D(s, a, F)$ are actually due to Hecke [27]; the results for the series $D^+(s, a, F)$ were obtained by Shimura in [46].

§4.3.2. For $q = 1$, the results of this subsection were obtained in [3] with some restrictions and finally in [4]. Here we follow the same ideas. The similar questions for the group $\Gamma^2(q)$ were considered in [19]. In [4] it is also proved that for $q = 1$, the function

$$\psi_F(s) = (2\pi)^{-s}\Gamma(s)\Gamma(s-k+2)\zeta(s,1,F),$$

where Γ is the gamma-function, can be analytically continued to a function meromorphic on the entire s-plane which is holomorphic everywhere with the possible exception of poles at $s = 0$, $k-2$, k, $2k-2$ and satisfies the functionsl equation

$$\psi_F(2k-2-s) = (-1)^k\psi_F(s),$$

where k is the weight of the eigenfunction F.

§4.3.3. Here we present the results obtained in [5] (the case $n = 2$) [7], and [8] (the general case). In [13] it is proved that the even zeta-function $\zeta^+(s, \lambda, F)$ with a Dirichlet character λ can be analytically continued to a function meromorphic on the entire s-plane and, with certain assumptions, satisfies a functional equation of usual form.

On Chapter 5

§ 5.1. The fact that, generally speaking, Hecke operators carry theta-series to linear combinations of theta-series with similar characteristics was first established without any explicit formulae by Freitag in [23]. In the local case the formulae of Theorem 5.1.15 were obtained in [8]. For the Siegel theorem see, for example, [47] [48].

§ 5.2. "Anzahlmatrizen" were considered by Eichler in [18].

§ 5.2.1. In the case $q = 1$, the formula (5.2.17) was originally obtained by Freitag in [24] by the method indicated in Exercise 5.2.4.

§ 5.2.2. For a more general situation, the formula (5.2.25) was obtained by an elementary method in [9]. The relations (5.2.28) were obtained in [24].

§5.2.3. The results are taken from [12].

§ 5.2.5. See also [10].

References

1. Andrianov, A. N.: Rationality theorems for Hecke series and zeta-functions of the groups GL_n and Sp_n over local fields. Izv. Akad. Nauk SSSR Ser. Mat. 33, 466–505 (1969); English transl. Math. USSR Izv. 3, 439–476 (1969).
2. Andrianov, A. N.: Spherical functions for GL_n over local fields and summation of Hecke series. Mat. Sb. Nov. Ser. 83 (125) No 3, 429–451 (1970); English transl. Math. USSR Sb. 12, No 3, 429–452 (1970).
3. Andrianov, A. N.: Dirichlet series with Euler product in the theory of Siegel modular forms of genus 2. Tr. Mat. Inst. Steklova 112, 73–94 (1971); English transl. Proc. Steklov Inst. Math. 112, 70–93 (1971).
4. Andrianov, A. N.: Euler products that correspond to Siegel modular forms of genus 2. Usp. Mat. Nauk 29, No 3 (177), 43–110 (1974); English transl. Russ. Math. Surv. 29 (1974).
5. Andrianov, A. N.: Symmetric squares of zeta-functions of Siegel modular forms of genus 2. Tr. Mat. Inst. Steklova 142, 22–45 (1976); English transl. Proc. Steklov Inst. Math. No 3 (142), 21–45 (1979).
6. Andrianov, A. N.: On factorization of Hecke polynomials for the symplectic group of genus n. Mat. Sb. Nov. Ser. 104 (146), No 3, 390–427 (1977) English transl. Math. USSR Sb. No 3, 343–373 (1977).
7. Andrianov, A. N.: Euler expansions of theta-transforms of Siegel modular forms of degree n. Mat. Sb. Nov. Ser. 105 (147), No 3, 291–341 (1978); English transl. Math. USSR Sb. 34, No 3, 259–300 (1978).
8. Andrianov, A. N.: The multiplicative arithmetic of Siegel modular forms. Usp. Mat. Nauk 34 No 1, 67–135 (1979); English transl. Russ. Math. Surv. 34 No 1, 75–148 (1979).
9. Andrianov, A. N.: Action of Hecke operator $T(p)$ on theta series. Math. Ann. 247, 245–254 (1980).
10. Andrianov, A. N.: Integral representations of quadratic forms by quadratic forms: Multiplicative properties. Proc. Int. Congress Math. August 16–24, 1983, Warszawa.
11. Andrianov, A. N.: Hecke operators and representations of binary quadratic forms. Tr. Mat. Inst. Steklova 165, 4–15 (1984); English transl. Proc. Steklov Inst. Math. 3, 3–15 (1985).
12. Andrianov, A. N.: Representations of even zeta-function on theta-series. Zap. Nauchn. Sem. Leningr. Otd. Mat. Inst. Steklova 134, 5–14 (1984).
13. Andrianov, A. N., Kalinin, V. L.: On analytic properties of standard zeta-functions of Siegel modular forms. Mat. Sb. Nov. Ser. 106 (148) No 3, 323–339 (1978); English transl. Math. USSR Sb. 35, No 1, 1–17 (1979).
14. Andrianov, A. N., Maloletkin, G. N.: The behaviour of theta-series of genus n under modular substitutions. Izv. Akad. Nauk SSSR, Ser. Mat. 39, No 2, 243–258 (1975); English transl. Math. USSR Izv. 9, No 2, 227–241 (1975).
15. Borevich, Z. I., Shafarevich, I. R.: Number theory. Moscow: Nauka 1966, 1972, 1985; English transl. New York London: Academic Press 1966.
16. Cassels, J. W. S.: Rational quadratic forms. London New York San Francisco: Academic Press 1978.
17. Eichler, M.: Einführung in die Theorie der algebraischen Zahlen und Funktionen. Basel Stuttgart: Birkhäuser 1963; English transl. Introduction to the theory of algebraic numbers and functions. New York London: Academic Press 1966.

18. Eichler, M.: Quadratische Formen und orthogonale Gruppen. Grundlehren Math. Wiss. 63. Berlin Heidelberg New York: Springer 1952, 1974.
19. Evdokimov, S. A.: Euler products for congruence subgroups of the Siegel group of genus 2. Mat. Sb. Nov. Ser. 99 (141), 483–513 (1976); English transl. Math. USSR Sb. 28 (1976).
20. Evdokimov, S. A.: Basis of eigenfunctions of Hecke operators in the theory of modular forms of degree n. Mat. Sb. Nov. Ser. 115 (157) No 3, 337–363 (1981); English transl. Math. USSR Sb. 44 (1981).
21. Evdokimov, S. A.: A letter to the editorial board. Mat. Sb. Nov. Ser. 116 (158) No 4 (1981); English transl. Math. USSR Sb. 45 (1981).
22. Freitag, E.: Holomorphe Differentialformen zu Kongruenzgruppen der Siegelschen Modulgruppe. Invent. Math. 30, 181–196 (1975).
23. Freitag, E.: Die Invarianz gewisser von Thetareihen erzeugter Vektorräume unter Heckeoperatoren. Math. Z. 156, 141–155 (1971).
24. Freitag, E.: Eine Bemerkung zu Andrianovs expliziten Formeln für die Wirkung der Heckeoperatoren auf Thetareihen. In E. B. Christofel, 336–351. Basel Aachen: Birkhäuser 1981.
25. Freitag, E.: Siegelsche Modulfunktionen. Grundlehren Math. Wiss. 254. Berlin Heidelberg New York: Springer 1983.
26. Hecke, E.: Theorie der Eisensteinschen Reihen höherer Stufe und ihre Anwendung auf Funktionentheorie und Arithmetik. Abh. Math. Semin. Univ. Hamb. 5, 199–224 (1927) (= Math. Werke, 461–486).
27. Hecke, E.: Über Modulfunktionene und die Dirichletschen Reihen mit Eulerscher Produktentwicklung. I, II. Math. Ann. 114, 1–28, 316–351 (1937) (= Math. Werke, 644–707).
28. Hecke, E.: Analytische Arithmetik der positiven quadratischen Formen. Danske Vidensk. Selsk. Mathem.-fys. Meddel XVII, 12. Copenhagen 1940 (= Math. Werke, 789–918).
29. Koecher, M.: Zur Theorie der Modulformen n-ten Grades. I, II. Math. Z. 59, No 4, 399–416 (1954); 61, No 4, 455–466 (1955).
30. Maass, H.: Die Primzahlen in der Theorie der Siegelschen Modulfunktionen. Math. Ann. 124, 87–122 (1951).
31. Maass, H.: Lectures on Siegel's modular functions. Bombay: Tata Inst. Fund. Research 1954–1955.
32. Maass, H.: Lectures on modular functions of one complex variable. Bombay: Tata Inst. Fund. Research 1964 (Revised 1983).
33. Maass, H.: Siegel's modular forms and Dirichlet series. Lect. Notes Math. 216. Berlin Heidelberg New York: Springer 1971.
34. Ogg, A.: Modular forms and Dirichlet series. New York Amsterdam: Benjamin 1969.
35. Petersson, H.: Konstruktion der sämtlichen Lösungen einer Riemannschen Funktionalgleichung durch Dirichletreihen mit Eulerscher Produktentwicklung I, II, III, Math. Ann. 116, 401–412 (1939), 117, 39–64 (1939), 117, 277–300 (1940/41).
36. Pfetzer, W.: Die Wirkung der Modulsubstitutionen auf mehrfache Thetareihen zu quadratischen Formen ungerader Variablenzahl. Arch. Math. 4, 448–454 (1953).
37. Raghavan, S.: Modular forms of degree n and representations by quadratic forms. Ann. Math. II Ser. 70, No 3, 446–477 (1959).
38. Rankin, R. A.: Contributions to the theory of Ramanujan's function $\tau(n)$ and similar arithmetical functions. I, II. Proc. Cambridge Phil. Soc. 35, 351–372 (1939).
39. Satake, I.: Theory of spherical functions on reductive algebraic groups over $\wp$-adic fields. Publ. Math. Inst. Hautes Etud. Sci. No 18, 5–69 (1963).
40. Schoeneberg, B.: Das Verhalten von mehrfachen Thetareihen bei Modulsubstitutionen. Math. Ann. 116, 511–523 (1939).
41. Séminaire H. Cartan, 10e année, 1957/58, Fonctions automorphes, vol. 1, 2. Secrétariat Mathématique, Paris 1958.
42. Serre, J-P.: Cours d'arithmetique. Paris: Presses Universitaires de France 1970.
43. Shimura, G.: On modular correspondences for $Sp(n, \mathbb{Z})$ and their congruence relations. Proc. Natl. Acad. Sci. USA 49, No 6, 824–828 (1963).

44. Shimura, G.: Introduction to the arithmetical theory of automorphic functions. Iwanami Shoten and Princeton Univ. Press 1971.
45. Shimura, G.: On modular forms of half integral weight. Ann. Math. II Ser. 97, No 3, 440–481 (1973).
46. Shimura, G.: On the holomorphy of certain Dirichlet series. Proc. Lond. Math. Soc. III Ser. 31, 79–98 (1975).
47. Siegel, C. L.: Über die analytische Theorie der quadratischen Formen. Ann. Math. II Ser. 36, 527–606 (1935) (= Gesammelte Abhandlungen, Bd. I, 326–405).
48. Siegel, C. L.: Lectures on the analytical theory of quadratic forms. Göttingen: Buchhandlung Robert Peppmüller 1963.
49. Sugawara, M.: On the transformation theory of Siegel's modular group of the n-th degree. Proc. Imp. Acad. Japan 13, 335–336 (1937).
50. Sugawara, M.: An invariant property of Siegel's modular functions. Proc. Imp. Acad. Japan 14, 1–3 (1938).
51. Tamagawa, T.: On the ζ-functions of a division algebra. Ann. Math. II Ser. 77, 387–405 (1963).
52. Witt, E.: Eine Identität zwischen Modulformen zweiten Grades. Abh. Math. Semin. Univ. Hamburg 14, 323–337 (1941).
53. Žarkovskaya, N. A.: The Siegel operator and Hecke operators. Funkts. Anal. Prilozh. 8, No 12, 30–38 (1974); English transl. Funct. Anal. Appl. 8, 113–120 (1974).
54. Žarkovskaya, N. A.: On the connection between the eigenvalues of Hecke operators and the Fourier coefficients of eigenfunctions for Siegel's modular forms of genus n. Mat. Sb. Nov. Ser. 96 (138), 584–593 (1975); English transl. Math. USSR Sb. 25 (1975).
55. Žuravlev, V. G.: Euler expansions of theta-transforms of Siegel modular forms of half integral weight and their analytic properties. Mat. Sb. Nov. Ser. 123 (165), No 2, 174–194 (1984).

Index of Terminology

Index of Notation

This list gives special symbols which have constant meaning in the whole book, or in substantial parts of it. There are symbols which are used with different meaning in different parts of the book, but we hoped that this will not cause confusion. The symbols below are listed in the order (1) latin letters (all founts), (2) greek letters (all founts), (3) other symbols as special uses of brackets, indices, and so on.

Note that special alphabets and typefaces are used in the following cases: special roman capitals as $\mathbb{C}$ are reserved for certain standard sets as the field of complex numbers; gothic capitals are only used for functional spaces; sanserif capitals as A, B are normally used for sets; sanserif lower case letters as q, i are reserved for quadratic forms and certain mappings; boldface letters are used to distinguish images of Hecke rings of the symplectic group, or those of their elements, in Hecke rings of the triangular subgroup from inverse images (starting from § 3.4); boldface lower case letters are also used for columns, rows, and vectors; script capitals as $\mathscr{L}$ are used for groups and rings related with the triangular subgroup of the symplectic group; discrete subgroups of the real general linear and symplectic groups are usually denoted by greek capitals.

Symbol	Meaning	Page
$\mathbb{A}_n$	The set of even positive semi-definite matrices of order n	3
$\mathbb{A}_n^+$	The set of positive definite matrices of $\mathbb{A}_n$	3
$\mathbb{A}_n(d, \delta, q)$		300
$\mathsf{B}(D) = \mathsf{B}_{\mathbb{Q}}(D)$	The set of rational square matrices B with ${}^tBD = {}^tDB$	143
$\underline{\mathsf{B}}(D)$	The subset of integral matrices of $\mathsf{B}(D)$	
$B(v) = B_p^n(v)$	$= \sum_{i=0}^{n} (-1)^i b_i^n(p) v^i$	208, 243
$\mathbb{C}$	The field of complex numbers	
$\mathscr{C}_-^n(q), \mathscr{C}_+^n(q)$	The centralizers of $\Pi_-^n(a)$, $\Pi_+^n(a)$ ($a \in \mathbb{N}_{(q)}$) in $\mathscr{L}^n(q)$	171
$\mathscr{C}_{-p}^n, \mathscr{C}_p^n$	The centralizers of Frobenius elements in $\mathscr{L}_p^n$	175
$\mathrm{ch}(v, \alpha)$	The characteristic polynomial of α	350
D_n	The fundamental domain for Γ^n on $\mathbb{H}_n$	58
$d(\mathsf{V}, f)$	The determinant of (V, f)	335
d^*Z	The invariant element of volume on $\mathbb{H}_n$	9
$d_i(M)$	Elementary divisors of M	117
$\mathsf{D}(\Gamma, \mathsf{S})$	The Hecke ring of Hecke pair (Γ, S) (over $\mathbb{Z}$)	108
$\mathsf{D}_{\mathsf{K}}(\Gamma, \mathsf{S})$	The Hecke ring of Hecke pair (Γ, S) over K	115
$D_i^n(p)$	$= \begin{pmatrix} E_{n-i} & 0 \\ 0 & pE_i \end{pmatrix}$	126
$D_{\alpha,\beta}^n(p)$	$= \begin{pmatrix} D_\alpha^{n-\beta}(p) & 0 \\ 0 & p^2E_\beta \end{pmatrix}$	126
$\det M$	The determinant of M	
$\mathrm{diag}(a_1, \ldots, a_n)$	The diagonal matrix with the diagonal entries $a_1, \ldots, a_n$	

Symbol	Meaning	Page
$E_n = (e_{\alpha\beta})$	The unit matrix of order n	
$\mathbb{E}_n$	The set of integral symmetric matrices of order n with even diagonal entries (even matrices)	
$\mathbf{e}_j$	Usually $\mathbf{e}_1, \ldots, \mathbf{e}_n$ are the columns of the unit matrix E_n, or a basis for a vector space	
$\mathsf{E}(A)$	The group of integral units of A	230
$e(A)$	The order of $\mathsf{E}(A)$	
$e\{A\}$	$= \exp(2\pi i \operatorname{Tr}(A))$	
$\operatorname{ed} g$	The matrix of elementary divisors of $g \in \mathsf{G}^n$	116
$\operatorname{ed}_p g$	The matrix of elementary p-divisors of $g \in \mathsf{G}^n$	121
$\mathsf{E}^n(q)$, $\mathsf{E}^n_p(q)$, $\mathbf{E}^n_p$	Even subrings of $\mathsf{L}^n(q)$, $\mathsf{L}^n_p(q)$, $\mathbf{L}^n_p$ respectively	283
F	An automorphic function or modular form	
F	A field	
F_n	The Minkowski domain	48
$\mathbb{F}_p$	The field of p elements	
$\mathfrak{F}^n_k(q, \chi)$, $\mathfrak{F}^n_\xi$	Spaces of Fourier coefficients	224
$\mathsf{GL}_n(\mathsf{K})$	The group of invertible matrices of order n with entries in K	
G^n	$= \mathsf{GL}_n(\mathbb{Q})$	
G^n_p	$= \mathsf{GL}_n(\mathbb{Z}[p^{-1}])$	
$\mathsf{GSp}^+_n(\mathsf{K})$	$= \{M \in \mathsf{GL}_{2n}(\mathsf{K}); {}^tMJ_nM = \mu(M)J_n, \mu(M) > 0\}$	28
$\mathsf{G}(Q)$, $\mathsf{G}(\mathsf{q})$	The genus of Q and that of q	348
g.c.d.	= The greatest common divisor	
$\mathbb{H}_n$	The Siegel upper half-plane of genus n	4
$\mathbb{H}_n(\varepsilon)$	$= \{X + iY \in \mathbb{H}_n, Y \geq \varepsilon E_n\}$	4
H^n	The global Hecke ring of the general linear group $= D_{\mathbb{Q}}(\Lambda^n, \mathsf{G}^n)$	116
H^n_p	A local subring of H^n, $= D_{\mathbb{Q}}(\Lambda^n, \mathsf{G}^n_p)$	121
$\underline{\mathsf{H}}^n_p$	The integral subring of H^n_p	123
$\mathsf{H}(\mathsf{O}_l) = \mathsf{H}(\Delta l^2)$	The group of classes of complete modules in $\mathbb{Q}(\Delta^{1/2})$ with the ring of multipliers O_l, or one of its realisation	355, 359
$h(X + iY)$	The height of a matrix of $\mathbb{H}_n$, i.e. $\det Y$	57
i	An index, or $(-1)^{1/2} \in \mathbb{H}_1$	
i	Standard imbedding of Hecke rings	110
$i(\mathsf{V}, f, r)$	The number of isotropic sets of r vectors in quadratic space (V, f)	343
J_n	$= \begin{pmatrix} 0 & E_n \\ -E_n & 0 \end{pmatrix}$	
j	Standard anti-isomorphism of Hecke rings	111
$j(M, Z)$	$= \det(CZ + D)\left(M = \begin{pmatrix} A & B \\ C & D \end{pmatrix}\right)$	28
K	A commutative ring with the unity	
$\mathsf{L}(\mathsf{K})$	$= \mathsf{D}_{\mathbb{Q}}(\mathsf{K}, \mathsf{S}(\mathsf{K}))$	137
$\mathsf{L}^n(q)$	$= \mathsf{L}(\Gamma^n_0(q))$-(global) Hecke ring of $\Gamma^n_0(q)$, $\mathsf{L}(q) = \mathsf{L}^1(q)$, $\mathsf{L}^n = \mathsf{L}^n(1)$.	139
$\mathsf{L}^n_p(q)$	A local subring of $\mathsf{L}^n(q)$, $\mathsf{L}^n_p = \mathsf{L}^n_p(1)$.	143
$\underline{\mathsf{L}}^n_p(q)$	The integral subring of $\mathsf{L}^n_p(q)$, $\underline{\mathsf{L}}^n_p = \underline{\mathsf{L}}^n_p(1)$.	146
$\mathbf{L}^n_p$	$= \mathsf{i}(\mathsf{L}^n_p) \subset \mathscr{L}^n_p$	152
$\mathscr{L}^n(q)$	$= \mathsf{D}_{\mathbb{Q}}(\Gamma^n_0, \mathscr{S}^n(q))$-(global) Hecke ring of Γ^n_0, $\mathscr{L}^n = \mathscr{L}^n(1)$	168
$\mathscr{L}^n_p$	$= \mathsf{D}_{\mathbb{Q}}(\Gamma^n_0, \mathscr{S}^n_p)$-a local subring of $\mathscr{L}^n$.	152
M	A module in an algebraic number field	
$\mathsf{M}_{m,n}(\mathsf{K})$	The set of $m \times n$-matrices with entries in K	
$\mathsf{M}_n(\mathsf{K})$	$= \mathsf{M}_{n,n}(\mathsf{K})$	
$\mathsf{M}_{m,n}$	$= \mathsf{M}_{m,n}(\mathbb{Z})$	
M_n	$= \mathsf{M}_n(\mathbb{Z})$	

Grundlehren der mathematischen Wissenschaften

A Series of Comprehensive Studies in Mathematics

A Selection

190. Faith: Algebra: Rings, Modules, and Categories I
191. Faith: Algebra II, Ring Theory
192. Mal'cev: Algebraic Systems
193. Pólya/Szegö: Problems and Theorems in Analysis I
194. Igusa: Theta Functions
195. Berberian: Baer*-Rings
196. Athreya/Ney: Branching Processes
197. Benz: Vorlesungen über Geometrie der Algebren
198. Gaal: Linear Analysis and Representation Theory
199. Nitsche: Vorlesungen über Minimalflächen
200. Dold: Lectures on Algebraic Topology
201. Beck: Continuous Flows in the Plane
202. Schmetterer: Introduction to Mathematical Statistics
203. Schoeneberg: Elliptic Modular Functions
204. Popov: Hyperstability of Control Systems
205. Nikol'skiĭ: Approximation of Functions of Several Variables and Imbedding Theorems
206. André: Homologie des Algébres Commutatives
207. Donoghue: Monotone Matrix Functions and Analytic Continuation
208. Lacey: The Isometric Theory of Classical Banach Spaces
209. Ringel: Map Color Theorem
210. Gihman/Skorohod: The Theory of Stochastic Processes I
211. Comfort/Negrepontis: The Theory of Ultrafilters
212. Switzer: Algebraic Topology - Homotopy and Homology
215. Schaefer: Banach Lattices and Positive Operators
217. Stenström: Rings of Quotients
218. Gihman/Skorohod: The Theory of Stochastic Processes II
219. Duvant/Lions: Inequalities in Mechanics and Physics
220. Kirillov: Elements of the Theory of Representations
221. Mumford: Algebraic Geometry I: Complex Projective Varieties
222. Lang: Introduction to Modular Forms
223. Bergh/Löfström: Interpolation Spaces. An Introduction
224. Gilbarg/Trudinger: Elliptic Partial Differential Equations of Second Order
225. Schütte: Proof Theory
226. Karoubi: K-Theory. An Introduction
227. Grauert/Remmert: Theorie der Steinschen Räume
228. Segal/Kunze: Integrals and Operators
229. Hasse: Number Theory
230. Klingenberg: Lectures on Closed Geodesics
231. Lang: Elliptic Curves: Diophantine Analysis
232. Gihman/Skorohod: The Theory of Stochastic Processes III
233. Stroock/Varadhan: Multidimensional Diffusion Processes
234. Aigner: Combinatorial Theory
235. Dynkin/Yushkevich: Controlled Markov Processes
236. Grauert/Remmert: Theory of Stein Spaces
237. Köthe: Topological Vector Spaces II
238. Graham/McGehee: Essays in Commutative Harmonic Analysis
239. Elliott: Probabilistic Number Theory I

240. Elliott: Probabilistic Number Theory II
241. Rudin: Function Theory in the Unit Ball of C^n
242. Huppert/Blackburn: Finite Groups II
243. Huppert/Blackburn: Finite Groups III
244. Kubert/Lang: Modular Units
245. Cornfeld/Fomin/Sinai: Ergodic Theory
246. Naimark/Štern: Theory of Group Representations
247. Suzuki: Group Theory I
248. Suzuki: Group Theory II
249. Chung: Lectures from Markov Processes to Brownian Motion
250. Arnold: Geometrical Methods in the Theory of Ordinary Differential Equations
251. Chow/Hale: Methods of Bifurcation Theory
252. Aubin: Nonlinear Analysis on Manifolds. Monge-Ampère Equations
253. Dwork: Lectures on p-adic Differential Equations
254. Freitag: Siegelsche Modulfunktionen
255. Lang: Complex Multiplication
256. Hörmander: The Analysis of Linear Partial Differential Operators I
257. Hörmander: The Analysis of Linear Partial Differential Operators II
258. Smoller: Shock Waves and Reaction-Diffusion Equations
259. Duren: Univalent Functions
260. Freidlin/Wentzell: Random Perturbations of Dynamical Systems
261. Bosch/Güntzer/Remmert: Non Archimedian Analysis – A Systematic Approach to Rigid Geometry
262. Doob: Classical Potential Theory and Its Probabilistic Counterpart
263. Krasnosel'skiĭ/Zabreĭko: Geometrical Methods of Nonlinear Analysis
264. Aubin/Cellina: Differential Inclusions
265. Grauert/Remmert: Coherent Analytic Sheaves
266. de Rham: Differentiable Manifolds
267. Arbarello/Cornalba/Griffiths/Harris: Geometry of Algebraic Curves, Vol. I
268. Arbarello/Cornalba/Griffiths/Harris: Geometry of Algebraic Curves, Vol. II
269. Schapira: Microdifferential Systems in the Complex Domain
270. Scharlau: Quadratic and Hermitian Forms
271. Ellis: Entropy, Large Deviations, and Statistical Mechanics
272. Elliott: Arithmetic Functions and Integer Products
273. Nikols'kii: Treatise on the Shift Operators
274. Hörmander: The Analysis of Linear Partial Differential Operators III
275. Hörmander: The Analysis of Linear Partial Differential Operators IV
276. Liggett: Interacting Particle Systems
277. Fulton/Lang: Riemann-Roch Algebra
278. Barr/Wells: Toposes, Triples and Theories
279. Bishop/Bridges: Constructive Analysis
280. Neukirch: Class Field Theory
281. Chandrasekharan: Elliptic Function
282. P. Leong, L. Gruman: Entire Functions of Several Complex Variables
283. K. Kodaira: Complex Manifolds and Deformation of Complex Structures
284. R. Finn: Equilibrium Capillary Surfaces
285. Y. D. Burago, V. A. Zalgaller: Geometric Inequalities
286. A. N. Andrianov: Quadratic Forms and Hecke Operators

Springer-Verlag
Berlin Heidelberg New York London Paris Tokyo

GPSR Compliance

The European Union's (EU) General Product Safety Regulation (GPSR) is a set of rules that requires consumer products to be safe and our obligations to ensure this.

If you have any concerns about our products, you can contact us on ProductSafety@springernature.com

In case Publisher is established outside the EU, the EU authorized representative is:

Springer Nature Customer Service Center GmbH
Europaplatz 3
69115 Heidelberg, Germany

Batch number: 10514799

Printed by Printforce, the Netherlands